M000250023

Differential Equations with Boundary-Value Problems

NINTH EDITION

Dennis G. Zill
Loyola Marymount University

Prepared by

Roberto Martinez
Loyola Marymount University

CENGAGE
Learning·

Australia • Brazil • Mexico • Singapore • United Kingdom • United States

For product information and technology assistance, contact us at
**Cengage Learning Customer & Sales Support,
1-800-354-9706**.

For permission to use material from this text or product, submit all requests online at **www.cengage.com/permissions**
Further permissions questions can be emailed to
permissionrequest@cengage.com.

ISBN: 978-1-305-96581-2

Cengage Learning
20 Channel Center Street
Boston, MA 02210
USA

Cengage Learning is a leading provider of customized learning solutions with office locations around the globe, including Singapore, the United Kingdom, Australia, Mexico, Brazil, and Japan. Locate your local office at: **www.cengage.com/global**.

Cengage Learning products are represented in Canada by Nelson Education, Ltd.

To learn more about Cengage Learning Solutions, visit **www.cengage.com**.

Purchase any of our products at your local college store or at our preferred online store **www.cengagebrain.com**.

Printed in the United States of America
Print Number: 01   Print Year: 2017

CONTENTS

Chapter 1

Introduction to Differential Equations

1.1	Definitions and Terminology

1. Second order; linear

3. Fourth order; linear

5. Second order; nonlinear because of $(dy/dx)^2$ or $\sqrt{1 + (dy/dx)^2}$

7. Third order; linear

9. Writing the differential equation in the form $x(dy/dx) + y^2 = 1$, we see that it is nonlinear in y because of y^2. However, writing it in the form $(y^2 - 1)(dx/dy) + x = 0$, we see that it is linear in x.

11. From $y = e^{-x/2}$ we obtain $y' = -\frac{1}{2}e^{-x/2}$. Then $2y' + y = -e^{-x/2} + e^{-x/2} = 0$.

13. From $y = e^{3x}\cos 2x$ we obtain $y' = 3e^{3x}\cos 2x - 2e^{3x}\sin 2x$ and $y'' = 5e^{3x}\cos 2x - 12e^{3x}\sin 2x$, so that $y'' - 6y' + 13y = 0$.

15. The domain of the function, found by solving $x + 2 \geq 0$, is $[-2, \infty)$. From $y' = 1 + 2(x+2)^{-1/2}$ we have

$$(y - x)y' = (y - x)[1 + (2(x + 2)^{-1/2}]$$

$$= y - x + 2(y - x)(x + 2)^{-1/2}$$

$$= y - x + 2[x + 4(x + 2)^{1/2} - x](x + 2)^{-1/2}$$

$$= y - x + 8(x + 2)^{1/2}(x + 2)^{-1/2} = y - x + 8.$$

An interval of definition for the solution of the differential equation is $(-2, \infty)$ because y' is not defined at $x = -2$.

17. The domain of the function is $\{x \mid 4 - x^2 \neq 0\}$ or $\{x \mid x \neq -2 \text{ and } x \neq 2\}$. From $y' = 2x/(4 - x^2)^2$ we have

$$y' = 2x\left(\frac{1}{4 - x^2}\right)^2 = 2xy^2.$$

1

An interval of definition for the solution of the differential equation is $(-2, 2)$. Other intervals are $(-\infty, -2)$ and $(2, \infty)$.

19. Writing $\ln(2X - 1) - \ln(X - 1) = t$ and differentiating implicitly we obtain

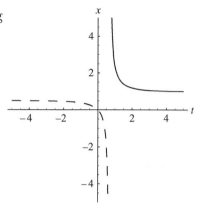

$$\frac{2}{2X - 1}\frac{dX}{dt} - \frac{1}{X - 1}\frac{dX}{dt} = 1$$

$$\left(\frac{2}{2X - 1} - \frac{1}{X - 1}\right)\frac{dX}{dt} = 1$$

$$\frac{2X - 2 - 2X + 1}{(2X - 1)(X - 1)}\frac{dX}{dt} = 1$$

$$\frac{dX}{dt} = -(2X - 1)(X - 1) = (X - 1)(1 - 2X).$$

Exponentiating both sides of the implicit solution we obtain

$$\frac{2X - 1}{X - 1} = e^t$$

$$2X - 1 = Xe^t - e^t$$

$$(e^t - 1) = (e^t - 2)X$$

$$X = \frac{e^t - 1}{e^t - 2}.$$

Solving $e^t - 2 = 0$ we get $t = \ln 2$. Thus, the solution is defined on $(-\infty, \ln 2)$ or on $(\ln 2, \infty)$. The graph of the solution defined on $(-\infty, \ln 2)$ is dashed, and the graph of the solution defined on $(\ln 2, \infty)$ is solid.

21. Differentiating $P = c_1 e^t / \left(1 + c_1 e^t\right)$ we obtain

$$\frac{dP}{dt} = \frac{\left(1 + c_1 e^t\right) c_1 e^t - c_1 e^t \cdot c_1 e^t}{\left(1 + c_1 e^t\right)^2} = \frac{c_1 e^t}{1 + c_1 e^t}\frac{\left[\left(1 + c_1 e^t\right) - c_1 e^t\right]}{1 + c_1 e^t}$$

$$= \frac{c_1 e^t}{1 + c_1 e^t}\left[1 - \frac{c_1 e^t}{1 + c_1 e^t}\right] = P(1 - P).$$

23. From $y = c_1 e^{2x} + c_2 x e^{2x}$ we obtain $\dfrac{dy}{dx} = (2c_1 + c_2)e^{2x} + 2c_2 x e^{2x}$ and $\dfrac{d^2y}{dx^2} = (4c_1 + 4c_2)e^{2x} + 4c_2 x e^{2x}$, so that

$$\frac{d^2y}{dx^2} - 4\frac{dy}{dx} + 4y = (4c_1 + 4c_2 - 8c_1 - 4c_2 + 4c_1)e^{2x} + (4c_2 - 8c_2 + 4c_2)xe^{2x} = 0.$$

In Problems 25–27, we use the Product Rule and the derivative of an integral ((12) of this section):

$$\frac{d}{dx}\int_a^x g(t)\,dt = g(x).$$

25. Differentiating $y = e^{3x}\int_1^x \frac{e^{-3t}}{t}\,dt$ we obtain $\frac{dy}{dx} = 3e^{3x}\int_1^x \frac{e^{-3t}}{t}\,dt + \frac{e^{-3x}}{x}\cdot e^{3x}$ or

$$\frac{dy}{dx} = 3e^{3x}\int_1^x \frac{e^{-3t}}{t}\,dt + \frac{1}{x},\text{ so that}$$

$$x\frac{dy}{dx} - 3xy = x\left(3e^{3x}\int_1^x \frac{e^{-3t}}{t}\,dt + \frac{1}{x}\right) - 3x\left(e^{3x}\int_1^x \frac{e^{-3t}}{t}\,dt\right)$$

$$= 3xe^{3x}\int_1^x \frac{e^{-3t}}{t}\,dt + 1 - 3xe^{3x}\int_1^x \frac{e^{-3t}}{t}\,dt = 1.$$

27. Differentiating $y = \frac{5}{x} + \frac{10}{x}\int_1^x \frac{\sin t}{t}\,dt$ we obtain $\frac{dy}{dx} = -\frac{5}{x^2} - \frac{10}{x^2}\int_1^x \frac{\sin t}{t}\,dt + \frac{\sin x}{x}\cdot\frac{10}{x}$ or

$$\frac{dy}{dx} = -\frac{5}{x^2} - \frac{10}{x^2}\int_1^x \frac{\sin t}{t}\,dt + \frac{10\sin x}{x^2},\text{ so that}$$

$$x^2\frac{dy}{dx} + xy = x^2\left(-\frac{5}{x^2} - \frac{10}{x^2}\int_1^x \frac{\sin t}{t}\,dt + \frac{10\sin x}{x^2}\right) + x\left(\frac{5}{x} + \frac{10}{x}\int_1^x \frac{\sin t}{t}\,dt\right)$$

$$= -5 - 10\int_1^x \frac{\sin t}{t}\,dt + 10\sin x + 5 + 10\int_1^x \frac{\sin t}{t}\,dt = 10\sin x.$$

29. From

$$y = \begin{cases} -x^2, & x < 0 \\ x^2, & x \geq 0 \end{cases}$$

we obtain

$$y' = \begin{cases} -2x, & x < 0 \\ 2x, & x \geq 0 \end{cases}$$

so that $xy' - 2y = 0$.

31. Substitute the function $y = e^{mx}$ into the equation $y' + 2y = 0$ to get

$$(e^{mx})' + 2(e^{mx}) = 0$$

$$me^{mx} + 2e^{mx} = 0$$

$$e^{mx}(m + 2) = 0$$

Now since $e^{mx} > 0$ for all values of x, we must have $m = -2$ and so $y = e^{-2x}$ is a solution.

33. Substitute the function $y = e^{mx}$ into the equation $y'' - 5y' + 6y = 0$ to get

$$(e^{mx})'' - 5(e^{mx})' + 6(e^{mx}) = 0$$

$$m^2 e^{mx} - 5me^{mx} + 6e^{mx} = 0$$

$$e^{mx}(m^2 - 5m + 6) = 0$$

$$e^{mx}(m - 2)(m - 3) = 0$$

Now since $e^{mx} > 0$ for all values of x, we must have $m = 2$ or $m = 3$ therefore $y = e^{2x}$ and $y = e^{3x}$ are solutions.

35. Substitute the function $y = x^m$ into the equation $xy'' + 2y' = 0$ to get

$$x \cdot (x^m)'' + 2(x^m)' = 0$$

$$x \cdot m(m-1)x^{m-2} + 2mx^{m-1} = 0$$

$$(m^2 - m)x^{m-1} + 2mx^{m-1} = 0$$

$$x^{m-1}[m^2 + m] = 0$$

$$x^{m-1}[m(m+1)] = 0$$

The last line implies that $m = 0$ or $m = -1$ therefore $y = x^0 = 1$ and $y = x^{-1}$ are solutions.

In Problems 37–39, we substitute $y = c$ into the differential equations and use $y' = 0$ and $y'' = 0$

37. Solving $5c = 10$ we see that $y = 2$ is a constant solution.

39. Since $1/(c-1) = 0$ has no solutions, the differential equation has no constant solutions.

41. From $x = e^{-2t} + 3e^{6t}$ and $y = -e^{-2t} + 5e^{6t}$ we obtain

$$\frac{dx}{dt} = -2e^{-2t} + 18e^{6t} \quad \text{and} \quad \frac{dy}{dt} = 2e^{-2t} + 30e^{6t}.$$

Then

$$x + 3y = (e^{-2t} + 3e^{6t}) + 3(-e^{-2t} + 5e^{6t}) = -2e^{-2t} + 18e^{6t} = \frac{dx}{dt}$$

and

$$5x + 3y = 5(e^{-2t} + 3e^{6t}) + 3(-e^{-2t} + 5e^{6t}) = 2e^{-2t} + 30e^{6t} = \frac{dy}{dt}.$$

43. $(y')^2 + 1 = 0$ has no real solutions because $(y')^2 + 1$ is positive for all differentiable functions $y = \phi(x)$.

45. The first derivative of $f(x) = e^x$ is e^x. The first derivative of $f(x) = e^{kx}$ is ke^{kx}. The differential equations are $y' = y$ and $y' = ky$, respectively.

47. We first note that $\sqrt{1 - y^2} = \sqrt{1 - \sin^2 x} = \sqrt{\cos^2 x} = |\cos x|$. This prompts us to consider values of x for which $\cos x < 0$, such as $x = \pi$. In this case

$$\left.\frac{dy}{dx}\right|_{x=\pi} = \left.\frac{d}{dx}(\sin x)\right|_{x=\pi} = \cos x\big|_{x=\pi} = \cos \pi = -1,$$

but

$$\sqrt{1 - y^2}\big|_{x=\pi} = \sqrt{1 - \sin^2 \pi} = \sqrt{1} = 1.$$

Thus, $y = \sin x$ will only be a solution of $y' = \sqrt{1 - y^2}$ when $\cos x > 0$. An interval of definition is then $(-\pi/2, \pi/2)$. Other intervals are $(3\pi/2, 5\pi/2)$, $(7\pi/2, 9\pi/2)$, and so on.

49. One solution is given by the upper portion of the graph with domain approximately $(0, 2.6)$. The other solution is given by the lower portion of the graph, also with domain approximately $(0, 2.6)$.

51. Differentiating $(x^3 + y^3)/xy = 3c$ we obtain

$$\frac{xy(3x^2 + 3y^2 y') - (x^3 + y^3)(xy' + y)}{x^2 y^2} = 0$$

$$3x^3 y + 3xy^3 y' - x^4 y' - x^3 y - xy^3 y' - y^4 = 0$$

$$(3xy^3 - x^4 - xy^3)y' = -3x^3 y + x^3 y + y^4$$

$$y' = \frac{y^4 - 2x^3 y}{2xy^3 - x^4} = \frac{y(y^3 - 2x^3)}{x(2y^3 - x^3)}.$$

53. The derivatives of the functions are $\phi_1'(x) = -x/\sqrt{25 - x^2}$ and $\phi_2'(x) = x/\sqrt{25 - x^2}$, neither of which is defined at $x = \pm 5$.

55. For the first-order differential equation integrate $f(x)$. For the second-order differential equation integrate twice. In the latter case we get $y = \int (\int f(x)dx)\, dx + c_1 x + c_2$.

57. The differential equation $yy' - xy = 0$ has normal form $dy/dx = x$. These are not equivalent because $y = 0$ is a solution of the first differential equation but not a solution of the second.

59. (a) Since e^{-x^2} is positive for all values of x, $dy/dx > 0$ for all x, and a solution, $y(x)$, of the differential equation must be increasing on any interval.

(b) $\displaystyle\lim_{x \to -\infty} \frac{dy}{dx} = \lim_{x \to -\infty} e^{-x^2} = 0$ and $\displaystyle\lim_{x \to \infty} \frac{dy}{dx} = \lim_{x \to \infty} e^{-x^2} = 0$. Since dy/dx approaches 0 as x approaches $-\infty$ and ∞, the solution curve has horizontal asymptotes to the left and to the right.

(c) To test concavity we consider the second derivative

$$\frac{d^2 y}{dx^2} = \frac{d}{dx}\left(\frac{dy}{dx}\right) = \frac{d}{dx}\left(e^{-x^2}\right) = -2xe^{-x^2}.$$

Since the second derivative is positive for $x < 0$ and negative for $x > 0$, the solution curve is concave up on $(-\infty, 0)$ and concave down on $(0, \infty)$.

(d)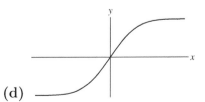

61. (a) The derivative of a constant solution is 0, so solving $y(a - by) = 0$ we see that $y = 0$ and $y = a/b$ are constant solutions.

(b) A solution is increasing where $dy/dx = y(a - by) = by(a/b - y) > 0$ or $0 < y < a/b$. A solution is decreasing where $dy/dx = by(a/b - y) < 0$ or $y < 0$ or $y > a/b$.

(c) Using implicit differentiation we compute

$$\frac{d^2y}{dx^2} = y(-by') + y'(a - by) = y'(a - 2by).$$

Thus $d^2y/dx^2 = 0$ when $y = a/2b$. Since $d^2y/dx^2 > 0$ for $0 < y < a/2b$ and $d^2y/dx^2 < 0$ for $a/2b < y < a/b$, the graph of $y = \phi(x)$ has a point of inflection at $y = a/2b$.

(d)

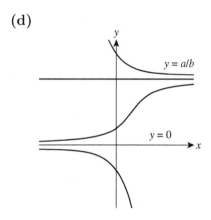

63. In *Mathematica* use

Clear[y]

y[x_]:= x Exp[5x] Cos[2x]

y[x]

y''''[x] − 20y'''[x] + 158y''[x] − 580y'[x] +841y[x]//Simplify

The output will show $y(x) = e^{5x}x \cos 2x$, which verifies that the correct function was entered, and 0, which verifies that this function is a solution of the differential equation.

1.2 | Initial-Value Problems

1. Solving $-1/3 = 1/(1 + c_1)$ we get $c_1 = -4$. The solution is $y = 1/(1 - 4e^{-x})$.

3. Letting $x = 2$ and solving $1/3 = 1/(4 + c)$ we get $c = -1$. The solution is $y = 1/(x^2 - 1)$. This solution is defined on the interval $(1, \infty)$.

5. Letting $x = 0$ and solving $1 = 1/c$ we get $c = 1$. The solution is $y = 1/(x^2 + 1)$. This solution is defined on the interval $(-\infty, \infty)$.

In Problems 7–9, we use $x = c_1 \cos t + c_2 \sin t$ and $x' = -c_1 \sin t + c_2 \cos t$ to obtain a system of two equations in the two unknowns c_1 and c_2.

7. From the initial conditions we obtain the system

$$c_1 = -1 \quad c_2 = 8$$

The solution of the initial-value problem is $x = -\cos t + 8 \sin t$.

9. From the initial conditions we obtain

$$\frac{\sqrt{3}}{2}c_1 + \frac{1}{2}c_2 = \frac{1}{2} - \frac{1}{2}c_2 + \frac{\sqrt{3}}{2} = 0$$

Solving, we find $c_1 = \sqrt{3}/4$ and $c_2 = 1/4$. The solution of the initial-value problem is

$$x = (\sqrt{3}/4) \cos t + (1/4) \sin t.$$

In Problems 11–13, we use $y = c_1 e^x + c_2 e^{-x}$ and $y' = c_1 e^x - c_2 e^{-x}$ to obtain a system of two equations in the two unknowns c_1 and c_2.

11. From the initial conditions we obtain

$$c_1 + c_2 = 1$$

$$c_1 - c_2 = 2.$$

Solving, we find $c_1 = \frac{3}{2}$ and $c_2 = -\frac{1}{2}$. The solution of the initial-value problem is $y = \frac{3}{2}e^x - \frac{1}{2}e^{-x}$.

13. From the initial conditions we obtain

$$e^{-1}c_1 + ec_2 = 5$$

$$e^{-1}c_1 - ec_2 = -5.$$

Solving, we find $c_1 = 0$ and $c_2 = 5e^{-1}$. The solution of the initial-value problem is $y = 5e^{-1}e^{-x} = 5e^{-1-x}$.

15. Two solutions are $y = 0$ and $y = x^3$.

17. For $f(x, y) = y^{2/3}$ we have $\dfrac{\partial f}{\partial y} = \dfrac{2}{3}y^{-1/3}$. Thus, the differential equation will have a unique solution in any rectangular region of the plane where $y \neq 0$.

19. For $f(x, y) = \dfrac{y}{x}$ we have $\dfrac{\partial f}{\partial y} = \dfrac{1}{x}$. Thus, the differential equation will have a unique solution in any region where $x \neq 0$.

21. For $f(x, y) = x^2/(4 - y^2)$ we have $\partial f/\partial y = 2x^2 y/(4 - y^2)^2$. Thus the differential equation will have a unique solution in any region where $y < -2$, $-2 < y < 2$, or $y > 2$.

23. For $f(x, y) = \dfrac{y^2}{x^2 + y^2}$ we have $\dfrac{\partial f}{\partial y} = \dfrac{2x^2 y}{(x^2 + y^2)^2}$. Thus, the differential equation will have a unique solution in any region not containing $(0, 0)$.

In Problems 25–27, we identify $f(x, y) = \sqrt{y^2 - 9}$ and $\partial f/\partial y = y/\sqrt{y^2 - 9}$. We see that f and $\partial f/\partial y$ are both continuous in the regions of the plane determined by $y < -3$ and $y > 3$ with no restrictions on x.

25. Since $4 > 3$, $(1, 4)$ is in the region defined by $y > 3$ and the differential equation has a unique solution through $(1, 4)$.

27. Since $(2, -3)$ is not in either of the regions defined by $y < -3$ or $y > 3$, there is no guarantee of a unique solution through $(2, -3)$.

29. (a) A one-parameter family of solutions is $y = cx$. Since $y' = c$, $xy' = xc = y$ and $y(0) = c \cdot 0 = 0$.

(b) Writing the equation in the form $y' = y/x$, we see that R cannot contain any point on the y-axis. Thus, any rectangular region disjoint from the y-axis and containing (x_0, y_0) will determine an interval around x_0 and a unique solution through (x_0, y_0). Since $x_0 = 0$ in part (a), we are not guaranteed a unique solution through $(0, 0)$.

(c) The piecewise-defined function which satisfies $y(0) = 0$ is not a solution since it is not differentiable at $x = 0$.

31. (a) Since $\dfrac{d}{dx}\left(-\dfrac{1}{x + c}\right) = \dfrac{1}{(x + c)^2} = y^2$, we see that $y = -\dfrac{1}{x + c}$ is a solution of the differential equation.

(b) Solving $y(0) = -1/c = 1$ we obtain $c = -1$ and $y = 1/(1 - x)$. Solving $y(0) = -1/c = -1$ we obtain $c = 1$ and $y = -1/(1 + x)$. Being sure to include $x = 0$, we see that the interval of existence of $y = 1/(1 - x)$ is $(-\infty, 1)$, while the interval of existence of $y = -1/(1 + x)$ is $(-1, \infty)$.

(c) By inspection we see that $y = 0$ is a solution on $(-\infty, \infty)$.

33. (a) Differentiating $3x^2 - y^2 = c$ we get $6x - 2yy' = 0$ or $yy' = 3x$.

(b) Solving $3x^2 - y^2 = 3$ for y we get

$$y = \phi_1(x) = \sqrt{3(x^2 - 1)}, \qquad 1 < x < \infty,$$

$$y = \phi_2(x) = -\sqrt{3(x^2 - 1)}, \qquad 1 < x < \infty,$$

$$y = \phi_3(x) = \sqrt{3(x^2 - 1)}, \qquad -\infty < x < -1,$$

$$y = \phi_4(x) = -\sqrt{3(x^2 - 1)}, \qquad -\infty < x < -1.$$

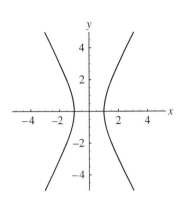

(c) Only $y = \phi_3(x)$ satisfies $y(-2) = 3$.

In Problems 35–37, we consider the points on the graphs with x-coordinates $x_0 = -1$, $x_0 = 0$, and $x_0 = 1$. The slopes of the tangent lines at these points are compared with the slopes given by $y'(x_0)$ in (a) through (f).

35. The graph satisfies the conditions in (b) and (f).

37. The graph satisfies the conditions in (c) and (d).

In Problems 39–43 $y = c_1 \cos 2x + c_2 \sin 2x$ is a two parameter family of solutions of the second-order differential equation $y'' + 4y = 0$. In some of the problems we will use the fact that $y' = -2c_1 \sin 2x + 2c_2 \cos 2x$.

39. From the boundary conditions $y(0) = 0$ and $y\left(\dfrac{\pi}{4}\right) = 3$ we obtain

$$y(0) = c_1 = 0$$

$$y\left(\frac{\pi}{4}\right) = c_1 \cos\left(\frac{\pi}{2}\right) + c_2 \sin\left(\frac{\pi}{2}\right) = c_2 = 3.$$

Thus, $c_1 = 0$, $c_2 = 3$, and the solution of the boundary-value problem is $y = 3\sin 2x$.

41. From the boundary conditions $y'(0) = 0$ and $y'\left(\frac{\pi}{6}\right) = 0$ we obtain

$$y'(0) = 2c_2 = 0$$

$$y'\left(\frac{\pi}{6}\right) = -2c_1 \sin\left(\frac{\pi}{3}\right) = -\sqrt{3}\, c_1 = 0.$$

Thus, $c_2 = 0$, $c_1 = 0$, and the solution of the boundary-value problem is $y = 0$.

43. From the boundary conditions $y(0) = 0$ and $y(\pi) = 2$ we obtain

$$y(0) = c_1 = 0$$

$$y(\pi) = c_1 = 2.$$

Since $0 \neq 2$, this is not possible and there is no solution.

45. Integrating $y' = 8e^{2x} + 6x$ we obtain

$$y = \int (8e^{2x} + 6x)\, dx = 4e^{2x} + 3x^2 + c.$$

Setting $x = 0$ and $y = 9$ we have $9 = 4 + c$ so $c = 5$ and $y = 4e^{2x} + 3x^2 + 5$.

47. When $x = 0$ and $y = \frac{1}{2}$, $y' = -1$, so the only plausible solution curve is the one with negative slope at $(0, \frac{1}{2})$, or the red curve.

49. If the solution is tangent to the x-axis at $(x_0, 0)$, then $y' = 0$ when $x = x_0$ and $y = 0$. Substituting these values into $y' + 2y = 3x - 6$ we get $0 + 0 = 3x_0 - 6$ or $x_0 = 2$.

51. When $y = \frac{1}{16}x^4$, $y' = \frac{1}{4}x^3 = x(\frac{1}{4}x^2) = xy^{1/2}$, and $y(2) = \frac{1}{16}(16) = 1$. When

$$y = \begin{cases} 0, & x < 0 \\ \dfrac{1}{16}x^4, & x \geq 0 \end{cases}$$

we have

$$y' = \begin{cases} 0, & x < 0 \\ \frac{1}{4}x^3, & x \geq 0 \end{cases} = x \begin{cases} 0, & x < 0 \\ \frac{1}{4}x^2, & x \geq 0 \end{cases} = xy^{1/2},$$

and $y(2) = \frac{1}{16}(16) = 1$. The two different solutions are the same on the interval $(0, \infty)$, which is all that is required by Theorem 1.2.1.

1.3 | Differential Equations as Mathematical Models

1. $\dfrac{dP}{dt} = kP + r; \qquad \dfrac{dP}{dt} = kP - r$

3. Let b be the rate of births and d the rate of deaths. Then $b = k_1 P$ and $d = k_2 P^2$. Since $dP/dt = b - d$, the differential equation is $dP/dt = k_1 P - k_2 P^2$.

5. From the graph in the text we estimate $T_0 = 180°$ and $T_m = 75°$. We observe that when $T = 85$, $dT/dt \approx -1$. From the differential equation we then have

$$k = \frac{dT/dt}{T - T_m} = \frac{-1}{85 - 75} = -0.1.$$

7. The number of students with the flu is x and the number not infected is $1000 - x$, so $dx/dt = kx(1000 - x)$.

9. The rate at which salt is leaving the tank is

$$R_{out}\ (3\ \text{gal/min}) \cdot \left(\frac{A}{300}\ \text{lb/gal}\right) = \frac{A}{100}\ \text{lb/min}.$$

Thus $dA/dt = A/100$. The initial amount is $A(0) = 50$.

11. The rate at which salt is entering the tank is

$$R_{in} = (3 \text{ gal/min}) \cdot (2 \text{ lb/gal}) = 6 \text{ lb/min}.$$

Since the tank loses liquid at the net rate of

$$3 \text{ gal/min} - 3.5 \text{ gal/min} = -0.5 \text{ gal/min},$$

after t minutes the number of gallons of brine in the tank is $300 - \frac{1}{2}t$ gallons. Thus the rate at which salt is leaving is

$$R_{out} = \left(\frac{A}{300 - t/2} \text{ lb/gal} \right) \cdot (3.5 \text{ gal/min}) = \frac{3.5A}{300 - t/2} \text{ lb/min} = \frac{7A}{600 - t} \text{ lb/min}.$$

The differential equation is

$$\frac{dA}{dt} = 6 - \frac{7A}{600 - t} \quad \text{or} \quad \frac{dA}{dt} + \frac{7}{600 - t} A = 6.$$

13. The volume of water in the tank at time t is $V = A_w h$. The differential equation is then

$$\frac{dh}{dt} = \frac{1}{A_w} \frac{dV}{dt} = \frac{1}{A_w} \left(-cA_h \sqrt{2gh} \right) = -\frac{cA_h}{A_w} \sqrt{2gh}.$$

Using $A_h = \pi \left(\frac{2}{12} \right)^2 = \frac{\pi}{36}$, $A_w = 10^2 = 100$, and $g = 32$, this becomes

$$\frac{dh}{dt} = -\frac{c\pi/36}{100} \sqrt{64h} = -\frac{c\pi}{450} \sqrt{h}.$$

15. Since $i = dq/dt$ and $L\, d^2q/dt^2 + R\, dq/dt = E(t)$, we obtain $L\, di/dt + Ri = E(t)$.

17. From Newton's second law we obtain $m\dfrac{dv}{dt} = -kv^2 + mg$.

19. The net force acting on the mass is

$$F = ma = m\frac{d^2x}{dt^2} = -k(s + x) + mg = -kx + mg - ks.$$

Since the condition of equilibrium is $mg = ks$, the differential equation is

$$m\frac{d^2x}{dt^2} = -kx.$$

21. As the rocket climbs (in the positive direction), it spends its amount of fuel and therefore the mass of the fuel changes with time. The air resistance acts in the opposite direction of the motion and the upward thrust R works in the same direction. Using Newton's second law we get

$$\frac{d}{dt}(mv) = -mg - kv + R$$

Now because the mass is variable, we must use the product rule to expand the left side of the equation. Doing so gives us the following:

$$\frac{d}{dt}(mv) = -mg - kv + R$$

$$v \times \frac{dm}{dt} + m \times \frac{dv}{dt} = -mg - kv + R$$

The last line is the differential equation we wanted to find.

23. From $g = k/R^2$ we find $k = gR^2$. Using $a = d^2r/dt^2$ and the fact that the positive direction is upward we get

$$\frac{d^2r}{dt^2} = -a = -\frac{k}{r^2} = -\frac{gR^2}{r^2} \qquad \text{or} \qquad \frac{d^2r}{dt^2} + \frac{gR^2}{r^2} = 0.$$

25. The differential equation is $\dfrac{dA}{dt} = k(M - A)$.

27. The differential equation is $x'(t) = r - kx(t)$ where $k > 0$.

29. We see from the figure that $2\theta + \alpha = \pi$. Thus

$$\frac{y}{-x} = \tan \alpha = \tan(\pi - 2\theta) = -\tan 2\theta = -\frac{2\tan\theta}{1 - \tan^2\theta}.$$

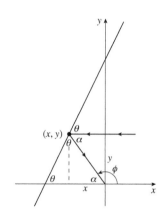

Since the slope of the tangent line is $y' = \tan\theta$ we have $y/x = 2y'[1-(y')^2]$ or $y - y(y')^2 = 2xy'$, which is the quadratic equation $y(y')^2 + 2xy' - y = 0$ in y'. Using the quadratic formula, we get

$$y' = \frac{-2x \pm \sqrt{4x^2 + 4y^2}}{2y} = \frac{-x \pm \sqrt{x^2 + y^2}}{y}.$$

Since $dy/dx > 0$, the differential equation is

$$\frac{dy}{dx} = \frac{-x + \sqrt{x^2 + y^2}}{y} \qquad \text{or} \qquad y\frac{dy}{dx} - \sqrt{x^2 + y^2} + x = 0.$$

31. The differential equation in (3) is $dT/dt = k(T - T_m)$. When the body is cooling, $T > T_m$, so $T - T_m > 0$. Since T is decreasing, $dT/dt < 0$ and $k < 0$. When the body is warming, $T < T_m$, so $T - T_m < 0$. Since T is increasing, $dT/dt > 0$ and $k < 0$.

33. This differential equation could describe a population that undergoes periodic fluctuations.

35. From Problem 23, $d^2r/dt^2 = -gR^2/r^2$. Since R is a constant, if $r = R + s$, then $d^2r/dt^2 = d^2s/dt^2$ and, using a Taylor series, we get

$$\frac{d^2s}{dt^2} = -g\frac{R^2}{(R+s)^2} = -gR^2(R+s)^{-2} \approx -gR^2[R^{-2} - 2sR^{-3} + \cdots] = -g + \frac{2gs}{R} + \cdots.$$

Thus, for R much larger than s, the differential equation is approximated by $d^2s/dt^2 = -g$.

37. We assume that the plow clears snow at a constant rate of k cubic miles per hour. Let t be the time in hours after noon, $x(t)$ the depth in miles of the snow at time t, and $y(t)$ the distance the plow has moved in t hours. Then dy/dt is the velocity of the plow and the assumption gives

$$wx\frac{dy}{dt} = k,$$

where w is the width of the plow. Each side of this equation simply represents the volume of snow plowed in one hour. Now let t_0 be the number of hours before noon when it started snowing and let s be the constant rate in miles per hour at which x increases. Then for $t > -t_0$, $x = s(t + t_0)$. The differential equation then becomes

$$\frac{dy}{dt} = \frac{k}{ws}\frac{1}{t + t_0}.$$

Integrating, we obtain

$$y = \frac{k}{ws}\left[\ln(t + t_0) + c\right]$$

where c is a constant. Now when $t = 0$, $y = 0$ so $c = -\ln t_0$ and

$$y = \frac{k}{ws}\ln\left(1 + \frac{t}{t_0}\right).$$

Finally, from the fact that when $t = 1$, $y = 2$ and when $t = 2$, $y = 3$, we obtain

$$\left(1 + \frac{2}{t_0}\right)^2 = \left(1 + \frac{1}{t_0}\right)^3.$$

Expanding and simplifying gives $t_0^2 + t_0 - 1 = 0$. Since $t_0 > 0$, we find $t_0 \approx 0.618$ hours ≈ 37 minutes. Thus it started snowing at about 11:23 in the morning.

39. Setting $A'(t) = -0.002$ and solving $A'(t) = -0.0004332A(t)$ for $A(t)$, we obtain

$$A(t) = \frac{A'(t)}{-0.0004332} = \frac{-0.002}{-0.0004332} \approx 4.6 \text{ grams.}$$

Chapter 1 in Review

1. $\dfrac{d}{dx} c_1 e^{10x} = 10\overbrace{c_1 e^{10x}}^{y};$ $\qquad \dfrac{dy}{dx} = 10y$

3. $\dfrac{d}{dx}(c_1 \cos kx + c_2 \sin kx) = -kc_1 \sin kx + kc_2 \cos kx;$

$$\frac{d^2}{dx^2}(c_1 \cos kx + c_2 \sin kx) = -k^2 c_1 \cos kx - k^2 c_2 \sin kx = -k^2\overbrace{(c_1 \cos kx + c_2 \sin kx)}^{y};$$

$\dfrac{d^2 y}{dx^2} = -k^2 y$ or $\dfrac{d^2 y}{dx^2} + k^2 y = 0$

5. $y = c_1 e^x + c_2 x e^x;$ $\quad y' = c_1 e^x + c_2 x e^x + c_2 e^x;$ $\quad y'' = c_1 e^x + c_2 x e^x + 2c_2 e^x;$

$y'' + y = 2(c_1 e^x + c_2 x e^x) + 2c_2 e^x = 2(c_1 e^x + c_2 x e^x + c_2 e^x) = 2y';$ $\quad y'' - 2y' + y = 0$

7. a, d   **9.** b   **11.** b

13. A few solutions are $y = 0$, $y = c$, and $y = e^x$.

15. The slope of the tangent line at (x, y) is y', so the differential equation is $y' = x^2 + y^2$.

17. (a) The domain is all real numbers.

(b) Since $y' = 2/3x^{1/3}$, the solution $y = x^{2/3}$ is undefined at $x = 0$. This function is a solution of the differential equation on $(-\infty, 0)$ and also on $(0, \infty)$.

19. Setting $x = x_0$ and $y = 1$ in $y = -2/x + x$, we get

$$1 = -\frac{2}{x_0} + x_0 \qquad \text{or} \qquad x_0^2 - x_0 - 2 = (x_0 - 2)(x_0 + 1) = 0.$$

Thus, $x_0 = 2$ or $x_0 = -1$. Since $x \neq 0$ in $y = -2/x + x$, we see that $y = -2/x + x$ is a solution of the initial-value problem $xy' + y = 2x$, $y(-1) = 1$, on the interval $(-\infty, 0)$ because $-1 < 0$, and $y = -2/x + x$ is a solution of the initial-value problem $xy' + y = 2x$, $y(2) = 1$, on the interval $(0, \infty)$ because $2 > 0$.

21. (a)

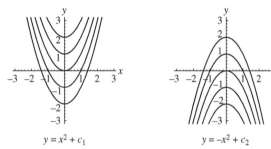

$$y = x^2 + c_1 \qquad\qquad y = -x^2 + c_2$$

(b) When $y = x^2 + c_1$, $y' = 2x$ and $(y')^2 = 4x^2$. When $y = -x^2 + c_2$, $y' = -2x$ and $(y')^2 = 4x^2$.

(c) Pasting together x^2, $x \geq 0$, and $-x^2$, $x \leq 0$, we get $y = \begin{cases} -x^2, & x \leq 0 \\ x^2, & x > 0. \end{cases}$

23. Differentiating $y = x \sin x + x \cos x$ we get

$$y' = x \cos x + \sin x - x \sin x + \cos x$$

and

$$y'' = -x \sin x + \cos x + \cos x - x \cos x - \sin x - \sin x$$

$$= -x \sin x - x \cos x + 2 \cos x - 2 \sin x.$$

Thus

$$y'' + y = -x \sin x - x \cos x + 2 \cos x - 2 \sin x + x \sin x + x \cos x = 2 \cos x - 2 \sin x.$$

An interval of definition for the solution is $(-\infty, \infty)$.

25. Differentiating $y = \sin(\ln x)$ we obtain $y' = \cos(\ln x)/x$ and $y'' = -[\sin(\ln x) + \cos(\ln x)]/x^2$.
Then

$$x^2 y'' + xy' + y = x^2 \left(-\frac{\sin(\ln x) + \cos(\ln x)}{x^2} \right) + x \frac{\cos(\ln x)}{x} + \sin(\ln x) = 0.$$

An interval of definition for the solution is $(0, \infty)$.

In Problems 27 - 29 we use (12) of Section 1.1 and the Product Rule.

27.

$$y = e^{\cos x} \int_0^x t e^{-\cos t}\, dt$$

$$\frac{dy}{dx} = e^{\cos x} \left(x e^{-\cos x} \right) - \sin x e^{\cos x} \int_0^x t e^{-\cos t}\, dt$$

$$\frac{dy}{dx} + (\sin x)\, y = e^{\cos x} x e^{-\cos x} - \sin x e^{\cos x} \int_0^x t e^{-\cos t}\, dt + \sin x \left(e^{\cos x} \int_0^x t e^{-\cos t}\, dt \right)$$

$$= x - \sin x e^{\cos x} \int_0^x t e^{-\cos t}\, dt + \sin x e^{\cos x} \int_0^x t e^{-\cos t}\, dt = x$$

29.

$$y = x \int_1^x \frac{e^{-t}}{t}\, dt$$

$$y' = x \frac{e^{-x}}{x} + \int_1^x \frac{e^{-t}}{t}\, dt = e^{-x} + \int_1^x \frac{e^{-t}}{t}\, dt$$

$$y'' = -e^{-x} + \frac{e^{-x}}{x}$$

$$x^2 y'' + \left(x^2 - x \right) y' + (1 - x)\, y = \left(-x^2 e^{-x} + x e^{-x} \right)$$

$$+ \left(x^2 e^{-x} + x^2 \int_1^x \frac{e^{-t}}{t}\, dt - x e^{-x} - x \int_1^x \frac{e^{-t}}{t}\, dt \right)$$

$$+ \left(x \int_1^x \frac{e^{-t}}{t}\, dt - x^2 \int_1^x \frac{e^{-t}}{t}\, dt \right) = 0$$

31. Using implicit differentiation we get

$$x^3 y^3 = x^3 + 1$$

$$3x^2 \cdot y^3 + x^3 \cdot 3y^2 \frac{dy}{dx} = 3x^2$$

$$\frac{3x^2 y^3}{3x^2 y^2} + \frac{x^3 3y^2}{3x^2 y^2} \frac{dy}{dx} = \frac{3x^2}{3x^2 y^2}$$

$$y + x \frac{dy}{dx} = \frac{1}{y^2}$$

33. Using implicit differentiation we get

$$y^3 + 3y = 1 - 3x$$

$$3y^2 y' + 3y' = -3$$

$$y^2 y' + y' = -1$$

$$(y^2 + 1)y' = -1$$

$$y' = \frac{-1}{y^2 + 1}$$

Differentiating the last line and remembering to use the quotient rule on the right side leads to

$$y'' = \frac{2yy'}{(y^2 + 1)^2}$$

Now since $y' = -1 / (y^2 + 1)$ we can write the last equation as

$$y'' = \frac{2y}{(y^2 + 1)^2} y' = \frac{2y}{(y^2 + 1)^2} \frac{-1}{(y^2 + 1)} = 2y \left(\frac{-1}{y^2 + 1} \right)^3 = 2y(y')^3$$

which is what we wanted to show.

In Problem 35–37, $y = c_1 e^{3x} + c_2 e^{-x} - 2x$ is given as a two-parameter family of solutions of the second-order differential equation $y'' - 2y' - 3y = 6x + 4$.

35. If $y(0) = 0$ and $y'(0) = 0$, then

$$c_1 + c_2 = 0$$
$$3c_1 - c_2 - 2 = 0$$

so $c_1 = \frac{1}{2}$ and $c_2 = -\frac{1}{2}$. Thus $y = \frac{1}{2} e^{3x} - half\ e^{-x} - 2x$.

37. If $y(1) = 4$ and $y'(1) = -2$, then

$$c_1 e^3 + c_2 e^{-1} - 2 = 4$$
$$3c_1 e^3 - c_2 e^{-1} - 2 = -2$$

so $c_1 = \frac{3}{2} e^{-3}$ and $c_2 = \frac{9}{2} e$. Thus $y = \frac{3}{2} e^{3x-3} + \frac{9}{2} e^{-x+1} - 2x$.

39. From the graph we see that estimates for y_0 and y_1 are $y_0 = -3$ and $y_1 = 0$.

Chapter 2

First-Order Differential Equations

2.1 | Solution Curves Without a Solution

1.

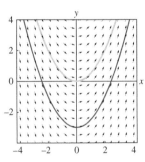

3.

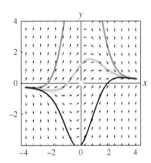

5.

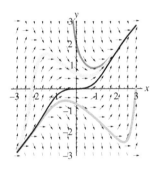

7.

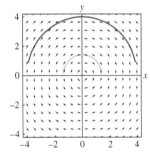

9.

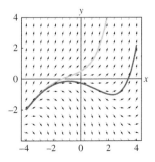

11.

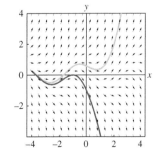

13.

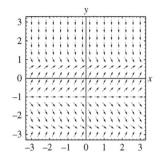

15. (a) The isoclines have the form $y = -x + c$, which are straight lines with slope -1.

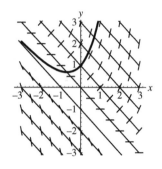

(b) The isoclines have the form $x^2 + y^2 = c$, which are circles centered at the origin.

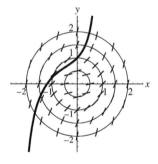

17. When $y < \frac{1}{2}x^2$, $y' = x^2 - 2y$ is positive and the portions of solution curves "outside" the nullcline parabola are increasing. When $y > \frac{1}{2}x^2$, $y' = x^2 - 2y$ is negative and the portions of the solution curves "inside" the nullcline parabola are decreasing.

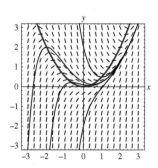

19. Writing the differential equation in the form $dy/dx = y(1-y)(1+y)$ we see that critical points are $y = -1$, $y = 0$, and $y = 1$. The phase portrait is shown at the right.

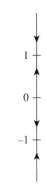

(a)

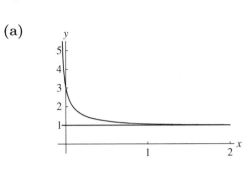

(b)

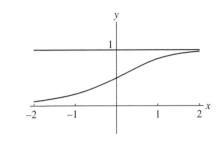

(c)

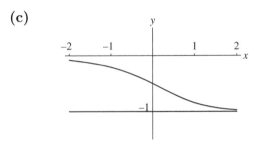

(d)

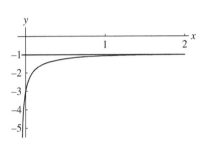

21. Solving $y^2 - 3y = y(y-3) = 0$ we obtain the critical points 0 and 3. From the phase portrait we see that 0 is asymptotically stable (attractor) and 3 is unstable (repeller).

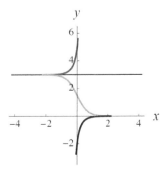

23. Solving $(y-2)^4 = 0$ we obtain the critical point 2. From the phase portrait we see that 2 is semi-stable.

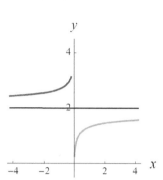

25. Solving $y^2(4-y^2) = y^2(2-y)(2+y) = 0$ we obtain the critical points -2, 0, and 2. From the phase portrait we see that 2 is asymptotically stable (attractor), 0 is semi-stable, and -2 is unstable (repeller).

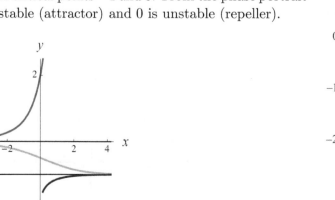

27. Solving $y\ln(y+2) = 0$ we obtain the critical points -1 and 0. From the phase portrait we see that -1 is asymptotically stable (attractor) and 0 is unstable (repeller).

29. The critical points are 0 and c because the graph of $f(y)$ is 0 at these points. Since $f(y) > 0$ for $y < 0$ and $y > c$, the graph of the solution is increasing on the y-intervals $(-\infty, 0)$ and (c, ∞). Since $f(y) < 0$ for $0 < y < c$, the graph of the solution is decreasing on the y-interval $(0, c)$.

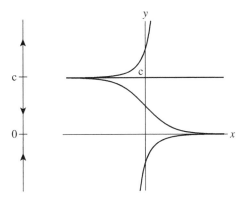

31. From the graphs of $z = \pi/2$ and $z = \sin y$ we see that $(2/\pi)y - \sin y = 0$ has only three solutions. By inspection we see that the critical points are $-\pi/2$, 0, and $\pi/2$.

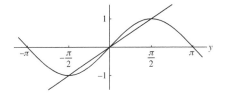

From the graph at the right we see that

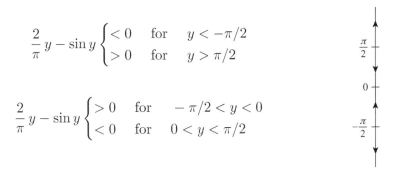

$$\frac{2}{\pi}y - \sin y \begin{cases} < 0 & \text{for} & y < -\pi/2 \\ > 0 & \text{for} & y > \pi/2 \end{cases}$$

$$\frac{2}{\pi}y - \sin y \begin{cases} > 0 & \text{for} & -\pi/2 < y < 0 \\ < 0 & \text{for} & 0 < y < \pi/2 \end{cases}$$

This enables us to construct the phase portrait shown at the right. From this portrait we see that $\pi/2$ and $-\pi/2$ are unstable (repellers), and 0 is asymptotically stable (attractor).

33. Recall that for $dy/dx = f(y)$ we are assuming that f and f' are continuous functions of y on some interval I. Now suppose that the graph of a nonconstant solution of the differential equation crosses the line $y = c$. If the point of intersection is taken as an initial condition we have two distinct solutions of the initial-value problem. This violates uniqueness, so the graph of any nonconstant solution must lie entirely on one side of any equilibrium solution. Since f is continuous it can only change signs at a point where it is 0. But this is a critical point. Thus, $f(y)$ is completely positive or completely negative in each region R_i. If $y(x)$ is oscillatory or has a relative extremum, then it must have a horizontal tangent line at some point (x_0, y_0). In this case y_0 would be a critical point of the differential equation, but we saw above that the graph of a nonconstant solution cannot intersect the graph of the equilibrium solution $y = y_0$.

35. Assuming the existence of the second derivative, points of inflection of $y(x)$ occur where $y''(x) = 0$. From $dy/dx = f(y)$ we have $d^2y/dx^2 = f'(y)\, dy/dx$. Thus, the y-coordinate of a point of inflection can be located by solving $f'(y) = 0$. (Points where $dy/dx = 0$ correspond to constant solutions of the differential equation.)

37. If (1) in the text has no critical points it has no constant solutions. The solutions have neither an upper nor lower bound. Since solutions are monotonic, every solution assumes all real values.

39. From the equation $dP/dt = k\,(P - h/k)$ we see that the only critical point of the autonomous differential equationis the positive number h/k. A phase portrait shows that this point is unstable, that is, h/k is a repeller. For any initial condition $P(0) = P_0$ for which $0 < P_0 < h/k$, $dP/dt < 0$ which means $P(t)$ is monotonic decreasing and so the graph of $P(t)$ must cross the t-axis or the line $P - 0$ at some time $t_1 > 0$. But $P(t_1) = 0$ means the population is extinct at time t_1.

41. Writing the differential equation in the form

$$\frac{dv}{dt} = \frac{k}{m}\left(\frac{mg}{k} - v^2\right) = \frac{k}{m}\left(\sqrt{\frac{mg}{k}} - v\right)\left(\sqrt{\frac{mg}{k}} + v\right)$$

we see that the only physically meaningful critical point is $\sqrt{mg/k}$.

From the phase portrait we see that $\sqrt{mg/k}$ is an asymptotically stable critical point. Thus, $\lim\limits_{t\to\infty} v = \sqrt{mg/k}$.

2.2 | Separable Variables

In many of the following problems we will encounter an expression of the form $\ln|g(y)| = f(x)+c$. To solve for $g(y)$ we exponentiate both sides of the equation. This yields $|g(y)| = e^{f(x)+c} = e^c e^{f(x)}$ which implies $g(y) = \pm e^c e^{f(x)}$. Letting $c_1 = \pm e^c$ we obtain $g(y) = c_1 e^{f(x)}$.

1. From $dy = \sin 5x\, dx$ we obtain $y = -\frac{1}{5}\cos 5x + c$.

3. From $dy = -e^{-3x}\, dx$ we obtain $y = \frac{1}{3}e^{-3x} + c$.

5. From $\dfrac{1}{y}\, dy = \dfrac{4}{x}\, dx$ we obtain $\ln|y| = 4\ln|x| + c$ or $y = c_1 x^4$.

7. From $e^{-2y} dy = e^{3x}\, dx$ we obtain $3e^{-2y} + 2e^{3x} = c$.

9. From $\left(y + 2 + \dfrac{1}{y}\right) dy = x^2 \ln x \, dx$ we obtain $\dfrac{y^2}{2} + 2y + \ln|y| = \dfrac{x^3}{3} \ln|x| - \dfrac{1}{9}x^3 + c.$

11. From $\dfrac{1}{\csc y} dy = -\dfrac{1}{\sec^2 x} dx$ or $\sin y \, dy = -\cos^2 x \, dx = -\frac{1}{2}(1 + \cos 2x) \, dx$ we obtain
$-\cos y = -\frac{1}{2}x - \frac{1}{4}\sin 2x + c$ or $4\cos y = 2x + \sin 2x + c_1.$

13. From $\dfrac{e^y}{(e^y + 1)^2} dy = \dfrac{-e^x}{(e^x + 1)^3} dx$ we obtain $-(e^y + 1)^{-1} = \frac{1}{2}(e^x + 1)^{-2} + c.$

15. From $\dfrac{1}{S} dS = k \, dr$ we obtain $S = ce^{kr}.$

17. From $\dfrac{1}{P - P^2} dP = \left(\dfrac{1}{P} + \dfrac{1}{1 - P}\right) dP = dt$ we obtain $\ln|P| - \ln|1 - P| = t + c$ so that
$\ln\left|\dfrac{P}{1 - P}\right| = t + c$ or $\dfrac{P}{1 - P} = c_1 e^t.$ Solving for P we have $P = \dfrac{c_1 e^t}{1 + c_1 e^t}.$

19. From $\dfrac{y - 2}{y + 3} dy = \dfrac{x - 1}{x + 4} dx$ or $\left(1 - \dfrac{5}{y + 3}\right) dy = \left(1 - \dfrac{5}{x + 4}\right) dx$ we obtain
$y - 5\ln|y + 3| = x - 5\ln|x + 4| + c$ or $\left(\dfrac{x + 4}{y + 3}\right)^5 = c_1 e^{x-y}.$

21. From $x \, dx = \dfrac{1}{\sqrt{1 - y^2}} dy$ we obtain $\frac{1}{2}x^2 = \sin^{-1} y + c$ or $y = \sin\left(\dfrac{x^2}{2} + c_1\right).$

23. From $\dfrac{1}{x^2 + 1} dx = 4 \, dt$ we obtain $\tan^{-1} x = 4t + c.$ Using $x(\pi/4) = 1$ we find $c = -3\pi/4.$ The
solution of the initial-value problem is $\tan^{-1} x = 4t - \dfrac{3\pi}{4}$ or $x = \tan\left(4t - \dfrac{3\pi}{4}\right).$

25. From $\dfrac{1}{y} dy = \dfrac{1 - x}{x^2} dx = \left(\dfrac{1}{x^2} - \dfrac{1}{x}\right) dx$ we obtain $\ln|y| = -\dfrac{1}{x} - \ln|x| = c$ or $xy = c_1 e^{-1/x}.$
Using $y(-1) = -1$ we find $c_1 = e^{-1}.$ The solution of the initial-value problem is $xy = e^{-1-1/x}$
or $y = e^{-(1+1/x)}/x.$

27. Separating variables and integrating we obtain

$$\dfrac{dx}{\sqrt{1 - x^2}} - \dfrac{dy}{\sqrt{1 - y^2}} = 0 \quad \text{and} \quad \sin^{-1} x - \sin^{-1} y = c.$$

Setting $x = 0$ and $y = \sqrt{3}/2$ we obtain $c = -\pi/3.$ Thus, an implicit solution of the initial-value problem is $\sin^{-1} x - \sin^{-1} y = \pi/3.$ Solving for y and using an addition formula from trigonometry, we get

$$y = \sin\left(\sin^{-1} x + \dfrac{\pi}{3}\right) = x\cos\dfrac{\pi}{3} + \sqrt{1 - x^2}\sin\dfrac{\pi}{3} = \dfrac{x}{2} + \dfrac{\sqrt{3}\sqrt{1 - x^2}}{2}.$$

29. Separating variables and then proceeding as in Example 5 we get

$$\frac{dy}{dx} = ye^{-x^2}$$

$$\frac{1}{y}\frac{dy}{dx} = e^{-x^2}$$

$$\int_4^x \frac{1}{y(t)}\frac{dy}{dt}\,dt = \int_4^x e^{-t^2}\,dt$$

$$\ln y(t)\Big|_4^x = \int_4^x e^{-t^2}\,dt$$

$$\ln y(x) - \ln y(4) = \int_4^x e^{-t^2}\,dt$$

$$\ln y(x) = \int_4^x e^{-t^2}\,dt$$

$$y(x) = e^{\int_4^x e^{-t^2}\,dt}$$

31. Separating variables we get

$$\frac{dy}{dx} = \frac{2x+1}{2y}$$

$$2y\,dy = (2x+1)\,dx$$

$$\int 2y\,dy = \int (2x+1)\,dx$$

$$y^2 = x^2 + x + c$$

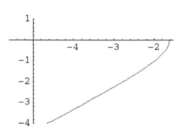

The condition $y(-2) = -1$ implies $c = -1$. Thus $y^2 = x^2 + x - 1$ and $y = -\sqrt{x^2 + x - 1}$ in order for y to be negative. Moreover for an interval containing -2 for values of x such that $x^2 + x - 1 > 0$ we get $\left(-\infty, -\frac{1}{2} - \frac{\sqrt{5}}{2}\right)$.

33. Separating variables we get

$$e^y\,dx - e^{-x}\,dy = 0$$

$$e^y\,dx = e^{-x}\,dy$$

$$e^x\,dx = e^{-y}\,dy$$

$$\int e^x\,dx = \int e^{-y}\,dy$$

$$e^x = -e^{-y} + c$$

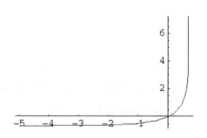

The condition $y(0) = 0$ implies $c = 2$. Thus $e^{-y} = 2 - e^x$. Therefore $y = -\ln(2 - e^x)$. Now we must have $2 - e^x > 0$ or $e^x < 2$. Since e^x is an increasing function this imples $x < \ln 2$ and so the interval of definition is $(-\infty, \ln 2)$.

35. (a) The equilibrium solutions $y(x) = 2$ and $y(x) = -2$ satisfy the initial conditions $y(0) = 2$ and $y(0) = -2$, respectively. Setting $x = \frac{1}{4}$ and $y = 1$ in $y = 2(1 + ce^{4x})/(1 - ce^{4x})$ we obtain

$$1 = 2\frac{1 + ce}{1 - ce}, \quad 1 - ce = 2 + 2ce, \quad -1 = 3ce, \quad \text{and} \quad c = -\frac{1}{3e}.$$

The solution of the corresponding initial-value problem is

$$y = 2\frac{1 - \frac{1}{3}e^{4x-1}}{1 + \frac{1}{3}e^{4x-1}} = 2\frac{3 - e^{4x-1}}{3 + e^{4x-1}}.$$

(b) Separating variables and integrating yields

$$\frac{1}{4}\ln|y - 2| - \frac{1}{4}\ln|y + 2| + \ln c_1 = x$$

$$\ln|y - 2| - \ln|y + 2| + \ln c = 4x$$

$$\ln\left|\frac{c(y - 2)}{y + 2}\right| = 4x$$

$$c\frac{y - 2}{y + 2} = e^{4x}.$$

Solving for y we get $y = 2(c + e^{4x})/(c - e^{4x})$. The initial condition $y(0) = -2$ implies $2(c + 1)/(c - 1) = -2$ which yields $c = 0$ and $y(x) = -2$. The initial condition $y(0) = 2$ does not correspond to a value of c, and it must simply be recognized that $y(x) = 2$ is a solution of the initial-value problem. Setting $x = \frac{1}{4}$ and $y = 1$ in $y = 2(c + e^{4x})/(c - e^{4x})$ leads to $c = -3e$. Thus, a solution of the initial-value problem is

$$y = 2\frac{-3e + e^{4x}}{-3e - e^{4x}} = 2\frac{3 - e^{4x-1}}{3 + e^{4x-1}}.$$

37. Singular solutions of $dy/dx = x\sqrt{1 - y^2}$ are $y = -1$ and $y = 1$. A singular solution of $(e^x + e^{-x})dy/dx = y^2$ is $y = 0$.

39. The singular solution $y = 1$ satisfies the initial-value problem.

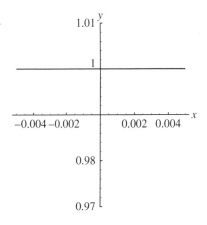

41. Separating variables we obtain $\dfrac{dy}{(y-1)^2+0.01}=dx$. Then

$$10\tan^{-1}10(y-1)=x+c \quad \text{and} \quad y=1+\frac{1}{10}\tan\frac{x+c}{10}\,.$$

Setting $x=0$ and $y=1$ we obtain $c=0$. The solution is

$$y=1+\frac{1}{10}\tan\frac{x}{10}\,.$$

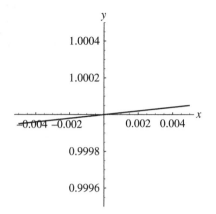

43. Separating variables, we have

$$\frac{dy}{y-y^3}=\frac{dy}{y(1-y)(1+y)}=\left(\frac{1}{y}+\frac{1/2}{1-y}-\frac{1/2}{1+y}\right)dy=dx.$$

Integrating, we get

$$\ln|y|-\frac{1}{2}\ln|1-y|-\frac{1}{2}\ln|1+y|=x+c.$$

When $y>1$, this becomes

$$\ln y-\frac{1}{2}\ln(y-1)-\frac{1}{2}\ln(y+1)=\ln\frac{y}{\sqrt{y^2-1}}=x+c.$$

Letting $x=0$ and $y=2$ we find $c=\ln(2/\sqrt{3})$. Solving for y we get $y_1(x)=2e^x/\sqrt{4e^{2x}-3}$, where $x>\ln(\sqrt{3}/2)$.

When $0<y<1$ we have

$$\ln y-\frac{1}{2}\ln(1-y)-\frac{1}{2}\ln(1+y)=\ln\frac{y}{\sqrt{1-y^2}}=x+c.$$

Letting $x=0$ and $y=\frac{1}{2}$ we find $c=\ln(1/\sqrt{3})$. Solving for y we get $y_2(x)=e^x/\sqrt{e^{2x}+3}$, where $-\infty<x<\infty$.

When $-1<y<0$ we have

$$\ln(-y)-\frac{1}{2}\ln(1-y)-\frac{1}{2}\ln(1+y)=\ln\frac{-y}{\sqrt{1-y^2}}=x+c.$$

Letting $x=0$ and $y=-\frac{1}{2}$ we find $c=\ln(1/\sqrt{3})$. Solving for y we get $y_3(x)=-e^x/\sqrt{e^{2x}+3}$, where $-\infty<x<\infty$.

When $y<-1$ we have

$$\ln(-y)-\frac{1}{2}\ln(1-y)-\frac{1}{2}\ln(-1-y)=\ln\frac{-y}{\sqrt{y^2-1}}=x+c.$$

Letting $x=0$ and $y=-2$ we find $c=\ln(2/\sqrt{3})$. Solving for y we get $y_4(x)=-2e^x/\sqrt{4e^{2x}-3}$, where $x>\ln(\sqrt{3}/2)$.

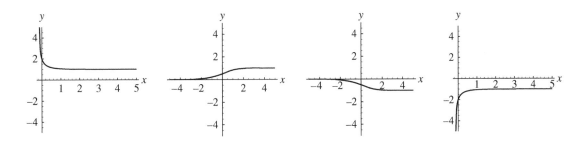

45. We separate variables and rationalize the denominator. Then

$$dy = \frac{1}{1+\sin x} \cdot \frac{1-\sin x}{1-\sin x}\, dx = \frac{1-\sin x}{1-\sin^2 x}\, dx = \frac{1-\sin x}{\cos^2 x}\, dx$$

$$= \left(\sec^2 x - \tan x \sec x\right)\, dx.$$

Integrating, we have $y = \tan x - \sec x + C$.

47. Separating variables we have $dy/\left(\sqrt{y}+y\right) = dx/\left(\sqrt{x}+x\right)$. To integrate $\int dx/\left(\sqrt{x}+x\right)$ we substitute $u^2 = x$ and get

$$\int \frac{2u}{u+u^2}\, du = \int \frac{2}{1+u}\, du = 2\ln|1+u| + c = 2\ln\left(1+\sqrt{x}\right) + c.$$

Integrating the separated differential equation we have

$$2\ln\left(1+\sqrt{y}\right) = 2\ln\left(1+\sqrt{x}\right) + c \quad \text{or} \quad \ln\left(1+\sqrt{y}\right) = \ln\left(1+\sqrt{x}\right) + \ln c_1.$$

Solving for y we get $y = \left[c_1\left(1+\sqrt{x}\right) - 1\right]^2$.

49. Separating variables we have $y\, dy = e^{\sqrt{x}}\, dx$. If $u = \sqrt{x}$, then $u^2 = x$ and $2u\, du = dx$. Thus, $\int e^{\sqrt{x}}\, dx = \int 2ue^u\, du$ and, using integration by parts, we find

$$\int y\, dy = \int e^{\sqrt{x}}\, dx \qquad \text{so} \qquad \frac{1}{2}y^2 = \int 2ue^u\, du = -2e^u + C = 2\sqrt{x}\, e^{\sqrt{x}} - 2e^{\sqrt{x}} + C,$$

and

$$y = 2\sqrt{\sqrt{x}\, e^{\sqrt{x}} - e^{\sqrt{x}} + C}\,.$$

To find C we solve $y(1) = 4$.

$$y(1) = 2\sqrt{\sqrt{1}\, e^{\sqrt{1}} - e^{\sqrt{1}} + C} = 2\sqrt{C} = 4 \qquad \text{so} \qquad C = 4.$$

and the solution of the intial-value problem is $y = 2\sqrt{\sqrt{x}\, e^{\sqrt{x}} - e^{\sqrt{x}} + 4}\,.$

51. (a) While $y_2(x) = -\sqrt{25 - x^2}$ is defined at $x = -5$ and $x = 5$, $y_2'(x)$ is not defined at these values, and so the interval of definition is the open interval $(-5, 5)$.

(b) At any point on the x-axis the derivative of $y(x)$ is undefined, so no solution curve can cross the x-axis. Since $-x/y$ is not defined when $y = 0$, the initial-value problem has no solution.

53. Separating variables we have $dy/\left(\sqrt{1+y^2} \, \sin^2 y \right) = dx$ which is not readily integrated (even by a CAS). We note that $dy/dx \geq 0$ for all values of x and y and that $dy/dx = 0$ when $y = 0$ and $y = \pi$, which are equilibrium solutions.

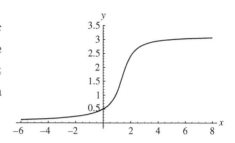

55. We are looking for a function $y(x)$ such that

$$y^2 + \left(\frac{dy}{dx} \right)^2 = 1.$$

Using the positive square root gives

$$\frac{dy}{dx} = \sqrt{1 - y^2}$$

$$\frac{dy}{\sqrt{1 - y^2}} = dx$$

$$\sin^{-1} y = x + c.$$

Thus a solution is $y = \sin(x + c)$. If we use the negative square root we obtain

$$y = \sin(c - x) = -\sin(x - c) = -\sin(x + c_1).$$

Note that when $c = c_1 = 0$ and when $c = c_1 = \pi/2$ we obtain the well known particular solutions $y = \sin x$, $y = -\sin x$, $y = \cos x$, and $y = -\cos x$. Note also that $y = 1$ and $y = -1$ are singular solutions.

57. Since the tension T_1 (or magnitude T_1) acts at the lowest point of the cable, we use symmetry to solve the problem on the interval $[0, L/2]$. The assumption that the roadbed is uniform (that is, weighs a constant ρ pounds per horizontal foot) implies $W = \rho x$, where x is measured in feet and $0 \leq x \leq L/2$. Therefore (10) becomes $dy/dx = (\rho/T_1)x$. This last equation is a separable equation of the form given in (1) of Section 2.2 in the text. Integrating and using the initial condition $y(0) = a$ shows that the shape of the cable is a parabola: $y(x) = (\rho/2T_1)x^2 + a$. In terms of the sag h of the cable and the span L, we see from Figure 2.2.5 in the text that $y(L/2) = h + a$. By applying this last condition to $y(x) = (\rho/2T_1)x^2 + a$ enables us to express $\rho/2T_1$ in terms of h and L: $y(x) = (4h/L^2)x^2 + a$. Since $y(x)$ is an even function of x, the solution is valid on $-L/2 \leq x \leq L/2$.

59. (a) An implicit solution of the differential equation $(2y + 2)dy - (4x^3 + 6x)\, dx = 0$ is

$$y^2 + 2y - x^4 - 3x^2 + c = 0.$$

The condition $y(0) = -3$ implies that $c = -3$. Therefore $y^2 + 2y - x^4 - 3x^2 - 3 = 0$.

(b) Using the quadratic formula we can solve for y in terms of x:

$$y = \frac{-2 \pm \sqrt{4 + 4(x^4 + 3x^2 + 3)}}{2}.$$

The explicit solution that satisfies the initial condition is then

$$y = -1 - \sqrt{x^4 + 3x^3 + 4}.$$

(c) From the graph of the function $f(x) = x^4 + 3x^3 + 4$ below we see that $f(x) \le 0$ on the approximate interval $-2.8 \le x \le -1.3$. Thus the approximate domain of the function

$$y = -1 - \sqrt{x^4 + 3x^3 + 4} = -1 - \sqrt{f(x)}$$

is $x \le -2.8$ or $x \ge -1.3$. The graph of this function is shown below.

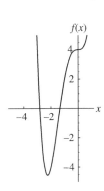

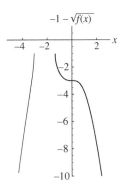

(d) Using the root finding capabilities of a CAS, the zeros of f are found to be -2.82202 and -1.3409. The domain of definition of the solution $y(x)$ is then $x > -1.3409$. The equality has been removed since the derivative dy/dx does not exist at the points where $f(x) = 0$. The graph of the solution $y = \phi(x)$ is given on the right.

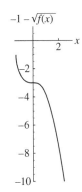

2.3 | Linear Equations

1. For $y' - 5y = 0$ an integrating factor is $e^{-\int 5\, dx} = e^{-5x}$ so that $\dfrac{d}{dx}\left[e^{-5x}y\right] = 0$ and $y = ce^{5x}$

for $-\infty < x < \infty$.

3. For $y' + y = e^{3x}$ an integrating factor is $e^{\int dx} = e^x$ so that $\dfrac{d}{dx}[e^x y] = e^{4x}$ and $y = \frac{1}{4}e^{3x} + ce^{-x}$ for $-\infty < x < \infty$. The transient term is ce^{-x}.

5. For $y' + 3x^2 y = x^2$ an integrating factor is $e^{\int 3x^2\, dx} = e^{x^3}$ so that $\dfrac{d}{dx}\left[e^{x^3} y\right] = x^2 e^{x^3}$ and $y = \frac{1}{3} + ce^{-x^3}$ for $-\infty < x < \infty$. The transient term is ce^{-x^3}.

7. For $y' + \dfrac{1}{x}y = \dfrac{1}{x^2}$ an integrating factor is $e^{\int (1/x)\, dx} = x$ so that $\dfrac{d}{dx}[xy] = \dfrac{1}{x}$ and $y = \dfrac{1}{x}\ln x + \dfrac{c}{x}$ for $0 < x < \infty$. The entire solution is transient.

9. For $y' - \dfrac{1}{x}y = x\sin x$ an integrating factor is $e^{-\int (1/x)\, dx} = \dfrac{1}{x}$ so that $\dfrac{d}{dx}\left[\dfrac{1}{x}y\right] = \sin x$ and $y = cx - x\cos x$ for $0 < x < \infty$. There is no transient term.

11. For $y' + \dfrac{4}{x}y = x^2 - 1$ an integrating factor is $e^{\int (4/x)dx} = x^4$ so that $\dfrac{d}{dx}\left[x^4 y\right] = x^6 - x^4$ and $y = \frac{1}{7}x^3 - \frac{1}{5}x + cx^{-4}$ for $0 < x < \infty$. The transient term is cx^{-4}.

13. For $y' + \left(1 + \dfrac{2}{x}\right)y = \dfrac{e^x}{x^2}$ an integrating factor is $e^{\int [1+(2/x)]dx} = x^2 e^x$ so that $\dfrac{d}{dx}[x^2 e^x y] = e^{2x}$ and $y = \dfrac{1}{2}\dfrac{e^x}{x^2} + \dfrac{ce^{-x}}{x^2}$ for $0 < x < \infty$. The transient term is $\dfrac{ce^{-x}}{x^2}$.

15. For $\dfrac{dx}{dy} - \dfrac{4}{y}x = 4y^5$ an integrating factor is $e^{-\int (4/y)\, dy} = e^{\ln y^{-4}} = y^{-4}$ so that $\dfrac{d}{dy}\left[y^{-4}x\right] = 4y$ and $x = 2y^6 + cy^4$ for $0 < y < \infty$. There is no transient term.

17. For $y' + (\tan x)y = \sec x$ an integrating factor is $e^{\int \tan x\, dx} = \sec x$ so that $\dfrac{d}{dx}[(\sec x)y] = \sec^2 x$ and $y = \sin x + c\cos x$ for $-\pi/2 < x < \pi/2$. There is no transient term.

19. For $y' + \dfrac{x+2}{x+1}y = \dfrac{2xe^{-x}}{x+1}$ an integrating factor is $e^{\int [(x+2)/(x+1)]dx} = (x+1)e^x$, so $\dfrac{d}{dx}[(x+1)e^x y] = 2x$ and $y = \dfrac{x^2}{x+1}e^{-x} + \dfrac{c}{x+1}e^{-x}$ for $-1 < x < \infty$. The entire solution is transient.

21. For $\dfrac{dr}{d\theta} + r\sec\theta = \cos\theta$ an integrating factor is $e^{\int \sec\theta\, d\theta} = e^{\ln|\sec x + \tan x|} = \sec\theta + \tan\theta$ so

that $\dfrac{d}{d\theta}[(\sec\theta + \tan\theta)r] = 1 + \sin\theta$ and $(\sec\theta + \tan\theta)r = \theta - \cos\theta + c$ for $-\pi/2 < \theta < \pi/2$.

There is no transient term.

23. For $y' + \left(3 + \dfrac{1}{x}\right)y = \dfrac{e^{-3x}}{x}$ an integrating factor is $e^{\int[3+(1/x)]dx} = xe^{3x}$ so that $\dfrac{d}{dx}\left[xe^{3x}y\right] = 1$

and $y = e^{-3x} + \dfrac{ce^{-3x}}{x}$ for $0 < x < \infty$. The transient term is ce^{-3x}/x.

25. For $y' - 5y = x$ an integrating factor is $e^{\int -5\,dx} = e^{-5x}$ so that $\dfrac{d}{dx}\left[e^{-5x}y\right] = xe^{-5x}$ and

$$y = e^{5x}\int xe^{-5x}\,dx = e^{5x}\left(-\frac{1}{5}xe^{-5x} - \frac{1}{25}e^{-5x} + c\right) = -\frac{1}{5}x - \frac{1}{25} + ce^{5x}.$$

If $y(0) = 3$ then $c = \dfrac{1}{25}$ and $y = -\dfrac{1}{5}x - \dfrac{1}{25} + \dfrac{76}{25}e^{5x}$. The solution is defined on $I = (-\infty, \infty)$.

27. For $y' + \dfrac{1}{x}y = \dfrac{1}{x}e^x$ an integrating factor is $e^{\int(1/x)dx} = x$ so that $\dfrac{d}{dx}[xy] = e^x$ and $y = \dfrac{1}{x}e^x + \dfrac{c}{x}$

for $0 < x < \infty$. If $y(1) = 2$ then $c = 2 - e$ and $y = \dfrac{1}{x}e^x + \dfrac{2 - e}{x}$. The solution is defined on

$I = (0, \infty)$.

29. For $\dfrac{di}{dt} + \dfrac{R}{L}i = \dfrac{E}{L}$ an integrating factor is $e^{\int(R/L)\,dt} = e^{Rt/L}$ so that $\dfrac{d}{dt}\left[e^{Rt/L}\,i\right] = \dfrac{E}{L}e^{Rt/L}$

and $i = \dfrac{E}{R} + ce^{-Rt/L}$ for $-\infty < t < \infty$. If $i(0) = i_0$ then $c = i_0 - E/R$ and $i = \dfrac{E}{R} +$

$\left(i_0 - \dfrac{E}{R}\right)e^{-Rt/L}$. The solution is defined on $I = (-\infty, \infty)$

31. For $y' + \dfrac{1}{x}y = 4 + \dfrac{1}{x}$ an integrating factor is $e^{\int(1/x)\,dx} = x$ so that $\dfrac{d}{dx}[xy] = 4x + 1$ and

$$y = \frac{1}{x}\int (4x + 1)\,dx = \frac{1}{x}\left(2x^2 + x + c\right) = 2x + 1 + \frac{c}{x}.$$

If $y(1) = 8$ then $c = 5$ and $y = 2x + 1 + \dfrac{5}{x}$. The solution is defined on $I = (0, \infty)$.

33. For $y' + \dfrac{1}{x+1}y = \dfrac{\ln x}{x+1}$ an integrating factor is $e^{\int[1/(x+1)]\,dx} = x+1$ so that $\dfrac{d}{dx}[(x+1)y] = \ln x$

and

$$y = \frac{x}{x+1}\ln x - \frac{x}{x+1} + \frac{c}{x+1} \qquad \text{for} \quad 0 < x < \infty.$$

If $y(1) = 10$ then $c = 21$ and $y = \dfrac{x}{x+1}\ln x - \dfrac{x}{x+1} + \dfrac{21}{x+1}$. The solution is defined on $I = (0, \infty)$.

35. For $y' - (\sin x)\, y = 2\sin x$ an integrating factor is $e^{\int(-\sin x)\,dx} = e^{\cos x}$ so that $\dfrac{d}{dx}\left[e^{\cos x}y\right] = 2(\sin x)\,e^{\cos x}$ and

$$y = e^{-\cos x}\int 2(\sin x)\,e^{\cos x}\,dx = e^{-\cos x}\left(-2e^{\cos x} + c\right) = -2 + ce^{-\cos x}.$$

If $y(\pi/2) = 1$ then $c = 3$ and $y = -2 + 3e^{-\cos x}$. The solution is defined on $I = (-\infty, \infty)$.

37. For $y' + 2y = f(x)$ an integrating factor is e^{2x} so that

$$ye^{2x} = \begin{cases} \dfrac{1}{2}e^{2x} + c_1, & 0 \le x \le 3 \\[2mm] c_2, & x > 3. \end{cases}$$

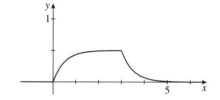

If $y(0) = 0$ then $c_1 = -1/2$ and for continuity we must have $c_2 = \frac{1}{2}e^6 - \frac{1}{2}$ so that

$$y = \begin{cases} \dfrac{1}{2}(1 - e^{-2x}), & 0 \le x \le 3 \\[2mm] \dfrac{1}{2}(e^6 - 1)e^{-2x}, & x > 3. \end{cases}$$

39. For $y' + 2xy = f(x)$ an integrating factor is e^{x^2} so that

$$ye^{x^2} = \begin{cases} \dfrac{1}{2}e^{x^2} + c_1, & 0 \le x \le 1 \\[2mm] c_2, & x > 1. \end{cases}$$

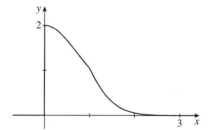

If $y(0) = 2$ then $c_1 = 3/2$ and for continuity we must have $c_2 = \frac{1}{2}e + \frac{3}{2}$ so that

$$y = \begin{cases} \dfrac{1}{2} + \dfrac{3}{2}e^{-x^2}, & 0 \le x \le 1 \\[2mm] \left(\dfrac{1}{2}e + \dfrac{3}{2}\right)e^{-x^2}, & x > 1. \end{cases}$$

41. We first solve the initial-value problem $y' + 2y = 4x$, $y(0) = 3$ on the interval $[0, 1]$. The integrating factor is $e^{\int 2\,dx} = e^{2x}$, so

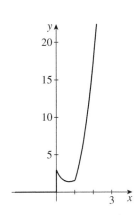

$$\frac{d}{dx}[e^{2x}y] = 4xe^{2x}$$

$$e^{2x}y = \int 4xe^{2x}\,dx = 2xe^{2x} - e^{2x} + c_1$$

$$y = 2x - 1 + c_1e^{-2x}.$$

Using the initial condition, we find $y(0) = -1 + c_1 = 3$, so $c_1 = 4$ and $y = 2x - 1 + 4e^{-2x}$,

$0 \le x \le 1$. Now, since $y(1) = 2 - 1 + 4e^{-2} = 1 + 4e^{-2}$, we solve the initial-value problem

$y' - (2/x)y = 4x$, $y(1) = 1 + 4e^{-2}$ on the interval $(1, \infty)$. The integrating factor is

$e^{\int (-2/x)\,dx} = e^{-2\ln x} = x^{-2}$, so

$$\frac{d}{dx}[x^{-2}y] = 4xx^{-2} = \frac{4}{x}$$

$$x^{-2}y = \int \frac{4}{x}\,dx = 4\ln x + c_2$$

$$y = 4x^2 \ln x + c_2x^2.$$

(We use $\ln x$ instead of $\ln |x|$ because $x > 1$.) Using the initial condition we find

$y(1) = c_2 = 1 + 4e^{-2}$, so $y = 4x^2 \ln x + (1 + 4e^{-2})x^2$, $x > 1$. Thus,

$$y = \begin{cases} 2x - 1 + 4e^{-2x}, & 0 \le x \le 1 \\ \\ 4x^2 \ln x + \left(1 + 4e^{-2}\right)x^2, & x > 1. \end{cases}$$

43. An integrating factor for $y' - 2xy = 1$ is e^{-x^2}. Thus

$$\frac{d}{dx}[e^{-x^2}y] = e^{-x^2}$$

$$e^{-x^2}y = \int_0^x e^{-t^2}\,dt = \frac{\sqrt{\pi}}{2}\operatorname{erf}(x) + c$$

$$y = \frac{\sqrt{\pi}}{2}e^{x^2}\operatorname{erf}(x) + ce^{x^2}.$$

From $y(1) = (\sqrt{\pi}/2)e\,\mathrm{erf}(1) + ce = 1$ we get $c = e^{-1} - \frac{\sqrt{\pi}}{2}\,\mathrm{erf}(1)$. The solution of the initial-value problem is

$$y = \frac{\sqrt{\pi}}{2}\,e^{x^2}\,\mathrm{erf}(x) + \left(e^{-1} - \frac{\sqrt{\pi}}{2}\,\mathrm{erf}(1)\right)e^{x^2}$$

$$= e^{x^2-1} + \frac{\sqrt{\pi}}{2}\,e^{x^2}\,(\mathrm{erf}(x) - \mathrm{erf}(1)).$$

45. For $y' + e^x y = 1$ an integrating factor is e^{e^x}. Thus

$$\frac{d}{dx}\left[e^{e^x} y\right] = e^{e^x} \quad\text{and}\quad e^{e^x} y = \int_0^x e^{e^t}\,dt + c.$$

From $y(0) = 1$ we get $c = e$, so $y = e^{-e^x}\int_0^x e^{e^t}\,dt + e^{1-e^x}$.

47. An integrating factor for

$$y' + \frac{2}{x}\,y = \frac{10\sin x}{x^3}$$

is x^2. Thus

$$\frac{d}{dx}\left[x^2 y\right] = 10\frac{\sin x}{x}$$

$$x^2 y = 10\int_0^x \frac{\sin t}{t}\,dt + c$$

$$y = 10x^{-2}\,\mathrm{Si}\,(x) + cx^{-2}.$$

From $y(1) = 0$ we get $c = -10\,\mathrm{Si}\,(1)$. Thus

$$y = 10x^{-2}\,\mathrm{Si}\,(x) - 10x^{-2}\,\mathrm{Si}\,(1) = 10x^{-2}\,(\mathrm{Si}\,(x) - \mathrm{Si}\,(1))\,.$$

49. We want 4 to be a critical point, so we use $y' = 4 - y$.

51. On the interval $(-3, 3)$ the integrating factor is

$$e^{\int x\,dx/(x^2-9)} = e^{-\int x\,dx/(9-x^2)} = e^{\frac{1}{2}\ln(9-x^2)} = \sqrt{9 - x^2}$$

and so

$$\frac{d}{dx}\left[\sqrt{9-x^2}\,y\right]=0 \quad \text{and} \quad y=\frac{c}{\sqrt{9-x^2}}.$$

53. The left-hand derivative of the function at $x=1$ is $1/e$ and the right-hand derivative at $x=1$ is $1-1/e$. Thus, y is not differentiable at $x=1$.

55. Since $e^{\int P(x)\,dx+c}=e^c e^{\int P(x)\,dx}=c_1 e^{\int P(x)\,dx}$, we would have

$$c_1 e^{\int P(x)\,dx}y=c_2+\int c_1 e^{\int P(x)\,dx}f(x)\,dx \quad \text{and} \quad e^{\int P(x)\,dx}y=c_3+\int e^{\int P(x)\,dx}f(x)\,dx,$$

which is the same as (4) in the text.

57. The solution of the first equation is $x=c_1 e^{-\lambda_1 t}$. From $x(0)=x_0$ we obtain $c_1=x_0$ and so $x=x_0 e^{-\lambda_1 t}$. The second equation then becomes

$$\frac{dy}{dt}=x_0\lambda_1 e^{-\lambda_1 t}-\lambda_2 y \quad \text{or} \quad \frac{dy}{dt}+\lambda_2 y=x_0\lambda_1 e^{-\lambda_1 t}$$

which is linear. An integrating factor is $e^{\lambda_2 t}$. Thus

$$\frac{d}{dt}\left[e^{\lambda_2 t}y\right]=x_0\lambda_1 e^{-\lambda_1 t}e^{\lambda_2 t}=x_0\lambda_1 e^{(\lambda_2-\lambda_1)t}$$

$$e^{\lambda_2 t}y=\frac{x_0\lambda_1}{\lambda_2-\lambda_1}e^{(\lambda_2-\lambda_1)t}+c_2$$

$$y=\frac{x_0\lambda_1}{\lambda_2-\lambda_1}e^{-\lambda_1 t}+c_2 e^{-\lambda_2 t}.$$

From $y(0)=y_0$ we obtain $c_2=(y_0\lambda_2-y_0\lambda_1-x_0\lambda_1)/(\lambda_2-\lambda_1)$. The solution is

$$y=\frac{x_0\lambda_1}{\lambda_2-\lambda_1}e^{-\lambda_1 t}+\frac{y_0\lambda_2-y_0\lambda_1-x_0\lambda_1}{\lambda_2-\lambda_1}e^{-\lambda_2 t}.$$

59. (a)

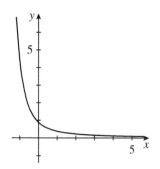

(b) Using a CAS we find $y(2) \approx 0.226339$.

61. (a)

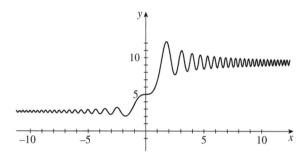

(b) From the graph we see that as $x \to \infty$, $y(x)$ oscillates with decreasing amplitudes approaching 9.35672. Since $\lim\limits_{x \to \infty} S(x) = \dfrac{1}{2}$, we have $\lim\limits_{x \to \infty} y(x) = 5e^{\sqrt{\pi/8}} \approx 9.357$, and since $\lim\limits_{x \to -\infty} S(x) = -\dfrac{1}{2}$, we have $\lim\limits_{x \to -\infty} y(x) = 5e^{-\sqrt{\pi/8}} \approx 2.672$.

(c) From the graph in part (b) we see that the absolute maximum occurs around $x = 1.7$ and the absolute minimum occurs around $x = -1.8$. Using the root-finding capability of a CAS and solving $y'(x) = 0$ for x, we see that the absolute maximum is $(1.772, 12.235)$ and the absolute minimum is $(-1.772, 2.044)$.

2.4 | Exact Equations

1. Let $M = 2x - 1$ and $N = 3y + 7$ so that $M_y = 0 = N_x$. From $f_x = 2x - 1$ we obtain $f = x^2 - x + h(y)$, $h'(y) = 3y + 7$, and $h(y) = \frac{3}{2}y^2 + 7y$. A solution is $x^2 - x + \frac{3}{2}y^2 + 7y = c$.

3. Let $M = 5x + 4y$ and $N = 4x - 8y^3$ so that $M_y = 4 = N_x$. From $f_x = 5x + 4y$ we obtain $f = \frac{5}{2}x^2 + 4xy + h(y)$, $h'(y) = -8y^3$, and $h(y) = -2y^4$. A solution is $\frac{5}{2}x^2 + 4xy - 2y^4 = c$.

5. Let $M = 2y^2x - 3$ and $N = 2yx^2 + 4$ so that $M_y = 4xy = N_x$. From $f_x = 2y^2x - 3$ we obtain $f = x^2y^2 - 3x + h(y)$, $h'(y) = 4$, and $h(y) = 4y$. A solution is $x^2y^2 - 3x + 4y = c$.

7. Let $M = x^2 - y^2$ and $N = x^2 - 2xy$ so that $M_y = -2y$ and $N_x = 2x - 2y$. The equation is not exact.

9. Let $M = y^3 - y^2 \sin x - x$ and $N = 3xy^2 + 2y \cos x$ so that $M_y = 3y^2 - 2y \sin x = N_x$. From $f_x = y^3 - y^2 \sin x - x$ we obtain $f = xy^3 + y^2 \cos x - \frac{1}{2}x^2 + h(y)$, $h'(y) = 0$, and $h(y) = 0$. A solution is $xy^3 + y^2 \cos x - \frac{1}{2}x^2 = c$.

11. Let $M = y \ln y - e^{-xy}$ and $N = 1/y + x \ln y$ so that $M_y = 1 + \ln y + xe^{-xy}$ and $N_x = \ln y$. The equation is not exact.

13. Let $M = y - 6x^2 - 2xe^x$ and $N = x$ so that $M_y = 1 = N_x$. From $f_x = y - 6x^2 - 2xe^x$ we obtain $f = xy - 2x^3 - 2xe^x + 2e^x + h(y)$, $h'(y) = 0$, and $h(y) = 0$. A solution is $xy - 2x^3 - 2xe^x + 2e^x = c$.

15. Let $M = x^2 y^3 - 1/\left(1 + 9x^2\right)$ and $N = x^3 y^2$ so that $M_y = 3x^2 y^2 = N_x$. From $f_x = x^2 y^3 - 1/\left(1 + 9x^2\right)$ we obtain $f = \frac{1}{3}x^3 y^3 - \frac{1}{3} \arctan(3x) + h(y)$, $h'(y) = 0$, and $h(y) = 0$. A solution is $x^3 y^3 - \arctan(3x) = c$.

17. Let $M = \tan x - \sin x \sin y$ and $N = \cos x \cos y$ so that $M_y = -\sin x \cos y = N_x$. From $f_x = \tan x - \sin x \sin y$ we obtain $f = \ln|\sec x| + \cos x \sin y + h(y)$, $h'(y) = 0$, and $h(y) = 0$. A solution is $\ln|\sec x| + \cos x \sin y = c$.

19. Let $M = 4t^3 y - 15t^2 - y$ and $N = t^4 + 3y^2 - t$ so that $M_y = 4t^3 - 1 = N_t$. From $f_t = 4t^3 y - 15t^2 - y$ we obtain $f = t^4 y - 5t^3 - ty + h(y)$, $h'(y) = 3y^2$, and $h(y) = y^3$. A solution is $t^4 y - 5t^3 - ty + y^3 = c$.

21. Let $M = x^2 + 2xy + y^2$ and $N = 2xy + x^2 - 1$ so that $M_y = 2(x + y) = N_x$. From $f_x = x^2 + 2xy + y^2$ we obtain $f = \frac{1}{3}x^3 + x^2 y + xy^2 + h(y)$, $h'(y) = -1$, and $h(y) = -y$. The solution is $\frac{1}{3}x^3 + x^2 y + xy^2 - y = c$. If $y(1) = 1$ then $c = 4/3$ and a solution of the initial-value problem is $\frac{1}{3}x^3 + x^2 y + xy^2 - y = \frac{4}{3}$.

23. Let $M = 4y + 2t - 5$ and $N = 6y + 4t - 1$ so that $M_y = 4 = N_t$. From $f_t = 4y + 2t - 5$ we obtain $f = 4ty + t^2 - 5t + h(y)$, $h'(y) = 6y - 1$, and $h(y) = 3y^2 - y$. The solution is $4ty + t^2 - 5t + 3y^2 - y = c$. If $y(-1) = 2$ then $c = 8$ and a solution of the initial-value problem is $4ty + t^2 - 5t + 3y^2 - y = 8$.

25. Let $M = y^2 \cos x - 3x^2 y - 2x$ and $N = 2y \sin x - x^3 + \ln y$ so that $M_y = 2y \cos x - 3x^2 = N_x$. From $f_x = y^2 \cos x - 3x^2 y - 2x$ we obtain $f = y^2 \sin x - x^3 y - x^2 + h(y)$, $h'(y) = \ln y$, and $h(y) = y \ln y - y$. The solution is $y^2 \sin x - x^3 y - x^2 + y \ln y - y = c$. If $y(0) = e$ then $c = 0$ and a solution of the initial-value problem is $y^2 \sin x - x^3 y - x^2 + y \ln y - y = 0$.

27. Equating $M_y = 3y^2 + 4kxy^3$ and $N_x = 3y^2 + 40xy^3$ we obtain $k = 10$.

29. Let $M = -x^2 y^2 \sin x + 2xy^2 \cos x$ and $N = 2x^2 y \cos x$ so that $M_y = -2x^2 y \sin x + 4xy \cos x = N_x$. From $f_y = 2x^2 y \cos x$ we obtain $f = x^2 y^2 \cos x + h(y)$, $h'(y) = 0$, and $h(y) = 0$. A solution of the differential equation is $x^2 y^2 \cos x = c$.

31. We note that $(M_y - N_x)/N = 1/x$, so an integrating factor is $e^{\int dx/x} = x$. Let $M = 2xy^2 + 3x^2$ and $N = 2x^2y$ so that $M_y = 4xy = N_x$. From $f_x = 2xy^2 + 3x^2$ we obtain $f = x^2y^2 + x^3 + h(y)$, $h'(y) = 0$, and $h(y) = 0$. A solution of the differential equation is $x^2y^2 + x^3 = c$.

33. We note that $(N_x - M_y)/M = 2/y$, so an integrating factor is $e^{\int 2\,dy/y} = y^2$. Let $M = 6xy^3$ and $N = 4y^3 + 9x^2y^2$ so that $M_y = 18xy^2 = N_x$. From $f_x = 6xy^3$ we obtain $f = 3x^2y^3 + h(y)$, $h'(y) = 4y^3$, and $h(y) = y^4$. A solution of the differential equation is $3x^2y^3 + y^4 = c$.

35. We note that $(M_y - N_x)/N = 3$, so an integrating factor is $e^{\int 3\,dx} = e^{3x}$. Let
$M = (10 - 6y + e^{-3x})e^{3x} = 10e^{3x} - 6ye^{3x} + 1$ and $N = -2e^{3x}$, so that $M_y = -6e^{3x} = N_x$.
From $f_x = 10e^{3x} - 6ye^{3x} + 1$ we obtain $f = \frac{10}{3}e^{3x} - 2ye^{3x} + x + h(y)$, $h'(y) = 0$, and $h(y) = 0$.
A solution of the differential equation is $\frac{10}{3}e^{3x} - 2ye^{3x} + x = c$.

37. We note that $(M_y - N_x)/N = 2x/(4 + x^2)$, so an integrating factor is
$e^{-2\int x\,dx/(4+x^2)} = 1/(4 + x^2)$. Let $M = x/(4 + x^2)$ and $N = (x^2y + 4y)/(4 + x^2) = y$, so
that $M_y = 0 = N_x$. From $f_x = x(4 + x^2)$ we obtain $f = \frac{1}{2}\ln(4 + x^2) + h(y)$, $h'(y) = y$, and
$h(y) = \frac{1}{2}y^2$. A solution of the differential equation is $\frac{1}{2}\ln(4 + x^2) + \frac{1}{2}y^2 = c$. Multiplying both
sides by 2 the last equation can be written as $e^{y^2}(x^2 + 4) = c_1$. Using the initial condition
$y(4) = 0$ we see that $c_1 = 20$. A solution of the initial-value problem is $e^{y^2}(x^2 + 4) = 20$.

39. (a) Implicitly differentiating $x^3 + 2x^2y + y^2 = c$ and solving for dy/dx we obtain

$$3x^2 + 2x^2\frac{dy}{dx} + 4xy + 2y\frac{dy}{dx} = 0 \quad \text{and} \quad \frac{dy}{dx} = -\frac{3x^2 + 4xy}{2x^2 + 2y}.$$

By writing the last equation in differential form we get $(4xy + 3x^2)dx + (2y + 2x^2)dy = 0$.

(b) Setting $x = 0$ and $y = -2$ in $x^3 + 2x^2y + y^2 = c$ we find $c = 4$, and setting $x = y = 1$ we also find $c = 4$. Thus, both initial conditions determine the same implicit solution.

(c) Solving $x^3 + 2x^2y + y^2 = 4$ for y we get

$$y_1(x) = -x^2 - \sqrt{4 - x^3 + x^4}$$

and

$$y_2(x) = -x^2 + \sqrt{4 - x^3 + x^4}.$$

Observe in the figure that $y_1(0) = -2$ and $y_2(1) = 1$.

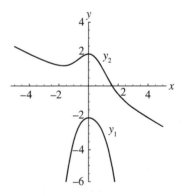

41. The explicit solution is $y = \sqrt{(3 + \cos^2 x)/(1 - x^2)}$. Since $3 + \cos^2 x > 0$ for all x we must have $1 - x^2 > 0$ or $-1 < x < 1$. Thus, the interval of definition is $(-1, 1)$.

43. First note that
$$d\left(\sqrt{x^2 + y^2}\right) = \frac{x}{\sqrt{x^2 + y^2}}\,dx + \frac{y}{\sqrt{x^2 + y^2}}\,dy.$$

Then $x\,dx + y\,dy = \sqrt{x^2+y^2}\,dx$ becomes

$$\frac{x}{\sqrt{x^2+y^2}}\,dx + \frac{y}{\sqrt{x^2+y^2}}\,dy = d\left(\sqrt{x^2+y^2}\right) = dx.$$

The left side is the total differential of $\sqrt{x^2+y^2}$ and the right side is the total differential of $x+c$. Thus $\sqrt{x^2+y^2} = x+c$ is a solution of the differential equation.

45. (a) In differential form

$$\left(v^2 - 32x\right)dx + xv\,dv = 0$$

This is not an exact equation, but $\mu(x) = x$ is an integrating factor. The new equation $\left(xv^2 - 32x^2\right)dx + x^2 v\,dv = 0$ is exact and solving yields $\frac{1}{2}x^2 v^2 - \frac{32}{3}x^3 = c$. When $x = 3$, $v = 0$ and so $c = -288$. Solving $\frac{1}{2}x^2 v^2 - \frac{32}{3}x^3 = -288$ for v yields the explicit solution

$$v(x) = 8\sqrt{\frac{x}{3} - \frac{9}{x^2}}\,.$$

(b) The chain leaves the platform when $x = 8$, and so

$$v(8) = 8\sqrt{\frac{8}{3} - \frac{9}{64}} \approx 12.7 \text{ ft/s}$$

2.5 Solutions by Substitutions

1. Letting $y = ux$ we have

$$(x - ux)\,dx + x(u\,dx + x\,du) = 0$$

$$dx + x\,du = 0$$

$$\frac{dx}{x} + du = 0$$

$$\ln|x| + u = c$$

$$x\ln|x| + y = cx.$$

3. Letting $x = vy$ we have

$$vy(v\,dy + y\,dv) + (y - 2vy)\,dy = 0$$

$$vy^2\,dv + y\left(v^2 - 2v + 1\right)dy = 0$$

$$\frac{v\,dv}{(v-1)^2} + \frac{dy}{y} = 0$$

$$\ln|v-1| - \frac{1}{v-1} + \ln|y| = c$$

$$\ln\left|\frac{x}{y} - 1\right| - \frac{1}{x/y - 1} + \ln|y| = c$$

$$(x - y)\ln|x - y| - y = c(x - y).$$

5. Letting $y = ux$ we have

$$\left(u^2 x^2 + ux^2\right) dx - x^2(u\, dx + x\, du) = 0$$

$$u^2\, dx - x\, du = 0$$

$$\frac{dx}{x} - \frac{du}{u^2} = 0$$

$$\ln|x| + \frac{1}{u} = c$$

$$\ln|x| + \frac{x}{y} = c$$

$$y \ln|x| + x = cy.$$

7. Letting $y = ux$ we have

$$(ux - x)\, dx - (ux + x)(u\, dx + x\, du) = 0$$

$$\left(u^2 + 1\right) dx + x(u + 1)\, du = 0$$

$$\frac{dx}{x} + \frac{u+1}{u^2+1}\, du = 0$$

$$\ln|x| + \frac{1}{2}\ln\left(u^2 + 1\right) + \tan^{-1} u = c$$

$$\ln x^2 \left(\frac{y^2}{x^2} + 1\right) + 2\tan^{-1}\frac{y}{x} = c_1$$

$$\ln\left(x^2 + y^2\right) + 2\tan^{-1}\frac{y}{x} = c_1$$

9. Letting $y = ux$ we have

$$-ux\, dx + (x + \sqrt{u}\, x)(u\, dx + x\, du) = 0$$

$$\left(x^2 + x^2\sqrt{u}\,\right) du + xu^{3/2}\, dx = 0$$

$$\left(u^{-3/2} + \frac{1}{u}\right) du + \frac{dx}{x} = 0$$

$$-2u^{-1/2} + \ln|u| + \ln|x| = c$$

$$\ln|y/x| + \ln|x| = 2\sqrt{x/y} + c$$

$$y(\ln|y| - c)^2 = 4x.$$

11. Letting $y = ux$ we have

$$\left(x^3 - u^3 x^3\right) dx + u^2 x^3 (u\, dx + x\, du) = 0$$

$$dx + u^2 x\, du = 0$$

$$\frac{dx}{x} + u^2\, du = 0$$

$$\ln|x| + \frac{1}{3}u^3 = c$$

$$3x^3 \ln|x| + y^3 = c_1 x^3.$$

Using $y(1) = 2$ we find $c_1 = 8$. The solution of the initial-value problem is $3x^3 \ln|x| + y^3 = 8x^3$.

13. Letting $y = ux$ we have

$$(x + uxe^u)\, dx - xe^u (u\, dx + x\, du) = 0$$

$$dx - xe^u\, du = 0$$

$$\frac{dx}{x} - e^u\, du = 0$$

$$\ln|x| - e^u = c$$

$$\ln|x| - e^{y/x} = c.$$

Using $y(1) = 0$ we find $c = -1$. The solution of the initial-value problem is $\ln|x| = e^{y/x} - 1$.

15. From $y' + \dfrac{1}{x}y = \dfrac{1}{x}y^{-2}$ and $w = y^3$ we obtain $\dfrac{dw}{dx} + \dfrac{3}{x}w = \dfrac{3}{x}$. An integrating factor is x^3 so that $x^3 w = x^3 + c$ or $y^3 = 1 + cx^{-3}$.

17. From $y' + y = xy^4$ and $w = y^{-3}$ we obtain $\dfrac{dw}{dx} - 3w = -3x$. An integrating factor is e^{-3x} so that $e^{-3x}w = xe^{-3x} + \frac{1}{3}e^{-3x} + c$ or $y^{-3} = x + \frac{1}{3} + ce^{3x}$.

19. From $y' - \dfrac{1}{t}y = -\dfrac{1}{t^2}y^2$ and $w = y^{-1}$ we obtain $\dfrac{dw}{dt} + \dfrac{1}{t}w = \dfrac{1}{t^2}$. An integrating factor is t so that $tw = \ln t + c$ or $y^{-1} = \dfrac{1}{t}\ln t + \dfrac{c}{t}$. Writing this in the form $\dfrac{t}{y} = \ln t + c$, we see that the solution can also be expressed in the form $e^{t/y} = c_1 t$.

21. From $y' - \dfrac{2}{x}y = \dfrac{3}{x^2}y^4$ and $w = y^{-3}$ we obtain $\dfrac{dw}{dx} + \dfrac{6}{x}w = -\dfrac{9}{x^2}$. An integrating factor is x^6 so that $x^6 w = -\frac{9}{5}x^5 + c$ or $y^{-3} = -\frac{9}{5}x^{-1} + cx^{-6}$. If $y(1) = \frac{1}{2}$ then $c = \frac{49}{5}$ and $y^{-3} = -\frac{9}{5}x^{-1} + \frac{49}{5}x^{-6}$.

23. Let $u = x + y + 1$ so that $du/dx = 1 + dy/dx$. Then $\dfrac{du}{dx} - 1 = u^2$ or $\dfrac{1}{1 + u^2}\, du = dx$. Thus $\tan^{-1} u = x + c$ or $u = \tan(x + c)$, and $x + y + 1 = \tan(x + c)$ or $y = \tan(x + c) - x - 1$.

25. Let $u = x + y$ so that $du/dx = 1 + dy/dx$. Then $\dfrac{du}{dx} - 1 = \tan^2 u$ or $\cos^2 u \, du = dx$. Thus $\frac{1}{2}u + \frac{1}{4}\sin 2u = x + c$ or $2u + \sin 2u = 4x + c_1$, and $2(x+y) + \sin 2(x+y) = 4x + c_1$ or $2y + \sin 2(x+y) = 2x + c_1$.

27. Let $u = y - 2x + 3$ so that $du/dx = dy/dx - 2$. Then $\dfrac{du}{dx} + 2 = 2 + \sqrt{u}$ or $\dfrac{1}{\sqrt{u}}\, du = dx$. Thus $2\sqrt{u} = x + c$ and $2\sqrt{y - 2x + 3} = x + c$.

29. Let $u = x + y$ so that $du/dx = 1 + dy/dx$. Then $\dfrac{du}{dx} - 1 = \cos u$ and $\dfrac{1}{1 + \cos u}\, du = dx$. Now

$$\frac{1}{1 + \cos u} = \frac{1 - \cos u}{1 - \cos^2 u} = \frac{1 - \cos u}{\sin^2 u} = \csc^2 u - \csc u \cot u$$

so we have $\int (\csc^2 u - \csc u \cot u)\, du = \int dx$ and $-\cot u + \csc u = x + c$. Thus $-\cot(x+y) + \csc(x+y) = x + c$. Setting $x = 0$ and $y = \pi/4$ we obtain $c = \sqrt{2} - 1$. The solution is

$$\csc(x+y) - \cot(x+y) = x + \sqrt{2} - 1.$$

31. We write the differential equation $M(x, y)dx + N(x, y)dy = 0$ as $dy/dx = f(x, y)$ where

$$f(x, y) = -\frac{M(x, y)}{N(x, y)}.$$

The function $f(x, y)$ must necessarily be homogeneous of degree 0 when M and N are homogeneous of degree α. Since M is homogeneous of degree α, $M(tx, ty) = t^\alpha M(x, y)$, and letting $t = 1/x$ we have

$$M(1, y/x) = \frac{1}{x^\alpha} M(x, y) \quad \text{or} \quad M(x, y) = x^\alpha M(1, y/x).$$

Thus

$$\frac{dy}{dx} = f(x, y) = -\frac{x^\alpha M(1, y/x)}{x^\alpha N(1, y/x)} = -\frac{M(1, y/x)}{N(1, y/x)} = F\left(\frac{y}{x}\right).$$

33. **(a)** By inspection $y = x$ and $y = -x$ are solutions of the differential equation and not members of the family $y = x\sin(\ln x + c_2)$.

(b) Letting $x = 5$ and $y = 0$ in $\sin^{-1}(y/x) = \ln x + c_2$ we get $\sin^{-1} 0 = \ln 5 + c_2$ or $c_2 = -\ln 5$. Then $\sin^{-1}(y/x) = \ln x - \ln 5 = \ln(x/5)$. Because the range of the arcsine function is $[-\pi/2, \pi/2]$ we must have

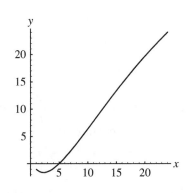

$$-\frac{\pi}{2} \leq \ln\frac{x}{5} \leq \frac{\pi}{2}$$

$$e^{-\pi/2} \leq \frac{x}{5} \leq e^{\pi/2}$$

$$5e^{-\pi/2} \leq x \leq 5e^{\pi/2}$$

The interval of definition of the solution is approximately $[1.04, 24.05]$.

35. (a) The substitutions $y = y_1 + u$ and

$$\frac{dy}{dx} = \frac{dy_1}{dx} + \frac{du}{dx}$$

lead to

$$\frac{dy_1}{dx} + \frac{du}{dx} = P + Q(y_1 + u) + R(y_1 + u)^2$$

$$= P + Qy_1 + Ry_1^2 + Qu + 2y_1Ru + Ru^2$$

or

$$\frac{du}{dx} - (Q + 2y_1R)u = Ru^2.$$

This is a Bernoulli equation with $n = 2$ which can be reduced to the linear equation

$$\frac{dw}{dx} + (Q + 2y_1R)w = -R$$

by the substitution $w = u^{-1}$.

(b) Identify $P(x) = -4/x^2$, $Q(x) = -1/x$, and $R(x) = 1$. Then $\dfrac{dw}{dx} + \left(-\dfrac{1}{x} + \dfrac{4}{x}\right)w = -1$.

An integrating factor is x^3 so that $x^3w = -\frac{1}{4}x^4 + c$ or $u = \left[-\frac{1}{4}x + cx^{-3}\right]^{-1}$. Thus,

$$y = \frac{2}{x} + u \quad \text{or} \quad y = \frac{2}{x} + \left(-\frac{1}{4}x + cx^{-3}\right)^{-1}$$

37. Write the differential equation as

$$\frac{dv}{dx} + \frac{1}{x}v = 32v^{-1},$$

and let $u = v^2$ or $v = u^{1/2}$. Then

$$\frac{dv}{dx} = \frac{1}{2}u^{-1/2}\frac{du}{dx},$$

and substituting into the differential equation, we have

$$\frac{1}{2}u^{-1/2}\frac{du}{dx} + \frac{1}{x}u^{1/2} = 32u^{-1/2} \quad \text{or} \quad \frac{du}{dx} + \frac{2}{x}u = 64.$$

The latter differential equation is linear with integrating factor $e^{\int (2/x)\,dx} = x^2$, so

$$\frac{d}{dx}\left[x^2u\right] = 64x^2$$

and

$$x^2u = \frac{64}{3}x^3 + c \quad \text{or} \quad v^2 = \frac{64}{3}x + \frac{c}{x^2}.$$

2.6   A Numerical Method

1. We identify $f(x,y) = 2x - 3y + 1$. Then, for $h = 0.1$,

$$y_{n+1} = y_n + 0.1(2x_n - 3y_n + 1) = 0.2x_n + 0.7y_n + 0.1,$$

and

$$y(1.1) \approx y_1 = 0.2(1) + 0.7(5) + 0.1 = 3.8$$
$$y(1.2) \approx y_2 = 0.2(1.1) + 0.7(3.8) + 0.1 = 2.98$$

For $h = 0.05$,

$$y_{n+1} = y_n + 0.05(2x_n - 3y_n + 1) = 0.1x_n + 0.85y_n + 0.1,$$

and

$$y(1.05) \approx y_1 = 0.1(1) + 0.85(5) + 0.1 = 4.4$$
$$y(1.1) \approx y_2 = 0.1(1.05) + 0.85(4.4) + 0.1 = 3.895$$
$$y(1.15) \approx y_3 = 0.1(1.1) + 0.85(3.895) + 0.1 = 3.47075$$
$$y(1.2) \approx y_4 = 0.1(1.15) + 0.85(3.47075) + 0.1 = 3.11514$$

3. Separating variables and integrating, we have

$$\frac{dy}{y} = dx \quad \text{and} \quad \ln|y| = x + c.$$

Thus $y = c_1 e^x$ and, using $y(0) = 1$, we find $c = 1$, so $y = e^x$ is the solution of the initial-value problem.

$h = 0.1$

x_n	y_n	Actual Value	Abs. Error	% Rel. Error
0.00	1.0000	1.0000	0.0000	0.00
1.10	1.1000	1.1052	0.0052	0.47
0.20	1.2100	1.2214	0.0114	0.93
0.30	1.3310	1.3499	0.0189	1.40
0.40	1.4641	1.4918	0.0277	1.86
0.50	1.6105	1.6487	0.0382	2.32
0.60	1.7716	1.8221	0.0506	2.77
0.70	1.9487	2.0138	0.0650	3.23
0.80	2.1436	2.2255	0.0820	3.68
0.90	2.3579	2.4596	0.1017	4.13
1.00	2.5937	2.7183	0.1245	4.58

$h = 0.05$

x_n	y_n	Actual Value	Abs. Error	% Rel. Error
0.00	1.0000	1.0000	0.0000	0.00
0.05	1.0500	1.0513	0.0013	0.12
0.10	1.1025	1.1052	0.0027	0.24
0.15	1.1576	1.1618	0.0042	0.36
0.20	1.2155	1.2214	0.0059	0.48
0.25	1.2763	1.2840	0.0077	0.60
0.30	1.3401	1.3499	0.0098	0.72
0.35	1.4071	1.4191	0.0120	0.84
0.40	1.4775	1.4918	0.0144	0.96
0.45	1.5513	1.5683	0.0170	1.08
0.50	1.6289	1.6487	0.0198	1.20
0.55	1.7103	1.7333	0.0229	1.32
0.60	1.7959	1.8221	0.0263	1.44
0.65	1.8856	1.9155	0.0299	1.56
0.70	1.9799	2.0138	0.0338	1.68
0.75	2.0789	2.1170	0.0381	1.80
0.80	2.1829	2.2255	0.0427	1.92
0.85	2.2920	2.3396	0.0476	2.04
0.90	2.4066	2.4596	0.0530	2.15
0.95	2.5270	2.5857	0.0588	2.27
1.00	2.6533	2.7183	0.0650	2.39

5.

$h = 0.1$

x_n	y_n
0.00	0.0000
0.10	0.1000
0.20	0.1905
0.30	0.2731
0.40	0.3492
0.50	0.4198

$h = 0.05$

x_n	y_n
0.00	0.0000
0.05	0.0500
0.10	0.0976
0.15	0.1429
0.20	0.1863
0.25	0.2278
0.30	0.2676
0.35	0.3058
0.40	0.3427
0.45	0.3782
0.50	0.4124

7.

$h = 0.1$

x_n	y_n
0.00	0.5000
0.10	0.5250
0.20	0.5431
0.30	0.5548
0.40	0.5613
0.50	0.5639

$h = 0.05$

x_n	y_n
0.00	0.5000
0.05	0.5125
0.10	0.5232
0.15	0.5322
0.20	0.5395
0.25	0.5452
0.30	0.5496
0.35	0.5527
0.40	0.5547
0.45	0.5559
0.50	0.5565

9.

$h = 0.1$

x_n	y_n
1.00	1.0000
1.10	1.0000
1.20	1.0191
1.30	1.0588
1.40	1.1231
1.50	1.2194

$h = 0.05$

x_n	y_n
1.00	1.0000
1.05	1.0000
1.10	1.0049
1.15	1.0147
1.20	1.0298
1.25	1.0506
1.30	1.0775
1.35	1.1115
1.40	1.1538
1.45	1.2057
1.50	1.2696

11. Tables of values were computed using the Euler and RK4 methods. The resulting points were plotted and joined using **ListPlot** in *Mathematica.*

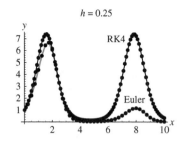

$h = 0.25$

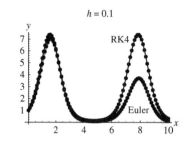

$h = 0.1$

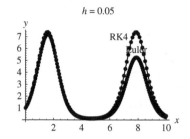

$h = 0.05$

13. Using separation of variables we find that the solution of the differential equation is $y = 1/(1 - x^2)$, which is undefined at $x = 1$, where the graph has a vertical asymptote. Because the actual solution of the differential equation becomes unbounded at x approaches 1, very small changes in the inputs x will result in large changes in the corresponding outputs y. This can be expected to have a serious effect on numerical procedures. The graphs below were obtained as described in Problem 11.

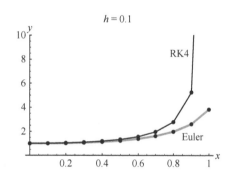

$h = 0.1$

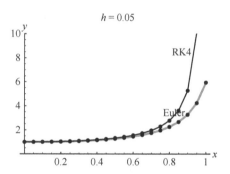

$h = 0.05$

Chapter 2 in Review

1. Writing the differential equation in the form $y' = k(y + A/k)$ we see that the critical point $-A/k$ is a repeller for $k > 0$ and an attractor for $k < 0$.

3. True; $y = k_2/k_1$ is always a solution for $k_1 \neq 0$.

5. $\dfrac{d^3 y}{dx^3} = x \sin y$ (There are many answers.)

7. True

9. $\dfrac{dy}{y} = e^x \, dx$

   $\ln y = e^x + c$

   $y = e^{e^x + c} = e^c e^{e^x}$ or $y = c_1 e^{e^x}$

11. $y = e^{\cos x} \displaystyle\int_0^x t e^{-\cos t}\, dt$

$$\frac{dy}{dx} = e^{\cos x} x e^{-\cos x} + (-\sin x)\, e^{\cos x} \int_0^x t e^{-\cos t}\, dt$$

$$\frac{dy}{dx} = x - (\sin x)\, y \quad \text{or} \quad \frac{dy}{dx} + (\sin x)\, y = x.$$

13. $\dfrac{dy}{dx} = (y-1)^2 (y-3)^2$

15. When n is odd, $x^n < 0$ for $x < 0$ and $x^n > 0$ for $x > 0$. In this case 0 is unstable. When n is even, $x^n > 0$ for $x < 0$ and for $x > 0$. In this case 0 is semi-stable.

When n is odd, $-x^n > 0$ for $x < 0$ and $-x^n < 0$ for $x > 0$. In this case 0 is asymptotically stable. When n is even, $-x^n < 0$ for $x < 0$ and for $x > 0$. In this case 0 is semi-stable.

17.

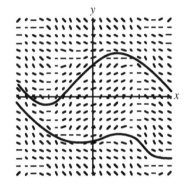

19. Separating variables and using the identity $\cos^2 x = \frac{1}{2}(1 + \cos 2x)$, we have

$$\cos^2 x\, dx = \frac{y}{y^2 + 1}\, dy,$$

$$\frac{1}{2} x + \frac{1}{4}\sin 2x = \frac{1}{2}\ln\left(y^2 + 1\right) + c,$$

and

$$2x + \sin 2x = 2\ln\left(y^2 + 1\right) + c.$$

21. The differential equation

$$\frac{dy}{dx} + \frac{2}{6x + 1}y = -\frac{3x^2}{6x + 1}y^{-2}$$

is Bernoulli. Using $w = y^3$, we obtain the linear equation

$$\frac{dw}{dx} + \frac{6}{6x + 1}w = -\frac{9x^2}{6x + 1}.$$

An integrating factor is $6x + 1$, so

$$\frac{d}{dx}\left[(6x + 1)w\right] = -9x^2,$$

$$w = -\frac{3x^3}{6x + 1} + \frac{c}{6x + 1},$$

and

$$(6x + 1)y^3 = -3x^3 + c.$$

(Note: The differential equation is also exact.)

23. Write the equation in the form

$$\frac{dQ}{dt} + \frac{1}{t}Q = t^3 \ln t.$$

An integrating factor is $e^{\ln t} = t$, so

$$\frac{d}{dt}[tQ] = t^4 \ln t$$

$$tQ = -\frac{1}{25}t^5 + \frac{1}{5}t^5 \ln t + c$$

and

$$Q = -\frac{1}{25}t^4 + \frac{1}{5}t^4 \ln t + \frac{c}{t}.$$

25. Write the equation in the form

$$\frac{dy}{dx} + \frac{8x}{x^2 + 4}y = \frac{2x}{x^2 + 4}.$$

An integrating factor is $(x^2 + 4)^4$, so

$$\frac{d}{dx}\left[(x^2 + 4)^4 y\right] = 2x(x^2 + 4)^3$$

$$(x^2 + 4)^4 y = \frac{1}{4}(x^2 + 4)^4 + c$$

and

$$y = \frac{1}{4} + c(x^2 + 4)^{-4}.$$

27. We put the equation $\frac{dy}{dx} + 4(\cos x)y = x$ in the standard form $\frac{dy}{dx} + 2(\cos x)y = \frac{1}{2}x$ then the integrating factor is $e^{\int 2\cos x\, dx} = e^{2\sin x}$. Therefore

$$\frac{d}{dx}\left[e^{2\sin x}y\right] = \frac{1}{2}xe^{2\sin x}$$

$$\int_0^x \frac{d}{dt}\left[e^{2\sin t}y(t)\right] dt = \frac{1}{2}\int_0^x te^{2\sin t}\, dt$$

$$e^{2\sin x}y(x) - e^0 \overset{1}{\overbrace{y(0)}} = \frac{1}{2}\int_0^x te^{2\sin t}\, dt$$

$$e^{2\sin x}y(x) - 1 = \frac{1}{2}\int_0^x te^{2\sin t}\, dy$$

$$y(x) = e^{-2\sin x} + \frac{1}{2}e^{-2\sin x}\int_0^x te^{2\sin t}\, dt$$

29. We put the equation $x\frac{dy}{dx}+2y = xe^{x^2}$ into standard form $\frac{dy}{dx}+\frac{2}{x}y = e^{x^2}$. Then the integrating factor is $e^{\int \frac{2}{x}\,dx} = e^{\ln x^2} = x^2$. Therefore

$$x^2 \frac{dy}{dx} + 2xy = x^2 e^{x^2}$$

$$\frac{d}{dx}\left[x^2 y\right] = x^2 e^{x^2}$$

$$\int_1^x \frac{d}{dt}\left[t^2 y(t)\right]\,dt = \int_1^x t^2 e^{t^2}\,dt$$

$$x^2 y(x) - \overbrace{y(1)}^{3} = \int_1^x t^2 e^{t^2}\,dt$$

$$y(x) = \frac{3}{x^2} + \frac{1}{x^2}\int_1^x t^2 e^{t^2}\,dt$$

31.

$$\frac{dy}{dx} + y = f(x), \quad y(0) = 5, \quad \text{where} \quad f(x) = \begin{cases} e^{-x}, & 0 \le x < 1 \\ 0, & x \ge 1 \end{cases}$$

For $0 \le x < 1$,

$$\frac{d}{dx}\left[e^x y\right] = 1$$

$$e^x y = x + c_1$$

$$y = xe^{-x} + c_1 e^{-x}$$

Using $y(0) = 5$, we have $c_1 = 5$. Therefore $y = xe^{-x} + 5e^{-x}$. Then for $x \ge 1$,

$$\frac{d}{dx}\left[e^x y\right] = 0$$

$$e^x y = c_2$$

$$y = c_2 e^{-x}$$

Requiring that $y(x)$ be continuous at $x = 1$ yields

$$c_2 e^{-1} = e^{-1} + 5e^{-1}$$

$$c_2 = 6$$

Therefore

$$y(x) = \begin{cases} xe^{-x} + 5e^{-x}, & 0 \le x < 1 \\ 6e^{-x}, & x \ge 1 \end{cases}$$

33. The differential equation has the form $(d/dx)\left[(\sin x)y\right] = 0$. Integrating, we have $(\sin x)y = c$ or $y = c/\sin x$. The initial condition implies $c = -2\sin(7\pi/6) = 1$. Thus, $y = 1/\sin x$, where the interval $(\pi, 2\pi)$ is chosen to include $x = 7\pi/6$.

35. (a) For $y < 0$, $\sqrt{y}$ is not a real number.

(b) Separating variables and integrating we have

$$\frac{dy}{\sqrt{y}} = dx \quad \text{and} \quad 2\sqrt{y} = x + c.$$

Letting $y(x_0) = y_0$ we get $c = 2\sqrt{y_0} - x_0$, so that

$$2\sqrt{y} = x + 2\sqrt{y_0} - x_0 \quad \text{and} \quad y = \frac{1}{4}(x + 2\sqrt{y_0} - x_0)^2.$$

Since $\sqrt{y} > 0$ for $y \neq 0$, we see that $dy/dx = \frac{1}{2}(x + 2\sqrt{y_0} - x_0)$ must be positive. Thus, the interval on which the solution is defined is $(x_0 - 2\sqrt{y_0}, \infty)$.

37. The graph of $y_1(x)$ is the portion of the closed blue curve lying in the fourth quadrant. Its interval of definition is approximately $(0.7, 4.3)$. The graph of $y_2(x)$ is the portion of the left-hand blue curve lying in the third quadrant. Its interval of definition is $(-\infty, 0)$.

39. Since the differential equation is autonomous, all lineal elements on a given horizontal line have the same slope. The direction field is then as shown in the figure at the right. It appears from the figure that the differential equation has critical points at -2 (an attractor) and at 2 (a repeller). Thus, -2 is an aymptotically stable critical point and 2 is an unstable critical point.

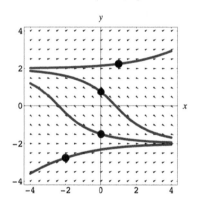

Chapter 3

Modeling with First-Order Differential Equations

3.1 | Linear Models

1. Let $P = P(t)$ be the population at time t, and P_0 the initial population. From $dP/dt = kP$ we obtain $P = P_0 e^{kt}$. Using $P(5) = 2P_0$ we find $k = \frac{1}{5} \ln 2$ and $P = P_0 e^{(\ln 2)t/5}$. Setting $P(t) = 3P_0$ we have $3 = e^{(\ln 2)t/5}$, so

$$\ln 3 = \frac{(\ln 2)t}{5} \quad \text{and} \quad t = \frac{5 \ln 3}{\ln 2} \approx 7.9 \text{ years.}$$

Setting $P(t) = 4P_0$ we have $4 = e^{(\ln 2)t/5}$, so

$$\ln 4 = \frac{(\ln 2)t}{5} \quad \text{and} \quad t = 10 \text{ years.}$$

3. Let $P = P(t)$ be the population at time t. Then $dP/dt = kP$ and $P = ce^{kt}$. From $P(0) = c = 500$ we see that $P = 500e^{kt}$. Since 15% of 500 is 75, we have $P(10) = 500e^{10k} = 575$. Solving for k, we get $k = \frac{1}{10} \ln \frac{575}{500} = \frac{1}{10} \ln 1.15$. When $t = 30$,

$$P(30) = 500e^{(1/10)(\ln 1.15)30} = 500e^{3 \ln 1.15} \approx 760 \text{ years}$$

and

$$P'(30) = kP(30) \approx \frac{1}{10}(\ln 1.15)760 \approx 10.62 \text{ persons/year.}$$

5. Let $A = A(t)$ be the amount of lead present at time t. From $dA/dt = kA$ and $A(0) = 1$ we obtain $A = e^{kt}$. Using $A(3.3) = 1/2$ we find $k = \frac{1}{3.3} \ln (1/2)$. When 90% of the lead has decayed, 0.1 grams will remain. Setting $A(t) = 0.1$ we have $e^{t(1/3.3) \ln (1/2)} = 0.1$, so

$$\frac{t}{3.3} \ln \frac{1}{2} = \ln 0.1 \quad \text{and} \quad t = \frac{3.3 \ln 0.1}{\ln (1/2)} \approx 10.96 \text{ hours.}$$

7. Setting $A(t) = 50$ in Problem 6 we obtain $50 = 100e^{kt}$, so

$$kt = \ln \frac{1}{2} \quad \text{and} \quad t = \frac{\ln (1/2)}{(1/6) \ln 0.97} \approx 136.5 \text{ hours.}$$

51

9. Let $I = I(t)$ be the intensity, t the thickness, and $I(0) = I_0$. If $dI/dt = kI$ and $I(3) = 0.25I_0$, then $I = I_0 e^{kt}$, $k = \frac{1}{3}\ln 0.25$, and $I(15) = 0.00098I_0$.

11. Using 5730 years as the half-life of C-14 we have from Example 3 in the text $A(t) = A_0 e^{-0.00012097t}$. Since 85.5% of the C-14 has decayed, $1 - 0.855 = 0.145$ times the original amount is now present, so

$$0.145A_0 = A_0 e^{-0.00012097t}, \quad e^{-0.00012097t} = 0.145, \quad \text{and} \quad t = -\frac{\ln 0.145}{0.00012097} \approx 15{,}968 \text{ years}$$

is the approximate age.

13. Assume that $dT/dt = k(T - 10)$ so that $T = 10 + ce^{kt}$. If $T(0) = 70°$ and $T(1/2) = 50°$ then $c = 60$ and $k = 2\ln(2/3)$ so that $T(1) = 36.67°$. If $T(t) = 15°$ then $t = 3.06$ minutes.

15. Assume that $dT/dt = k(T - 100)$ so that $T = 100 + ce^{kt}$. If $T(0) = 20°$ and $T(1) = 22°$, then $c = -80$ and $k = \ln(39/40)$ so that $T(t) = 90°$, which implies $t = 82.1$ seconds. If $T(t) = 98°$ then $t = 145.7$ seconds.

17. Using separation of variables to solve $dT/dt = k(T - T_m)$ we get $T(t) = T_m + ce^{kt}$. Using $T(0) = 70$ we find $c = 70 - T_m$, so $T(t) = T_m + (70 - T_m)e^{kt}$. Using the given observations, we obtain

$$T\left(\frac{1}{2}\right) = T_m + (70 - T_m)e^{k/2} = 110$$

$$T(1) = T_m + (70 - T_m)e^k = 145.$$

Then, from the first equation, $e^{k/2} = (110 - T_m)/(70 - T_m)$ and

$$e^k = (e^{k/2})^2 = \left(\frac{110 - T_m}{70 - T_m}\right)^2 = \frac{145 - T_m}{70 - T_m} \frac{(110 - T_m)^2}{70 - T_m} \qquad = 145 - T_m$$

$$12100 - 220T_m + T_m^2 = 10150 - 250T_m + T_m^2$$

$$T_m = 390.$$

The temperature in the oven is $390°$.

19. According to Newton's Law of Cooling

$$\frac{dT}{dt} = k(T - T_m).$$

Separating variables we have

$$\frac{dT}{T - T_m} = k\,dt \qquad \text{so} \qquad \ln|T - T_m| = kt + c \qquad \text{and} \qquad T = T_m + c_1 e^{kt}.$$

Setting $T(0) = T_0$ we find $c_1 = T_0 - T_m$. Thus

$$T(t) = T_m + (T_0 - T_m)e^{kt}.$$

In this problem we use $T_0 = 98.6$ and $T_m = 70$. Now, let n denote the number of hours elapsed before the body was found. Then $T(n) = 85$ and $T(n+1) = 80$. Using this information, we have

$$70 + (98.6 - 70)e^{kn} = 85 \quad \text{and} \quad 70 + (98.6 - 70)e^{k(n+1)} = 80$$

or

$$28.6e^{kn} = 15 \quad \text{and} \quad 28.6e^{kn+k} = 28.6e^{kn}e^k = 10.$$

The last equation is the same as $15e^k = 10$. Solving for k, we have $k = \ln\frac{2}{3} \approx -0.4055$. Finally, solving $e^{-0.4055n} = 15/28.6$ for n, we have

$$-0.4055n = \ln\left(\frac{15}{28.6}\right)$$

$$n = \frac{1}{-0.4055}\ln\left(\frac{15}{28.6}\right) \approx 1.6.$$

Thus, about 1.6 hours elapsed before the body was found.

21. From $dA/dt = 4 - A/50$ we obtain $A = 200 + ce^{-t/50}$. If $A(0) = 30$ then $c = -170$ and $A = 200 - 170e^{-t/50}$.

23. From $dA/dt = 10 - A/100$ we obtain $A = 1000 + ce^{-t/100}$. If $A(0) = 0$ then $c = -1000$ and $A(t) = 1000 - 1000e^{-t/100}$.

25. From

$$\frac{dA}{dt} = 10 - \frac{10A}{500 - (10-5)t} = 10 - \frac{2A}{100 - t}$$

we obtain $A = 1000 - 10t + c(100 - t)^2$. If $A(0) = 0$ then $c = -\frac{1}{10}$. The tank is empty in 100 minutes.

27. From

$$\frac{dA}{dt} = 3 - \frac{4A}{100 + (6-4)t} = 3 - \frac{2A}{50 + t}$$

we obtain $A = 50 + t + c(50 + t)^{-2}$. If $A(0) = 10$ then $c = -100{,}000$ and $A(30) = 64.38$ pounds.

29. Assume $L\, di/dt + Ri = E(t)$, $L = 0.1$, $R = 50$, and $E(t) = 50$ so that $i = \frac{3}{5} + ce^{-500t}$. If $i(0) = 0$ then $c = -3/5$ and $\lim_{t\to\infty} i(t) = 3/5$.

31. Assume $R\, dq/dt + (1/C)q = E(t)$, $R = 200$, $C = 10^{-4}$, and $E(t) = 100$ so that $q = 1/100 + ce^{-50t}$. If $q(0) = 0$ then $c = -1/100$ and $i = \frac{1}{2}e^{-50t}$.

33. For $0 \le t \le 20$ the differential equation is $20\, di/dt + 2i = 120$. An integrating factor is $e^{t/10}$, so $(d/dt)[e^{t/10}i] = 6e^{t/10}$ and $i = 60 + c_1 e^{-t/10}$. If $i(0) = 0$ then $c_1 = -60$ and $i = 60 - 60e^{-t/10}$.

For $t > 20$ the differential equation is $20\,di/dt + 2i = 0$ and $i = c_2 e^{-t/10}$. At $t = 20$ we want $c_2 e^{-2} = 60 - 60e^{-2}$ so that $c_2 = 60\left(e^2 - 1\right)$. Thus

$$i(t) = \begin{cases} 60 - 60e^{-t/10}, & 0 \le t \le 20 \\ 60\left(e^2 - 1\right)e^{-t/10}, & t > 20 \end{cases}$$

35. (a) From $m\,dv/dt = mg - kv$ we obtain $v = mg/k + ce^{-kt/m}$. If $v(0) = v_0$ then $c = v_0 - mg/k$ and the solution of the initial-value problem is

$$v(t) = \frac{mg}{k} + \left(v_0 - \frac{mg}{k}\right)e^{-kt/m}.$$

(b) As $t \to \infty$ the limiting velocity is mg/k.

(c) From $ds/dt = v$ and $s(0) = 0$ we obtain

$$s(t) = \frac{mg}{k}t - \frac{m}{k}\left(v_0 - \frac{mg}{k}\right)e^{-kt/m} + \frac{m}{k}\left(v_0 - \frac{mg}{k}\right).$$

37. When air resistance is proportional to velocity, the model for the velocity is $m\,dv/dt = -mg - kv$ (using the fact that the positive direction is upward.) Solving the differential equation using separation of variables we obtain $v(t) = -mg/k + ce^{-kt/m}$. From $v(0) = 300$ we get

$$v(t) = -\frac{mg}{k} + \left(300 + \frac{mg}{k}\right)e^{-kt/m}.$$

Integrating and using $s(0) = 0$ we find

$$s(t) = -\frac{mg}{k}t + \frac{m}{k}\left(300 + \frac{mg}{k}\right)\left(1 - e^{-kt/m}\right).$$

Setting $k = 0.0025$, $m = 16/32 = 0.5$, and $g = 32$ we have

$$s(t) = 1{,}340{,}000 - 6{,}400t - 1{,}340{,}000e^{-0.005t}$$

and

$$v(t) = -6{,}400 + 6{,}700e^{-0.005t}.$$

The maximum height is attained when $v = 0$, that is, at $t_a = 9.162$. The maximum height will be $s(9.162) = 1363.79$ ft, which is less than the maximum height in Problem 36.

39. (a) With the values given in the text the initial-value problem becomes

$$\frac{dv}{dt} + \frac{2}{200 - t}v = -9.8 + \frac{2000}{200 - t}, \quad v(0) = 0.$$

This is a linear differential equationwith integrating factor

$$e^{\int [2/(200-t)]\,dt} = e^{-2\ln|200-t|} = (200 - t)^{-2}.$$

Then

$$\frac{d}{dt}\left[(200-t)^{-2}v\right] = -9.8(200-t)^{-2} + 2000(200-t)^{-3} \quad \leftarrow \text{integrate}$$

$$(200-t)^{-2}v = -9.8(200-t)^{-1} + 1000(200-t)^{-2} + c \quad \leftarrow \text{multiply by}(200-t)^2$$

$$v = -9.8(200-t) + 1000 + c(200-t)^2$$

$$= -960 + 9.8t + c(200-t)^2.$$

Using the initial condition we have

$$0 = v(0) = -960 + 40{,}000c \quad \text{so} \quad c = \frac{960}{40{,}000} = 0.024.$$

Thus

$$v(t) = -960 + 9.8t + 0.024(200-t)^2$$

$$= -960 + 9.8t + 0.024(40{,}000 - 400t + t^2)$$

$$= -960 + 9.8t + 960 - 9.6t + 0.024t^2$$

$$= 0.024t^2 + 0.2t.$$

(b) Integrating both sides of

$$\frac{ds}{dt} = v(t) = 0.024t^2 + 0.2t$$

we find

$$s(t) = 0.008t^3 + 0.1t^2 + c_1.$$

We assume that the height of the rocket is measured from $s = 0$, so that $s(0) = 0$ and $c_1 = 0$. Then the height of the rocket at time t is $s(t) = 0.008t^3 + 0.1t^2$.

41. (a) The differential equation is first-order and linear. Letting $b = k/\rho$, the integrating factor is $e^{\int 3b\,dt/(bt+r_0)} = (r_0 + bt)^3$. Then

$$\frac{d}{dt}[(r_0 + bt)^3 v] = g(r_0 + bt)^3 \quad \text{and} \quad (r_0 + bt)^3 v = \frac{g}{4b}(r_0 + bt)^4 + c.$$

The solution of the differential equation is $v(t) = (g/4b)(r_0 + bt) + c(r_0 + bt)^{-3}$. Using $v(0) = 0$ we find $c = -gr_0^4/4b$, so that

$$v(t) = \frac{g}{4b}(r_0 + bt) - \frac{gr_0^4}{4b(r_0 + bt)^3} = \frac{g\rho}{4k}\left(r_0 + \frac{k}{\rho}t\right) - \frac{g\rho r_0^4}{4k(r_0 + kt/\rho)^3}.$$

(b) Integrating $dr/dt = k/\rho$ we get $r = kt/\rho + c$. Using $r(0) = r_0$ we have $c = r_0$, so $r(t) = kt/\rho + r_0$.

(c) If $r = 0.007$ ft when $t = 10$ s, then solving $r(10) = 0.007$ for k/ρ, we obtain $k/\rho = -0.0003$ and $r(t) = 0.01 - 0.0003t$. Solving $r(t) = 0$ we get $t = 33.3$, so the raindrop will have evaporated completely at 33.3 seconds.

43. (a) From $dP/dt = (k_1 - k_2)P$ we obtain $P = P_0 e^{(k_1 - k_2)t}$ where $P_0 = P(0)$.

(b) If $k_1 > k_2$ then $P \to \infty$ as $t \to \infty$. If $k_1 = k_2$ then $P = P_0$ for every t. If $k_1 < k_2$ then $P \to 0$ as $t \to \infty$.

45. (a) Solving $r - kx = 0$ for x we find the equilibrium solution $x = r/k$. When $x < r/k$, $dx/dt > 0$ and when $x > r/k$, $dx/dt < 0$. From the phase portrait we see that $\lim\limits_{t \to \infty} x(t) = r/k$.

(b) From $dx/dt = r - kx$ and $x(0) = 0$ we obtain $x = r/k - (r/k)e^{-kt}$ so that $x \to r/k$ as $t \to \infty$. If $x(T) = r/2k$ then $T = (\ln 2)/k$.

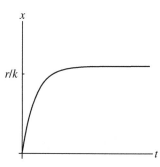

47. (a) For $0 \le t < 4$, $6 \le t < 10$ and $12 \le t < 16$, no voltage is applied to the heart and $E(t) = 0$. At the other times, the differential equation is $dE/dt = -E/RC$. Separating variables, integrating, and solving for e, we get $E = ke^{-t/RC}$, subject to $E(4) = E(10) = E(16) = 12$. These intitial conditions yield, respectively, $k = 12e^{4/RC}$, $k = 12e^{10/RC}$, $k = 12e^{16/RC}$, and $k = 12e^{22/RC}$. Thus

$$E(t) = \begin{cases} 0, & 0 \le t < 4,\ 6 \le t < 10,\ 12 \le t < 16 \\ 12e^{(4-t)/RC}, & 4 \le t < 6 \\ 12e^{(10-t)/RC}, & 10 \le t < 12 \\ 12e^{(16-t)/RC}, & 16 \le t < 18 \\ 12e^{(22-t)/RC}, & 22 \le t < 24 \end{cases}$$

(b)

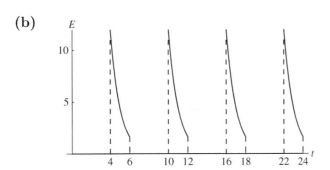

49. **(a)** (i) If $s(t)$ is distance measured down the plane from the highest point, then $ds/dt = v$. Integrating $ds/dt = 16t$ gives $s(t) = 8t^2 + c_2$. Using $s(0) = 0$ then gives $c_2 = 0$. Now the length L of the plane is $L = 50/\sin 30° = 100$ ft. The time it takes the box to slide completely down the plane is the solution of $s(t) = 100$ or $t^2 = 25/2$, so $t \approx 3.54$ s.

(ii) Integrating $ds/dt = 4t$ gives $s(t) = 2t^2 + c_2$. Using $s(0) = 0$ gives $c_2 = 0$, so $s(t) = 2t^2$ and the solution of $s(t) = 100$ is now $t \approx 7.07$ s.

(iii) Integrating $ds/dt = 48 - 48e^{-t/12}$ and using $s(0) = 0$ to determine the constant of integration, we obtain $s(t) = 48t + 576e^{-t/12} - 576$. With the aid of a CAS we find that the solution of $s(t) = 100$, or

$$100 = 48t + 576e^{-t/12} - 576 \qquad \text{or} \qquad 0 = 48t + 576e^{-t/12} - 676,$$

is now $t \approx 7.84$ s.

(b) The differential equation $m\, dv/dt = mg \sin\theta - \mu mg \cos\theta$ can be written

$$m\frac{dv}{dt} = mg\cos\theta(\tan\theta - \mu).$$

If $\tan\theta = \mu$, $dv/dt = 0$ and $v(0) = 0$ implies that $v(t) = 0$. If $\tan\theta < \mu$ and $v(0) = 0$, then integration implies $v(t) = g\cos\theta(\tan\theta - \mu)t < 0$ for all time t.

(c) Since $\tan 23° = 0.4245$ and $\mu = \sqrt{3}/4 = 0.4330$, we see that $\tan 23° < 0.4330$. The differential equation is $dv/dt = 32\cos 23°(\tan 23° - \sqrt{3}/4) = -0.251493$. Integration and the use of the initial condition gives $v(t) = -0.251493t + 1$. When the box stops, $v(t) = 0$ or $0 = -0.251493t + 1$ or $t = 3.976254$ s. From $s(t) = -0.125747t^2 + t$ we find $s(3.976254) = 1.988119$ ft.

(d) With $v_0 > 0$, $v(t) = -0.251493t + v_0$ and $s(t) = -0.125747t^2 + v_0 t$. Because two real positive solutions of the equation $s(t) = 100$, or $0 = -0.125747t^2 + v_0 t - 100$, would be physically meaningless, we use the quadratic formula and require that $b^2 - 4ac = 0$ or $v_0^2 - 50.2987 = 0$. From this last equality we find $v_0 \approx 7.092164$ ft/s. For the time it takes the box to traverse the entire inclined plane, we must have $0 = -0.125747t^2 + 7.092164t - 100$. *Mathematica* gives complex roots for the last equation: $t = 28.2001 \pm 0.0124458i$. But, for

$$0 = -0.125747t^2 + 7.092164691t - 100,$$

the roots are $t = 28.1999\,\mathrm{s}$ and $t = 28.2004\,\mathrm{s}$. So if $v_0 > 7.092164$, we are guaranteed that the box will slide completely down the plane.

3.2   Nonlinear Models

1. **(a)** Solving $N(1 - 0.0005N) = 0$ for N we find the equilibrium solutions $N = 0$ and $N = 2000$. When $0 < N < 2000$, $dN/dt > 0$. From the phase portrait we see that $\lim_{t \to \infty} N(t) = 2000$. A graph of the solution is shown in part (b).

(b) Separating variables and integrating we have

$$\frac{dN}{N(1 - 0.0005N)} = \left(\frac{1}{N} - \frac{1}{N - 2000} \right) dN = dt$$

and

$$\ln N - \ln|N - 2000| = t + c.$$

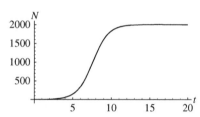

Solving for N we get $N(t) = 2000e^{c+t}/(1 + e^{c+t}) = 2000e^c e^t/(1 + e^c e^t)$. Using $N(0) = 1$ and solving for e^c we find $e^c = 1/1999$ and so $N(t) = 2000e^t/(1999 + e^t)$. Then $N(10) = 1833.59$, so 1834 companies are expected to adopt the new technology when $t = 10$.

3. From $dP/dt = P\left(10^{-1} - 10^{-7}P\right)$ and $P(0) = 5000$ we obtain $P = 500/(0.0005 + 0.0995e^{-0.1t})$ so that $P \to 1{,}000{,}000$ as $t \to \infty$. If $P(t) = 500{,}000$ then $t = 52.9\,\text{months}$.

5. **(a)** The differential equation is $dP/dt = P(5 - P) - 4$. Solving $P(5 - P) - 4 = 0$ for P we obtain equilibrium solutions $P = 1$ and $P = 4$. The phase portrait is shown on the right and solution curves are shown in part (b). We see that for $P_0 > 4$ and $1 < P_0 < 4$ the population approaches 4 as t increases. For $0 < P < 1$ the population decreases to 0 in finite time.

(b) The differential equation is

$$\frac{dP}{dt} = P(5 - P) - 4 = -(P^2 - 5P + 4) = -(P - 4)(P - 1).$$

Separating variables and integrating, we obtain

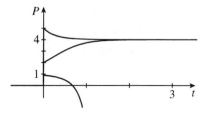

$$\frac{dP}{(P-4)(P-1)} = -dt$$

$$\left(\frac{1/3}{P-4} - \frac{1/3}{P-1}\right) dP = -dt$$

$$\frac{1}{3} \ln\left|\frac{P-4}{P-1}\right| = -t + c$$

$$\frac{P-4}{P-1} = c_1 e^{-3t}.$$

Setting $t = 0$ and $P = P_0$ we find $c_1 = (P_0 - 4)/(P_0 - 1)$. Solving for P we obtain

$$P(t) = \frac{4(P_0 - 1) - (P_0 - 4)e^{-3t}}{(P_0 - 1) - (P_0 - 4)e^{-3t}}.$$

(c) To find when the population becomes extinct in the case $0 < P_0 < 1$ we set $P = 0$ in

$$\frac{P-4}{P-1} = \frac{P_0 - 4}{P_0 - 1} e^{-3t}$$

from part (a) and solve for t. This gives the time of extinction

$$t = -\frac{1}{3} \ln \frac{4(P_0 - 1)}{P_0 - 4}.$$

7. Solving $P(5-P) - 7 = 0$ for P we obtain complex roots, so there are no equilibrium solutions. Since $dP/dt < 0$ for all values of P, the population becomes extinct for any initial condition. Using separation of variables to solve the initial-value problem, we get

$$P(t) = \frac{5}{2} + \frac{\sqrt{3}}{2} \tan\left[\tan^{-1}\left(\frac{2P_0 - 5}{\sqrt{3}}\right) - \frac{\sqrt{3}}{2}t\right].$$

Solving $P(t) = 0$ for t we see that the time of extinction is

$$t = \frac{2}{3}\left(\sqrt{3}\tan^{-1}(5/\sqrt{3}) + \sqrt{3}\tan^{-1}\left[(2P_0 - 5)/\sqrt{3}\right]\right).$$

9. Let $X = X(t)$ be the amount of C at time t and $dX/dt = k(120 - 2X)(150 - X)$. If $X(0) = 0$ and $X(5) = 10$, then

$$X(t) = \frac{150 - 150e^{180kt}}{1 - 2.5e^{180kt}},$$

where $k = .0001259$ and $X(20) = 29.3$ grams. Now by L'Hôpital's rule, $X \to 60$ as $t \to \infty$, so that the amount of $A \to 0$ and the amount of $B \to 30$ as $t \to \infty$.

11. (a) The initial-value problem is $dh/dt = -8A_h\sqrt{h}/A_w$, $h(0) = H$. Separating variables and integrating we have

$$\frac{dh}{\sqrt{h}} = -\frac{8A_h}{A_w}dt \quad \text{and} \quad 2\sqrt{h} = -\frac{8A_h}{A_w}t + c.$$

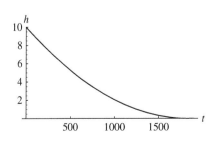

Using $h(0) = H$ we find $c = 2\sqrt{H}$, so the solution of the initial-value problem is $\sqrt{h(t)} = (A_w\sqrt{H} - 4A_h t)/A_w$, where $A_w\sqrt{H} - 4A_h t \geq 0$. Thus,

$$h(t) = (A_w\sqrt{H} - 4A_h t)^2/A_w^2 \quad \text{for} \quad 0 \leq t \leq A_w\sqrt{H}/4A_h.$$

(b) Identifying $H = 10$, $A_w = 4\pi$, and $A_h = \pi/576$ we have
$h(t) = t^2/331{,}776 - (\sqrt{5/2}/144)t + 10$. Solving $h(t) = 0$ we see that the tank empties in $576\sqrt{10}$ seconds or 30.36 minutes.

13. (a) Separating variables and integrating gives

$$6h^{3/2}\, dh = -5t \quad \text{and} \quad \frac{12}{5}h^{5/2} = -5t + c.$$

Using $h(0) = 20$ we find $c = 1920\sqrt{5}$, so the solution of the initial-value problem is $h(t) = \left(800\sqrt{5} - \frac{25}{12}t\right)^{2/5}$. Solving $h(t) = 0$ we see that the tank empties in $384\sqrt{5}$ seconds or 14.31 minutes.

(b) When the height of the water is h, the radius of the top of the water is $r = h\tan 30° = h/\sqrt{3}$ and $A_w = \pi h^2/3$. The differential equation is

$$\frac{dh}{dt} = -c\frac{A_h}{A_w}\sqrt{2gh} = -0.6\frac{\pi(2/12)^2}{\pi h^2/3}\sqrt{64h} = -\frac{2}{5h^{3/2}}.$$

Separating variables and integrating gives

$$5h^{3/2}\, dh = -2\, dt \quad \text{and} \quad 2h^{5/2} = -2t + c.$$

Using $h(0) = 9$ we find $c = 486$, so the solution of the initial-value problem is $h(t) = (243 - t)^{2/5}$. Solving $h(t) = 0$ we see that the tank empties in 243 seconds or 4.05 minutes.

15. (a) After separating variables we obtain

$$\frac{m\, dv}{mg - kv^2} = dt$$

$$\frac{1}{g}\frac{dv}{1 - (\sqrt{k}\,v/\sqrt{mg})^2} = dt$$

$$\frac{\sqrt{mg}}{\sqrt{k}\,g}\frac{\sqrt{k/mg}\, dv}{1 - (\sqrt{k}\,v/\sqrt{mg})^2} = dt$$

$$\sqrt{\frac{m}{kg}}\tanh^{-1}\frac{\sqrt{k}\,v}{\sqrt{mg}} = t + c$$

$$\tanh^{-1}\frac{\sqrt{k}\,v}{\sqrt{mg}} = \sqrt{\frac{kg}{m}}\,t + c_1.$$

Thus the velocity at time t is

$$v(t) = \sqrt{\frac{mg}{k}} \tanh\left(\sqrt{\frac{kg}{m}}\, t + c_1\right).$$

Setting $t = 0$ and $v = v_0$ we find $c_1 = \tanh^{-1}\left(\sqrt{k}\, v_0/\sqrt{mg}\,\right)$.

(b) Since $\tanh t \to 1$ as $t \to \infty$, we have $v \to \sqrt{mg/k}$ as $t \to \infty$.

(c) Integrating the expression for $v(t)$ in part (a) we obtain an integral of the form $\int du/u$:

$$s(t) = \sqrt{\frac{mg}{k}} \int \tanh\left(\sqrt{\frac{kg}{m}}\, t + c_1\right) dt = \frac{m}{k}\ln\left[\cosh\left(\sqrt{\frac{kg}{m}}\, t + c_1\right)\right] + c_2.$$

Setting $t = 0$ and $s = 0$ we find $c_2 = -(m/k)\ln(\cosh c_1)$, where c_1 is given in part (a).

17. (a) Let ρ be the weight density of the water and V the volume of the object. Archimedes' principle states that the upward buoyant force has magnitude equal to the weight of the water displaced. Taking the positive direction to be down, the differential equation is

$$m\frac{dv}{dt} = mg - kv^2 - \rho V.$$

(b) Using separation of variables we have

$$\frac{m\, dv}{(mg - \rho V) - kv^2} = dt$$

$$\frac{m}{\sqrt{k}}\frac{\sqrt{k}\, dv}{(\sqrt{mg - \rho V}\,)^2 - (\sqrt{k}\, v)^2} = dt$$

$$\frac{m}{\sqrt{k}}\frac{1}{\sqrt{mg - \rho V}}\tanh^{-1}\frac{\sqrt{k}\, v}{\sqrt{mg - \rho V}} = t + c.$$

Thus

$$v(t) = \sqrt{\frac{mg - \rho V}{k}} \tanh\left(\frac{\sqrt{kmg - k\rho V}}{m}\, t + c_1\right).$$

(c) Since $\tanh t \to 1$ as $t \to \infty$, the terminal velocity is $\sqrt{(mg - \rho V)/k}$.

19. (a) From $2W^2 - W^3 = W^2(2 - W) = 0$ we see that $W = 0$ and $W = 2$ are constant solutions.

(b) Separating variables and using a CAS to integrate we get

$$\frac{dW}{W\sqrt{4 - 2W}} = dx \quad \text{and} \quad -\tanh^{-1}\left(\frac{1}{2}\sqrt{4 - 2W}\right) = x + c.$$

Using the facts that the hyperbolic tangent is an odd function and $1 - \tanh^2 x = \mathrm{sech}^2\, x$ we have

$$\frac{1}{2}\sqrt{4 - 2W} = \tanh\left(-x - c\right) = -\tanh\left(x + c\right)$$

$$\frac{1}{4}(4 - 2W) = \tanh^2\left(x + c\right)$$

$$1 - \frac{1}{2}W = \tanh^2\left(x + c\right)$$

$$\frac{1}{2}W = 1 - \tanh^2\left(x + c\right) = \mathrm{sech}^2\left(x + c\right)$$

Thus, $W(x) = 2\,\mathrm{sech}^2\left(x + c\right)$.

(c) Letting $x = 0$ and $W = 2$ we find that $\mathrm{sech}^2\left(c\right) = 1$ and $c = 0$.

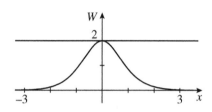

21. (a) With $c = 0.01$ the differential equation is $dP/dt = kP^{1.01}$. Separating variables and integrating we obtain

$$P^{-1.01}\, dP = k\, dt$$

$$\frac{P^{-0.01}}{-0.01} = kt + c_1$$

$$P^{-0.01} = -0.01kt + c_2$$

$$P(t) = (-0.01kt + c_2)^{-100}$$

$$P(0) = c_2^{-100} = 10$$

$$c_2 = 10^{-0.01}.$$

Then

$$P(t) = \frac{1}{\left(-0.01kt + 10^{-0.01}\right)^{100}}$$

and, since P doubles in 5 months from 10 to 20,

$$P(5) = \frac{1}{\left(-0.01k(5) + 10^{-0.01}\right)^{100}} = 20$$

so

$$\left(-0.01k(5) + 10^{-0.01}\right)^{100} = \frac{1}{20}$$

$$-0.01k = \frac{\left[\left(\frac{1}{20}\right)^{1/100} - \left(\frac{1}{10}\right)^{1/100}\right]}{5}$$

$$= -0.001350\,.$$

Thus $P(t) = 1/\left(-0.001350t + 10^{-0.01}\right)^{100}$.

(b) Define $T = \left(\frac{1}{10}\right)^{1/100}/0.001350 \approx 724$ months $= 60$ years. As $t \to 724$ (from the left), $P \to \infty$.

(c)
$$P(50) = 1/\left[-0.001350\,(50) + 10^{-0.01}\right]^{100} \approx 12{,}839 \quad \text{and}$$
$$P(100) = 1/\left[-0.001350\,(100) + 10^{-0.01}\right]^{100} \approx 28{,}630{,}966$$

23. (a)

t	P(t)	Q(t)
0	3.929	0.035
10	5.308	0.036
20	7.240	0.033
30	9.638	0.033
40	12.866	0.033
50	17.069	0.036
60	23.192	0.036
70	31.433	0.023
80	38.558	0.030
90	50.156	0.026
100	62.948	0.021
110	75.996	0.021
120	91.972	0.015
130	105.711	0.016
140	122.775	0.007
150	131.669	0.014
160	150.697	0.019
170	179.300	

(b) The regression line is $Q = 0.0348391 - 0.000168222P$.

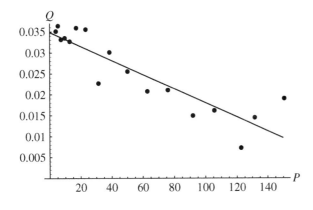

(c) The solution of the logistic equation is given in Equation (5) in the text. Identifying $a = 0.0348391$ and $b = 0.000168222$ we have
$$P(t) = \frac{aP_0}{bP_0 + (a - bP_0)e^{-at}}.$$

(d) With $P_0 = 3.929$ the solution becomes
$$P(t) = \frac{0.136883}{0.000660944 + 0.0341781e^{-0.0348391t}}.$$

(e)

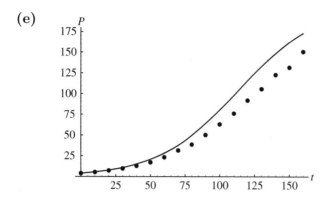

(f) We identify $t = 180$ with 1970, $t = 190$ with 1980, and $t = 200$ with 1990. The model predicts $P(180) = 188.661$, $P(190) = 193.735$, and $P(200) = 197.485$. The actual population figures for these years are 203.303, 226.542, and 248.765 millions. As $t \to \infty$, $P(t) \to a/b = 207.102$.

25. To find t_d we solve

$$m\frac{dv}{dt} = mg - kv^2, \qquad v(0) = 0$$

using separation of variables. This gives

$$v(t) = \sqrt{\frac{mg}{k}}\,\tanh\sqrt{\frac{kg}{m}}\,t.$$

Integrating and using $s(0) = 0$ gives

$$s(t) = \frac{m}{k}\ln\left(\cosh\sqrt{\frac{kg}{m}}\,t\right).$$

To find the time of descent we solve $s(t) = 823.84$ and find $t_d = 7.77882$. The impact velocity is $v(t_d) = 182.998$, which is positive because the positive direction is downward.

27. While the object is in the air its velocity is modeled by the linear differential equation $m\,dv/dt = mg - kv$. Using $m = 160$, $k = \frac{1}{4}$, and $g = 32$, the differential equation becomes $dv/dt + (1/640)v = 32$. The integrating factor is $e^{\int dt/640} = e^{t/640}$ and the solution of the differential equation is $e^{t/640}v = \int 32e^{t/640}\,dt = 20{,}480e^{t/640} + c$. Using $v(0) = 0$ we see that $c = -20{,}480$ and $v(t) = 20{,}480 - 20{,}480e^{-t/640}$. Integrating we get $s(t) = 20{,}480t + 13{,}107{,}200e^{-t/640} + c$. Since $s(0) = 0$, $c = -13{,}107{,}200$ and $s(t) = -13{,}107{,}200 + 20{,}480t + 13{,}107{,}200e^{-t/640}$. To find when the object hits the liquid we solve $s(t) = 500 - 75 = 425$, obtaining $t_a = 5.16018$. The velocity at the time of impact with the liquid is $v_a = v(t_a) = 164.482$. When the object is in the liquid its velocity is modeled by the nonlinear differential equation $m\,dv/dt = mg - kv^2$. Using $m = 160$, $g = 32$, and $k = 0.1$ this becomes $dv/dt = (51{,}200 - v^2)/1600$. Separating variables and integrating we have

$$\frac{dv}{51{,}200 - v^2} = \frac{dt}{1600} \quad \text{and} \quad \frac{\sqrt{2}}{640}\ln\left|\frac{v - 160\sqrt{2}}{v + 160\sqrt{2}}\right| = \frac{1}{1600}t + c.$$

Solving $v(0) = v_a = 164.482$ we obtain $c = -0.00407537$. Then, for $v < 160\sqrt{2} = 226.274$,

$$\left| \frac{v - 160\sqrt{2}}{v + 160\sqrt{2}} \right| = e^{\sqrt{2}t/5 - 1.8443} \quad \text{or} \quad -\frac{v - 160\sqrt{2}}{v + 160\sqrt{2}} = e^{\sqrt{2}t/5 - 1.8443}.$$

Solving for v we get

$$v(t) = \frac{13964.6 - 2208.29e^{\sqrt{2}t/5}}{61.7153 + 9.75937e^{\sqrt{2}t/5}}.$$

Integrating we find

$$s(t) = 226.275t - 1600\ln\left(6.3237 + e^{\sqrt{2}t/5}\right) + c.$$

Solving $s(0) = 0$ we see that $c = 3185.78$, so

$$s(t) = 3185.78 + 226.275t - 1600\ln\left(6.3237 + e^{\sqrt{2}t/5}\right).$$

To find when the object hits the bottom of the tank we solve $s(t) = 75$, obtaining $t_b = 0.466273$. The time from when the object is dropped from the helicopter to when it hits the bottom of the tank is $t_a + t_b = 5.62708$ seconds.

29. **(a)** With $k = v_r/v_s$,

$$\frac{dy}{dx} = \frac{y - k\sqrt{x^2 + y^2}}{x}$$

is a first-order homogeneous differential equation (see Section 2.5). Substituting $y = ux$ into the differential equation gives

$$u + x\frac{du}{dx} = u - k\sqrt{1 + u^2} \quad \text{or} \quad \frac{du}{dx} = -k\sqrt{1 + u^2}.$$

Separating variables and integrating we obtain

$$\int \frac{du}{\sqrt{1 + u^2}} = -\int k\,dx \quad \text{or} \quad \ln\left(u + \sqrt{1 + u^2}\right) = -k\ln x + \ln c.$$

This implies

$$\ln x^k\left(u + \sqrt{1 + u^2}\right) = \ln c \quad \text{or} \quad x^k\left(\frac{y}{x} + \frac{\sqrt{x^2 + y^2}}{x}\right) = c.$$

The condition $y(1) = 0$ gives $c = 1$ and so $y + \sqrt{x^2 + y^2} = x^{1-k}$. Solving for y gives

$$y(x) = \frac{1}{2}\left(x^{1-k} - x^{1+k}\right).$$

(b) If $k = 1$, then $v_s = v_r$ and $y = \frac{1}{2}(1 - x^2)$. Since $y(0) = \frac{1}{2}$, the swimmer lands on the west beach at $(0, \frac{1}{2})$. That is, $\frac{1}{2}$ mile north of $(0,0)$. If $k > 1$, then $v_r > v_s$ and $1 - k < 0$. This means $\lim_{x\to 0^+} y(x)$ becomes infinite, since $\lim_{x\to 0^+} x^{1-k}$ becomes infinite. The swimmer never makes it to the west beach and is swept northward with the current. If $0 < k < 1$, then $v_s > v_r$ and $1 - k > 0$. The value of $y(x)$ at $x = 0$ is $y(0) = 0$. The swimmer has made it to the point $(0,0)$.

31. The differential equation

$$\frac{dy}{dx} = -\frac{30x(1-x)}{2}$$

separates into $dy = 15(-x + x^2)dx$. Integration gives $y(x) = -\frac{15}{2}x^2 + 5x^3 + c$. The condition $y(1) = 0$ gives $c = \frac{5}{2}$ and so $y(x) = \frac{1}{2}(-15x^2 + 10x^3 + 5)$. Since $y(0) = \frac{5}{2}$, the swimmer has to walk 2.5 miles back down the west beach to reach $(0, 0)$.

33. (a) Letting $c = 0.6$, $A_h = \pi(\frac{1}{32} \cdot \frac{1}{12})^2$, $A_w = \pi \cdot 1^2 = \pi$, and $g = 32$, the differential equation in Problem 12 becomes $dh/dt = -0.00003255\sqrt{h}$. Separating variables and integrating, we get $2\sqrt{h} = -0.00003255t + c$, so $h = (c_1 - 0.00001628t)^2$. Setting $h(0) = 2$, we find $c = \sqrt{2}$, so $h(t) = (\sqrt{2} - 0.00001628t)^2$, where h is measured in feet and t in seconds.

(b) One hour is 3,600 seconds, so the hour mark should be placed at

$$h(3600) = [\sqrt{2} - 0.00001628(3600)]^2 \approx 1.838 \,\text{ft} \approx 22.0525 \,\text{in}.$$

up from the bottom of the tank. The remaining marks corresponding to the passage of 2, 3, 4, ..., 12 hours are placed at the values shown in the table. The marks are not evenly spaced because the water is not draining out at a uniform rate; that is, $h(t)$ is not a linear function of time.

time (seconds)	height (inches)
0	24.0000
1	22.0520
2	20.1864
3	18.4033
4	16.7026
5	15.0844
6	13.5485
7	12.0952
8	10.7242
9	9.4357
10	8.2297
11	7.1060
12	6.0648

35. If we let r_h denote the radius of the hole and $A_w = \pi[f(h)]^2$, then the differential equation $dh/dt = -k\sqrt{h}$, where $k = cA_h\sqrt{2g}/A_w$, becomes

$$\frac{dh}{dt} = -\frac{c\pi r_h^2 \sqrt{2g}}{\pi[f(h)]^2}\sqrt{h} = -\frac{8cr_h^2\sqrt{h}}{[f(h)]^2}.$$

For the time marks to be equally spaced, the rate of change of the height must be a constant; that is, $dh/dt = -a$. (The constant is negative because the height is decreasing.) Thus

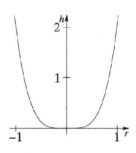

$$-a = -\frac{8cr_h^2\sqrt{h}}{[f(h)]^2}, \qquad [f(h)]^2 = \frac{8cr_h^2\sqrt{h}}{a}, \qquad \text{and} \qquad r = f(h) = 2r_h\sqrt{\frac{2c}{a}}\,h^{1/4}.$$

Solving for h, we have

$$h = \frac{a^2}{64c^2 r_h^4}r^4.$$

The shape of the tank with $c = 0.6$, $a = 2$ ft/12 hr = 1 ft/21,600 s, and $r_h = 1/32(12) = 1/384$ is shown in the above figure.

3.3 | Modeling with Systems of First-Order DEs

1. The linear equation $dx/dt = -\lambda_1 x$ can be solved by either separation of variables or by an integrating factor. Integrating both sides of $dx/x = -\lambda_1 dt$ we obtain $\ln|x| = -\lambda_1 t + c$ from which we get $x = c_1 e^{-\lambda_1 t}$. Using $x(0) = x_0$ we find $c_1 = x_0$ so that $x = x_0 e^{-\lambda_1 t}$. Substituting this result into the second differential equation we have

$$\frac{dy}{dt} + \lambda_2 y = \lambda_1 x_0 e^{-\lambda_1 t}$$

which is linear. An integrating factor is $e^{\lambda_2 t}$ so that

$$\frac{d}{dt}\left[e^{-\lambda_2 t} y\right] + \lambda_2 y = \lambda_1 x_0 e^{(\lambda_2 - \lambda_1)t}$$

$$y = \frac{\lambda_1 x_0}{\lambda_2 - \lambda_1} e^{(\lambda_2 - \lambda_1)t} e^{-\lambda_2 t} + c_2 e^{-\lambda_2 t} = \frac{\lambda_1 x_0}{\lambda_2 - \lambda_1} e^{-\lambda_1 t} + c_2 e^{-\lambda_2 t}.$$

Using $y(0) = 0$ we find $c_2 = -\lambda_1 x_0 / (\lambda_2 - \lambda_1)$. Thus

$$y = \frac{\lambda_1 x_0}{\lambda_2 - \lambda_1}\left(e^{-\lambda_1 t} - e^{-\lambda_2 t}\right).$$

Substituting this result into the third differential equation we have

$$\frac{dz}{dt} = \frac{\lambda_1 \lambda_2 x_0}{\lambda_2 - \lambda_1}\left(e^{-\lambda_1 t} - e^{-\lambda_2 t}\right).$$

Integrating we find

$$z = -\frac{\lambda_2 x_0}{\lambda_2 - \lambda_1} e^{-\lambda_1 t} + \frac{\lambda_1 x_0}{\lambda_2 - \lambda_1} e^{-\lambda_2 t} + c_3.$$

Using $z(0) = 0$ we find $c_3 = x_0$. Thus

$$z = x_0\left(1 - \frac{\lambda_2}{\lambda_2 - \lambda_1} e^{-\lambda_1 t} + \frac{\lambda_1}{\lambda_2 - \lambda_1} e^{-\lambda_2 t}\right).$$

3. The amounts x and y are the same at about $t = 5$ days. The amounts x and z are the same at about $t = 20$ days. The amounts y and z are the same at about $t = 147$ days. The time when y and z are the same makes sense because most of A and half of B are gone, so half of C should have been formed.

5. (a) Since the third equation in the system is linear and containts only the variable $K(t)$ we have

$$\frac{dK}{dt} = -(\lambda_1 + \lambda_2)P$$

$$K(t) = c_1 e^{-(\lambda_1 + \lambda_2)t}$$

Using $K(0) = K_0$ yields $K(t) = K_0 e^{-(\lambda_1 + \lambda_2)t}$.

We can now solve for $A(t)$ and $C(t)$:

$$\frac{dA}{dt} = \lambda_2 P(t) = \lambda_2 K_0 e^{-(\lambda_1+\lambda_2)t}$$

$$A(t) = -\frac{\lambda_2}{\lambda_1+\lambda_2} K_0 e^{-(\lambda_1+\lambda_2)t} + c_2$$

Using $A(0) = 0$ implies $c_2 = \frac{\lambda_2}{\lambda_1+\lambda_2} K_0$. Therefore,

$$A(t) = \frac{\lambda_2}{\lambda_1+\lambda_2} K_0 \left[1 - e^{-(\lambda_1+\lambda_2)t}\right].$$

We use the same approach to solve for $C(t)$:

$$\frac{dC}{dt} = \lambda_1 K(t) = \lambda_1 K_0 e^{-(\lambda_1+\lambda_2)t}$$

$$C(t) = -\frac{\lambda_1}{\lambda_1+\lambda_2} K_0 e^{-(\lambda_1+\lambda_2)t} + c_3$$

Using $C(0) = 0$ implies $c_3 = \frac{\lambda_1}{\lambda_1+\lambda_2} K_0$. Therefore,

$$C(t) = \frac{\lambda_1}{\lambda_1+\lambda_2} K_0 \left[1 - e^{-(\lambda_1+\lambda_2)t}\right].$$

(b) It is known that $\lambda_1 = 4.7526 \times 10^{-10}$ and $\lambda_2 = 0.5874 \times 10^{-10}$ so

$$\lambda_1 + \lambda_2 = 5.34 \times 10^{-10}$$

$$K(t) = K_0 e^{-0.000000000534t}$$

$$K(t) = \frac{1}{2}K_0$$

$$t = \frac{\ln\frac{1}{2}}{-0.000000000534} \approx 1.3 \times 10^9 \text{ years}$$

or the half-life of K-40 is about 1.3 billion years.

(c) Using the solutions $A(t)$, $C(t)$, and the values of λ_1 and λ_2 from part (b) we see that

$$\lim_{t\to\infty} A(t) = \frac{\lambda_2}{\lambda_1+\lambda_2} \lim_{t\to\infty} \left[K_0\left(1 - e^{-(\lambda_1+\lambda_2)t}\right)\right]$$

$$= \frac{\lambda_2}{\lambda_1+\lambda_2} K_0 = \frac{0.5874 \times 10^{-10}}{5.34 \times 10^{-10}} K_0 = 0.11 K_0 \text{ or } 11\% \text{ of } K_0$$

$$\lim_{t\to\infty} C(t) = \frac{\lambda_1}{\lambda_1+\lambda_2} \lim_{t\to\infty} \left[K_0\left(1 - e^{-(\lambda_1+\lambda_2)t}\right)\right]$$

$$= \frac{\lambda_1}{\lambda_1+\lambda_2} K_0 = \frac{4.7526 \times 10^{-10}}{5.34 \times 10^{-10}} K_0 = 0.89 K_0 \text{ or } 89\% \text{ of } K_0$$

7. The system is

$$x_1' = 2 \cdot 3 + \frac{1}{50}x_2 - \frac{1}{50}x_1 \cdot 4 = -\frac{2}{25}x_1 + \frac{1}{50}x_2 + 6$$

$$x_2' = \frac{1}{50}x_1 \cdot 4 - \frac{1}{50}x_2 - \frac{1}{50}x_2 \cdot 3 = \frac{2}{25}x_1 - \frac{2}{25}x_2.$$

9. (a) A model is

$$\frac{dx_1}{dt} = 3 \cdot \frac{x_2}{100 - t} - 2 \cdot \frac{x_1}{100 + t}, \qquad x_1(0) = 100$$

$$\frac{dx_2}{dt} = 2 \cdot \frac{x_1}{100 + t} - 3 \cdot \frac{x_2}{100 - t}, \qquad x_2(0) = 50.$$

(b) Since the system is closed, no salt enters or leaves the system and $x_1(t) + x_2(t) = 100 + 50 = 150$ for all time. Thus $x_1 = 150 - x_2$ and the second equation in part (a) becomes

$$\frac{dx_2}{dt} = \frac{2(150 - x_2)}{100 + t} - \frac{3x_2}{100 - t} = \frac{300}{100 + t} - \frac{2x_2}{100 + t} - \frac{3x_2}{100 - t}$$

or

$$\frac{dx_2}{dt} + \left(\frac{2}{100 + t} + \frac{3}{100 - t}\right)x_2 = \frac{300}{100 + t},$$

which is linear in x_2. An integrating factor is

$$e^{2\ln(100+t) - 3\ln(100-t)} = (100 + t)^2(100 - t)^{-3}$$

so

$$\frac{d}{dt}[(100 + t)^2(100 - t)^{-3}x_2] = 300(100 + t)(100 - t)^{-3}.$$

Using integration by parts, we obtain

$$(100 + t)^2(100 - t)^{-3}x_2 = 300\left[\frac{1}{2}(100 + t)(100 - t)^{-2} - \frac{1}{2}(100 - t)^{-1} + c\right].$$

Thus

$$x_2 = \frac{300}{(100 + t)^2}\left[c(100 - t)^3 - \frac{1}{2}(100 - t)^2 + \frac{1}{2}(100 + t)(100 - t)\right]$$

$$= \frac{300}{(100 + t)^2}[c(100 - t)^3 + t(100 - t)].$$

Using $x_2(0) = 50$ we find $c = 5/3000$. At $t = 30$, $x_2 = (300/130^2)(70^3 c + 30 \cdot 70) \approx 47.4$ lbs.

11. Zooming in on the graph it can be seen that the populations are first equal at about $t = 5.6$. The approximate periods of x and y are both 45.

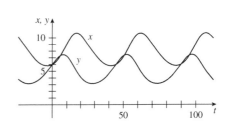

13. (a)

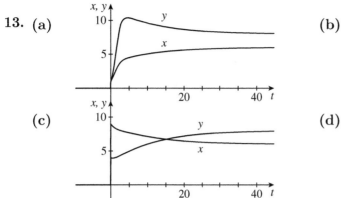

(b)

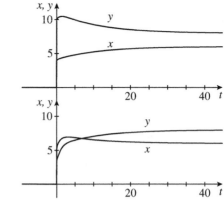

(c)

(d)

In each case the population $x(t)$ approaches 6,000, while the population $y(t)$ approaches 8,000.

15. By Kirchhoff's first law we have $i_1 = i_2 + i_3$. Applying Kirchhoff's second law to each loop we obtain

$$E(t) = i_1 R_1 + L_1 \frac{di_2}{dt} + i_2 R_2$$

and

$$E(t) = i_1 R_1 + L_2 \frac{di_3}{dt} + i_3 R_3.$$

Combining the three equations, we obtain the system

$$L_1 \frac{di_2}{dt} + (R_1 + R_2)i_2 + R_1 i_3 = E$$

$$L_2 \frac{di_3}{dt} + R_1 i_2 + (R_1 + R_3)i_3 = E.$$

17. We first note that $s(t) + i(t) + r(t) = n$. Now the rate of change of the number of susceptible persons, $s(t)$, is proportional to the number of contacts between the number of people infected and the number who are susceptible; that is, $ds/dt = -k_1 si$. We use $-k_1 < 0$ because $s(t)$ is decreasing. Next, the rate of change of the number of persons who have recovered is proportional to the number infected; that is, $dr/dt = k_2 i$ where $k_2 > 0$ since r is increasing. Finally, to obtain di/dt we use

$$\frac{d}{dt}(s + i + r) = \frac{d}{dt} n = 0.$$

This gives

$$\frac{di}{dt} = -\frac{dr}{dt} - \frac{ds}{dt} = -k_2 i + k_1 si.$$

The system of differential equations is then

$$\frac{ds}{dt} = -k_1 si$$

$$\frac{di}{dt} = -k_2 i + k_1 si$$

$$\frac{dr}{dt} = k_2 i.$$

A reasonable set of initial conditions is $i(0) = i_0$, the number of infected people at time 0, $s(0) = n - i_0$, and $r(0) = 0$.

19. Since $x_0 > y_0 > 0$ we have $x(t) > y(t)$ and $y - x < 0$. Thus $dx/dt < 0$ and $dy/dt > 0$. We conclude that $x(t)$ is decreasing and $y(t)$ is increasing. As $t \to \infty$ we expect that $x(t) \to C$ and $y(t) \to C$, where C is a constant common equilibrium concentration.

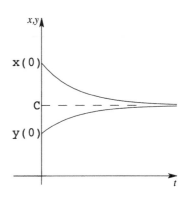

21. Since there are initially 25 pounds of salt in tank A and none in tank B, and since furthermore only pure water is being pumped into tank A, we would expect that $x_1(t)$ would steadily decrease over time. On the other hand, since salt is being added to tank B from tank A, we would expect $x_2(t)$ to increase over time. However, since pure water is being added to the system at a constant rate and a mixed solution is being pumped out of the system, it makes sense that the amount of salt in both tanks would approach 0 over time.

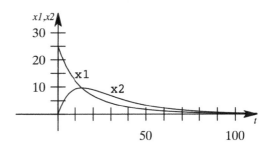

Chapter 3 in Review

1. The differential equation is $dP/dt = 0.15P$.

3. From $\dfrac{dP}{dt} = 0.018P$ and $P(0) = 4$ billion we obtain $P = 4e^{0.018t}$ so that $P(45) = 8.99$ billion.

5. The starting point is $A(t) = A_0 e^{-0.00012097t}$. With $A(t) = 0.53A_0$ we have

$$-0.00012097t = \ln 0.53 \quad \text{or} \quad t = \frac{\ln 0.53}{-0.00012097} \approx 5248 \text{ years.}$$

This represents the iceman's age in 1991, so the approximate date of his death would be

$$1991 - 5248 = -3257 \quad \text{or} \quad 3257 \text{ BC.}$$

7. Separating variables, we have

$$\frac{\sqrt{a^2 - y^2}}{y}\, dy = -dx.$$

Substituting $y = s \sin \theta$, this becomes

$$\frac{\sqrt{a^2 - a^2 \sin^2 \theta}}{a \sin \theta} (a \cos \theta) \, d\theta = -dx$$

$$a \int \frac{\cos^2 \theta}{\sin \theta} \, d\theta = - \int dx$$

$$a \int \frac{1 - \sin^2 \theta}{\sin \theta} \, d\theta = -x + c$$

$$a \int (\csc \theta - \sin \theta) \, d\theta = -x + c$$

$$a \ln |\csc \theta - \cot \theta| + a \cos \theta = -x + c$$

$$a \ln \left| \frac{a}{y} - \frac{\sqrt{a^2 - y^2}}{y} \right| + a \frac{\sqrt{a^2 - y^2}}{a} = -x + c.$$

Letting $s = 10$, this is

$$10 \ln \left| \frac{10}{y} - \frac{\sqrt{100 - y^2}}{y} \right| + \sqrt{100 - y^2} = -x + c.$$

Letting $x = 0$ and $y = 10$ we determine that $c = 0$, so the solution is

$$10 \ln \left| \frac{10}{y} - \frac{\sqrt{100 - y^2}}{y} \right| + \sqrt{100 - y^2} = -x.$$

9. **(a)** The differential equation

$$\frac{dT}{dt} = k(T - T_m) = k[T - T_2 - B(T_1 - T)]$$

$$= k[(1 + B)T - (BT_1 + T_2)] = k(1 + B) \left(T - \frac{BT_1 + T_2}{1 + B} \right)$$

is autonomous and has the single critical point $(BT_1 + T_2)/(1 + B)$. Since $k < 0$ and $B > 0$, by phase-line analysis it is found that the critical point is an attractor and

$$\lim_{t \to \infty} T(t) = \frac{BT_1 + T_2}{1 + B}.$$

Moreover,

$$\lim_{t \to \infty} T_m(t) = \lim_{t \to \infty} [T_2 + B(T_1 - T)] = T_2 + B \left(T_1 - \frac{BT_1 + T_2}{1 + B} \right) = \frac{BT_1 + T_2}{1 + B}.$$

(b) The differential equation is

$$\frac{dT}{dt} = k(T - T_m) = k(T - T_2 - BT_1 + BT)$$

or

$$\frac{dT}{dt} - k(1+B)T = -k(BT_1 + T_2).$$

This is linear and has integrating factor $e^{-\int k(1+B)dt} = e^{-k(1+B)t}$. Thus,

$$\frac{d}{dt}[e^{-k(1+B)t}T] = -k(BT_1 + T_2)e^{-k(1+B)t}$$

$$e^{-k(1+B)t}T = \frac{BT_1 + T_2}{1+B}e^{-k(1+B)t} + c$$

$$T(t) = \frac{BT_1 + T_2}{1+B} + ce^{k(1+B)t}.$$

Since $T(0) = T_1$ we find, $T(t) = \frac{BT_1 + T_2}{1+B} + \frac{T_1 - T_2}{1+B}e^{k(1+B)t}$.

(c) The temperature $T(t)$ decreases to $(BT_1+T_2)/(1+B)$, whereas $T_m(t)$ increases to $(BT_1+T_2)/(1+B)$ as $t \to \infty$. Thus, the temperature $(BT_1 + T_2)/(1+B)$, (which is a weighted average,

$$\frac{B}{1+B}T_1 + \frac{1}{1+B}T_2,$$

of the two initial temperatures), can be interpreted as an equilibrium temperature. The body cannot get cooler than this value whereas the medium cannot get hotter than this value.

11. Separating variables, we obtain

$$\frac{dq}{E_0 - q/C} = \frac{dt}{k_1 + k_2 t}$$

$$-C\ln\left|E_0 - \frac{q}{C}\right| = \frac{1}{k_2}\ln|k_1 + k_2 t| + c_1$$

$$\frac{(E_0 - q/C)^{-C}}{(k_1 + k_2 t)^{1/k_2}} = c_2.$$

Setting $q(0) = q_0$ we find $c_2 = (E_0 - q_0/C)^{-C}/k_1^{1/k_2}$, so

$$\frac{(E_0 - q/C)^{-C}}{(k_1 + k_2 t)^{1/k_2}} = \frac{(E_0 - q_0/C)^{-C}}{k_1^{1/k_2}}$$

$$\left(E_0 - \frac{q}{C}\right)^{-C} = \left(E_0 - \frac{q_0}{C}\right)^{-C}\left(\frac{k_1}{k_1 + k_2 t}\right)^{-1/k_2}$$

$$E_0 - \frac{q}{C} = \left(E_0 - \frac{q_0}{C}\right)\left(\frac{k_1}{k_1 + k_2 t}\right)^{1/Ck_2}$$

$$q = E_0 C + (q_0 - E_0 C)\left(\frac{k_1}{k_1 + k_2 t}\right)^{1/Ck_2}.$$

13. From $dx/dt = k_1 x(\alpha - x)$ we obtain

$$\left(\frac{1/\alpha}{x} + \frac{1/\alpha}{\alpha - x}\right) dx = k_1\, dt$$

so that $x = \alpha c_1 e^{\alpha k_1 t}/(1 + c_1 e^{\alpha k_1 t})$. From $dy/dt = k_2 xy$ we obtain

$$\ln|y| = \frac{k_2}{k_1} \ln\left|1 + c_1 e^{\alpha k_1 t}\right| + c \quad \text{or} \quad y = c_2 \left(1 + c_1 e^{\alpha k_1 t}\right)^{k_2/k_1}.$$

15.

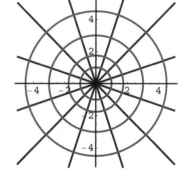

$$y = c_1 x$$

$$\frac{dy}{dx} = c_1$$

$$\frac{dy}{dx} = \frac{y}{x}$$

Therefore the differential equation of the orthogonal family is

$$\frac{dy}{dx} = -\frac{x}{y}$$

$$y\, dy + x\, dx = 0$$

$$x^2 + y^2 = c_2$$

which is a family of circles $(c_2 > 0)$ centered at the origin.

17. From $y = c_1 e^x$ we obtain $y' = y$ so that the differential equation of the orthogonal family is

$$\frac{dy}{dx} = -\frac{1}{y}.$$

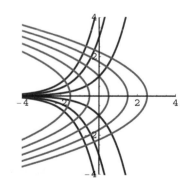

Separating variables and integrating we get

$$y\, dy = -\, dx$$

$$\frac{1}{2} y^2 = -x + c$$

$$y^2 + 2x = c_2$$

19. Critical points of the equation

$$\frac{dP}{dt} = rP\left(1 - \frac{P}{K}\right)\left(\frac{P}{A} - 1\right) \quad r > 0,$$

are 0, A, and K. Here A is called the **Allee threshold** and satisfies $0 < A < K$. From the accompanying phase portrait we see that K and 0 are attractors, or asymptotically stable, but A is a repeller, or unstable. Thus, for an initial value $P_0 < A$ the population decreases over time, that is, $P \to 0$ as $t \to \infty$.

21. The piecewise-defined function $w(x)$ is now

$$w(x) = \begin{cases} x, & 0 \le x \le \dfrac{\sqrt{2}}{2} \\ \sqrt{2} - x, & \dfrac{\sqrt{}}{2} < x \le \sqrt{2} \end{cases}$$

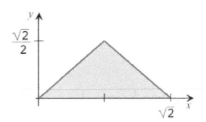

First, we solve

$$x\frac{dx}{dt} = 1, \qquad x(0) = 0$$

by separation of variables. This yield $x(t) = \sqrt{2t}$. The time interval corresponding to $0 \le x \le \frac{\sqrt{2}}{2}$ is defined by $0 \le t \le \frac{1}{4}$. Second, we solve

$$\left(\sqrt{2} - x\right)\frac{dx}{dt} = 1, \qquad x\left(\frac{1}{4}\right) = \frac{\sqrt{2}}{2}.$$

This gives $x^2 - 2\sqrt{2}\,x + 2t + 1 = 0$. Using the quadratic formula, we have $x(t) = \sqrt{2} - \sqrt{1 - 2t}$. The time interval corresponding to $\frac{\sqrt{2}}{2} < x \le \sqrt{2}$ is defined by $\frac{1}{4} \le t \le \frac{1}{2}$. Thus,

$$x(t) = \begin{cases} \sqrt{2t}, & 0 \le t \le \frac{1}{4} \\ \sqrt{2} - \sqrt{1 - 2t}, & \frac{1}{4} < t \le \frac{1}{2}. \end{cases}$$

The time that it takes the saw to cut through the piece of wood is then $t = \frac{1}{2}$.

Chapter 4

Higher-Order Differential Equations

4.1 | Preliminary Theory - Linear Equations

1. From $y = c_1 e^x + c_2 e^{-x}$ we find $y' = c_1 e^x - c_2 e^{-x}$. Then $y(0) = c_1 + c_2 = 0$, $y'(0) = c_1 - c_2 = 1$ so that $c_1 = \frac{1}{2}$ and $c_2 = -\frac{1}{2}$. The solution is $y = \frac{1}{2} e^x - \frac{1}{2} e^{-x}$.

3. From $y = c_1 x + c_2 x \ln x$ we find $y' = c_1 + c_2(1 + \ln x)$. Then $y(1) = c_1 = 3$, $y'(1) = c_1 + c_2 = -1$ so that $c_1 = 3$ and $c_2 = -4$. The solution is $y = 3x - 4x \ln x$.

5. From $y = c_1 + c_2 x^2$ we find $y' = 2c_2 x$. Then $y(0) = c_1 = 0$, $y'(0) = 2c_2 \cdot 0 = 0$ and hence $y'(0) = 1$ is not possible. Since $a_2(x) = x$ is 0 at $x = 0$, Theorem 4.1.1 is not violated.

7. From $x(0) = x_0 = c_1$ we see that $x(t) = x_0 \cos \omega t + c_2 \sin \omega t$ and $x'(t) = -x_0 \sin \omega t + c_2 \omega \cos \omega t$. Then $x'(0) = x_1 = c_2 \omega$ implies $c_2 = x_1/\omega$. Thus

$$x(t) = x_0 \cos \omega t + \frac{x_1}{\omega} \sin \omega t.$$

9. Since $a_2(x) = x - 2$ and $x_0 = 0$ the problem has a unique solution for $-\infty < x < 2$.

11. (a) We have $y(0) = c_1 + c_2 = 0$, $y''(1) = c_1 e + c_2 e^{-1} = 1$ so that $c_1 = e/(e^2 - 1)$ and $c_2 = -e/(e^2 - 1)$. The solution is $y = e(e^x - e^{-x})/(e^2 - 1)$.

(b) We have $y(0) = c_3 \cosh 0 + c_4 \sinh 0 = c_3 = 0$ and $y(1) = c_3 \cosh 1 + c_4 \sinh 1 = c_4 \sinh 1 = 1$, so $c_3 = 0$ and $c_4 = 1/\sinh 1$. The solution is $y = (\sinh x)/(\sinh 1)$.

(c) Starting with the solution in part (b) we have

$$y = \frac{1}{\sinh 1} \sinh x = \frac{2}{e^1 - e^{-1}} \frac{e^x - e^{-x}}{2} = \frac{e^x - e^{-x}}{e - 1/e} = \frac{e}{e^2 - 1} (e^x - e^{-x}).$$

13. From $y = c_1 e^x \cos x + c_2 e^x \sin x$ we find $y' = c_1 e^x(-\sin x + \cos x) + c_2 e^x(\cos x + \sin x)$.

(a) We have $y(0) = c_1 = 1$, $y'(0) = c_1 + c_2 = 0$ so that $c_1 = 1$ and $c_2 = -1$. The solution is $y = e^x \cos x - e^x \sin x$.

(b) We have $y(0) = c_1 = 1$, $y(\pi) = -e^\pi = -1$, which is not possible.

(c) We have $y(0) = c_1 = 1$, $y(\pi/2) = c_2 e^{\pi/2} = 1$ so that $c_1 = 1$ and $c_2 = e^{-\pi/2}$. The solution is $y = e^x \cos x + e^{-\pi/2} e^x \sin x$.

(d) We have $y(0) = c_1 = 0$, $y(\pi) = c_2 e^\pi \sin \pi = 0$ so that $c_1 = 0$ and c_2 is arbitrary. Solutions are $y = c_2 e^x \sin x$, for any real numbers c_2.

15. Since $(-4)x + (3)x^2 + (1)(4x - 3x^2) = 0$ the set of functions is linearly dependent.

17. Since $(-1/5)5 + (1)\cos^2 x + (1)\sin^2 x = 0$ the set of functions is linearly dependent.

19. Since $(-4)x + (3)(x - 1) + (1)(x + 3) = 0$ the set of functions is linearly dependent.

21. Suppose $c_1(1 + x) + c_2 x + c_3 x^2 = 0$. Then $c_1 + (c_1 + c_2)x + c_3 x^2 = 0$ and so $c_1 = 0$, $c_1 + c_2 = 0$, and $c_3 = 0$. Since $c_1 = 0$ we also have $c_2 = 0$. Thus, the set of functions is linearly independent.

23. The functions satisfy the differential equation and are linearly independent since

$$W\left(e^{-3x}, e^{4x}\right) = 7e^x \neq 0$$

for $-\infty < x < \infty$. The general solution is

$$y = c_1 e^{-3x} + c_2 e^{4x}.$$

25. The functions satisfy the differential equation and are linearly independent since

$$W\left(e^x \cos 2x, e^x \sin 2x\right) = 2e^{2x} \neq 0$$

for $-\infty < x < \infty$. The general solution is $y = c_1 e^x \cos 2x + c_2 e^x \sin 2x$.

27. The functions satisfy the differential equation and are linearly independent since

$$W\left(x^3, x^4\right) = x^6 \neq 0$$

for $0 < x < \infty$. The general solution is

$$y = c_1 x^3 + c_2 x^4.$$

29. The functions satisfy the differential equation and are linearly independent since

$$W\left(x, x^{-2}, x^{-2} \ln x\right) = 9x^{-6} \neq 0$$

for $0 < x < \infty$. The general solution is

$$y = c_1 x + c_2 x^{-2} + c_3 x^{-2} \ln x.$$

31. The functions $y_1 = e^{2x}$ and $y_2 = e^{5x}$ form a fundamental set of solutions of the associated homogeneous equation, and $y_p = 6e^x$ is a particular solution of the nonhomogeneous equation.

33. The functions $y_1 = e^{2x}$ and $y_2 = xe^{2x}$ form a fundamental set of solutions of the associated homogeneous equation, and $y_p = x^2 e^{2x} + x - 2$ is a particular solution of the nonhomogeneous equation.

35. (a) We have $y'_{p_1} = 6e^{2x}$ and $y''_{p_1} = 12e^{2x}$, so

$$y''_{p_1} - 6y'_{p_1} + 5y_{p_1} = 12e^{2x} - 36e^{2x} + 15e^{2x} = -9e^{2x}.$$

Also, $y'_{p_2} = 2x + 3$ and $y''_{p_2} = 2$, so

$$y''_{p_2} - 6y'_{p_2} + 5y_{p_2} = 2 - 6(2x + 3) + 5(x^2 + 3x) = 5x^2 + 3x - 16.$$

(b) By the superposition principle for nonhomogeneous equations a particular solution of $y'' - 6y' + 5y = 5x^2 + 3x - 16 - 9e^{2x}$ is $y_p = x^2 + 3x + 3e^{2x}$. A particular solution of the second equation is

$$y_p = -2y_{p_2} - \frac{1}{9}y_{p_1} = -2x^2 - 6x - \frac{1}{3}e^{2x}.$$

37. (a) Since $D^2 x = 0$, x and 1 are solutions of $y'' = 0$. Since they are linearly independent, the general solution is $y = c_1 x + c_2$.

(b) Since $D^3 x^2 = 0$, x^2, x, and 1 are solutions of $y''' = 0$. Since they are linearly independent, the general solution is $y = c_1 x^2 + c_2 x + c_3$.

(c) Since $D^4 x^3 = 0$, x^3, x^2, x, and 1 are solutions of $y^{(4)} = 0$. Since they are linearly independent, the general solution is $y = c_1 x^3 + c_2 x^2 + c_3 x + c_4$.

(d) By part (a), the general solution of $y'' = 0$ is $y_c = c_1 x + c_2$. Since $D^2 x^2 = 2! = 2$, $y_p = x^2$ is a particular solution of $y'' = 2$. Thus, the general solution is $y = c_1 x + c_2 + x^2$.

(e) By part (b), the general solution of $y''' = 0$ is $y_c = c_1 x^2 + c_2 x + c_3$. Since $D^3 x^3 = 3! = 6$, $y_p = x^3$ is a particular solution of $y''' = 6$. Thus, the general solution is $y = c_1 x^2 + c_2 x + c_3 + x^3$.

(f) By part (c), the general solution of $y^{(4)} = 0$ is $y_c = c_1 x^3 + c_2 x^2 + c_3 x + c_4$. Since $D^4 x^4 = 4! = 24$, $y_p = x^4$ is a particular solution of $y^{(4)} = 24$. Thus, the general solution is $y = c_1 x^3 + c_2 x^2 + c_3 x + c_4 + x^4$.

39. (a) From the graphs of $y_1 = x^3$ and $y_2 = |x|^3$ we see that the functions are linearly independent since they cannot be multiples of each other. It is easily shown that $y_1 = x^3$ is a solution of $x^2 y'' - 4xy' + 6y = 0$. To show that $y_2 = |x|^3$ is a solution let $y_2 = x^3$ for $x \geq 0$ and let $y_2 = -x^3$ for $x < 0$.

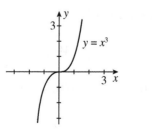
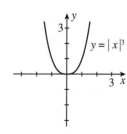

(b) If $x \geq 0$ then $y_2 = x^3$ and

$$W(y_1, y_2) = \begin{vmatrix} x^3 & x^3 \\ 3x^2 & 3x^2 \end{vmatrix} = 0$$

If $x < 0$ then $y_2 = -x^3$ and

$$W(y_1, y_2) = \begin{vmatrix} x^3 & -x^3 \\ 3x^2 & -3x^2 \end{vmatrix} = 0$$

This does not violate Theorem 4.1.3 since $a_2(x) = x^2$ is zero at $x = 0$.

(c) The functions $Y_1 = x^3$ and $Y_2 = x^2$ are solutions of $x^2 y'' - 4xy' + 6y = 0$ on the interval $(-\infty, \infty)$ because we have, in turn,

$$x^2 Y_1'' - 4x Y_1' + 6 Y_1 = x^2 (6x) - 4x (3x^2) + 6x^3 = 0$$
$$x^2 Y_2'' - 4x Y_2' + 6 Y_2 = x^2 (2) - 4x (2x) + 6x^2 = 0.$$

The solutions $Y_1 = x^3$ and $Y_2 = x^2$ are also linearly independent on the interval $(-\infty, \infty)$. In order for $c_1 x^3 + c_2 x^2 = 0$ for *every* real number x it is necessary that $c_1 = c_2 = 0$. To see this, observe that if either c_1 or c_2 were not 0, then the equation $c_1 c^3 + c_2 x^2 = 0$ or $x^2 (c_1 x + c_2) = 0$ would hold for at most two real numbers.

(d) Since the linear differential equation is homogeneous, the superposition principle indicates that $y = x^3 + x^2$ is a solution of the equation. It is also clear that $y = x^3 + x^2$ satisfies the initial conditions $y(0) = 0$, $y'(0) = 0$.

(e) Neither is the general solution on $(-\infty, \infty)$ since we form a general solution on an interval for which $a_2(x) \neq 0$ for every x in the interval.

41. Since $0y_1 + 0y_2 + \cdots + 0y_k + 1y_{k+1} = 0$, the set of solutions is linearly dependent.

4.2 | Reduction of Order

In Problems 1–7 we use reduction of order to find a second solution. In Problems 9–16 we use formula (5) from the text.

1. Define $y = u(x)e^{2x}$ so

$$y' = 2ue^{2x} + u'e^{2x}, \quad y'' = e^{2x}u'' + 4e^{2x}u' + 4e^{2x}u, \quad \text{and} \quad y'' - 4y' + 4y = e^{2x}u'' = 0.$$

Therefore $u'' = 0$ and $u = c_1 x + c_2$. Taking $c_1 = 1$ and $c_2 = 0$ we see that a second solution is $y_2 = xe^{2x}$.

3. Define $y = u(x)\cos 4x$ so

$$y' = -4u\sin 4x + u'\cos 4x, \quad y'' = u''\cos 4x - 8u'\sin 4x - 16u\cos 4x$$

and

$$y'' + 16y = (\cos 4x)u'' - 8(\sin 4x)u' = 0 \quad \text{or} \quad u'' - 8(\tan 4x)u' = 0.$$

If $w = u'$ we obtain the linear first-order equation $w' - 8(\tan 4x)w = 0$ which has the integrating factor $e^{-8\int \tan 4x\, dx} = \cos^2 4x$. Now

$$\frac{d}{dx}[(\cos^2 4x)w] = 0 \quad \text{gives} \quad (\cos^2 4x)w = c.$$

Therefore $w = u' = c\sec^2 4x$ and $u = c_1\tan 4x$. A second solution is
$y_2 = \tan 4x\cos 4x = \sin 4x$.

5. Define $y = u(x)\cosh x$ so

$$y' = u\sinh x + u'\cosh x, \quad y'' = u''\cosh x + 2u'\sinh x + u\cosh x$$

and

$$y'' - y = (\cosh x)u'' + 2(\sinh x)u' = 0 \quad \text{or} \quad u'' + 2(\tanh x)u' = 0.$$

If $w = u'$ we obtain the linear first-order equation $w' + 2(\tanh x)w = 0$ which has the integrating factor $e^{2\int \tanh x\, dx} = \cosh^2 x$. Now

$$\frac{d}{dx}[(\cosh^2 x)w] = 0 \quad \text{gives} \quad (\cosh^2 x)w = c.$$

Therefore $w = u' = c\operatorname{sech}^2 x$ and $u = c\tanh x$. A second solution is
$y_2 = \tanh x\cosh x = \sinh x$.

7. Define $y = u(x)e^{2x/3}$ so

$$y' = \frac{2}{3}e^{2x/3}u + e^{2x/3}u', \quad y'' = e^{2x/3}u'' + \frac{4}{3}e^{2x/3}u' + \frac{4}{9}e^{2x/3}u$$

and

$$9y'' - 12y' + 4y = 9e^{2x/3}u'' = 0.$$

Therefore $u'' = 0$ and $u = c_1 x + c_2$. Taking $c_1 = 1$ and $c_2 = 0$ we see that a second solution
is $y_2 = xe^{2x/3}$.

9. Identifying $P(x) = -7/x$ we have

$$y_2 = x^4 \int \frac{e^{-\int (-7/x)\, dx}}{x^8}\, dx = x^4 \int \frac{1}{x}\, dx = x^4 \ln|x|.$$

A second solution is $y_2 = x^4 \ln|x|$.

11. Identifying $P(x) = 1/x$ we have

$$y_2 = \ln x \int \frac{e^{-\int dx/x}}{(\ln x)^2} \, dx = \ln x \int \frac{dx}{x(\ln x)^2} = \ln x \left(-\frac{1}{\ln x} \right) = -1.$$

A second solution is $y_2 = 1$.

13. Identifying $P(x) = -1/x$ we have

$$y_2 = x \sin (\ln x) \int \frac{e^{-\int -dx/x}}{x^2 \sin^2 (\ln x)} \, dx = x \sin (\ln x) \int \frac{x}{x^2 \sin^2 (\ln x)} \, dx$$

$$= x \sin (\ln x) \int \frac{\csc^2 (\ln x)}{x} \, dx = [x \sin (\ln x)] \, [-\cot (\ln x)] = -x \cos (\ln x).$$

A second solution is $y_2 = x \cos(\ln x)$.

15. Identifying $P(x) = 2(1+x)/\left(1 - 2x - x^2\right)$ we have

$$y_2 = (x+1) \int \frac{e^{-\int 2(1+x)\,dx/\left(1-2x-x^2\right)}}{(x+1)^2} \, dx = (x+1) \int \frac{e^{\ln \left(1-2x-x^2\right)}}{(x+1)^2} \, dx$$

$$= (x+1) \int \frac{1 - 2x - x^2}{(x+1)^2} \, dx = (x+1) \int \left[\frac{2}{(x+1)^2} - 1 \right] dx$$

$$= (x+1) \left[-\frac{2}{x+1} - x \right] = -2 - x^2 - x.$$

A second solution is $y_2 = x^2 + x + 2$.

17. Define $y = u(x)e^{-2x}$ so

$$y' = -2ue^{-2x} + u'e^{-2x}, \quad y'' = u''e^{-2x} - 4u'e^{-2x} + 4ue^{-2x}$$

and

$$y'' - 4y = e^{-2x}u'' - 4e^{-2x}u' = \left(u'' - 4u' \right) e^{-2x} = 2.$$

If $w = u'$ we obtain the linear first-order equation $w' - 4w = 2e^{2x}$ which has the integrating factor $e^{-4 \int dx} = e^{-4x}$. Now

$$\frac{d}{dx}[e^{-4x}w] = 2e^{-2x} \quad \text{gives} \quad e^{-4x}w = -e^{-2x} + c_1.$$

Therefore $w = u' = -e^{2x} + c_1 e^{4x}$ and

$$u = -\frac{1}{2}e^{2x} + \frac{1}{4}c_1 e^{4x} + c_2$$

$$y = -\frac{1}{2} + \frac{1}{4}c_1 e^{2x} + c_2 e^{-2x}.$$

From the last equation we see that a second solution is $y_2 = e^{2x}$ and $y_p = -\frac{1}{2}$.

19. Define $y = u(x)e^x$ so

$$y' = ue^x + u'e^x, \quad y'' = u''e^x + 2u'e^x + ue^x$$

and

$$y'' - 3y' + 2y = e^x u'' - e^x u' = 5e^{3x}.$$

If $w = u'$ we obtain the linear first-order equation $w' - w = 5e^{2x}$ which has the integrating factor $e^{-\int dx} = e^{-x}$. Now

$$\frac{d}{dx}[e^{-x}w] = 5e^x \quad \text{gives} \quad e^{-x}w = 5e^x + c_1.$$

Therefore $w = u' = 5e^{2x} + c_1 e^x$ and

$$u = \frac{5}{2}e^{2x} + c_1 e^x + c_2$$

$$y = \frac{5}{2}e^{3x} + c_1 e^{2x} + c_2 e^x$$

From the last equation we see that a second solution is $y_2 = e^{2x}$ and $y_p = \frac{5}{2}e^{3x}$.

21. Dividing by x^2 we have

$$x^2 y'' + (x^2 - x)\, y' + (1 - x)\, y = 0$$

$$y'' + \left(1 - \frac{1}{x}\right) y' + \left(\frac{1}{x^2} - \frac{1}{x}\right) y = 0$$

Using $P(x) = 1 - \dfrac{1}{x}$ and formula (5) in the text we have

$$y_2(x) = x \int_{x_0}^{x} \frac{e^{-(1-1/t)\,dt}}{t^2}\, dt = x \int_{x_0}^{x} \frac{e^{-t+\ln t}}{t^2}\, dt = x \int_{x_0}^{x} \frac{e^{-t}e^{\ln t}}{t^2}\, dt = x \int_{x_0}^{x} \frac{e^{-t}t}{t^2}\, dt$$

Therefore $y_2(x) = x \displaystyle\int_{x_0}^{x} \frac{e^{-t}}{t}\, dt$, $x + 0 > 0$.

23. (a) For m_1 constant, let $y_1 = e^{m_1 x}$. Then $y_1' = m_1 e^{m_1 x}$ and $y_1'' = m_1^2 e^{m_1 x}$. Substituting into the differential equation we obtain

$$ay_1'' + by_1' + cy_1 = am_1^2 e^{m_1 x} + bm_1 e^{m_1 x} + ce^{m_1 x}$$
$$= e^{m_1 x}(am_1^2 + bm_1 + c) = 0.$$

Thus, $y_1 = e^{m_1 x}$ will be a solution of the differential equation whenever $am_1^2 + bm_1 + c = 0$. Since a quadratic equation always has at least one real or complex root, the differential equation must have a solution of the form $y_1 = e^{m_1 x}$.

(b) Write the differential equation in the form

$$y'' + \frac{b}{a}y' + \frac{c}{a}y = 0,$$

and let $y_1 = e^{m_1 x}$ be a solution. Then a second solution is given by

$$y_2 = e^{m_1 x} \int \frac{e^{-bx/a}}{e^{2m_1 x}}\,dx$$

$$= e^{m_1 x} \int e^{-(b/a+2m_1)x}\,dx$$

$$= -\frac{1}{b/a + 2m_1}\,e^{m_1 x}e^{-(b/a+2m_1)x} \qquad (m_1 \neq -b/2a)$$

$$= -\frac{1}{b/a + 2m_1}\,e^{-(b/a+m_1)x}.$$

Thus, when $m_1 \neq -b/2a$, a second solution is given by $y_2 = e^{m_2 x}$ where $m_2 = -b/a - m_1$. When $m_1 = -b/2a$ a second solution is given by

$$y_2 = e^{m_1 x} \int dx = xe^{m_1 x}.$$

(c) The functions

$$\sin x = \frac{1}{2i}(e^{ix} - e^{-ix}) \qquad\qquad \cos x = \frac{1}{2}(e^{ix} + e^{-ix})$$

$$\sinh x = \frac{1}{2}(e^x - e^{-x}) \qquad\qquad \cosh x = \frac{1}{2}(e^x + e^{-x})$$

are all expressible in terms of exponential functions.

25. (a) We have $y' = y'' = e^x$, so

$$xy'' - (x+10)y' + 10y = xe^x - (x+10)e^x + 10e^x = 0,$$

and $y = e^x$ is a solution of the differential equation.

(b) By (5) in the text a second solution is

$$y_2 = y_1 \int \frac{e^{-\int P(x)\,dx}}{y_1^2}\,dx = e^x \int \frac{e^{\int \frac{x+10}{x}\,dx}}{e^{2x}}\,dx = e^x \int \frac{e^{\int (1+10/x)\,dx}}{e^{2x}}\,dx$$

$$= e^x \int \frac{e^{x+\ln x^{10}}}{e^{2x}}\,dx = e^x \int x^{10}e^{-x}\,dx$$

$$= e^x(-3{,}628{,}800 - 3{,}628{,}800x - 1{,}814{,}400x^2 - 604{,}800x^3 - 151{,}200x^4$$
$$- 30{,}240x^5 - 5{,}040x^6 - 720x^7 - 90x^8 - 10x^9 - x^{10})e^{-x}$$

$$= -3{,}628{,}800 - 3{,}628{,}800x - 1{,}814{,}400x^2 - 604{,}800x^3 - 151{,}200x^4$$
$$- 30{,}240x^5 - 5{,}040x^6 - 720x^7 - 90x^8 - 10x^9 - x^{10}.$$

(c) By Corollary (a) of Theorem 3.1.2, $-\dfrac{1}{10!}\,y_2 = \displaystyle\sum_{n=0}^{10}\frac{1}{n!}x^n$ is a solution.

4.3	Homogeneous Linear Equations with Constant Coefficients

1. From $4m^2 + m = 0$ we obtain $m_1 = 0$ and $m_2 = -1/4$ so that $y = c_1 + c_2 e^{-x/4}$.

3. From $m^2 - m - 6 = 0$ we obtain $m_1 = 3$ and $m_2 = -2$ so that $y = c_1 e^{3x} + c_2 e^{-2x}$.

5. From $m^2 + 8m + 16 = 0$ we obtain $m_1 = -4$ and $m_2 = -4$ so that $y = c_1 e^{-4x} + c_2 x e^{-4x}$.

7. From $12m^2 - 5m - 2 = 0$ we obtain $m_1 = -1/4$ and $m_2 = 2/3$ so that $y = c_1 e^{-x/4} + c_2 e^{2x/3}$.

9. From $m^2 + 9 = 0$ we obtain $m_1 = 3i$ and $m_2 = -3i$ so that $y = c_1 \cos 3x + c_2 \sin 3x$.

11. From $m^2 - 4m + 5 = 0$ we obtain $m = 2 \pm i$ so that $y = e^{2x}(c_1 \cos x + c_2 \sin x)$.

13. From $3m^2 + 2m + 1 = 0$ we obtain $m = -1/3 \pm \sqrt{2}\,i/3$ so that

$$y = e^{-x/3}[c_1 \cos\left(\sqrt{2}\,x/3\right) + c_2 \sin\left(\sqrt{2}\,x/3\right)].$$

15. From $m^3 - 4m^2 - 5m = 0$ we obtain $m_1 = 0$, $m_2 = 5$, and $m_3 = -1$ so that

$$y = c_1 + c_2 e^{5x} + c_3 e^{-x}.$$

17. From $m^3 - 5m^2 + 3m + 9 = 0$ we obtain $m_1 = -1$, $m_2 = 3$, and $m_3 = 3$ so that

$$y = c_1 e^{-x} + c_2 e^{3x} + c_3 x e^{3x}.$$

19. From $m^3 + m^2 - 2 = 0$ we obtain $m_1 = 1$ and $m_2 = -1 \pm i$ so that

$$u = c_1 e^t + e^{-t}(c_2 \cos t + c_3 \sin t).$$

21. From $m^3 + 3m^2 + 3m + 1 = 0$ we obtain $m_1 = -1$, $m_2 = -1$, and $m_3 = -1$ so that

$$y = c_1 e^{-x} + c_2 x e^{-x} + c_3 x^2 e^{-x}.$$

23. From $m^4 + m^3 + m^2 = 0$ we obtain $m_1 = 0$, $m_2 = 0$, and $m_3 = -1/2 \pm \sqrt{3}\,i/2$ so that

$$y = c_1 + c_2 x + e^{-x/2}[c_3 \cos\left(\sqrt{3}\,x/2\right) + c_4 \sin\left(\sqrt{3}\,x/2\right)].$$

25. From $16m^4 + 24m^2 + 9 = 0$ we obtain $m_1 = \pm\sqrt{3}\,i/2$ and $m_2 = \pm\sqrt{3}\,i/2$ so that

$$y = c_1 \cos\left(\sqrt{3}\,x/2\right) + c_2 \sin\left(\sqrt{3}\,x/2\right) + c_3 x \cos\left(\sqrt{3}\,x/2\right) + c_4 x \sin\left(\sqrt{3}\,x/2\right).$$

27. From $m^5 + 5m^4 - 2m^3 - 10m^2 + m + 5 = 0$ we obtain $m_1 = -1$, $m_2 = -1$, $m_3 = 1$, and $m_4 = 1$, and $m_5 = -5$ so that

$$u = c_1 e^{-r} + c_2 r e^{-r} + c_3 e^r + c_4 r e^r + c_5 e^{-5r}.$$

29. From $m^2 + 16 = 0$ we obtain $m = \pm 4i$ so that $y = c_1 \cos 4x + c_2 \sin 4x$. If $y(0) = 2$ and $y'(0) = -2$ then $c_1 = 2$, $c_2 = -1/2$, and $y = 2\cos 4x - \frac{1}{2}\sin 4x$.

31. From $m^2 - 4m - 5 = 0$ we obtain $m_1 = -1$ and $m_1 = 5$, so that $y = c_1 e^{-t} + c_2 e^{5t}$. If $y(1) = 0$ and $y'(1) = 2$, then $c_1 e^{-1} + c_2 e^5 = 0$, $-c_1 e^{-1} + 5c_2 e^5 = 2$, so $c_1 = -e/3$, $c_2 = e^{-5}/3$, and $y = -\frac{1}{3}e^{1-t} + \frac{1}{3}e^{5t-5}$.

33. From $m^2 + m + 2 = 0$ we obtain $m = -1/2 \pm \sqrt{7}\, i/2$ so that $y = e^{-x/2}[c_1 \cos(\sqrt{7}\, x/2) + c_2 \sin(\sqrt{7}\, x/2)]$. If $y(0) = 0$ and $y'(0) = 0$ then $c_1 = 0$ and $c_2 = 0$ so that $y = 0$.

35. From $m^3 + 12m^2 + 36m = 0$ we obtain $m_1 = 0$, $m_2 = -6$, and $m_3 = -6$ so that $y = c_1 + c_2 e^{-6x} + c_3 x e^{-6x}$. If $y(0) = 0$, $y'(0) = 1$, and $y''(0) = -7$ then

$$c_1 + c_2 = 0, \quad -6c_2 + c_3 = 1, \quad 36c_2 - 12c_3 = -7,$$

so $c_1 = 5/36$, $c_2 = -5/36$, $c_3 = 1/6$, and $y = \frac{5}{36} - \frac{5}{36}e^{-6x} + \frac{1}{6}x e^{-6x}$.

37. From $m^2 - 10m + 25 = 0$ we obtain $m_1 = 5$ and $m_2 = 5$ so that $y = c_1 e^{5x} + c_2 x e^{5x}$. If $y(0) = 1$ and $y(1) = 0$ then $c_1 = 1$, $c_1 e^5 + c_2 e^5 = 0$, so $c_1 = 1$, $c_2 = -1$, and $y = e^{5x} - x e^{5x}$.

39. From $m^2 + 1 = 0$ we obtain $m = \pm i$ so that $y = c_1 \cos x + c_2 \sin x$ and $y' = -c_1 \sin x + c_2 \cos x$. From $y'(0) = c_1(0) + c_2(1) = c_2 = 0$ and $y'(\pi/2) = -c_1(1) = 0$ we find $c_1 = c_2 = 0$. A solution of the boundary-value problem is $y = 0$.

41. The auxiliary equation is $m^2 - 3 = 0$ which has roots $-\sqrt{3}$ and $\sqrt{3}$. By (10) the general solution is $y = c_1 e^{\sqrt{3}\, x} + c_2 e^{-\sqrt{3}\, x}$. By (11) the general solution is $y = c_1 \cosh \sqrt{3}\, x + c_2 \sinh \sqrt{3}\, x$. For $y = c_1 e^{\sqrt{3}\, x} + c_2 e^{-\sqrt{3}\, x}$ the initial conditions imply $c_1 + c_2 = 1$, $\sqrt{3}\, c_1 - \sqrt{3}\, c_2 = 5$. Solving for c_1 and c_2 we find $c_1 = \frac{1}{2}(1 + 5\sqrt{3})$ and $c_2 = \frac{1}{2}(1 - 5\sqrt{3})$ so $y = \frac{1}{2}(1 + 5\sqrt{3})e^{\sqrt{3}\, x} + \frac{1}{2}(1 - 5\sqrt{3})e^{-\sqrt{3}\, x}$. For $y = c_1 \cosh \sqrt{3}\, x + c_2 \sinh \sqrt{3}\, x$ the initial conditions imply $c_1 = 1$, $\sqrt{3}\, c_2 = 5$. Solving for c_1 and c_2 we find $c_1 = 1$ and $c_2 = \frac{5}{3}\sqrt{3}$ so $y = \cosh \sqrt{3}\, x + \frac{5}{3}\sqrt{3} \sinh \sqrt{3}\, x$.

43. The auxiliary equation should have two positive roots, so that the solution has the form $y = c_1 e^{k_1 x} + c_2 e^{k_2 x}$. Thus, the differential equation is (f).

45. The auxiliary equation should have a pair of complex roots $\alpha \pm \beta i$ where $\alpha < 0$, so that the solution has the form $e^{\alpha x}(c_1 \cos \beta x + c_2 \sin \beta x)$. Thus, the differential equation is (e).

47. The differential equation should have the form $y'' + k^2 y = 0$ where $k = 1$ so that the period of the solution is 2π. Thus, the differential equation is (d).

49. We have $(m-1)(m-5) = m^2 - 6m + 5$, so the differential equation is $y'' - 6y' + 5y = 0$.

51. We have $m(m-2) = m^2 - 2m$, so the differential equation is $y'' - 2y' = 0$.

53. We have $(m-3i)(m+3i) = m^2 + 9$, so the differential equation is $y'' + 9y = 0$.

55. We have $[m - (-1 + i)][m - (-1 - i)] = m^2 + 2m + 2$, so the differential equation is $y'' + 2y' + 2y = 0$.

57. We have $m^2(m - 8) = m^3 - 8m^2$, so the differential equation is $y''' - 8y'' = 0$.

59. A third root must be $m_3 = 3 - i$ and the auxiliary equation is

$$\left(m + \frac{1}{2}\right)[m - (3 + i)][m - (3 - i)] = \left(m + \frac{1}{2}\right)(m^2 - 6m + 10) = m^3 - \frac{11}{2}m^2 + 7m + 5.$$

The differential equation is

$$y''' - \frac{11}{2}y'' + 7y' + 5y = 0.$$

61. From the solution $y_1 = e^{-4x}\cos x$ we conclude that $m_1 = -4 + i$ and $m_2 = -4 - i$ are roots of the auxiliary equation. Hence another solution must be $y_2 = e^{-4x}\sin x$. Now dividing the polynomial $m^3 + 6m^2 + m - 34$ by $[m - (-4 + i)][m - (-4 - i)] = m^2 + 8m + 17$ gives $m - 2$. Therefore $m_3 = 2$ is the third root of the auxiliary equation, and the general solution of the differential equation is

$$y = c_1 e^{-4x}\cos x + c_2 e^{-4x}\sin x + c_3 e^{2x}.$$

63. Using the definition of $\sinh x$ and the formula for the cosine of the sum of two angles, we have

$$y = \sinh x - 2\cos(x + \pi/6)$$
$$= \frac{1}{2}e^x - \frac{1}{2}e^{-x} - 2\left[(\cos x)\left(\cos\frac{\pi}{6}\right) - (\sin x)\left(\sin\frac{\pi}{6}\right)\right]$$
$$= \frac{1}{2}e^x - \frac{1}{2}e^{-x} - 2\left(\frac{\sqrt{3}}{2}\cos x - \frac{1}{2}\sin x\right)$$
$$= \frac{1}{2}e^x - \frac{1}{2}e^{-x} - \sqrt{3}\cos x + \sin x.$$

This form of the solution can be obtained from the general solution $y = c_1 e^x + c_2 e^{-x} + c_3\cos x + c_4\sin x$ by choosing $c_1 = \frac{1}{2}$, $c_2 = -\frac{1}{2}$, $c_3 = -\sqrt{3}$, and $c_4 = 1$.

65. Using a CAS to solve the auxiliary equation $m^3 - 6m^2 + 2m + 1$ we find $m_1 = -0.270534$, $m_2 = 0.658675$, and $m_3 = 5.61186$. The general solution is

$$y = c_1 e^{-0.270534x} + c_2 e^{0.658675x} + c_3 e^{5.61186x}.$$

67. Using a CAS to solve the auxiliary equation $3.15m^4 - 5.34m^2 + 6.33m - 2.03 = 0$ we find $m_1 = -1.74806$, $m_2 = 0.501219$, $m_3 = 0.62342 + 0.588965i$, and $m_4 = 0.62342 - 0.588965i$. The general solution is

$$y = c_1 e^{-1.74806x} + c_2 e^{0.501219x} + e^{0.62342x}(c_3\cos 0.588965x + c_4\sin 0.588965x).$$

69. From $2m^4 + 3m^3 - 16m^2 + 15m - 4 = 0$ we obtain $m_1 = -4$, $m_2 = \frac{1}{2}$, $m_3 = 1$, and $m_4 = 1$, so that $y = c_1 e^{-4x} + c_2 e^{x/2} + c_3 e^x + c_4 x e^x$. If $y(0) = -2$, $y'(0) = 6$, $y''(0) = 3$, and $y'''(0) = \frac{1}{2}$, then

$$c_1 + c_2 + c_3 = -2$$

$$-4c_1 + \frac{1}{2}c_2 + c_3 + c_4 = 6$$

$$16c_1 + \frac{1}{4}c_2 + c_3 + 2c_4 = 3$$

$$-64c_1 + \frac{1}{8}c_2 + c_3 + 3c_4 = \frac{1}{2},$$

so $c_1 = -\frac{4}{75}$, $c_2 = -\frac{116}{3}$, $c_3 = \frac{918}{25}$, $c_4 = -\frac{58}{5}$, and

$$y = -\frac{4}{75}e^{-4x} - \frac{116}{3}e^{x/2} + \frac{918}{25}e^x - \frac{58}{5}xe^x.$$

4.4 Undetermined Coefficients - Superposition Approach

1. From $m^2 + 3m + 2 = 0$ we find $m_1 = -1$ and $m_2 = -2$. Then $y_c = c_1 e^{-x} + c_2 e^{-2x}$ and we assume $y_p = A$. Substituting into the differential equation we obtain $2A = 6$. Then $A = 3$, $y_p = 3$ and

$$y = c_1 e^{-x} + c_2 e^{-2x} + 3.$$

3. From $m^2 - 10m + 25 = 0$ we find $m_1 = m_2 = 5$. Then $y_c = c_1 e^{5x} + c_2 x e^{5x}$ and we assume $y_p = Ax + B$. Substituting into the differential equation we obtain $25A = 30$ and $-10A + 25B = 3$. Then $A = \frac{6}{5}$, $B = \frac{3}{5}$, $y_p = \frac{6}{5}x + \frac{3}{5}$, and

$$y = c_1 e^{5x} + c_2 x e^{5x} + \frac{6}{5}x + \frac{3}{5}.$$

5. From $\frac{1}{4}m^2 + m + 1 = 0$ we find $m_1 = m_2 = -2$. Then $y_c = c_1 e^{-2x} + c_2 x e^{-2x}$ and we assume $y_p = Ax^2 + Bx + C$. Substituting into the differential equation we obtain $A = 1$, $2A + B = -2$, and $\frac{1}{2}A + B + C = 0$. Then $A = 1$, $B = -4$, $C = \frac{7}{2}$, $y_p = x^2 - 4x + \frac{7}{2}$, and

$$y = c_1 e^{-2x} + c_2 x e^{-2x} + x^2 - 4x + \frac{7}{2}.$$

7. From $m^2 + 3 = 0$ we find $m_1 = \sqrt{3}\,i$ and $m_2 = -\sqrt{3}\,i$. Then $y_c = c_1 \cos\sqrt{3}\,x + c_2 \sin\sqrt{3}\,x$ and we assume $y_p = (Ax^2 + Bx + C)e^{3x}$. Substituting into the differential equation we obtain $2A + 6B + 12C = 0$, $12A + 12B = 0$, and $12A = -48$. Then $A = -4$, $B = 4$, $C = -\frac{4}{3}$, $y_p = \left(-4x^2 + 4x - \frac{4}{3}\right)e^{3x}$ and

$$y = c_1 \cos\sqrt{3}\,x + c_2 \sin\sqrt{3}\,x + \left(-4x^2 + 4x - \frac{4}{3}\right)e^{3x}.$$

9. From $m^2 - m = 0$ we find $m_1 = 1$ and $m_2 = 0$. Then $y_c = c_1 e^x + c_2$ and we assume $y_p = Ax$. Substituting into the differential equation we obtain $-A = -3$. Then $A = 3$, $y_p = 3x$ and $y = c_1 e^x + c_2 + 3x$.

11. From $m^2 - m + \frac{1}{4} = 0$ we find $m_1 = m_2 = \frac{1}{2}$. Then $y_c = c_1 e^{x/2} + c_2 x e^{x/2}$ and we assume $y_p = A + Bx^2 e^{x/2}$. Substituting into the differential equation we obtain $\frac{1}{4}A = 3$ and $2B = 1$. Then $A = 12$, $B = \frac{1}{2}$, $y_p = 12 + \frac{1}{2}x^2 e^{x/2}$, and

$$y = c_1 e^{x/2} + c_2 x e^{x/2} + 12 + \frac{1}{2}x^2 e^{x/2}.$$

13. From $m^2 + 4 = 0$ we find $m_1 = 2i$ and $m_2 = -2i$. Then $y_c = c_1 \cos 2x + c_2 \sin 2x$ and we assume $y_p = Ax \cos 2x + Bx \sin 2x$. Substituting into the differential equation we obtain $4B = 0$ and $-4A = 3$. Then $A = -\frac{3}{4}$, $B = 0$, $y_p = -\frac{3}{4}x \cos 2x$, and

$$y = c_1 \cos 2x + c_2 \sin 2x - \frac{3}{4}x \cos 2x.$$

15. From $m^2 + 1 = 0$ we find $m_1 = i$ and $m_2 = -i$. Then $y_c = c_1 \cos x + c_2 \sin x$ and we assume $y_p = (Ax^2 + Bx) \cos x + (Cx^2 + Dx) \sin x$. Substituting into the differential equation we obtain $4C = 0$, $2A + 2D = 0$, $-4A = 2$, and $-2B + 2C = 0$. Then $A = -\frac{1}{2}$, $B = 0$, $C = 0$, $D = \frac{1}{2}$, $y_p = -\frac{1}{2}x^2 \cos x + \frac{1}{2}x \sin x$, and

$$y = c_1 \cos x + c_2 \sin x - \frac{1}{2}x^2 \cos x + \frac{1}{2}x \sin x.$$

17. From $m^2 - 2m + 5 = 0$ we find $m_1 = 1 + 2i$ and $m_2 = 1 - 2i$. Then $y_c = e^x(c_1 \cos 2x + c_2 \sin 2x)$ and we assume $y_p = Axe^x \cos 2x + Bxe^x \sin 2x$. Substituting into the differential equation we obtain $4B = 1$ and $-4A = 0$. Then $A = 0$, $B = \frac{1}{4}$, $y_p = \frac{1}{4}xe^x \sin 2x$, and

$$y = e^x(c_1 \cos 2x + c_2 \sin 2x) + \frac{1}{4}xe^x \sin 2x.$$

19. From $m^2 + 2m + 1 = 0$ we find $m_1 = m_2 = -1$. Then $y_c = c_1 e^{-x} + c_2 x e^{-x}$ and we assume $y_p = A \cos x + B \sin x + C \cos 2x + D \sin 2x$. Substituting into the differential equation we obtain $2B = 0$, $-2A = 1$, $-3C + 4D = 3$, and $-4C - 3D = 0$. Then $A = -\frac{1}{2}$, $B = 0$, $C = -\frac{9}{25}$, $D = \frac{12}{25}$, $y_p = -\frac{1}{2}\cos x - \frac{9}{25}\cos 2x + \frac{12}{25}\sin 2x$, and

$$y = c_1 e^{-x} + c_2 x e^{-x} - \frac{1}{2}\cos x - \frac{9}{25}\cos 2x + \frac{12}{25}\sin 2x.$$

21. From $m^3 - 6m^2 = 0$ we find $m_1 = m_2 = 0$ and $m_3 = 6$. Then $y_c = c_1 + c_2 x + c_3 e^{6x}$ and we assume $y_p = Ax^2 + B \cos x + C \sin x$. Substituting into the differential equation we obtain $-12A = 3$, $6B - C = -1$, and $B + 6C = 0$. Then $A = -\frac{1}{4}$, $B = -\frac{6}{37}$, $C = \frac{1}{37}$, $y_p = -\frac{1}{4}x^2 - \frac{6}{37}\cos x + \frac{1}{37}\sin x$, and

$$y = c_1 + c_2 x + c_3 e^{6x} - \frac{1}{4}x^2 - \frac{6}{37}\cos x + \frac{1}{37}\sin x.$$

23. From $m^3 - 3m^2 + 3m - 1 = 0$ we find $m_1 = m_2 = m_3 = 1$. Then $y_c = c_1 e^x + c_2 x e^x + c_3 x^2 e^x$ and we assume $y_p = Ax + B + Cx^3 e^x$. Substituting into the differential equation we obtain $-A = 1$, $3A - B = 0$, and $6C = -4$. Then $A = -1$, $B = -3$, $C = -\frac{2}{3}$, $y_p = -x - 3 - \frac{2}{3}x^3 e^x$, and

$$y = c_1 e^x + c_2 x e^x + c_3 x^2 e^x - x - 3 - \frac{2}{3}x^3 e^x.$$

25. From $m^4 + 2m^2 + 1 = 0$ we find $m_1 = m_3 = i$ and $m_2 = m_4 = -i$. Then $y_c = c_1 \cos x + c_2 \sin x + c_3 x \cos x + c_4 x \sin x$ and we assume $y_p = Ax^2 + Bx + C$. Substituting into the differential equation we obtain $A = 1$, $B = -2$, and $4A + C = 1$. Then $A = 1$, $B = -2$, $C = -3$, $y_p = x^2 - 2x - 3$, and

$$y = c_1 \cos x + c_2 \sin x + c_3 x \cos x + c_4 x \sin x + x^2 - 2x - 3.$$

27. We have $y_c = c_1 \cos 2x + c_2 \sin 2x$ and we assume $y_p = A$. Substituting into the differential equation we find $A = -\frac{1}{2}$. Thus $y = c_1 \cos 2x + c_2 \sin 2x - \frac{1}{2}$. From the initial conditions we obtain $c_1 = 0$ and $c_2 = \sqrt{2}$, so

$$y = \sqrt{2} \sin 2x - \frac{1}{2}.$$

29. We have $y_c = c_1 e^{-x/5} + c_2$ and we assume $y_p = Ax^2 + Bx$. Substituting into the differential equation we find $A = -3$ and $B = 30$. Thus $y = c_1 e^{-x/5} + c_2 - 3x^2 + 30x$. From the initial conditions we obtain $c_1 = 200$ and $c_2 = -200$, so

$$y = 200 e^{-x/5} - 200 - 3x^2 + 30x.$$

31. We have $y_c = e^{-2x}(c_1 \cos x + c_2 \sin x)$ and we assume $y_p = Ae^{-4x}$. Substituting into the differential equation we find $A = 7$. Thus $y = e^{-2x}(c_1 \cos x + c_2 \sin x) + 7e^{-4x}$. From the initial conditions we obtain $c_1 = -10$ and $c_2 = 9$, so

$$y = e^{-2x}(-10 \cos x + 9 \sin x) + 7e^{-4x}.$$

33. We have $x_c = c_1 \cos \omega t + c_2 \sin \omega t$ and we assume $x_p = At \cos \omega t + Bt \sin \omega t$. Substituting into the differential equation we find $A = -F_0/2\omega$ and $B = 0$. Thus $x = c_1 \cos \omega t + c_2 \sin \omega t - (F_0/2\omega)t \cos \omega t$. From the initial conditions we obtain $c_1 = 0$ and $c_2 = F_0/2\omega^2$, so

$$x = (F_0/2\omega^2) \sin \omega t - (F_0/2\omega)t \cos \omega t.$$

35. We have $y_c = c_1 + c_2 e^x + c_3 x e^x$ and we assume $y_p = Ax + Bx^2 e^x + Ce^{5x}$. Substituting into the differential equation we find $A = 2$, $B = -12$, and $C = \frac{1}{2}$. Thus

$$y = c_1 + c_2 e^x + c_3 x e^x + 2x - 12x^2 e^x + \frac{1}{2}e^{5x}.$$

From the initial conditions we obtain $c_1 = 11$, $c_2 = -11$, and $c_3 = 9$, so

$$y = 11 - 11e^x + 9x e^x + 2x - 12x^2 e^x + \frac{1}{2}e^{5x}.$$

37. We have $y_c = c_1 \cos x + c_2 \sin x$ and we assume $y_p = Ax^2 + Bx + C$. Substituting into the differential equation we find $A = 1$, $B = 0$, and $C = -1$. Thus $y = c_1 \cos x + c_2 \sin x + x^2 - 1$. From $y(0) = 5$ and $y(1) = 0$ we obtain

$$c_1 - 1 = 5$$

$$(\cos 1)c_1 + (\sin 1)c_2 = 0.$$

Solving this system we find $c_1 = 6$ and $c_2 = -6 \cot 1$. The solution of the boundary-value problem is

$$y = 6 \cos x - 6(\cot 1) \sin x + x^2 - 1.$$

39. The general solution of the differential equation $y'' + 3y = 6x$ is $y = c_1 \cos \sqrt{3}x + c_2 \sin \sqrt{3}x + 2x$. The condition $y(0) = 0$ implies $c_1 = 0$ and so $y = c_2 \sin \sqrt{3}x + 2x$. The condition $y(1) + y'(1) = 0$ implies $c_2 \sin \sqrt{3} + 2 + c_2 \sqrt{3} \cos \sqrt{3} + 2 = 0$ so $c_2 = -4/(\sin \sqrt{3} + \sqrt{3} \cos \sqrt{3})$. The solution is

$$y = \frac{-4 \sin \sqrt{3}x}{\sin \sqrt{3} + \sqrt{3} \cos \sqrt{3}} + 2x.$$

41. We have $y_c = c_1 \cos 2x + c_2 \sin 2x$ and we assume $y_p = A \cos x + B \sin x$ on $[0, \pi/2]$. Substituting into the differential equation we find $A = 0$ and $B = \frac{1}{3}$. Thus $y = c_1 \cos 2x + c_2 \sin 2x + \frac{1}{3} \sin x$ on $[0, \pi/2]$. On $(\pi/2, \infty)$ we have $y = c_3 \cos 2x + c_4 \sin 2x$. From $y(0) = 1$ and $y'(0) = 2$ we obtain

$$c_1 = 1$$

$$\frac{1}{3} + 2c_2 = 2.$$

Solving this system we find $c_1 = 1$ and $c_2 = \frac{5}{6}$. Thus $y = \cos 2x + \frac{5}{6} \sin 2x + \frac{1}{3} \sin x$ on $[0, \pi/2]$. Now continuity of y at $x = \pi/2$ implies

$$\cos \pi + \frac{5}{6} \sin \pi + \frac{1}{3} \sin \frac{\pi}{2} = c_3 \cos \pi + c_4 \sin \pi$$

or $-1 + \frac{1}{3} = -c_3$. Hence $c_3 = \frac{2}{3}$. Continuity of y' at $x = \pi/2$ implies

$$-2 \sin \pi + \frac{5}{3} \cos \pi + \frac{1}{3} \cos \frac{\pi}{2} = -2c_3 \sin \pi + 2c_4 \cos \pi$$

or $-\frac{5}{3} = -2c_4$. Then $c_4 = \frac{5}{6}$ and so

$$y(x) = \begin{cases} \cos 2x + \dfrac{5}{6} \sin 2x + \dfrac{1}{3} \sin x, & 0 \leq x \leq \pi/2 \\[2mm] \dfrac{2}{3} \cos 2x + \dfrac{5}{6} \sin 2x, & x > \pi/2 \end{cases}$$

43. (a) From $y_p = Ae^{kx}$ we find $y_p' = Ake^{kx}$ and $y_p'' = Ak^2 e^{kx}$. Substituting into the differential equation we get

$$aAk^2 e^{kx} + bAke^{kx} + cAe^{kx} = (ak^2 + bk + c)Ae^{kx} = e^{kx},$$

so $(ak^2 + bk + c)A = 1$. Since k is not a root of $am^2 + bm + c = 0$, $A = 1/(ak^2 + bk + c)$.

(b) From $y_p = Axe^{kx}$ we find $y_p' = Akxe^{kx} + Ae^{kx}$ and $y_p'' = Ak^2xe^{kx} + 2Ake^{kx}$. Substituting into the differential equation we get

$$aAk^2xe^{kx} + 2aAke^{kx} + bAkxe^{kx} + bAe^{kx} + cAxe^{kx}$$
$$= (ak^2 + bk + c)Axe^{kx} + (2ak + b)Ae^{kx}$$
$$= (0)Axe^{kx} + (2ak + b)Ae^{kx} = (2ak + b)Ae^{kx} = e^{kx}$$

where $ak^2 + bk + c = 0$ because k is a root of the auxiliary equation. Now, the roots of the auxiliary equation are $-b/2a \pm \sqrt{b^2 - 4ac}/2a$, and since k is a root of multiplicity one, $k \neq -b/2a$ and $2ak + b \neq 0$. Thus $(2ak + b)A = 1$ and $A = 1/(2ak + b)$.

(c) If k is a root of multiplicity two, then, as we saw in part (b), $k = -b/2a$ and $2ak + b = 0$. From $y_p = Ax^2e^{kx}$ we find $y_p' = Akx^2e^{kx} + 2Axe^{kx}$ and $y_p'' = Ak^2x^2e^{kx} + 4Akxe^{kx} = 2Ae^{kx}$. Substituting into the differential equation, we get

$$aAk^2x^2e^{kx} + 4aAkxe^{kx} + 2aAe^{kx} + bAkx^2e^{kx} + 2bAxe^{kx} + cAx^2e^{kx}$$
$$= (ak^2 + bk + c)Ax^2e^{kx} + 2(2ak + b)Axe^{kx} + 2aAe^{kx}$$
$$= (0)Ax^2e^{kx} + 2(0)Axe^{kx} + 2aAe^{kx} = 2aAe^{kx} = e^{kx}.$$

Since the differential equation is second order, $a \neq 0$ and $A = 1/(2a)$.

45. $f(x) = e^x \sin x$. We see that $y_p \to \infty$ as $x \to \infty$ and $y_p \to 0$ as $x \to -\infty$.

47. $f(x) = \sin 2x$. We see that y_p is sinusoidal.

49. The complementary function is $y_c = e^{2x}(c_1 \cos 2x + c_2 \sin 2x)$. We assume a particular solution of the form $y_p = (Ax^3 + Bx^2 + Cx)e^{2x} \cos 2x + (Dx^3 + Ex^2 + F)e^{2x} \sin 2x$. Substituting into the differential equation and using a CAS to simplify yields

$$[12Dx^2 + (6A + 8E)x + (2B + 4F)]e^{2x} \cos 2x$$
$$+ [-12Ax^2 + (-8B + 6D)x + (-4C + 2E)]e^{2x} \sin 2x$$
$$= (2x^2 - 3x)e^{2x} \cos 2x + (10x^2 - x - 1)e^{2x} \sin 2x$$

This gives the system of equations

$12D = 2,$	$6A + 8E = -3$	$2B + 4F = 0,$
$-12A = 10,$	$-8B + 6D = -1$	$-4C + 2E = -1,$

from which we find $A = -\frac{5}{6}$, $B = \frac{1}{4}$, $C = \frac{3}{8}$, $D = \frac{1}{6}$, $E = \frac{1}{4}$, and $F = -\frac{1}{8}$. Thus, a particular solution of the differential equation is

$$y_p = \left(-\frac{5}{6}x^3 + \frac{1}{4}x^2 + \frac{3}{8}x\right)e^{2x} \cos 2x + \left(\frac{1}{6}x^3 + \frac{1}{4}x^2 - \frac{1}{8}x\right)e^{2x} \sin 2x.$$

4.5 | Undetermined Coefficients - Superposition Approach

1. $(9D^2 - 4)y = (3D - 2)(3D + 2)y = \sin x$

3. $(D^2 - 4D - 12)y = (D - 6)(D + 2)y = x - 6$

5. $(D^3 + 10D^2 + 25D)y = D(D + 5)^2 y = e^x$

7. $(D^3 + 2D^2 - 13D + 10)y = (D - 1)(D - 2)(D + 5)y = xe^{-x}$

9. $(D^4 + 8D)y = D(D + 2)(D^2 - 2D + 4)y = 4$

11. $D^4 y = D^4(10x^3 - 2x) = D^3(30x^2 - 2) = D^2(60x) = D(60) = 0$

13. $(D - 2)(D + 5)(e^{2x} + 3e^{-5x}) = (D - 2)(2e^{2x} - 15e^{-5x} + 5e^{2x} + 15e^{-5x}) = (D - 2)7e^{2x} = 14e^{2x} - 14e^{2x} = 0$

15. D^4 because of x^3

17. $D(D - 2)$ because of 1 and e^{2x}

19. $D^2 + 4$ because of $\cos 2x$

21. $D^3(D^2 + 16)$ because of x^2 and $\sin 4x$

23. $(D + 1)(D - 1)^3$ because of e^{-x} and $x^2 e^x$

25. $D(D^2 - 2D + 5)$ because of 1 and $e^x \cos 2x$

27. $1,\ x,\ x^2,\ x^3,\ x^4$

29. $e^{6x},\ e^{-3x/2}$

31. $\cos \sqrt{5}\, x,\ \sin \sqrt{5}\, x$

33. $D^3 - 10D^2 + 25D = D(D - 5)^2;\quad 1,\ e^{5x},\ xe^{5x}$

35. Applying D to the differential equation we obtain

$$D(D^2 - 9)y = 0.$$

Then

$$y = \underbrace{c_1 e^{3x} + c_2 e^{-3x}}_{y_c} + c_3$$

and $y_p = A$. Substituting y_p into the differential equation yields $-9A = 54$ or $A = -6$. The general solution is

$$y = c_1 e^{3x} + c_2 e^{-3x} - 6.$$

37. Applying D to the differential equation we obtain

$$D(D^2 + D)y = D^2(D+1)y = 0.$$

Then

$$y = \underbrace{c_1 + c_2e^{-x}}_{y_c} + c_3x$$

and $y_p = Ax$. Substituting y_p into the differential equation yields $A = 3$. The general solution is

$$y = c_1 + c_2e^{-3x} + 3x.$$

39. Applying D^2 to the differential equation we obtain

$$D^2(D^2 + 4D + 4)y = D^2(D+2)^2y = 0.$$

Then

$$y = \underbrace{c_1e^{-2x} + c_2xe^{-2x}}_{y_c} + c_3 + c_4x$$

and $y_p = Ax+B$. Substituting y_p into the differential equation yields $4Ax+(4A+4B) = 2x+6$. Equating coefficients gives

$$4A = 2$$

$$4A + 4B = 6.$$

Then $A = 1/2$, $B = 1$, and the general solution is

$$y = c_1e^{-2x} + c_2xe^{-2x} + \frac{1}{2}x + 1.$$

41. Applying D^3 to the differential equation we obtain

$$D^3(D^3 + D^2)y = D^5(D+1)y = 0.$$

Then

$$y = \underbrace{c_1 + c_2x + c_3e^{-x}}_{y_c} + c_4x^4 + c_5x^3 + c_6x^2$$

and $y_p = Ax^4 + Bx^3 + Cx^2$. Substituting y_p into the differential equation yields

$$12Ax^2 + (24A + 6B)x + (6B + 2C) = 8x^2.$$

Equating coefficients gives

$$12A = 8$$

$$24A + 6B = 0$$

$$6B + 2C = 0.$$

Then $A = 2/3$, $B = -8/3$, $C = 8$, and the general solution is

$$y = c_1 + c_2x + c_3e^{-x} + \frac{2}{3}x^4 - \frac{8}{3}x^3 + 8x^2.$$

43. Applying $D - 4$ to the differential equation we obtain

$$(D - 4)(D^2 - D - 12)y = (D - 4)^2(D + 3)y = 0.$$

Then

$$y = \underbrace{c_1 e^{4x} + c_2 e^{-3x}}_{y_c} + c_3 x e^{4x}$$

and $y_p = Axe^{4x}$. Substituting y_p into the differential equation yields $7Ae^{4x} = e^{4x}$. Equating coefficients gives $A = 1/7$. The general solution is

$$y = c_1 e^{4x} + c_2 e^{-3x} + \frac{1}{7}xe^{4x}.$$

45. Applying $D(D - 1)$ to the differential equation we obtain

$$D(D - 1)(D^2 - 2D - 3)y = D(D - 1)(D + 1)(D - 3)y = 0.$$

Then

$$y = \underbrace{c_1 e^{3x} + c_2 e^{-x}}_{y_c} + c_3 e^{x} + c_4$$

and $y_p = Ae^{x} + B$. Substituting y_p into the differential equation yields $-4Ae^{x} - 3B = 4e^{x} - 9$. Equating coefficients gives $A = -1$ and $B = 3$. The general solution is

$$y = c_1 e^{3x} + c_2 e^{-x} - e^{x} + 3.$$

47. Applying $D^2 + 1$ to the differential equation we obtain

$$(D^2 + 1)(D^2 + 25)y = 0.$$

Then

$$y = \underbrace{c_1 \cos 5x + c_2 \sin 5x}_{y_c} + c_3 \cos x + c_4 \sin x$$

and $y_p = A \cos x + B \sin x$. Substituting y_p into the differential equation yields

$$24A \cos x + 24B \sin x = 6 \sin x.$$

Equating coefficients gives $A = 0$ and $B = 1/4$. The general solution is

$$y = c_1 \cos 5x + c_2 \sin 5x + \frac{1}{4} \sin x.$$

49. Applying $(D - 4)^2$ to the differential equation we obtain

$$(D - 4)^2(D^2 + 6D + 9)y = (D - 4)^2(D + 3)^2 y = 0.$$

Then

$$y = \underbrace{c_1 e^{-3x} + c_2 x e^{-3x}}_{y_c} + c_3 x e^{4x} + c_4 e^{4x}$$

and $y_p = Axe^{4x} + Be^{4x}$. Substituting y_p into the differential equation yields

$$49Axe^{4x} + (14A + 49B)e^{4x} = -xe^{4x}.$$

Equating coefficients gives

$$49A = -1$$
$$14A + 49B = 0.$$

Then $A = -1/49$, $B = 2/343$, and the general solution is

$$y = c_1 e^{-3x} + c_2 x e^{-3x} - \frac{1}{49}xe^{4x} + \frac{2}{343}e^{4x}.$$

51. Applying $D(D-1)^3$ to the differential equation we obtain

$$D(D-1)^3(D^2-1)y = D(D-1)^4(D+1)y = 0.$$

Then

$$y = \underbrace{c_1 e^x + c_2 e^{-x}}_{y_c} + c_3 x^3 e^x + c_4 x^2 e^x + c_5 x e^x + c_6$$

and $y_p = Ax^3 e^x + Bx^2 e^x + Cxe^x + E$. Substituting y_p into the differential equation yields

$$6Ax^2 e^x + (6A + 4B)xe^x + (2B + 2C)e^x - E = x^2 e^x + 5.$$

Equating coefficients gives

$$6A = 1$$
$$6A + 4B = 0$$
$$2B + 2C = 0$$
$$-E = 5.$$

Then $A = 1/6$, $B = -1/4$, $C = 1/4$, $E = -5$, and the general solution is

$$y = c_1 e^x + c_2 e^{-x} + \frac{1}{6}x^3 e^x - \frac{1}{4}x^2 e^x + \frac{1}{4}xe^x - 5.$$

53. Applying $D^2 - 2D + 2$ to the differential equation we obtain

$$(D^2 - 2D + 2)(D^2 - 2D + 5)y = 0.$$

Then

$$y = \underbrace{e^x(c_1 \cos 2x + c_2 \sin 2x)}_{y_c} + e^x(c_3 \cos x + c_4 \sin x)$$

and $y_p = Ae^x \cos x + Be^x \sin x$. Substituting y_p into the differential equation yields

$$3Ae^x \cos x + 3Be^x \sin x = e^x \sin x.$$

Equating coefficients gives $A = 0$ and $B = 1/3$. The general solution is

$$y = e^x(c_1 \cos 2x + c_2 \sin 2x) + \frac{1}{3}e^x \sin x.$$

55. Applying $D^2 + 25$ to the differential equation we obtain

$$(D^2 + 25)(D^2 + 25) = (D^2 + 25)^2 = 0.$$

Then

$$y = \underbrace{c_1 \cos 5x + c_2 \sin 5x}_{y_c} + c_3 x \cos 5x + c_4 x \sin 5x$$

and $y_p = Ax \cos 5x + Bx \sin 5x$. Substituting y_p into the differential equation yields

$$10B \cos 5x - 10A \sin 5x = 20 \sin 5x.$$

Equating coefficients gives $A = -2$ and $B = 0$. The general solution is

$$y = c_1 \cos 5x + c_2 \sin 5x - 2x \cos 5x.$$

57. Applying $(D^2 + 1)^2$ to the differential equation we obtain

$$(D^2 + 1)^2(D^2 + D + 1) = 0.$$

Then

$$y = e^{-x/2}\underbrace{\left[c_1 \cos \frac{\sqrt{3}}{2}x + c_2 \sin \frac{\sqrt{3}}{2}x\right] + c_3 \cos x + c_4 \sin x + c_5 x \cos x + c_6 x \sin x}_{y_c}$$

and $y_p = A \cos x + B \sin x + Cx \cos x + Ex \sin x$. Substituting y_p into the differential equation yields

$$(B + C + 2E) \cos x + Ex \cos x + (-A - 2C + E) \sin x - Cx \sin x = x \sin x.$$

Equating coefficients gives

$$B + C + 2E = 0$$
$$E = 0$$
$$-A - 2C + E = 0$$
$$-C = 1.$$

Then $A = 2$, $B = 1$, $C = -1$, and $E = 0$, and the general solution is

$$y = e^{-x/2}\left[c_1 \cos \frac{\sqrt{3}}{2}x + c_2 \sin \frac{\sqrt{3}}{2}x\right] + 2\cos x + \sin x - x\cos x.$$

59. Applying D^3 to the differential equation we obtain

$$D^3(D^3 + 8D^2) = D^5(D + 8) = 0.$$

Then

$$y = \underbrace{c_1 + c_2 x + c_3 e^{-8x}}_{y_c} + c_4 x^2 + c_5 x^3 + c_6 x^4$$

and $y_p = Ax^2 + Bx^3 + Cx^4$. Substituting y_p into the differential equation yields

$$16A + 6B + (48B + 24C)x + 96Cx^2 = 2 + 9x - 6x^2.$$

Equating coefficients gives

$$16A + 6B = 2$$
$$48B + 24C = 9$$
$$96C = -6.$$

Then $A = 11/256$, $B = 7/32$, and $C = -1/16$, and the general solution is

$$y = c_1 + c_2 x + c_3 e^{-8x} + \frac{11}{256}x^2 + \frac{7}{32}x^3 - \frac{1}{16}x^4.$$

61. Applying $D^2(D - 1)$ to the differential equation we obtain

$$D^2(D - 1)(D^3 - 3D^2 + 3D - 1) = D^2(D - 1)^4 = 0.$$

Then

$$y = \underbrace{c_1 e^x + c_2 x e^x + c_3 x^2 e^x}_{y_c} + c_4 + c_5 x + c_6 x^3 e^x$$

and $y_p = A + Bx + Cx^3 e^x$. Substituting y_p into the differential equation yields

$$(-A + 3B) - Bx + 6Ce^x = 16 - x + e^x.$$

Equating coefficients gives

$$-A + 3B = 16$$
$$-B = -1$$
$$6C = 1.$$

Then $A = -13$, $B = 1$, and $C = 1/6$, and the general solution is

$$y = c_1 e^x + c_2 x e^x + c_3 x^2 e^x - 13 + x + \frac{1}{6}x^3 e^x.$$

63. Applying $D(D-1)$ to the differential equation we obtain

$$D(D-1)(D^4 - 2D^3 + D^2) = D^3(D-1)^3 = 0.$$

Then

$$y = \underbrace{c_1 + c_2 x + c_3 e^x + c_4 x e^x}_{y_c} + c_5 x^2 + c_6 x^2 e^x$$

and $y_p = Ax^2 + Bx^2 e^x$. Substituting y_p into the differential equation yields $2A + 2Be^x = 1 + e^x$. Equating coefficients gives $A = 1/2$ and $B = 1/2$. The general solution is

$$y = c_1 + c_2 x + c_3 e^x + c_4 x e^x + \frac{1}{2}x^2 + \frac{1}{2}x^2 e^x.$$

65. The complementary function is $y_c = c_1 e^{8x} + c_2 e^{-8x}$. Using D to annihilate 16 we find $y_p = A$. Substituting y_p into the differential equation we obtain $-64A = 16$. Thus $A = -1/4$ and

$$y = c_1 e^{8x} + c_2 e^{-8x} - \frac{1}{4}$$

$$y' = 8c_1 e^{8x} - 8c_2 e^{-8x}.$$

The initial conditions imply

$$c_1 + c_2 = \frac{5}{4}$$

$$8c_1 - 8c_2 = 0.$$

Thus $c_1 = c_2 = 5/8$ and

$$y = \frac{5}{8}e^{8x} + \frac{5}{8}e^{-8x} - \frac{1}{4}.$$

67. The complementary function is $y_c = c_1 + c_2 e^{5x}$. Using D^2 to annihilate $x - 2$ we find $y_p = Ax + Bx^2$. Substituting y_p into the differential equation we obtain $(-5A + 2B) - 10Bx = -2 + x$. Thus $A = 9/25$ and $B = -1/10$, and

$$y = c_1 + c_2 e^{5x} + \frac{9}{25}x - \frac{1}{10}x^2$$

$$y' = 5c_2 e^{5x} + \frac{9}{25} - \frac{1}{5}x.$$

The initial conditions imply

$$c_1 + c_2 = 0$$

$$c_2 = \frac{41}{125}.$$

Thus $c_1 = -41/125$ and $c_2 = 41/125$, and

$$y = -\frac{41}{125} + \frac{41}{125}e^{5x} + \frac{9}{25}x - \frac{1}{10}x^2.$$

69. The complementary function is $y_c = c_1 \cos x + c_2 \sin x$. Using $(D^2 + 1)(D^2 + 4)$ to annihilate $8 \cos 2x - 4 \sin x$ we find $y_p = Ax \cos x + Bx \sin x + C \cos 2x + E \sin 2x$. Substituting y_p into the differential equation we obtain $2B \cos x - 3C \cos 2x - 2A \sin x - 3E \sin 2x = 8 \cos 2x - 4 \sin x$. Thus $A = 2$, $B = 0$, $C = -8/3$, and $E = 0$, and

$$y = c_1 \cos x + c_2 \sin x + 2x \cos x - \frac{8}{3} \cos 2x$$

$$y' = -c_1 \sin x + c_2 \cos x + 2 \cos x - 2x \sin x + \frac{16}{3} \sin 2x.$$

The initial conditions imply

$$c_2 + \frac{8}{3} = -1$$

$$-c_1 - \pi = 0.$$

Thus $c_1 = -\pi$ and $c_2 = -11/3$, and

$$y = -\pi \cos x - \frac{11}{3} \sin x + 2x \cos x - \frac{8}{3} \cos 2x.$$

71. The complementary function is $y_c = e^{2x}(c_1 \cos 2x + c_2 \sin 2x)$. Using D^4 to annihilate x^3 we find $y_p = A + Bx + Cx^2 + Ex^3$. Substituting y_p into the differential equation we obtain $(8A - 4B + 2C) + (8B - 8C + 6E)x + (8C - 12E)x^2 + 8Ex^3 = x^3$. Thus $A = 0$, $B = 3/32$, $C = 3/16$, and $E = 1/8$, and

$$y = e^{2x}(c_1 \cos 2x + c_2 \sin 2x) + \frac{3}{32}x + \frac{3}{16}x^2 + \frac{1}{8}x^3$$

$$y' = e^{2x}[c_1(2 \cos 2x - 2 \sin 2x) + c_2(2 \cos 2x + 2 \sin 2x)] + \frac{3}{32} + \frac{3}{8}x + \frac{3}{8}x^2.$$

The initial conditions imply

$$c_1 = 2$$

$$2c_1 + 2c_2 + \frac{3}{32} = 4.$$

Thus $c_1 = 2$, $c_2 = -3/64$, and

$$y = e^{2x}\left(2 \cos 2x - \frac{3}{64} \sin 2x\right) + \frac{3}{32}x + \frac{3}{16}x^2 + \frac{1}{8}x^3.$$

73. To see in this case that the factors of L do not commute consider the operators $(xD-1)(D+4)$ and $(D+4)(xD-1)$. Applying the operators to the function x we find

$$(xD - 1)(D + 4)x = (xD^2 + 4xD - D - 4)x$$

$$= xD^2 x + 4xDx - Dx - 4x$$

$$= x(0) + 4x(1) - 1 - 4x = -1$$

and

$$(D + 4)(xD - 1)x = (D + 4)(xDx - x)$$

$$= (D + 4)(x \cdot 1 - x) = 0.$$

Thus, the operators are not the same.

4.6 | Variation of Parameters

The particular solution, $y_p = u_1 y_1 + u_2 y_2$, in the following problems can take on a variety of forms, especially where trigonometric functions are involved. The validity of a particular form can best be checked by substituting it back into the differential equation.

1. The auxiliary equation is $m^2 + 1 = 0$, so $y_c = c_1 \cos x + c_2 \sin x$ and

$$W = \begin{vmatrix} \cos x & \sin x \\ -\sin x & \cos x \end{vmatrix} = 1$$

Identifying $f(x) = \sec x$ we obtain

$$u_1' = -\frac{\sin x \sec x}{1} = -\tan x$$

$$u_2' = \frac{\cos x \sec x}{1} = 1.$$

Then $u_1 = \ln|\cos x|$, $u_2 = x$, and

$$y = c_1 \cos x + c_2 \sin x + \cos x \ln|\cos x| + x \sin x.$$

3. The auxiliary equation is $m^2 + 1 = 0$, so $y_c = c_1 \cos x + c_2 \sin x$ and

$$W = \begin{vmatrix} \cos x & \sin x \\ -\sin x & \cos x \end{vmatrix} = 1$$

Identifying $f(x) = \sin x$ we obtain

$$u_1' = -\sin^2 x$$

$$u_2' = \cos x \sin x.$$

Then

$$u_1 = \frac{1}{4}\sin 2x - \frac{1}{2}x = \frac{1}{2}\sin x \cos x - \frac{1}{2}x$$

$$u_2 = -\frac{1}{2}\cos^2 x.$$

and

$$y = c_1 \cos x + c_2 \sin x + \frac{1}{2}\sin x \cos^2 x - \frac{1}{2}x \cos x - \frac{1}{2}\cos^2 x \sin x$$

$$= c_1 \cos x + c_2 \sin x - \frac{1}{2}x \cos x.$$

5. The auxiliary equation is $m^2 + 1 = 0$, so $y_c = c_1 \cos x + c_2 \sin x$ and

$$W = \begin{vmatrix} \cos x & \sin x \\ -\sin x & \cos x \end{vmatrix} = 1$$

Identifying $f(x) = \cos^2 x$ we obtain

$$u_1' = -\sin x \cos^2 x$$
$$u_2' = \cos^3 x = \cos x \left(1 - \sin^2 x\right).$$

Then $u_1 = \frac{1}{3}\cos^3 x$, $u_2 = \sin x - \frac{1}{3}\sin^3 x$, and

$$y = c_1 \cos x + c_2 \sin x + \frac{1}{3}\cos^4 x + \sin^2 x - \frac{1}{3}\sin^4 x$$

$$= c_1 \cos x + c_2 \sin x + \frac{1}{3}\left(\cos^2 x + \sin^2 x\right)\left(\cos^2 x - \sin^2 x\right) + \sin^2 x$$

$$= c_1 \cos x + c_2 \sin x + \frac{1}{3}\cos^2 x + \frac{2}{3}\sin^2 x$$

$$= c_1 \cos x + c_2 \sin x + \frac{1}{3} + \frac{1}{3}\sin^2 x.$$

7. The auxiliary equation is $m^2 - 1 = 0$, so $y_c = c_1 e^x + c_2 e^{-x}$ and

$$W = \begin{vmatrix} e^x & e^{-x} \\ e^x & -e^{-x} \end{vmatrix} = -2$$

Identifying $f(x) = \cosh x = \frac{1}{2}(e^{-x} + e^x)$ we obtain

$$u_1' = \frac{1}{4}e^{-2x} + \frac{1}{4}$$

$$u_2' = -\frac{1}{4} - \frac{1}{4}e^{2x}.$$

Then

$$u_1 = -\frac{1}{8}e^{-2x} + \frac{1}{4}x$$

$$u_2 = -\frac{1}{8}e^{2x} - \frac{1}{4}x$$

and

$$y = c_1 e^x + c_2 e^{-x} - \frac{1}{8}e^{-x} + \frac{1}{4}xe^x - \frac{1}{8}e^x - \frac{1}{4}xe^{-x}$$

$$= c_3 e^x + c_4 e^{-x} + \frac{1}{4}x(e^x - e^{-x})$$

$$= c_3 e^x + c_4 e^{-x} + \frac{1}{2}x \sinh x.$$

9. The auxiliary equation is $m^2 - 9 = 0$, so $y_c = c_1 e^{3x} + c_2 e^{-3x}$ and

$$W = \begin{vmatrix} e^{3x} & e^{-3x} \\ 3e^{3x} & -3e^{-3x} \end{vmatrix} = -6$$

Identifying $f(x) = 9x/e^{3x}$ we obtain $u_1' = \frac{3}{2}xe^{-6x}$ and $u_2' = -\frac{3}{2}x$. Then

$$u_1 = -\frac{1}{24}e^{-6x} - \frac{1}{4}xe^{-6x},$$

$$u_2 = -\frac{3}{4}x^2$$

and

$$y = c_1 e^{3x} + c_2 e^{-3x} - \frac{1}{24}e^{-3x} - \frac{1}{4}xe^{-3x} - \frac{3}{4}x^2 e^{-3x}$$

$$= c_1 e^{3x} + c_3 e^{-3x} - \frac{1}{4}xe^{-3x}(1 + 3x).$$

11. The auxiliary equation is $m^2 + 3m + 2 = (m+1)(m+2) = 0$, so $y_c = c_1 e^{-x} + c_2 e^{-2x}$ and

$$W = \begin{vmatrix} e^{-x} & e^{-2x} \\ -e^{-x} & -2e^{-2x} \end{vmatrix} = -e^{-3x}$$

Identifying $f(x) = 1/(1 + e^x)$ we obtain

$$u_1' = \frac{e^x}{1 + e^x}$$

$$u_2' = -\frac{e^{2x}}{1 + e^x} = \frac{e^x}{1 + e^x} - e^x.$$

Then $u_1 = \ln(1 + e^x)$, $u_2 = \ln(1 + e^x) - e^x$, and

$$y = c_1 e^{-x} + c_2 e^{-2x} + e^{-x} \ln(1 + e^x) + e^{-2x} \ln(1 + e^x) - e^{-x}$$

$$= c_3 e^{-x} + c_2 e^{-2x} + (1 + e^{-x})e^{-x} \ln(1 + e^x).$$

13. The auxiliary equation is $m^2 + 3m + 2 = (m+1)(m+2) = 0$, so $y_c = c_1 e^{-x} + c_2 e^{-2x}$ and

$$W = \begin{vmatrix} e^{-x} & e^{-2x} \\ -e^{-x} & = 2e^{-2} \end{vmatrix} = -e^{-3x}$$

Identifying $f(x) = \sin e^x$ we obtain

$$u_1' = \frac{e^{-2x} \sin e^x}{e^{-3x}} = e^x \sin e^x$$

$$u_2' = \frac{e^{-x} \sin e^x}{-e^{-3x}} = -e^{2x} \sin e^x.$$

Then $u_1 = -\cos e^x$, $u_2 = e^x \cos x - \sin e^x$, and

$$y = c_1 e^{-x} + c_2 e^{-2x} - e^{-x} \cos e^x + e^{-x} \cos e^x - e^{-2x} \sin e^x$$

$$= c_1 e^{-x} + c_2 e^{-2x} - e^{-2x} \sin e^x.$$

15. The auxiliary equation is $m^2 + 2m + 1 = (m+1)^2 = 0$, so $y_c = c_1 e^{-t} + c_2 t e^{-t}$ and

$$W = \begin{vmatrix} e^{-t} & te^{-t} \\ -e^{-t} & -te^{-t} + e^{-t} \end{vmatrix} = e^{-2t}$$

Identifying $f(t) = e^{-t} \ln t$ we obtain

$$u_1' = -\frac{te^{-t}e^{-t} \ln t}{e^{-2t}} = -t \ln t$$

$$u_2' = \frac{e^{-t}e^{-t} \ln t}{e^{-2t}} = \ln t.$$

Then

$$u_1 = -\frac{1}{2}t^2 \ln t + \frac{1}{4}t^2$$

$$u_2 = t \ln t - t$$

and

$$y = c_1 e^{-t} + c_2 te^{-t} - \frac{1}{2}t^2 e^{-t} \ln t + \frac{1}{4}t^2 e^{-t} + t^2 e^{-t} \ln t - t^2 e^{-t}$$

$$= c_1 e^{-t} + c_2 te^{-t} + \frac{1}{2}t^2 e^{-t} \ln t - \frac{3}{4}t^2 e^{-t}.$$

17. The auxiliary equation is $3m^2 - 6m + 6 = 0$, so $y_c = e^x (c_1 \cos x + c_2 \sin x)$ and

$$W = \begin{vmatrix} e^x \cos x & e^x \sin x \\ e^x \cos x - e^x \sin x & e^x \cos x + e^x \sin x \end{vmatrix} = e^{2x}$$

Identifying $f(x) = \frac{1}{3}e^x \sec x$ we obtain

$$u_1' = -\frac{(e^x \sin x)(e^x \sec x)/3}{e^{2x}} = -\frac{1}{3}\tan x$$

$$u_2' = \frac{(e^x \cos x)(e^x \sec x)/3}{e^{2x}} = \frac{1}{3}.$$

Then $u_1 = \frac{1}{3}\ln|\cos x|$, $u_2 = \frac{1}{3}x$, and

$$y = c_1 e^x \cos x + c_2 e^x \sin x + \frac{1}{3}\ln|\cos x|e^x \cos x + \frac{1}{3}xe^x \sin x.$$

19. The auxiliary equation is $4m^2 - 1 = (2m - 1)(2m + 1) = 0$, so $y_c = c_1 e^{x/2} + c_2 e^{-x/2}$ and

$$W = \begin{vmatrix} e^{x/2} & e^{-x/2} \\ \frac{1}{2}e^{x/2} & -\frac{1}{2}e^{-x/2} \end{vmatrix} = -1$$

Identifying $f(x) = xe^{x/2}/4$ we obtain $u_1' = x/4$ and $u_2' = -xe^x/4$. Then $u_1 = x^2/8$ and $u_2 = -xe^x/4 + e^x/4$. Thus

$$y = c_1 e^{x/2} + c_2 e^{-x/2} + \frac{1}{8}x^2 e^{x/2} - \frac{1}{4}xe^{x/2} + \frac{1}{4}e^{x/2}$$

$$= c_3 e^{x/2} + c_2 e^{-x/2} + \frac{1}{8}x^2 e^{x/2} - \frac{1}{4}xe^{x/2}$$

and

$$y' = \frac{1}{2}c_3 e^{x/2} - \frac{1}{2}c_2 e^{-x/2} + \frac{1}{16}x^2 e^{x/2} + \frac{1}{8}xe^{x/2} - \frac{1}{4}e^{x/2}.$$

The initial conditions imply

$$c_3 + c_2 = 0$$

$$\frac{1}{2}c_3 - \frac{1}{2}c_2 - \frac{1}{4} = 0$$

Thus $c_3 = 3/4$ and $c_2 = 1/4$, and

$$y = \frac{3}{4}e^{x/2} + \frac{1}{4}e^{-x/2} + \frac{1}{8}x^2 e^{x/2} - \frac{1}{4}xe^{x/2}.$$

21. The auxiliary equation is $m^2 + 2m - 8 = (m-2)(m+4) = 0$, so $y_c = c_1 e^{2x} + c_2 e^{-4x}$ and

$$W = \begin{vmatrix} e^{2x} & e^{-4x} \\ 2e^{2x} & -4e^{-4x} \end{vmatrix} = -6e^{-2x}$$

Identifying $f(x) = 2e^{-2x} - e^{-x}$ we obtain

$$u_1' = \frac{1}{3}e^{-4x} - \frac{1}{6}e^{-3x}$$

$$u_2' = \frac{1}{6}e^{3x} - \frac{1}{3}e^{2x}.$$

Then

$$u_1 = -\frac{1}{12}e^{-4x} + \frac{1}{18}e^{-3x}$$

$$u_2 = \frac{1}{18}e^{3x} - \frac{1}{6}e^{2x}.$$

Thus

$$y = c_1 e^{2x} + c_2 e^{-4x} - \frac{1}{12}e^{-2x} + \frac{1}{18}e^{-x} + \frac{1}{18}e^{-x} - \frac{1}{6}e^{-2x}$$

$$= c_1 e^{2x} + c_2 e^{-4x} - \frac{1}{4}e^{-2x} + \frac{1}{9}e^{-x}$$

and

$$y' = 2c_1 e^{2x} - 4c_2 e^{-4x} + \frac{1}{2}e^{-2x} - \frac{1}{9}e^{-x}.$$

The initial conditions imply

$$c_1 + c_2 - \frac{5}{36} = 1$$

$$2c_1 - 4c_2 + \frac{7}{18} = 0$$

Thus $c_1 = 25/36$ and $c_2 = 4/9$, and

$$y = \frac{25}{36}e^{2x} + \frac{4}{9}e^{-4x} - \frac{1}{4}e^{-2x} + \frac{1}{9}e^{-x}.$$

23. The auxiliary equation is $m^2 + 1 = 0$, so $y_c = c_1 \cos x + c_2 \sin x$ and

$$W = \begin{vmatrix} \cos x & \sin x \\ -\sin x & \cos x \end{vmatrix} = 1$$

Identifying $f(x) = e^{x^2}$ we obtain $u_1' = -e^{x^2} \sin x$ and $u_2' = e^{x^2} \cos x$. Then

$$u_1 = -\int_{x_0}^x e^{t^2} \sin t\, dt$$

$$u_2 = \int_{x_0}^x e^{t^2} \cos t\, dt$$

and

$$y = c_1 \cos x + c_2 \sin x - \cos x \int_{x_0}^x e^{t^2} \sin t\, dt + \sin x \int_{x_0}^x e^{t^2} \cos t\, dt.$$

25. The auxiliary equation is $m^2 + m - 2 = 0$, so $y_c = c_1 e^{-2x} + c_2 e^x$ and

$$W = \begin{vmatrix} e^{-2x} & e^x \\ -2e^{-2x} & e^x \end{vmatrix} = 3e^{-x}$$

Identifying $f(x) = \ln x$ we obtain $u_1' = -\frac{1}{3}e^{2x} \ln x$ and $u_2' = \frac{1}{3}e^{-x} \ln x$. Then

$$u_1 = -\frac{1}{3}\int_{x_0}^x e^{2t} \ln t\, dt,$$

$$u_2 = \frac{1}{3}\int_{x_0}^x e^{-t} \ln t\, dt$$

and

$$y = c_1 e^{-2x} + c_2 e^x - \frac{1}{3}e^{-2x}\int_{x_0}^x e^{2t} \ln t\, dt + \frac{1}{3}e^x\int_{x_0}^x e^{-t} \ln t\, dt, \qquad x_0 > 0.$$

27. Write the equation in the form

$$y'' + \frac{1}{x}y' + \left(1 - \frac{1}{4x^2}\right)y = x^{-1/2}$$

and identify $f(x) = x^{-1/2}$. From $y_1 = x^{-1/2} \cos x$ and $y_2 = x^{-1/2} \sin x$ we compute

$$W = \begin{vmatrix} x^{-1/2} \cos x & x^{-1/2} \sin x \\ -x^{-1/2} \sin x - \dfrac{1}{2} x^{-3/2} \cos x & x^{-1/2} \cos x - \dfrac{1}{2} x^{-3/2} \sin x \end{vmatrix} = \dfrac{1}{x}$$

Now

$$u_1' = -\sin x \quad \text{so} \quad u_1 = \cos x,$$

and

$$u_2' = \cos x \quad \text{so} \quad u_2 = \sin x.$$

Thus a particular solution is

$$y_p = x^{-1/2} \cos^2 x + x^{-1/2} \sin^2 x,$$

and the general solution is

$$\begin{aligned} y &= c_1 x^{-1/2} \cos x + c_2 x^{-1/2} \sin x + x^{-1/2} \cos^2 x + x^{-1/2} \sin^2 x \\ &= c_1 x^{-1/2} \cos x + c_2 x^{-1/2} \sin x + x^{-1/2}. \end{aligned}$$

29. The auxiliary equation is $m^3 + m = m(m^2 + 1) = 0$, so $y_c = c_1 + c_2 \cos x + c_3 \sin x$ and

$$W = \begin{vmatrix} 1 & \cos x & \sin x \\ 0 & -\sin x & \cos x \\ 0 & -\cos x & -\sin x \end{vmatrix} = 1$$

Identifying $f(x) = \tan x$ we obtain

$$u_1' = W_1 = \begin{vmatrix} 0 & \cos x & \sin x \\ 0 & -\sin x & \cos x \\ \tan x & -\cos x & -\sin x \end{vmatrix} = \tan x$$

$$u_2' = W_2 = \begin{vmatrix} 1 & 0 & \sin x \\ 0 & 0 & \cos x \\ 0 & \tan x & -\sin x \end{vmatrix} = -\sin x$$

$$u_3' = W_3 = \begin{vmatrix} 1 & \cos x & 0 \\ 0 & -\sin x & 0 \\ 0 & -\cos x & \tan x \end{vmatrix} = -\sin x \tan x = \dfrac{\cos^2 x - 1}{\cos x} = \cos x - \sec x$$

Then

$$u_1 = -\ln|\cos x|$$

$$u_2 = \cos x$$

$$u_3 = \sin x - \ln|\sec x + \tan x|$$

and

$$y = c_1 + c_2 \cos x + c_3 \sin x - \ln|\cos x| + \cos^2 x$$
$$+ \sin^2 x - \sin x \ln|\sec x + \tan x|$$
$$= c_4 + c_2 \cos x + c_3 \sin x - \ln|\cos x| - \sin x \ln|\sec x + \tan x|$$

for $-\pi/2 < x < \pi/2$.

31. The auxiliary equation is $m^3 - 2m^2 - m + 2 = 0$ so $y_c = c_1 e^x + c_2 e^{-x} + c_3 e^{2x}$

$$W = \begin{vmatrix} 1 & e^x & e^{-x} & e^{2x} \\ e^x & -e^{-x} & 2e^{2x} \\ e^x & e^{-x} & 4e^{2x} \end{vmatrix} = -6e^{2x}$$

Identifying $f(x) = e^{4x}$ we obtain

$$u_1' = \frac{1}{-6e^{2x}} W_1 = \frac{1}{-6e^{2x}} \begin{vmatrix} 0 & e^{-x} & e^{2x} \\ 0 & -e-x & 2e^{2x} \\ e^{4x} & e^{-x} & 4e^{2x} \end{vmatrix} = \frac{1}{-6e^{2x}} \cdot 3e^{5x} = -\frac{1}{2} e^{3x}$$

$$u_2' = \frac{1}{-6e^{2x}} W_2 = \frac{1}{-6e^{2x}} \begin{vmatrix} e^x & 0 & e^{2x} \\ e^x & 0 & 2e^{2x} \\ e^x & e^{4x} & 4e^{2x} \end{vmatrix} = \frac{1}{-6e^{2x}} \cdot (-e^{7x}) = \frac{1}{6} e^{5x}$$

$$u_3' = \frac{1}{-6e^{2x}} W_3 = \frac{1}{-6e^{2x}} \begin{vmatrix} e^x & e^{-x} & 0 \\ e^x & -e-x & 0 \\ e^x & e^{-x} & e^{4x} \end{vmatrix} = \frac{1}{-6e^{2x}} \cdot (-2e^{4x}) = \frac{1}{3} e^{4x}$$

Then

$$u_1 = -\frac{1}{6} e^{3x}$$
$$u_2 = \frac{1}{30} e^{5x}$$
$$u_3 = \frac{1}{6} e^{2x}$$

Thus $y_p = -\frac{1}{6} e^{3x} \cdot e^x + \frac{1}{30} e^{5x} \cdot e^{-x} + \frac{1}{6} e^{2x} \cdot e^{2x} = \frac{1}{30} e^{4x}$ and $y = c_1 e^x + c_2 e^{-x} + c_3 e^{2x} + \frac{1}{30} e^{4x}$ on $(-\infty, \infty)$.

33. The auxiliary equation is $3m^2 - 6m + 30 = 0$, which has roots $1 \pm 3i$, so $y_c = e^x(c_1 \cos 3x + c_2 \sin 3x)$. We consider first the differential equation $3y'' - 6y' + 30y = 15 \sin x$, which can be solved using undetermined coefficients. Letting $y_{p_1} = A \cos x + B \sin x$ and substituting into the differential equation we get

$$(27A - 6B) \cos x + (6A + 27B) \sin x = 15 \sin x.$$

Then

$$27A - 6B = 0 \quad \text{and} \quad 6A + 27B = 15,$$

so $A = \frac{2}{17}$ and $B = \frac{9}{17}$. Thus, $y_{p_1} = \frac{2}{17}\cos x + \frac{9}{17}\sin x$. Next, we consider the differential equation $3y'' - 6y' + 30y$, for which a particular solution y_{p_2} can be found using variation of parameters. The Wronskian is

$$W = \begin{vmatrix} e^x \cos 3x & e^x \sin 3x \\ e^x \cos 3x - 3e^x \sin 3x & 3e^x \cos 3x + e^x \sin 3x \end{vmatrix} = 3e^{2x}$$

Identifying $f(x) = \frac{1}{3}e^x \tan x$ we obtain

$$u_1' = -\frac{1}{9}\sin 3x \tan 3x = -\frac{1}{9}\left(\frac{\sin^2 3x}{\cos 3x}\right) = -\frac{1}{9}\left(\frac{1 - \cos^2 3x}{\cos 3x}\right) = -\frac{1}{9}(\sec 3x - \cos 3x)$$

so

$$u_1 = -\frac{1}{27}\ln|\sec 3x + \tan 3x| + \frac{1}{27}\sin 3x.$$

Next

$$u_2' = \frac{1}{9}\sin 3x \quad \text{so} \quad u_2 = -\frac{1}{27}\cos 3x.$$

Thus

$$y_{p_2} = -\frac{1}{27}e^x \cos 3x(\ln|\sec 3x + \tan 3x| - \sin 3x) - \frac{1}{27}e^x \sin 3x \cos 3x$$

$$= -\frac{1}{27}e^x(\cos 3x)\ln|\sec 3x + \tan 3x|$$

and the general solution of the original differential equation is

$$y = e^x(c_1 \cos 3x + c_2 \sin 3x) + y_{p_1}(x) + y_{p_2}(x).$$

35. The interval of definition for Problem 1 is $(-\pi/2, \pi/2)$, for Problem 7 is $(-\infty, \infty)$, for Problem 15 is $(0, \infty)$, and for Problem 18 is $(-1, 1)$. In Problem 28 the general solution is

$$y = c_1 \cos(\ln x) + c_2 \sin(\ln x) + \cos(\ln x)\ln|\cos(\ln x)| + (\ln x)\sin(\ln x)$$

for $-\pi/2 < \ln x < \pi/2$ or $e^{-\pi/2} < x < e^{\pi/2}$. The bounds on $\ln x$ are due to the presence of $\sec(\ln x)$ in the differential equation.

4.7   Cauchy–Euler Equation

1. The auxiliary equation is $m^2 - m - 2 = (m+1)(m-2) = 0$ so that $y = c_1 x^{-1} + c_2 x^2$.

3. The auxiliary equation is $m^2 = 0$ so that $y = c_1 + c_2 \ln x$.

5. The auxiliary equation is $m^2 + 4 = 0$ so that $y = c_1 \cos(2\ln x) + c_2 \sin(2\ln x)$.

7. The auxiliary equation is $m^2 - 4m - 2 = 0$ so that $y = c_1 x^{2-\sqrt{6}} + c_2 x^{2+\sqrt{6}}$.

9. The auxiliary equation is $25m^2 + 1 = 0$ so that $y = c_1 \cos\left(\frac{1}{5}\ln x\right) + c_2 \sin\left(\frac{1}{5}\ln x\right)$.

11. The auxiliary equation is $m^2 + 4m + 4 = (m+2)^2 = 0$ so that $y = c_1 x^{-2} + c_2 x^{-2}\ln x$.

13. The auxiliary equation is $3m^2 + 3m + 1 = 0$ so that

$$y = x^{-1/2}\left[c_1 \cos\left(\frac{\sqrt{3}}{6}\ln x\right) + c_2 \sin\left(\frac{\sqrt{3}}{6}\ln x\right)\right].$$

15. Assuming that $y = x^m$ and substituting into the differential equation we obtain

$$m(m-1)(m-2) - 6 = m^3 - 3m^2 + 2m - 6 = (m-3)(m^2+2) = 0.$$

Thus

$$y = c_1 x^3 + c_2 \cos\left(\sqrt{2}\ln x\right) + c_3 \sin\left(\sqrt{2}\ln x\right).$$

17. Assuming that $y = x^m$ and substituting into the differential equation we obtain

$$m(m-1)(m-2)(m-3) + 6m(m-1)(m-2) = m^4 - 7m^2 + 6m = m(m-1)(m-2)(m+3) = 0.$$

Thus

$$y = c_1 + c_2 x + c_3 x^2 + c_4 x^{-3}.$$

19. The auxiliary equation is $m^2 - 5m = m(m-5) = 0$ so that $y_c = c_1 + c_2 x^5$ and

$$W\left(1, x^5\right) = \begin{vmatrix} 1 & x^5 \\ 0 & 5x^4 \end{vmatrix} = 5x^4.$$

Identifying $f(x) = x^3$ we obtain $u_1' = -\frac{1}{5}x^4$ and $u_2' = 1/5x$. Then $u_1 = -\frac{1}{25}x^5$, $u_2 = \frac{1}{5}\ln x$, and

$$y = c_1 + c_2 x^5 - \frac{1}{25}x^5 + \frac{1}{5}x^5\ln x = c_1 + c_3 x^5 + \frac{1}{5}x^5\ln x.$$

21. The auxiliary equation is $m^2 - 2m + 1 = (m-1)^2 = 0$ so that $y_c = c_1 x + c_2 x\ln x$ and

$$W\left(x, x^2\right) = \begin{vmatrix} x & x\ln x \\ 1 & 1+\ln x \end{vmatrix} = x$$

Identifying $f(x) = 2/x$ we obtain $u_1' = -2(\ln x)/x$ and $u_2' = 2/x$. Then $u_1 = -(\ln x)^2$, $u_2 = 2\ln x$, and

$$y = c_1 x + c_2 x\ln x - x(\ln x)^2 + 2x(\ln x)^2$$
$$= c_1 x + c_2 x\ln x + x(\ln x)^2, \qquad x > 0.$$

23. The auxiliary equation $m(m-1) + m - 1 = m^2 - 1 = 0$ has roots $m_1 = -1$, $m_2 = 1$, so $y_c = c_1 x^{-1} + c_2 x$. With $y_1 = x^{-1}$, $y_2 = x$, and the identification $f(x) = \ln x/x^2$, we get

$$W = 2x^{-1}, \qquad W_1 = -(\ln x)/x, \qquad \text{and} \qquad W_2 = (\ln x)/x^3.$$

Then $u_1' = W_1/W = -(\ln x)/2$, $u_2' = W_2/W = (\ln x)/2x^2$, and integration by parts gives

$$u_1 = \frac{1}{2}x - \frac{1}{2}x\ln x$$

$$u_2 = -\frac{1}{2}x^{-1}\ln x - \frac{1}{2}x^{-1},$$

so

$$y_p = u_1 y_1 + u_2 y_2 = \left(\frac{1}{2}x - \frac{1}{2}x\ln x\right)x^{-1} + \left(-\frac{1}{2}x^{-1}\ln x - \frac{1}{2}x^{-1}\right)x = -\ln x$$

and

$$y = y_c + y_p = c_1 x^{-1} + c_2 x - \ln x, \qquad x > 0.$$

25. The auxiliary equation is $m^2 + 2m = m(m + 2) = 0$, so that $y = c_1 + c_2 x^{-2}$ and $y' = -2c_2 x^{-3}$. The initial conditions imply

$$c_1 + c_2 = 0$$

$$-2c_2 = 4.$$

Thus, $c_1 = 2$, $c_2 = -2$, and $y = 2 - 2x^{-2}$. The graph is given to the right.

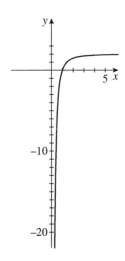

27. The auxiliary equation is $m^2 + 1 = 0$, so that

$$y = c_1 \cos(\ln x) + c_2 \sin(\ln x)$$

and

$$y' = -c_1 \frac{1}{x}\sin(\ln x) + c_2 \frac{1}{x}\cos(\ln x).$$

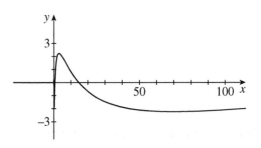

The initial conditions imply $c_1 = 1$ and $c_2 = 2$. Thus $y = \cos(\ln x) + 2\sin(\ln x)$. The graph is given to the right.

29. The auxiliary equation is $m^2 = 0$ so that $y_c = c_1 + c_2 \ln x$ and

$$W(1, \ln x) = \begin{vmatrix} 1 & \ln x \\ 0 & \dfrac{1}{x} \end{vmatrix} = \dfrac{1}{x}$$

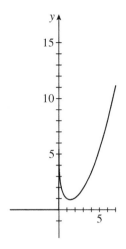

Identifying $f(x) = 1$ we obtain $u_1' = -x \ln x$ and $u_2' = x$. Then $u_1 = \frac{1}{4}x^2 - \frac{1}{2}x^2 \ln x$, $u_2 = \frac{1}{2}x^2$, and

$$y = c_1 + c_2 \ln x + \frac{1}{4}x^2 - \frac{1}{2}x^2 \ln x + \frac{1}{2}x^2 \ln x = c_1 + c_2 \ln x + \frac{1}{4}x^2.$$

The initial conditions imply $c_1 + \frac{1}{4} = 1$ and $c_2 + \frac{1}{2} = -\frac{1}{2}$. Thus, $c_1 = \frac{3}{4}$, $c_2 = -1$, and $y = \frac{3}{4} - \ln x + \frac{1}{4}x^2$. The graph is given to the right.

31. Substituting $x = e^t$ into the differential equation we obtain

$$\frac{d^2 y}{dt^2} + 8\frac{dy}{dt} - 20y = 0.$$

The auxiliary equation is $m^2 + 8m - 20 = (m + 10)(m - 2) = 0$ so that

$$y = c_1 e^{-10t} + c_2 e^{2t} \quad \text{or} \quad y = c_1 x^{-10} + c_2 x^2.$$

33. Substituting $x = e^t$ into the differential equation we obtain

$$\frac{d^2 y}{dt^2} + 9\frac{dy}{dt} + 8y = e^{2t}.$$

The auxiliary equation is $m^2 + 9m + 8 = (m+1)(m+8) = 0$ so that $y_c = c_1 e^{-t} + c_2 e^{-8t}$. Using undetermined coefficients we try $y_p = Ae^{2t}$. This leads to $30Ae^{2t} = e^{2t}$, so that $A = 1/30$ and

$$y = c_1 e^{-t} + c_2 e^{-8t} + \frac{1}{30}e^{2t} \quad \text{or} \quad y = c_1 x^{-1} + c_2 x^{-8} + \frac{1}{30}x^2.$$

35. Substituting $x = e^t$ into the differential equation we obtain

$$\frac{d^2 y}{dt^2} - 4\frac{dy}{dt} + 13y = 4 + 3e^t.$$

The auxiliary equation is $m^2 - 4m + 13 = 0$ so that $y_c = e^{2t}(c_1 \cos 3t + c_2 \sin 3t)$. Using undetermined coefficients we try $y_p = A + Be^t$. This leads to $13A + 10Be^t = 4 + 3e^t$, so that $A = 4/13$, $B = 3/10$, and

$$y = e^{2t}(c_1 \cos 3t + c_2 \sin 3t) + \frac{4}{13} + \frac{3}{10}e^t$$

or

$$y = x^2 [c_1 \cos(3\ln x) + c_2 \sin(3\ln x)] + \frac{4}{13} + \frac{3}{10}x.$$

In the next problem we use the substitution $t = -x$ since the initial conditions are on the interval $(-\infty, 0)$. In this case

$$\frac{dy}{dt} = \frac{dy}{dx}\frac{dx}{dt} = -\frac{dy}{dx}$$

and

$$\frac{d^2y}{dt^2} = \frac{d}{dt}\left(\frac{dy}{dt}\right) = \frac{d}{dt}\left(-\frac{dy}{dx}\right) = -\frac{d}{dt}(y') = -\frac{dy'}{dx}\frac{dx}{dt} = -\frac{d^2y}{dx^2}\frac{dx}{dt} = \frac{d^2y}{dx^2}.$$

37. The differential equation and initial conditions become

$$4t^2\frac{d^2y}{dt^2} + y = 0; \quad y(t)\Big|_{t=1} = 2, \quad y'(t)\Big|_{t=1} = -4.$$

The auxiliary equation is $4m^2 - 4m + 1 = (2m-1)^2 = 0$, so that

$$y = c_1 t^{1/2} + c_2 t^{1/2}\ln t \quad \text{and} \quad y' = \frac{1}{2}c_1 t^{-1/2} + c_2\left(t^{-1/2} + \frac{1}{2}t^{-1/2}\ln t\right).$$

The initial conditions imply $c_1 = 2$ and $1 + c_2 = -4$. Thus

$$y = 2t^{1/2} - 5t^{1/2}\ln t \quad \text{or} \quad y = 2(-x)^{1/2} - 5(-x)^{1/2}\ln(-x), \quad x < 0.$$

39. If we force the function $y = (x-(-3))^m = (x+3)^m$ into the equation we get

$$(x+3)^2 y'' - 8(x+3)y' + 14y = 0$$
$$(x+3)^2 m(m-1)(x+3)^{m-2} - 8(x+3)m(x+3)^{m-1} + 14(x+3)^m = 0$$
$$(x+3)^m[m^2 - 9m + 14] = 0$$
$$(x+3)^m(m-2)(m-7) = 0$$

The solutions to the auxiliary equation are therefore $m = 2$ and $m = 7$ from which we get the general solution $y = c_1(x+3)^2 + c_2(x+3)^7$.

41. Letting $t = x + 2$ we obtain $dy/dx = dy/dt$ and, using the Chain Rule,

$$\frac{d^2y}{dx^2} = \frac{d}{dx}\left(\frac{dy}{dt}\right) = \frac{d^2y}{dt^2}\frac{dt}{dx} = \frac{d^2y}{dt^2}(1) = \frac{d^2y}{dt^2}.$$

Substituting into the differential equation we obtain

$$t^2\frac{d^2y}{dt^2} + t\frac{dy}{dt} + y = 0.$$

The auxiliary equation is $m^2 + 1 = 0$ so that

$$y = c_1\cos(\ln 2t) + c_2\sin(\ln 2t) = c_2\cos[\ln(x+2)] + c_2\sin[\ln(x+2)].$$

43. Since the leading coefficient $a_2(x) = (x-4)^2 \neq 0$ for every value of x satisfying $x > 4$ we may take the interval of definition of the general solution to be $(4, \infty)$.

45. For $x^2 y'' = 0$ the auxiliary equation is $m(m-1) = 0$ and the general solution is $y = c_1 + c_2 x$. The initial conditions imply $c_1 = y_0$ and $c_2 = y_1$, so $y = y_0 + y_1 x$. The initial conditions are satisfied for all real values of y_0 and y_1.

For $x^2 y'' - 2xy' + 2y = 0$ the auxiliary equation is $m^2 - 3m + 2 = (m-1)(m-2) = 0$ and the general solution is $y = c_1 x + c_2 x^2$. The initial condition $y(0) = y_0$ implies $0 = y_0$ and the condition $y'(0) = y_1$ implies $c_1 = y_1$. Thus, the initial conditions are satisfied for $y_0 = 0$ and for all real values of y_1.

For $x^2 y'' - 4xy' + 6y = 0$ the auxiliary equation is $m^2 - 5m + 6 = (m-2)(m-3) = 0$ and the general solution is $y = c_1 x^2 + c_2 x^3$. The initial conditions imply $y(0) = 0 = y_0$ and $y'(0) = 0$. Thus, the initial conditions are satisfied only for $y_0 = y_1 = 0$.

47. (a) Multiplying by r^3 we have

$$\frac{d^3 w}{dr^3} + \frac{1}{r}\frac{d^2 w}{dr^2} - \frac{1}{r^2}\frac{dw}{dr} = \frac{q}{2D} r$$

$$r^3 \frac{d^3 w}{dr^3} + r^2 \frac{d^2 w}{dr^2} - r\frac{dw}{dr} = \frac{q}{2D} r^4$$

This is a nonhomogeneous third-order Cauchy-Euler differential equation. We first solve the homogenous equation. The auxiliary equation is

$$m(m-1)(m-2) + m(m-1) - m = 0$$

$$m^3 - 2m^2 = 0$$

$$m^2(m-2) = 0$$

Therefore $m_1 = m_2 = 0$ and $m_3 = 2$. Then the complementary function is $w_c(r) = c_1 + c_2 \ln r + c_3 r^2$. To use variation of parameters we must use the first form of the differential equation and identify $f(r) = qr/2D$. By (10) of Section 3.5 we identify $y_1 = 1$, $y_2 = \ln r$, and $y_3 = r^2$, and expand each of the four determinants by the first column:

$$W_1 = \begin{vmatrix} 0 & \ln r & r^2 \\ 0 & \dfrac{1}{r} & 2r \\ \dfrac{q}{2D} r & -\dfrac{1}{r^2} & 2 \end{vmatrix} = \frac{q}{2D} r \begin{vmatrix} \ln r & r^2 \\ \dfrac{1}{r} & 2r \end{vmatrix} = \frac{q}{2D}\left(2r^2 \ln r - r^2\right)$$

$$W_2 = \begin{vmatrix} 1 & 0 & r^2 \\ 0 & 0 & 2r \\ 0 & \dfrac{q}{2D} r & 2 \end{vmatrix} = 1 \cdot \begin{vmatrix} 0 & 2r \\ \dfrac{q}{2D} r & 2 \end{vmatrix} = -\frac{q}{D} r^2$$

$$W_3 = \begin{vmatrix} 1 & \ln r & 0 \\ 0 & \dfrac{1}{r} & 0 \\ 0 & -\dfrac{1}{r^2} & \dfrac{q}{2D} r \end{vmatrix} = 1 \cdot \begin{vmatrix} \dfrac{1}{r} & 0 \\ -\dfrac{1}{r^2} & \dfrac{q}{2D} r \end{vmatrix} = \frac{q}{2D}$$

$$W = \begin{vmatrix} 1 & \ln r & r^2 \\ 0 & \dfrac{1}{r} & 2r \\ 0 & -\dfrac{1}{r^2} & 2 \end{vmatrix} = 1 \cdot \begin{vmatrix} \dfrac{1}{r} & 2r \\ -\dfrac{1}{r^2} & 2 \end{vmatrix} = \dfrac{2}{r} - \left(-\dfrac{2}{r}\right) = \dfrac{4}{r}$$

Then

$$u_1' = \frac{W_1}{W} = \frac{\dfrac{q}{2D}\, r\,(2r\ln r - r)}{\dfrac{4}{r}} = \frac{q}{4D}\, r^3 \ln r - \frac{q}{8D}\, r^3$$

$$u_2' = \frac{W_2}{W} = \frac{-\dfrac{q}{D}\, r^2}{\dfrac{4}{r}} = -\frac{q}{4D}\, r^3$$

$$u_3' = \frac{W_3}{W} = \frac{\dfrac{q}{2D}}{\dfrac{4}{r}} = \frac{q}{8D}\, r$$

and

$$u_1 = \frac{q}{16D}\, r^4 \ln r - \frac{3q}{64D}\, r^4 \qquad \text{(integration by parts)}$$

$$u_2 = -\frac{q}{16D}\, r^4$$

$$u_3 = \frac{q}{16D}\, r^2$$

Thus a particular solution is

$$w_p(r) = \left(\frac{q}{16D}\, r^4 \ln r - \frac{3q}{64D}\, r^4\right) \cdot 1 + \left(-\frac{q}{16D}\, r^4\right)\ln r + \left(\frac{4q}{64D}\, r^2\right) r^2 = \frac{q}{64D}\, r^4$$

Therefore the general solution is $w(r) = w_c(r) + w_p(r)$ or

$$w(r) = c_1 + c_2 \ln r + c_3 r^2 + \frac{q}{64D}\, r^4$$

(b) Taking the derivative of w we have

$$w'(r) - \frac{c_2}{r} + 2c_3 r + \frac{q}{16D}\, r^3 .$$

The condition $w'(0) = 0$ requires $c_2 = 0$. Using the condition $w'(a) = 0$ gives

$$0 = 2c_3 a + \frac{q}{16D}\, a^3$$

$$c_3 = -\frac{q}{32D}\, a^2 .$$

So,

$$w(r) = c_1 - \frac{q}{32D}\, a^2 r^2 + \frac{q}{64D}\, r^4 .$$

Using the condition $w(a) = 0$ gives

$$0 = c_1 - \frac{q}{32D} a^4 + \frac{q}{64D} a^4$$

$$c_1 = \frac{q}{64D} a^4.$$

Therefore

$$w(r) = \frac{q}{64D} a^4 - \frac{q}{32D} a^2 r^2 + \frac{q}{64D} r^4 = \frac{q}{64D} \left(a^2 - r^2\right)^2.$$

49. The auxiliary equation is $2m(m-1)(m-2) - 10.98m(m-1) + 8.5m + 1.3 = 0$, so that $m_1 = -0.053299$, $m_2 = 1.81164$, $m_3 = 6.73166$, and

$$y = c_1 x^{-0.053299} + c_2 x^{1.81164} + c_3 x^{6.73166}.$$

51. The auxiliary equation is

$$m(m-1)(m-2)(m-3) + 6m(m-1)(m-2) + 3m(m-1) - 3m + 4 = 0,$$

so that $m_1 = m_2 = \sqrt{2}$ and $m_3 = m_4 = -\sqrt{2}$. The general solution of the differential equation is

$$y = c_1 x^{\sqrt{2}} + c_2 x^{\sqrt{2}} \ln x + c_3 x^{-\sqrt{2}} + c_4 x^{-\sqrt{2}} \ln x.$$

53. First solve the associated homogeneous equation. From the auxiliary equation $(m-3)(m-2)(m+1) = 0$ we find $y_c = c_1 x^3 + c_2 x^2 + c_3 x^{-1}$. Before finding the particular solution using variation of parameters, first put the differential equation in standard form by dividing through by x^3. We therefore identify $f(x) = x^{-1}$. Now with $y_1 = x^3$, $y_2 = x^2$, and $y_3 = x^{-1}$ we get

$$W = \begin{vmatrix} x^3 & x^2 & x^{-1} \\ 3x^2 & 2x & -x^{-2} \\ 6x & 2 & 2x^{-3} \end{vmatrix} = -12x, \qquad W_1 = \begin{vmatrix} 0 & x^2 & x^{-1} \\ 0 & 2x & -x^{-2} \\ x^{-1} & 2 & 2x^{-3} \end{vmatrix} = -3x,$$

$$W_2 = \begin{vmatrix} x^3 & 0 & x^{-1} \\ 3x^2 & 0 & -x^{-2} \\ 6x & x^{-1} & 2x^{-3} \end{vmatrix} = 4, \qquad W_3 = \begin{vmatrix} x^3 & x^2 & 0 \\ 3x^2 & 2x & 0 \\ 6x & 2 & x^{-1} \end{vmatrix} = -x^3.$$

Therefore $u_1' = \dfrac{W_1}{W} = \dfrac{1}{4x^2}$, $u_2' = \dfrac{W_2}{W} = \dfrac{-1}{3x}$, and $u_3' = \dfrac{W_3}{W} = \dfrac{x^2}{12}$. The integrals of these three functions are straight forward: $u_1 = -1/4x$, $u_2 = -(\ln x)/3$, $u_3 = x^3/36$. The particular solution is

$$y_p = u_1 y_1 + u_2 y_2 + u_3 y_3 = -\frac{x^2}{4} - \frac{x^2 \ln x}{3} + \frac{x^2}{36}$$

$$= -\frac{2x^2}{9} - \frac{x^2 \ln x}{3}.$$

Finally, because $-2x^2/9$ can be absorbed in the $c_2 x^2$ term the general solutions is

$$y = y_c + y_p = c_1 x^3 + c_2 x^2 + c_3 x^{-1} - \frac{x^2 \ln x}{3}, \qquad x > 0.$$

4.8 | Green's Functions

1.

$$y'' - 16y = f(x)$$

$$y'' - 16y = 0$$

$$y_1 = e^{-4x}, \quad y_2 = e^{4x}$$

$$W\left(e^{-4x}, e^{4x}\right) = \begin{vmatrix} e^{-4x} & e^{4x} \\ -4e^{-4x} & 4e^{4x} \end{vmatrix} = 8$$

$$G(x, t) = \frac{e^{-4t}e^{4x} - e^{-4x}e^{4t}}{8} = \frac{1}{8}\left[e^{4(x-t)} - e^{-4(x-t)}\right]$$

$$y_p(x) = \frac{1}{8}\int_{x_0}^{x}\left[e^{4(x-t)} - e^{-4(x-t)}\right]f(t)\,dt = \frac{1}{4}\int_{x_0}^{x}\sinh 4(x - t)\,f(t)\,dt$$

3.

$$y'' + 2y' + y = f(x)$$

$$y'' + 2y' + y = 0$$

$$y_1 = e^{-x}, \quad y_2 = xe^{-x}$$

$$W\left(e^{-x}, xe^{-x}\right) = \begin{vmatrix} e^{-x} & xe^{-x} \\ -e^{-x} & -xe^{-x} + e^{-x} \end{vmatrix} = e^{-2x}$$

$$G(x, t) = \frac{e^{-t}xe^{-x} - e^{-x}te^{-t}}{e^{-2t}} = (x - t)e^{-(x-t)}$$

$$y_p(x) = \int_{x_0}^{x}(x - t)e^{-(x-t)}f(t)\,dt$$

5.

$$y'' + 9y = f(x)$$

$$y'' + 9y = 0$$

$$y_1 = \cos 3x, \quad y_2 = \sin 3x$$

$$W\left(\cos 3x, \sin 3x\right) = \begin{vmatrix} \cos 3x & \sin 3x \\ -3\sin 3x & 3\cos 3x \end{vmatrix} = 3$$

$$G(x, t) = \frac{\cos 3t \sin 3x - \cos 3x \sin 3t}{3} = \frac{1}{3}\sin 3(x - t)$$

$$y_p(x) = \frac{1}{3}\int_{x_0}^{x}\sin 3(x - t)f(t)\,dt$$

7.

$$y'' - 16y = xe^{-2x}$$

$$y = y_c + y_p$$

$$y = c_1 e^{-4x} + c_2 e^{4x} + y_p$$

$$y = c_1 e^{-4x} + c_2 e^{4x} + \frac{1}{4} \int_{x_0}^{x} \sinh 4(x-t) t e^{-2t} \, dt$$

9.

$$y'' + 2y' + y = e^{-x}$$

$$y = y_c + y_p$$

$$y = c_1 e^{-x} + c_2 x e^{-x} + y_p$$

$$y = c_1 e^{-x} + c_2 x e^{-x} + \int_{x_0}^{x} (x-t) e^{-(x-t)} e^{-t} \, dt$$

11.

$$y'' + 9y = x + \sin x$$

$$y = y_c + y_p$$

$$y = c_1 \cos 3x + c_2 \sin 3x + y_p$$

$$y = c_1 \cos 3x + c_2 \sin 3x + \frac{1}{3} \int_{x_0}^{x} \sin 3(x-t)(t + \sin t) \, dt$$

13. The initial-value problem is $y'' - 4y = e^{2x}$, $y(0) = 0$, $y'(0) = 0$. Then we find that

$$y_1 = e^{-2x}, \quad y_2 = e^{2x}$$

$$W\left(e^{-2x}, e^{2x}\right) = \begin{vmatrix} e^{-2x} & e^{2x} \\ -2e^{-2x} & 2e^{2x} \end{vmatrix} = 4.$$

Then

$$G(x, t) = \frac{e^{-2t} e^{2x} - e^{-2x} e^{2t}}{4} = \frac{1}{4}\left[e^{2(x-t)} - e^{-2(x-t)}\right]$$

and the solution of the initial-value problem is

$$y_p(x) = \frac{1}{4} \int_0^x \left[e^{2(x-t)} - e^{-2(x-t)}\right] e^{2t} \, dt$$

$$= \frac{1}{4} e^{2x} \int_0^x dt - \frac{1}{4} e^{-2x} \int_0^x e^{4t} \, dt$$

$$= \frac{1}{4} x e^{2x} - \frac{1}{16} e^{2x} + \frac{1}{16} e^{-2x}.$$

15. The initial-value problem is $y'' - 10y' + 25y = e^{5x}$, $y(0) = 0$, $y'(0) = 0$. Then we find that

$$y_1 = e^{5x}, \quad y_2 = xe^{5x}$$

$$W\left(e^{5x}, xe^{5x}\right) = \begin{vmatrix} e^{5x} & xe^{5x} \\ 5e^{5x} & 5xe^{5x} + e^{5x} \end{vmatrix} = e^{10x}.$$

Then

$$G(x,t) = \frac{e^{5t}xe^{5x} - e^{5x}te^{5t}}{e^{10t}} = (x-t)e^{5(x-t)}$$

and the solution of the initial-value problem is

$$y_p(x) = \int_0^x (x-t)e^{5(x-t)}e^{5t}\, dt$$

$$= xe^{5x}\int_0^x dt - e^{5x}\int_0^x t\, dt$$

$$= x^2 e^{5x} - \frac{1}{2}x^2 e^{5x}$$

$$= \frac{1}{2}x^2 e^{5x}.$$

17. The initial-value problem is $y'' + y = \csc x \cot x$, $y(\pi/2) = 0$, $y'(\pi/2) = 0$. Then we find that

$$y_1 = \cos x, \quad y_2 = \sin x$$

$$W\left(\cos x, \sin x\right) = \begin{vmatrix} \cos x & \sin x \\ -\sin x & \cos x \end{vmatrix} = 1.$$

Then

$$G(x,t) = \cos t \sin x - \cos x \sin t = \sin(x-t)$$

and the solution of the initial-value problem is

$$y_p(x) = \int_{\pi/2}^x (\cos t \sin x - \cos x \sin t)\csc t \cot t\, dt$$

$$= \sin x \int_{\pi/2}^x \cot^2 t\, dt - \cos x \int_{\pi/2}^x \cot t\, dt$$

$$= \sin x \int_{\pi/2}^x (\csc^2 t - 1)\, dt - \cos x \int_{\pi/2}^x \cot t\, dt$$

$$= \sin x \left(-\cot x - x + \frac{\pi}{2}\right) - \cos x \ln|\sin x|$$

$$= -\cos x - x \sin x + \frac{\pi}{2}\sin x - \cos x \ln|\sin x|$$

$$= -\cos x + \frac{\pi}{2}\sin x - x \sin x - \cos x \ln|\sin x|.$$

19. The initial-value problem is $y'' - 4y = e^{2x}$, $y(0) = 1$, $y'(0) = -4$, so $y(x) = c_1 e^{-2x} + c_2 e^{2x}$. The initial conditions give

$$c_1 + c_2 = 4$$
$$-2c_1 + 2c_2 = -4,$$

so $c_1 = \frac{3}{2}$ and $c_2 = -\frac{1}{2}$, which implies that $y_h = \frac{3}{2}e^{-2x} - \frac{1}{2}e^{2x}$. Now, y_p found in the solution of Problem 13 in this section gives

$$y = y_h + y_p = \frac{3}{2}e^{-2x} - \frac{1}{2}e^{2x} + \left(\frac{1}{4}xe^{2x} - \frac{1}{16}e^{2x} + \frac{1}{16}e^{-2x}\right)$$
$$= \frac{25}{16}e^{-2x} - \frac{9}{16}e^{2x} + \frac{1}{4}xe^{2x}.$$

21. The initial-value problem is $y'' - 10y' + 25y = e^{5x}$, $y(0) = -1$, $y'(0) = 1$, so $y(x) = c_1 e^{5x} + c_2 xe^{5x}$. The initial conditions give

$$c_1 = -1$$
$$5c_1 + c_2 = 1,$$

so $c_1 = -1$ and $c_2 = 6$, which implies that $y_h = -e^{5x} + 6xe^{5x}$. Now, y_p found in the solution of Problem 15 in this section gives

$$y = y_h + y_p = -e^{5x} + 6xe^{5x} + \frac{1}{2}x^2 e^{5x}.$$

23. The initial-value problem is $y'' + y = \csc x \cot x$, $y(\pi/2) = -\frac{\pi}{2}$, $y'(\pi/2) = -1$, so $y(x) = c_1 \cos x + c_2 \sin x$. The initial conditions give

$$c_2 = -\frac{\pi}{2}$$
$$-c_1 = -1,$$

so $c_1 = 1$ and $c_2 = -\frac{\pi}{2}$, which implies that $y_h = \cos x - \frac{\pi}{2}\sin x$. Now, y_p found in the solution of Problem 17 in this section gives

$$y = y_h + y_p = \cos x - \frac{\pi}{2}\sin x + \left(-\cos x + \frac{\pi}{2}\sin x - x\sin x - \cos x \ln|\sin x|\right)$$
$$= -x\sin x - \cos x \ln|\sin x|.$$

25. The initial-value problem is $y'' + 3y' + 2y = \sin e^x$, $y(0) = -1$, $y'(0) = 0$, so $y(x) = c_1 e^{-x} + c_2 e^{-2x}$. The initial conditions give

$$c_1 + c_2 = -1$$
$$-c_1 - 2c_2 = 0,$$

so $c_1 = -2$ and $c_2 = 1$, which implies that $y_h = -2e^{-x} + e^{-2x}$. The Wronskian is

$$W\left(e^{-x}, e^{-2x}\right) = \begin{vmatrix} e^{-x} & e^{-2x} \\ -e^{-x} & -2e^{-2x} \end{vmatrix} = -e^{-3x}.$$

Then $G(x,t) = \dfrac{e^{-t}e^{-2x} - e^{-x}e^{-2t}}{-e^{-3t}} = e^{-x}e^{t} - e^{2t}e^{-2x}$ so

$$y_p(x) = \int_0^x \left(e^{-x}e^{t} - e^{2t}e^{-2x}\right)\sin e^{t}\, dt$$

$$= e^{-x}\int_0^x e^{t}\sin e^{t}\, dt - e^{-2x}\int_0^x e^{2t}\sin e^{t}\, dt$$

$$= e^{-x}\left(-\cos e^{x} + \cos 1\right) - e^{-2x}\left(-e^{x}\cos e^{x} + \sin e^{x} + \cos 1 - \sin 1\right)$$

$$= e^{-x}\cos 1 + (\sin 1 - \cos 1)e^{-2x} - e^{-2x}\sin e^{x}.$$

The solution of the initial-value problem is

$$y = y_h + y_p = -2e^{-x} + e^{-2x} + \left(e^{-x}\cos 1 + e^{-2x}(\sin 1 - \cos 1) - e^{-2x}\sin e^{x}\right)$$

$$= (\cos 1 - 2)e^{-x} + (1 + \sin 1 - \cos 1)e^{-2x} - e^{-2x}\sin e^{x}.$$

27. The Cauchy-Euler initial-value problem $x^2 y'' - 2xy' + 2y = x$, $y(1) = 2$, $y'(1) = -1$, has auxiliary equation $m(m-1) - 2m + 2 = m^2 - 3m + 2 = (m-1)(m-2) = 0$ so $m_1 = 1$, $m_2 = 2$, $y(x) = c_1 x + c_2 x^2$, and $y' = c_1 + 2c_2 x$. The initial conditions give

$$\begin{aligned} c_1 + c_2 &= 2 \\ c_1 + 2c_2 &= -1, \end{aligned}$$

so $c_1 = 5$ and $c_2 = -3$, which implies that $y_h = 5x - 3x^2$. The Wronskian is

$$W\left(x, x^2\right) = \begin{vmatrix} x & x^2 \\ 1 & 2x \end{vmatrix} = x^2.$$

Then $G(x,t) = \dfrac{tx^2 - xt^2}{t^2} = \dfrac{x(x-t)}{t}$. From the standard form of the differential equation we identify the forcing function $f(t) = \dfrac{1}{t}$. Then, for $x > 1$,

$$y_p(x) = \int_1^x \frac{x(x-t)}{t}\frac{1}{t}\, dt$$

$$= x^2 \int_1^x \frac{1}{t^2}\, dt - x \int_1^x \frac{1}{t}\, dt$$

$$= x^2\left(-\frac{1}{x} + 1\right) - x(\ln x - \ln 1)$$

$$= x^2 - x - x\ln x.$$

The solution of the initial-value problem is

$$y = y_h + y_p = (5x - 3x^2) + (x^2 - x - x\ln x) = 4x - 2x^2 - x\ln x.$$

29. The Cauchy-Euler initial-value problem $x^2 y'' - 6y = \ln x$, $y(1) = 1$, $y'(1) = 3$, has auxiliary equation $m(m-1) - 6 = m^2 - m - 6 = (m-3)(m+2) = 0$ so $m_1 = 3$, $m_2 = -2$, $y(x) = c_1 x^3 + c_2 x^{-2}$, and $y' = 3c_1 x^2 - 2c_2 x^{-3}$. The initial conditions give

$$c_1 + c_2 = 1$$
$$3c_1 - 2c_2 = 3,$$

so $c_1 = 1$ and $c_2 = 0$, which implies that $y_h = x^3$. The Wronskian is

$$W\left(x^3, x^{-2}\right) = \begin{vmatrix} x^3 & x^{-2} \\ 3x^2 & -2x^{-3} \end{vmatrix} = -5.$$

Then $G(x,t) = \dfrac{t^3 x^{-2} - x^3 t^{-2}}{-5} = -\dfrac{1}{5}\left(\dfrac{t^3}{x^2} - \dfrac{x^3}{t^2}\right)$. From the standard form of the differential

equation we identify the forcing function $f(t) = \dfrac{\ln t}{t^2}$. Then, for $x > 1$,

$$y_p(x) = \int_1^x \frac{1}{5}\left(\frac{t^3}{x^2} - \frac{x^3}{t^2}\right)\frac{\ln t}{t^2}\,dt$$

$$= -\frac{1}{5}x^{-2}\int_1^x t\ln t\,dt + \frac{1}{5}x^3\int_1^x t^{-4}\ln t\,dt$$

$$= -\frac{1}{5}x^{-2}\left(-\frac{x^2}{4} + \frac{1}{2}x^2\ln x + \frac{1}{4}\right) + \frac{1}{5}x^3\left(-\frac{1}{9x^3} - \frac{\ln x}{3x^3} + \frac{1}{9}\right)$$

$$= \frac{1}{20} - \frac{1}{10}\ln x - \frac{1}{20}x^{-2} - \frac{1}{45} - \frac{1}{15}\ln x + \frac{1}{45}x^3$$

$$= \frac{1}{36} - \frac{1}{6}\ln x + \frac{1}{45}x^3 - \frac{1}{20}x^{-2}.$$

The solution of the initial-value problem is

$$y = y_h + y_p = x^3 + \left(\frac{1}{36} - \frac{1}{6}\ln x + \frac{1}{45}x^3 - \frac{1}{20}x^{-2}\right) = \frac{46}{45}x^3 - \frac{1}{20}x^{-2} + \frac{1}{36} - \frac{1}{6}\ln x.$$

31. The initial-value problem is $y'' - y = f(x)$, $y(0) = 8$, $y'(0) = 2$, where

$$f(x) = \begin{cases} -1, & x < 0 \\ 1, & x \geq 0. \end{cases}$$

We first find

$$y_h(x) = 5e^x + 3e^{-x} \quad \text{and} \quad y_p(x) = \frac{1}{2}\int_0^x \left[e^{x-t} - e^{-(x-t)}\right]f(t)\,dt.$$

Then for $x < 0$,

$$y_p(x) = -\frac{1}{2}\int_0^x \left[e^{x-t} - e^{-(x-t)}\right]dt = -\frac{1}{2}e^x\int_0^x e^{-t}\,dt + \frac{1}{2}e^{-x}\int_0^x e^t\,dt$$

$$= -\frac{1}{2}e^x(1 - e^{-x}) + \frac{1}{2}e^{-x}(e^x - 1) = -\frac{1}{2}e^x + \frac{1}{2} + \frac{1}{2} - \frac{1}{2}e^{-x}$$

$$= 1 - \frac{1}{2}e^x - \frac{1}{2}e^{-x},$$

and for $x \geq 0$

$$y_p(x) = \frac{1}{2}\int_0^x \left[e^{x-t} - e^{-(x-t)}\right]dt = \frac{1}{2}e^x\int_0^x e^{-t}\,dt - \frac{1}{2}e^{-x}\int_0^x e^t\,dt$$

$$= \frac{1}{2}e^x(1 - e^{-x}) - \frac{1}{2}e^{-x}(e^x - 1) = \frac{1}{2}e^x - \frac{1}{2} - \frac{1}{2} + \frac{1}{2}e^{-x}$$

$$= -1 + \frac{1}{2}e^x + \frac{1}{2}e^{-x}.$$

The solution is

$$y(x) = y_h(x) + y_p(x) = 5e^x + 3e^{-x} + y_p(x),$$

where

$$y_p(x) = \begin{cases} 1 - \dfrac{1}{2}e^x - \dfrac{1}{2}e^{-x}, & x < 0 \\ -1 + \dfrac{1}{2}e^x + \dfrac{1}{2}e^{-x}, & x \geq 0 \end{cases} = \begin{cases} 1 - \cosh x, & x < 0 \\ -1 + \cosh x, & x \geq 0 \end{cases}$$

33. The initial-value problem is $y'' + y = f(x)$, $y(0) = 1$, $y'(0) = -1$, where

$$f(x) = \begin{cases} 0, & x < 0 \\ 10, & 0 \leq x \leq 3\pi \\ 0, & x > 3\pi. \end{cases}$$

We first find

$$y_h(x) = \cos x - \sin x \qquad \text{and} \qquad y_p(x) = \int_0^x \sin(x - t)f(t)\,dt.$$

Then for $x < 0$

$$y_p(x) = \int_0^x \sin(x - t)0\,dt = 0,$$

for $0 \leq x \leq 3\pi$

$$y_p(x) = 10\int_0^x \sin(x - t)\,dt = 10 - 10\cos x,$$

and for $x > 3\pi$

$$y_p(x) = 10\int_0^{3\pi} \sin(x - t)\,dt + \int_{3\pi}^x \sin(x - t)0\,dt = -20\cos x.$$

The solution is

$$y(x) = y_h(x) + y_p(x) = \cos x - \sin x + y_p(x),$$

where

$$y_p(x) = \begin{cases} 0, & x < 0 \\ 10 - 10\cos x, & 0 \leq x \leq 3\pi \\ -20\cos x, & x > 3\pi. \end{cases}$$

35. The boundary-value problem is $y'' = f(x), \quad y(0) = 0, \; y(1) = 0.$ The solution of the associated homogeneous equation is $y = c_1 + c_2 x$.

(a) To satisfy $y(0) = 0$ we take $y_1(x) = x$ and to satisfy $y(1) = 0$ we take $y_2(x) = x - 1$. The Wronskian of y_1 and y_2 is

$$W(y_1, y_2) = \begin{vmatrix} x & x-1 \\ 1 & 1 \end{vmatrix} = 1,$$

so

$$G(x, t) = \begin{cases} t(x-1), & 0 \le t \le x \\ x(t-1), & x \le t \le 1. \end{cases}$$

Therefore

$$y_p(x) = \int_0^1 G(x, t) f(t)\, dt = (x-1) \int_0^x t f(t)\, dt + x \int_x^1 (t-1) f(t)\, dt.$$

(b) By the Product Rule and the Fundamental Theorem of Calculus, the first two derivative of $y_p(x)$ are

$$y_p'(x) = (x-1) x f(x) + \int_0^x t f(t)\, dt + x[-(x-1)f(x)] + \int_x^1 (t-1) f(t)\, dt$$

$$\begin{aligned} y_p''(x) &= (x-1)[x f'(x) + f(x)] + x f(x) + x f(x) \\ &\quad - [(x^2 - x) f'(x) + (2x-1) f(x)] - (x-1) f(x) \\ &= x^2 f'(x) + x f(x) - x f'(x) - f(x) + x f(x) + x f(x) - x^2 f(x) \\ &\quad + x f'(x) - 2x f(x) + f(x) - x f(x) + f(x) = f(x) \end{aligned}$$

Thus, $y_p(x)$ satisfies the differential equation. To see that the boundary conditions are satisfied we compute

$$y_p(0) = (0-1) \int_0^0 t f(t)\, dt + 0 \cdot \int_0^1 (t-1) f(t)\, dt = 0$$

and

$$y_p(1) = (1-1) \int_0^1 t f(t)\, dt + 1 \cdot \int_1^1 (t-1) f(t)\, dt = 0.$$

37. If $f(x) = 1$ in Problem 35, then

$$y_p(x) = (x-1) \int_0^x t\, dt + x \int_x^1 (t-1)\, dt = \frac{1}{2} x^2 - \frac{1}{2} x.$$

39. The boundary-value problem is $y'' + y = 1, \quad y(0) = 0, \; y(1) = 0.$ The solution of the associated homogeneous equation is $y = c_1 \cos x + c_2 \sin x$. Since $y(0) = c_1 \cos 0 + c_2 \sin 0 =$

$c_1 = 0$, we take $y_1(x) = \sin x$. To satisfy $y(1) = 0$ we note that $y(1) = c_1 \cos 1 + c_2 \sin 1 = 0$ which implies that $c_1 = -c_2 \sin 1 / \cos 1$ so

$$y(x) = -c_2 \frac{\sin 1}{\cos 1} \cos x + c_2 \sin x = -\frac{c_2}{\cos 1} (\sin x \cos 1 - \cos x \sin 1).$$

Taking $c_2 = -\cos 1$, we have

$$y_2(x) = \sin x \cos 1 - \cos x \sin 1 = \sin(x - 1).$$

The Wronskian of y_1 and y_2 is

$$W(y_1, y_2) = \begin{vmatrix} \sin x & \sin(x-1) \\ \cos x & \cos(x-1) \end{vmatrix} = \sin x \cos(x-1) - \cos x \sin(x-1) = \sin[x - (x-1)] = \sin 1,$$

so

$$G(x,t) = \begin{cases} \dfrac{\sin t \sin(x-1)}{\sin 1}, & 0 \le t \le x \\[3mm] \dfrac{\sin x \sin(t-1)}{\sin 1}, & x \le t \le 1. \end{cases}$$

Therefore, taking $f(t) = 1$,

$$y_p(x) = \int_0^1 G(x,t)\,dt = \frac{\sin(x-1)}{\sin 1} \int_0^x \sin t\,dt + \frac{\sin x}{\sin 1} \int_x^1 \sin(t-1)\,dt$$

$$= \frac{\sin(x-1)}{\sin 1}(-\cos x + 1) + \frac{\sin x}{\sin 1}[-1 + \cos(x-1)] = \frac{\sin(x-1)}{\sin 1} - \frac{\sin x}{\sin 1} + 1.$$

41. The boundary-value problem is $y'' - 2y' + 2y = e^x$, $y(0) = 0$, $y(\pi/2) = 0$. The auxiliary equation is

$$m^2 - 2m + 2 = 0 \quad \text{so} \quad m = \frac{2 \pm \sqrt{4 - 8}}{2} = 1 \pm i.$$

The solution of the associated homogeneous equation is then $y = e^x(c_1 \cos x + c_2 \sin x)$. It is easily seen that

$$y_1(x) = e^x \sin x \quad \text{and} \quad y_2(x) = e^x \cos x$$

satisfy the boundary conditions. The Wronskian of y_1 and y_2 is

$$W(y_1, y_2) = \begin{vmatrix} e^x \sin x & e^x \cos x \\ e^x \cos x + e^x \sin x & -e^x \sin x + e^x \cos x \end{vmatrix}$$

$$= e^{2x}\left[-\sin^2 x + \sin x \cos x - (\cos^2 x + \sin x \cos x)\right] = -e^{2x},$$

so

$$G(x,t) = \begin{cases} \dfrac{e^t \sin t\, e^x \cos x}{-e^{2t}}, & 0 \le t \le x \\[3mm] \dfrac{e^x \sin x\, e^t \cos t}{-e^{2t}}, & x \le t \le \pi/2 \end{cases}$$

Therefore, taking $f(t) = e^t$,

$$y_p(x) = \int_0^{\pi/2} G(x,t)e^t \, dt = -e^x \cos x \int_0^x \sin t \, dt - e^x \sin x \int_x^{\pi/2} \cos t \, dt$$

$$= -e^x \cos x(1 - \cos x) - e^x \sin x(1 - \sin x) = -e^x \cos x - e^x \sin x + e^x.$$

43. The Cauchy-Euler boundary-value problem $x^2 y'' + xy' = 1$, $y(e^{-1}) = 0$, $y(1) = 0$ has auxiliary equation $m(m-1)+m = m^2 = 0$ so $y(x) = c_1 + c_2 \ln x$. Since $y(e^{-1}) = c_1 + c_2 \ln e^{-1} = c_1 - c_2 = 0$, $c_1 = c_2$ and $y(x) = c_2 + c_2 \ln x = c_2(1 + \ln x)$. Taking $c_2 = 1$ we have $y_1(x) = 1 + \ln x$. To satisfy $y(1) = 0$ we note that $y(1) = c_1 + c_2 \ln 1 = c_1 = 0$ which implies that $y(x) = c_2 \ln x$. Taking $c_2 = 1$ we find $y_2(x) = \ln x$. The Wronskian of y_1 and y_2 is

$$W(y_1, y_2) = \begin{vmatrix} 1 + \ln x & \ln x \\ 1/x & 1/x \end{vmatrix} = \frac{1}{x},$$

so

$$G(x,t) = \begin{cases} \dfrac{(1 + \ln t)(\ln x)}{1/t}, & 0 \le t \le x \\[3mm] \dfrac{(1 + \ln x)(\ln t)}{1/t}, & x \le t \le 1. \end{cases}$$

From the standard form of the differential equation we identify the forcing function $f(t) = \dfrac{1}{t^2}$. Then,

$$y_p(x) = \int_{e^{-1}}^1 G(x,t)\frac{1}{t^2} \, dt = \ln x \int_{e^{-1}}^x \left(\frac{1}{t} + \frac{\ln t}{t} \right) dt + (1 + \ln x) \int_x^1 \frac{\ln t}{t} \, dt$$

$$= (\ln x)\left(\ln x + \frac{1}{2}(\ln x)^2 + \frac{1}{2} \right) + (1 + \ln x)\left(-\frac{1}{2}(\ln x)^2 \right) = \frac{1}{2}(\ln x)^2 + \frac{1}{2}\ln x.$$

4.9   Solving Systems of Linear Equations

1. From $Dx = 2x - y$ and $Dy = x$ we obtain $y = 2x - Dx$, $Dy = 2Dx - D^2x$, and $(D^2 - 2D + 1)x = 0$. The solution is

$$x = c_1 e^t + c_2 t e^t$$
$$y = (c_1 - c_2)e^t + c_2 t e^t.$$

3. From $Dx = -y + t$ and $Dy = x - t$ we obtain $y = t - Dx$, $Dy = 1 - D^2x$, and $(D^2 + 1)x = 1 + t$. The solution is

$$x = c_1 \cos t + c_2 \sin t + 1 + t$$
$$y = c_1 \sin t - c_2 \cos t + t - 1.$$

5. From $(D^2 + 5)x - 2y = 0$ and $-2x + (D^2 + 2)y = 0$ we obtain $y = \frac{1}{2}(D^2 + 5)x$, $D^2y = \frac{1}{2}(D^4 + 5D^2)x$, and $(D^2 + 1)(D^2 + 6)x = 0$. The solution is

$$x = c_1 \cos t + c_2 \sin t + c_3 \cos \sqrt{6}\,t + c_4 \sin \sqrt{6}\,t$$

$$y = 2c_1 \cos t + 2c_2 \sin t - \frac{1}{2}c_3 \cos \sqrt{6}\,t - \frac{1}{2}c_4 \sin \sqrt{6}\,t.$$

7. From $D^2x = 4y + e^t$ and $D^2y = 4x - e^t$ we obtain $y = \frac{1}{4}D^2x - \frac{1}{4}e^t$, $D^2y = \frac{1}{4}D^4x - \frac{1}{4}e^t$, and $(D^2 + 4)(D - 2)(D + 2)x = -3e^t$. The solution is

$$x = c_1 \cos 2t + c_2 \sin 2t + c_3 e^{2t} + c_4 e^{-2t} + \frac{1}{5}e^t$$

$$y = -c_1 \cos 2t - c_2 \sin 2t + c_3 e^{2t} + c_4 e^{-2t} - \frac{1}{5}e^t.$$

9. From $Dx + D^2y = e^{3t}$ and $(D + 1)x + (D - 1)y = 4e^{3t}$ we obtain $D(D^2 + 1)x = 34e^{3t}$ and $D(D^2 + 1)y = -8e^{3t}$. The solution is

$$y = c_1 + c_2 \sin t + c_3 \cos t - \frac{4}{15}e^{3t}$$

$$x = c_4 + c_5 \sin t + c_6 \cos t + \frac{17}{15}e^{3t}.$$

Substituting into $(D + 1)x + (D - 1)y = 4e^{3t}$ gives

$$(c_4 - c_1) + (c_5 - c_6 - c_3 - c_2)\sin t + (c_6 + c_5 + c_2 - c_3)\cos t = 0$$

so that $c_4 = c_1$, $c_5 = c_3$, $c_6 = -c_2$, and

$$x = c_1 - c_2 \cos t + c_3 \sin t + \frac{17}{15}e^{3t}.$$

11. From $(D^2 - 1)x - y = 0$ and $(D - 1)x + Dy = 0$ we obtain $y = (D^2 - 1)x$, $Dy = (D^3 - D)x$, and $(D - 1)(D^2 + D + 1)x = 0$. The solution is

$$x = c_1 e^t + e^{-t/2}\left[c_2 \cos \frac{\sqrt{3}}{2}t + c_3 \sin \frac{\sqrt{3}}{2}t\right]$$

$$y = \left(-\frac{3}{2}c_2 - \frac{\sqrt{3}}{2}c_3\right)e^{-t/2}\cos \frac{\sqrt{3}}{2}t + \left(\frac{\sqrt{3}}{2}c_2 - \frac{3}{2}c_3\right)e^{-t/2}\sin \frac{\sqrt{3}}{2}t.$$

13. From $(2D - 5)x + Dy = e^t$ and $(D - 1)x + Dy = 5e^t$ we obtain $Dy = (5 - 2D)x + e^t$ and $(4 - D)x = 4e^t$. Then

$$x = c_1 e^{4t} + \frac{4}{3}e^t$$

and $Dy = -3c_1 e^{4t} + 5e^t$ so that

$$y = -\frac{3}{4}c_1 e^{4t} + c_2 + 5e^t.$$

15. Multiplying the first equation by $D+1$ and the second equation by D^2+1 and subtracting we obtain $(D^4 - D^2)x = 1$. Then

$$x = c_1 + c_2 t + c_3 e^t + c_4 e^{-t} - \frac{1}{2}t^2.$$

Multiplying the first equation by $D+1$ and subtracting we obtain $D^2(D+1)y = 1$. Then

$$y = c_5 + c_6 t + c_7 e^{-t} - \frac{1}{2}t^2.$$

Substituting into $(D-1)x + (D^2+1)y = 1$ gives

$$(-c_1 + c_2 + c_5 - 1) + (-2c_4 + 2c_7)e^{-t} + (-1 - c_2 + c_6)t = 1$$

so that $c_5 = c_1 - c_2 + 2$, $c_6 = c_2 + 1$, and $c_7 = c_4$. The solution of the system is

$$x = c_1 + c_2 t + c_3 e^t + c_4 e^{-t} - \frac{1}{2}t^2$$

$$y = (c_1 - c_2 + 2) + (c_2 + 1)t + c_4 e^{-t} - \frac{1}{2}t^2.$$

17. From $Dx = y$, $Dy = z$. and $Dz = x$ we obtain $x = D^2 y = D^3 x$ so that $(D-1)(D^2 + D + 1)x = 0$,

$$x = c_1 e^t + e^{-t/2}\left[c_2 \sin\frac{\sqrt{3}}{2}t + c_3 \cos\frac{\sqrt{3}}{2}t\right],$$

$$y = c_1 e^t + \left(-\frac{1}{2}c_2 - \frac{\sqrt{3}}{2}c_3\right)e^{-t/2}\sin\frac{\sqrt{3}}{2}t + \left(\frac{\sqrt{3}}{2}c_2 - \frac{1}{2}c_3\right)e^{-t/2}\cos\frac{\sqrt{3}}{2}t,$$

and

$$z = c_1 e^t + \left(-\frac{1}{2}c_2 + \frac{\sqrt{3}}{2}c_3\right)e^{-t/2}\sin\frac{\sqrt{3}}{2}t + \left(-\frac{\sqrt{3}}{2}c_2 - \frac{1}{2}c_3\right)e^{-t/2}\cos\frac{\sqrt{3}}{2}t.$$

19. Write the system in the form

$$Dx - 6y = 0$$
$$x - Dy + z = 0$$
$$x + y - Dz = 0.$$

Multiplying the second equation by D and adding to the third equation we obtain $(D+1)x - (D^2 - 1)y = 0$. Eliminating y between this equation and $Dx - 6y = 0$ we find

$$(D^3 - D - 6D - 6)x = (D+1)(D+2)(D-3)x = 0.$$

Thus

$$x = c_1 e^{-t} + c_2 e^{-2t} + c_3 e^{3t},$$

and, successively substituting into the first and second equations, we get

$$y = -\frac{1}{6}c_1 e^{-t} - \frac{1}{3}c_2 e^{-2t} + \frac{1}{2}c_3 e^{3t}$$

$$z = -\frac{5}{6}c_1 e^{-t} - \frac{1}{3}c_2 e^{-2t} + \frac{1}{2}c_3 e^{3t}.$$

21. From $(D+5)x + y = 0$ and $4x - (D+1)y = 0$ we obtain $y = -(D+5)x$ so that $Dy = -(D^2 + 5D)x$. Then $4x + (D^2 + 5D)x + (D+5)x = 0$ and $(D+3)^2 x = 0$. Thus

$$x = c_1 e^{-3t} + c_2 t e^{-3t}$$

$$y = -(2c_1 + c_2)e^{-3t} - 2c_2 t e^{-3t}.$$

Using $x(1) = 0$ and $y(1) = 1$ we obtain

$$c_1 e^{-3} + c_2 e^{-3} = 0$$

$$-(2c_1 + c_2)e^{-3} - 2c_2 e^{-3} = 1$$

or

$$c_1 + c_2 = 0$$

$$2c_1 + 3c_2 = -e^3.$$

Thus $c_1 = e^3$ and $c_2 = -e^3$. The solution of the initial-value problem is

$$x = e^{-3t+3} - t e^{-3t+3}$$

$$y = -e^{-3t+3} + 2t e^{-3t+3}.$$

23. Equating Newton's law with the net forces in the x- and y-directions gives $m\, d^2x/dt^2 = 0$ and $m\, d^2y/dt^2 = -mg$, respectively. From $mD^2x = 0$ we obtain $x(t) = c_1 t + c_2$, and from $mD^2y = -mg$ or $D^2y = -g$ we obtain $y(t) = -\frac{1}{2}gt^2 + c_3 t + c_4$.

25. Multiplying the first equation by $D + 1$ and the second equation by D we obtain

$$D(D+1)x - 2D(D+1)y = 2t + t^2$$

$$D(D+1)x - 2D(D+1)y = 0.$$

This leads to $2t + t^2 = 0$, so the system has no solutions.

27. (a) Separating variables in the first equation, we have $dx_1/x_1 = -dt/50$, so $x_1 = c_1 e^{-t/50}$. From $x_1(0) = 15$ we get $c_1 = 15$. The second differential equation then becomes

$$\frac{dx_2}{dt} = \frac{15}{50}e^{-t/50} - \frac{2}{75}x_2 \qquad \text{or} \qquad \frac{dx_2}{dt} + \frac{2}{75}x_2 = \frac{3}{10}e^{-t/50}.$$

This differential equation is linear and has the integrating factor $e^{\int 2\,dt/75} = e^{2t/75}$. Then

$$\frac{d}{dt}\left[e^{2t/75}x_2\right] = \frac{3}{10}e^{-t/50+2t/75} = \frac{3}{10}e^{t/150}$$

so

$$e^{2t/75}x_2 = 45e^{t/150} + c_2$$

and

$$x_2 = 45e^{-t/50} + c_2e^{-2t/75}.$$

From $x_2(0) = 10$ we get $c_2 = -35$. The third differential equation then becomes

$$\frac{dx_3}{dt} = \frac{90}{75}e^{-t/50} - \frac{70}{75}e^{-2t/75} - \frac{1}{25}x_3$$

or

$$\frac{dx_3}{dt} + \frac{1}{25}x_3 = \frac{6}{5}e^{-t/50} - \frac{14}{15}e^{-2t/75}.$$

This differential equation is linear and has the integrating factor $e^{\int dt/25} = e^{t/25}$. Then

$$\frac{d}{dt}\left[e^{t/25}x_3\right] = \frac{6}{5}e^{-t/50+t/25} - \frac{14}{15}e^{-2t/75+t/25} = \frac{6}{5}e^{t/50} - \frac{14}{15}e^{t/75},$$

so

$$e^{t/25}x_3 = 60e^{t/50} - 70e^{t/75} + c_3$$

and

$$x_3 = 60e^{-t/50} - 70e^{-2t/75} + c_3e^{-t/25}.$$

From $x_3(0) = 5$ we get $c_3 = 15$. The solution of the initial-value problem is

$$x_1(t) = 15e^{-t/50}$$
$$x_2(t) = 45e^{-t/50} - 35e^{-2t/75}$$
$$x_3(t) = 60e^{-t/50} - 70e^{-2t/75} + 15e^{-t/25}.$$

(b)

(c) Solving $x_1(t) = \frac{1}{2}$, $x_2(t) = \frac{1}{2}$, and $x_3(t) = \frac{1}{2}$, **FindRoot** gives, respectively, $t_1 = 170.06\,\text{min}$, $t_2 = 214.7\,\text{min}$, and $t_3 = 224.4\,\text{min}$. Thus, all three tanks will contain less than or equal to 0.5 pounds of salt after 224.4 minutes.

4.10 | Nonlinear Differential Equations

1. We have $y_1' = y_1'' = e^x$, so

$$(y_1'')^2 = (e^x)^2 = e^{2x} = y_1^2.$$

Also, $y_2' = -\sin x$ and $y_2'' = -\cos x$, so

$$(y_2'')^2 = (-\cos x)^2 = \cos^2 x = y_2^2.$$

However, if $y = c_1 y_1 + c_2 y_2$, we have $(y'')^2 = (c_1 e^x - c_2 \cos x)^2$ and $y^2 = (c_1 e^x + c_2 \cos x)^2$. Thus $(y'')^2 \neq y^2$.

3. Let $u = y'$ so that $u' = y''$. The equation becomes $u' = -u - 1$ which is separable. Thus

$$\frac{du}{u^2 + 1} = -dx$$

$$\tan^{-1} u = -x + c_1$$

$$y' = \tan(c_1 - x)$$

$$y = \ln|\cos(c_1 - x)| + c_2.$$

5. Let $u = y'$ so that $u' = y''$. The equation becomes $x^2 u' + u^2 = 0$. Separating variables we obtain

$$\frac{du}{u^2} = -\frac{dx}{x^2}$$

$$-\frac{1}{u} = \frac{1}{x} + c_1 = \frac{c_1 x + 1}{x}$$

$$u = -\frac{x}{c_1 x + 1} = \frac{1}{c_1}\left(\frac{1}{c_1 x + 1} - 1\right)$$

$$y = \frac{1}{c_1^2}\ln|c_1 x + 1| - \frac{1}{c_1}x + c_2.$$

7. Let $u = y'$ so that $y'' = u\,du/dy$. The equation becomes $yu\,du/dy + u^2 + 1 = 0$. Separating variables we obtain

$$\frac{u\,du}{u^2 + 1} + \frac{dy}{y} = 0$$

$$\frac{1}{2}\ln\left|u^2 + 1\right| + \ln|y| = \ln c_1$$

$$\left(u^2 + 1\right)y^2 = c_1^2$$

$$u^2 = \frac{c_1^2 - y^2}{y^2}$$

$$\frac{dy}{dx} = \pm\frac{\sqrt{c_1^2 - y^2}}{y}$$

$$\pm\frac{y}{\sqrt{c_1^2 - y^2}}\,dy = dx$$

$$\mp\sqrt{c_1^2 - y^2} = x + c_2$$

$$(x + c_2)^2 + y^2 = c_1^2$$

9. Let $u = y'$ so that $y'' = u\,du/dy$. The equation becomes $u\,du/dy + 2yu^3 = 0$. Separating variables we obtain

$$\frac{du}{u^2} + 2y\,dy = 0$$

$$-\frac{1}{u} + y^2 = c$$

$$u = \frac{1}{y^2 + c_1}$$

$$\left(y^2 + c_1\right)dy = dx$$

$$\frac{1}{3}y^3 + c_1 y = x + c_2.$$

11. Letting $u = y'$ we have

$$y'' = \frac{du}{dx} = \frac{du}{dy}\frac{dy}{du} = u\frac{du}{dy} \qquad \text{so} \qquad 2y'y'' = 1 \qquad \text{becomes} \qquad 2u^2\frac{du}{dy} = 1.$$

Then separating variables, integrating, and simplifying, we have

$$2u^2 \, du = dy$$

$$\frac{2}{3}u^3 = y + c_1$$

$$u = \left(\frac{3}{2}y + c_2\right)^{1/3} = \frac{dy}{dx}$$

$$\left(\frac{3}{2}y + c_2\right)^{-1/3} dy = dx$$

$$\left(\frac{3}{2}y + c_2\right)^{2/3} = x + c_3$$

$$\frac{3}{2}y + c_2 = (x + c_3)^{3/2} \ .$$

Now

$$y(0) = 2 \quad \text{implies} \quad 3 + c_2 = c_3^{3/2} \quad \text{and} \quad y'(0) = 1 \quad \text{implies} \quad c_3 = 1.$$

Thus $c_2 = -2$ and $\dfrac{3}{2}y - 2 = (x + 1)^{3/2}$. The solution of the initial-value problem is

$$y = \frac{2}{3}(x + 1)^{3/2} + \frac{4}{3}.$$

13. (a)

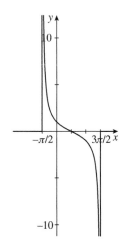

(b) Let $u = y'$ so that $y'' = u \, du/dy$. The equation becomes $u \, du/dy + yu = 0$. Separating variables we obtain

$$du = -y \, dy$$

$$u = -\frac{1}{2}y^2 + c_1$$

$$y' = -\frac{1}{2}y^2 + c_1.$$

When $x = 0$, $y = 1$ and $y' = -1$ so $-1 = -1/2 + c_1$ and $c_1 = -1/2$. Then

$$\frac{dy}{dx} = -\frac{1}{2}y^2 - \frac{1}{2}$$

$$\frac{dy}{y^2 + 1} = -\frac{1}{2}\,dx$$

$$\tan^{-1} y = -\frac{1}{2}x + c_2$$

$$y = \tan\left(-\frac{1}{2}x + c_2\right).$$

When $x = 0$, $y = 1$ so $1 = \tan c_2$ and $c_2 = \pi/4$. The solution of the initial-value problem is

$$y = \tan\left(\frac{\pi}{4} - \frac{1}{2}x\right).$$

The graph is shown in part (a).

(c) The interval of definition is $-\pi/2 < \pi/4 - x/2 < \pi/2$ or $-\pi/2 < x < 3\pi/2$.

15. Let $u = y'$ so that $u' = y''$. The equation becomes $u' - (1/x)u = (1/x)u^3$, which is Bernoulli. Using $w = u^{-2}$ we obtain $dw/dx + (2/x)w = -2/x$. An integrating factor is x^2, so

$$\frac{d}{dx}[x^2 w] = -2x$$

$$x^2 w = -x^2 + c_1$$

$$w = -1 + \frac{c_1}{x^2}$$

$$u^{-2} = -1 + \frac{c_1}{x^2}$$

$$u = \frac{x}{\sqrt{c_1 - x^2}}$$

$$\frac{dy}{dx} = \frac{x}{\sqrt{c_1 - x^2}}$$

$$y = -\sqrt{c_1 - x^2} + c_2$$

$$c_1 - x^2 = (c_2 - y)^2$$

$$x^2 + (c_2 - y)^2 = c_1.$$

In Problems 17–19 the thinner curve is obtained using a numerical solver, while the thicker curve is the graph of the Taylor polynomial.

17. We look for a solution of the form

$$y(x) = y(0) + y'(0)x + \frac{1}{2!}y''(0)x^2 + \frac{1}{3!}y'''(0)x^3 + \frac{1}{4!}y^{(4)}(0)x^4 + \frac{1}{5!}y^{(5)}(0)x^5.$$

From $y''(x) = x + y^2$ we compute

$$y'''(x) = 1 + 2yy'$$
$$y^{(4)}(x) = 2yy'' + 2(y')^2$$
$$y^{(5)}(x) = 2yy''' + 6y'y''.$$

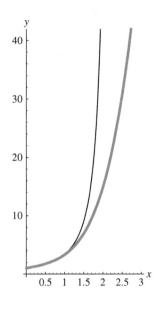

Using $y(0) = 1$ and $y'(0) = 1$ we find

$$y''(0) = 1, \quad y'''(0) = 3, \quad y^{(4)}(0) = 4, \quad y^{(5)}(0) = 12.$$

An approximate solution is

$$y(x) = 1 + x + \frac{1}{2}x^2 + \frac{1}{2}x^3 + \frac{1}{6}x^4 + \frac{1}{10}x^5.$$

19. We look for a solution of the form

$$y(x) = y(0) + y'(0)x + \frac{1}{2!}y''(0)x^2 + \frac{1}{3!}y'''(0)x^3 + \frac{1}{4!}y^{(4)}(0)x^4 + \frac{1}{5!}y^{(5)}(0)x^5.$$

From $y''(x) = x^2 + y^2 - 2y'$ we compute

$$y'''(x) = 2x + 2yy' - 2y''$$
$$y^{(4)}(x) = 2 + 2(y')^2 + 2yy'' - 2y'''$$
$$y^{(5)}(x) = 6y'y'' + 2yy''' - 2y^{(4)}.$$

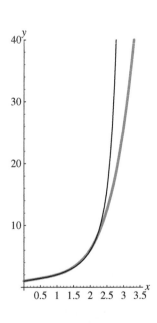

Using $y(0) = 1$ and $y'(0) = 1$ we find

$$y''(0) = -1, \quad y'''(0) = 4, \quad y^{(4)}(0) = -6, \quad y^{(5)}(0) = 14.$$

An approximate solution is

$$y(x) = 1 + x - \frac{1}{2}x^2 + \frac{2}{3}x^3 - \frac{1}{4}x^4 + \frac{7}{60}x^5.$$

21. We need to solve $[1 + (y')^2]^{3/2} = y''$. Let $u = y'$ so that $u' = y''$. The equation becomes $(1 + u^2)^{3/2} = u'$ or $(1 + u^2)^{3/2} = du/dx$. Separating variables and using the substitution

$u = \tan\theta$ we have

$$\frac{du}{(1+u^2)^{3/2}} = dx$$

$$\int \frac{\sec^2\theta}{(1+\tan^2\theta)^{3/2}} \, d\theta = x$$

$$\int \frac{\sec^2\theta}{\sec^3\theta} \, d\theta = x$$

$$\int \cos\theta \, d\theta = x$$

$$\sin\theta = x$$

$$\frac{u}{\sqrt{1+u^2}} = x$$

$$\frac{y'}{\sqrt{1+(y')^2}} = x$$

$$(y')^2 = x^2 \left[1 + (y')^2\right] = x^2 + x^2(y')^2$$

$$(y')^2 = \frac{x^2}{1-x^2}$$

$$y' = \frac{x}{\sqrt{1-x^2}} \quad (\text{ for } x > 0)$$

$$y = -\sqrt{1-x^2}.$$

23. Letting $u = y''$, separating variables, and integrating we have

$$\frac{du}{dx} = \sqrt{1+u^2}, \qquad \frac{du}{\sqrt{1+u^2}} = dx, \quad \text{and} \quad \sinh^{-1} u = x + c_1.$$

Then

$$u = y'' = \sinh(x+c_1), \quad y' = \cosh(x+c_1) + c_2, \quad \text{and} \quad y = \sinh(x+c_1) + c_2 x + c_3.$$

25. Let $u = dx/dt$ so that $d^2x/dt^2 = u\,du/dx$. The equation becomes $u\,du/dx = -k^2/x^2$. Separating variables we obtain

$$u\,du = -\frac{k^2}{x^2}\,dx$$

$$\frac{1}{2}u^2 = \frac{k^2}{x} + c$$

$$\frac{1}{2}v^2 = \frac{k^2}{x} + c.$$

When $t = 0$, $x = x_0$ and $v = 0$ so $0 = (k^2/x_0) + c$ and $c = -k^2/x_0$. Then

$$\frac{1}{2}v^2 = k^2 \left(\frac{1}{x} - \frac{1}{x_0}\right) \quad \text{and} \quad \frac{dx}{dt} = -k\sqrt{2}\sqrt{\frac{x_0 - x}{xx_0}}.$$

Separating variables we have

$$-\sqrt{\frac{xx_0}{x_0 - x}}\, dx = k\sqrt{2}\, dt$$

$$t = -\frac{1}{k}\sqrt{\frac{x_0}{2}} \int \sqrt{\frac{x}{x_0 - x}}\, dx.$$

Using *Mathematica* to integrate we obtain

$$t = -\frac{1}{k}\sqrt{\frac{x_0}{2}}\left[-\sqrt{x(x_0 - x)} - \frac{x_0}{2}\tan^{-1}\left(\frac{(x_0 - 2x)}{2x}\sqrt{\frac{x}{x_0 - x}}\right)\right]$$

$$= \frac{1}{k}\sqrt{\frac{x_0}{2}}\left[\sqrt{x(x_0 - x)} + \frac{x_0}{2}\tan^{-1}\frac{x_0 - 2x}{2\sqrt{x(x_0 - x)}}\right].$$

Chapter 4 in Review

1. Since, simply by substitution, $y = 0$ is seen to be a solution of the given initial-value problem by Theorem 4.1.1 in the text, it is the only solution

3. False; it is not true unless the differential equation is homogeneous. For example, $y_1 = x$ is a solution of $y'' + y = x$, but $y_2 = 5x$ is not.

5. The auxiliary equation is second-order and has $5i$ as a root. Thus, the other root of the auxiliary equation must be $-5i$, since complex roots of polynomials with real coefficients must occur in conjugate pairs. Thus, a second solution of the differential equation must be $\cos 5x$ and the general solution is $y = c_1 \cos 5x + c_2 \sin 5x$.

7. If $y = c_1 x^2 + c_2 x^2 \ln x$, $x > 0$, is the general solution of a Cauchy-Euler differential equation, then the roots of its auxiliary equation are $m_1 = m_2 = 2$ and the auxiliary equation is

$$m^2 - 4m + 4 = m(m - 1) - 3m + 4 = 0.$$

Thus, the Cauchy-Euler differential equation is $x^2 y'' - 3xy' + 4y = 0$.

9. By the superposition principle for nonhomogeneous equations a particular solution is

$$y_p = y_{p_1} + y_{p_2} = x + x^2 - 2 = x^2 + x - 2.$$

11. The set is linearly independent over $(-\infty, 0)$ and linearly dependent over $(0, \infty)$.

13. **(a)** The auxiliary equation is $(m-3)(m+5)(m-1) = m^3 + m^2 - 17m + 15 = 0$, so the differential equation is $y''' + y'' - 17y' + 15y = 0$.

(b) The form of the auxiliary equation is

$$m(m-1)(m-2) + bm(m-1) + cm + d = m^3 + (b-3)m^2 + (c-b+2)m + d = 0.$$

Since $(m-3)(m+5)(m-1) = m^3 + m^2 - 17m + 15 = 0$, we have $b - 3 = 1$, $c - b + 2 = -17$, and $d = 15$. Thus, $b = 4$ and $c = -15$, so the differential equation is $y''' + 4y'' - 15y' + 15y = 0$.

15. The auxiliary equation is $am(m-1) + bm + c = am^2 + (b-a)m + c = 0$. If the roots are 3 and -1, then we want $(m-3)(m+1) = m^2 - 2m - 3 = 0$. Thus, let $a = 1$, $b = -1$, and $c = -3$, so that the differential equation is $x^2 y'' - xy' - 3y = 0$.

17. From $m^2 - 2m - 2 = 0$ we obtain $m = 1 \pm \sqrt{3}$ so that

$$y = c_1 e^{(1+\sqrt{3})x} + c_2 e^{(1-\sqrt{3})x}.$$

19. From $m^3 + 10m^2 + 25m = 0$ we obtain $m = 0$, $m = -5$, and $m = -5$ so that

$$y = c_1 + c_2 e^{-5x} + c_3 x e^{-5x}.$$

21. From $3m^3 + 10m^2 + 15m + 4 = 0$ we obtain $m = -1/3$ and $m = -3/2 \pm (\sqrt{7}/2)i$ so that

$$y = c_1 e^{-x/3} + e^{-3x/2}\left(c_2 \cos \frac{\sqrt{7}}{2} x + c_3 \sin \frac{\sqrt{7}}{2} x\right).$$

23. Applying D^4 to the differential equation we obtain $D^4(D^2 - 3D + 5) = 0$. Then

$$y = \underbrace{e^{3x/2}\left(c_1 \cos \frac{\sqrt{11}}{2} x + c_2 \sin \frac{\sqrt{11}}{2} x\right)}_{y_c} + c_3 + c_4 x + c_5 x^2 + c_6 x^3$$

and $y_p = A + Bx + Cx^2 + Dx^3$. Substituting y_p into the differential equation yields

$$(5A - 3B + 2C) + (5B - 6C + 6D)x + (5C - 9D)x^2 + 5Dx^3 = -2x + 4x^3.$$

Equating coefficients gives $A = -222/625$, $B = 46/125$, $C = 36/25$, and $D = 4/5$. The general solution is

$$y = e^{3x/2}\left(c_1 \cos \frac{\sqrt{11}}{2} x + c_2 \sin \frac{\sqrt{11}}{2} x\right) - \frac{222}{625} + \frac{46}{125}x + \frac{36}{25}x^2 + \frac{4}{5}x^3.$$

25. Applying $D(D^2 + 1)$ to the differential equation we obtain

$$D(D^2 + 1)(D^3 - 5D^2 + 6D) = D^2(D^2 + 1)(D - 2)(D - 3) = 0.$$

Then

$$y = \underbrace{c_1 + c_2 e^{2x} + c_3 e^{3x}}_{y_c} + c_4 x + c_5 \cos x + c_6 \sin x$$

and $y_p = Ax + B \cos x + C \sin x$. Substituting y_p into the differential equation yields

$$6A + (5B + 5C) \cos x + (-5B + 5C) \sin x = 8 + 2 \sin x.$$

Equating coefficients gives $A = 4/3$, $B = -1/5$, and $C = 1/5$. The general solution is

$$y = c_1 + c_2 e^{2x} + c_3 e^{3x} + \frac{4}{3} x - \frac{1}{5} \cos x + \frac{1}{5} \sin x.$$

27. The auxiliary equation is $m^2 - 2m + 2 = [m - (1 + i)][m - (1 - i)] = 0$, so $y_c = c_1 e^x \sin x + c_2 e^x \cos x$ and

$$W = \begin{vmatrix} e^x \sin x & e^x \cos x \\ e^x \cos x + e^x \sin x & -e^x \sin x + e^x \cos x \end{vmatrix} = -e^{2x}.$$

Identifying $f(x) = e^x \tan x$ we obtain

$$u_1' = -\frac{(e^x \cos x)(e^x \tan x)}{-e^{2x}} = \sin x$$

$$u_2' = \frac{(e^x \sin x)(e^x \tan x)}{-e^{2x}} = -\frac{\sin^2 x}{\cos x} = \cos x - \sec x.$$

Then $u_1 = -\cos x$, $u_2 = \sin x - \ln|\sec x + \tan x|$, and

$$y = c_1 e^x \sin x + c_2 e^x \cos x - e^x \sin x \cos x + e^x \sin x \cos x - e^x \cos x \ln|\sec x + \tan x|$$

$$= c_1 e^x \sin x + c_2 e^x \cos x - e^x \cos x \ln|\sec x + \tan x|.$$

29. The auxiliary equation is $6m^2 - m - 1 = 0$ so that

$$y = c_1 x^{1/2} + c_2 x^{-1/3}.$$

31. The auxiliary equation is $m^2 - 5m + 6 = (m - 2)(m - 3) = 0$ and a particular solution is $y_p = x^4 - x^2 \ln x$ so that

$$y = c_1 x^2 + c_2 x^3 + x^4 - x^2 \ln x.$$

33. The auxiliary equation is $m^2 + \omega^2 = 0$, so $y_c = c_1 \cos \omega t + c_2 \sin \omega t$. When $\omega \neq \alpha$, $y_p = A \cos \alpha t + B \sin \alpha t$ and

$$y = c_1 \cos \omega t + c_2 \sin \omega t + A \cos \alpha t + B \sin \alpha t.$$

When $\omega = \alpha$, $y_p = At \cos \omega t + Bt \sin \omega t$ and

$$y = c_1 \cos \omega t + c_2 \sin \omega t + At \cos \omega t + Bt \sin \omega t.$$

35. If $y = \sin x$ is a solution then so is $y = \cos x$ and $m^2 + 1$ is a factor of the auxiliary equation $m^4 + 2m^3 + 11m^2 + 2m + 10 = 0$. Dividing by $m^2 + 1$ we get $m^2 + 2m + 10$, which has roots $-1 \pm 3i$. The general solution of the differential equation is

$$y = c_1 \cos x + c_2 \sin x + e^{-x}(c_3 \cos 3x + c_4 \sin 3x).$$

37. (a) The auxiliary equation is $m^4 - 2m^2 + 1 = (m^2 - 1)^2 = 0$, so the general solution of the differential equation is

$$y = c_1 \sinh x + c_2 \cosh x + c_3 x \sinh x + c_4 x \cosh x.$$

(b) Since both $\sinh x$ and $x \sinh x$ are solutions of the associated homogeneous differential equation , a particular solution of $y^{(4)} - 2y'' + y = \sinh x$ has the form $y_p = Ax^2 \sinh x + Bx^2 \cosh x$.

39. The auxiliary equation is $m^2 - 2m + 2 = 0$ so that $m = 1 \pm i$ and $y = e^x(c_1 \cos x + c_2 \sin x)$. Setting $y(\pi/2) = 0$ and $y(\pi) = -1$ we obtain $c_1 = e^{-\pi}$ and $c_2 = 0$. Thus, $y = e^{x-\pi} \cos x$.

41. The auxiliary equation is $m^2 - 1 = (m - 1)(m + 1) = 0$ so that $m = \pm 1$ and $y = c_1 e^x + c_2 e^{-x}$. Assuming $y_p = Ax + B + C \sin x$ and substituting into the differential equation we find $A = -1$, $B = 0$, and $C = -\frac{1}{2}$. Thus $y_p = -x - \frac{1}{2} \sin x$ and

$$y = c_1 e^x + c_2 e^{-x} - x - \frac{1}{2} \sin x.$$

Setting $y(0) = 2$ and $y'(0) = 3$ we obtain

$$c_1 + c_2 = 2$$
$$c_1 - c_2 - \frac{3}{2} = 3.$$

Solving this system we find $c_1 = \frac{13}{4}$ and $c_2 = -\frac{5}{4}$. The solution of the initial-value problem is

$$y = \frac{13}{4} e^x - \frac{5}{4} e^{-x} - x - \frac{1}{2} \sin x.$$

43. Let $u = y'$ so that $u' = y''$. The equation becomes $u\, du/dx = 4x$. Separating variables we obtain

$$u\, du = 4x\, dx$$
$$\frac{1}{2} u^2 = 2x^2 + c_1$$
$$u^2 = 4x^2 + c_2.$$

When $x = 1$, $y' = u = 2$, so $4 = 4 + c_2$ and $c_2 = 0$. Then

$$u^2 = 4x^2$$

$$\frac{dy}{dx} = 2x \quad \text{or} \quad \frac{dy}{dx} = -2x$$

$$y = x^2 + c_3 \quad \text{or} \quad y = -x^2 + c_4.$$

When $x = 1$, $y = 5$, so $5 = 1 + c_3$ and $5 = -1 + c_4$. Thus $c_3 = 4$ and $c_4 = 6$. We have $y = x^2 + 4$ and $y = -x^2 + 6$. Note however that when $y = -x^2 + 6$, $y' = -2x$ and $y'(1) = -2 \neq 2$. Thus, the solution of the initial-value problem is $y = x^2 + 4$.

45. (a) The auxiliary equation is $12m^4 + 64m^3 + 59m^2 - 23m - 12 = 0$ and has roots -4, $-\frac{3}{2}$, $-\frac{1}{3}$, and $\frac{1}{2}$. The general solution is

$$y = c_1 e^{-4x} + c_2 e^{-3x/2} + c_3 e^{-x/3} + c_4 e^{x/2}.$$

(b) The system of equations is

$$c_1 + c_2 + c_3 + c_4 = -1$$

$$-4c_1 - \frac{3}{2}c_2 - \frac{1}{3}c_3 + \frac{1}{2}c_4 = 2$$

$$16c_1 + \frac{9}{4}c_2 + \frac{1}{9}c_3 + \frac{1}{4}c_4 = 5$$

$$-64c_1 - \frac{27}{8}c_2 - \frac{1}{27}c_3 + \frac{1}{8}c_4 = 0.$$

Using a CAS we find $c_1 = -\frac{73}{495}$, $c_2 = \frac{109}{35}$, $c_3 = -\frac{3726}{385}$, and $c_4 = \frac{257}{45}$. The solution of the initial-value problem is

$$y = -\frac{73}{495}e^{-4x} + \frac{109}{35}e^{-3x/2} - \frac{3726}{385}e^{-x/3} + \frac{257}{45}e^{x/2}.$$

47. From $(D-2)x + (D-2)y = 1$ and $Dx + (2D-1)y = 3$ we obtain $(D-1)(D-2)y = -6$ and $Dx = 3 - (2D-1)y$. Then

$$y = c_1 e^{2t} + c_2 e^t - 3 \quad \text{and} \quad x = -c_2 e^t - \frac{3}{2}c_1 e^{2t} + c_3.$$

Substituting into $(D-2)x + (D-2)y = 1$ gives $c_3 = \frac{5}{2}$ so that

$$x = -c_2 e^t - \frac{3}{2}c_1 e^{2t} + \frac{5}{2}.$$

49. From $(D-2)x - y = -e^t$ and $-3x + (D-4)y = -7e^t$ we obtain $(D-1)(D-5)x = -4e^t$ so that

$$x = c_1 e^t + c_2 e^{5t} + te^t.$$

Then

$$y = (D-2)x + e^t = -c_1 e^t + 3c_2 e^{5t} - te^t + 2e^t.$$

Chapter 5

Modeling with Higher-Order Differential Equations

<table>
<tr><td>**5.1**</td><td>**Linear Models: Initial-Value Problems**</td></tr>
</table>

1. From $\frac{1}{8}x'' + 16x = 0$ we obtain

$$x = c_1 \cos 8\sqrt{2}\,t + c_2 \sin 8\sqrt{2}\,t$$

so that the period of motion is $2\pi/8\sqrt{2} = \sqrt{2}\,\pi/8$ seconds.

3. From $\frac{3}{4}x'' + 72x = 0$, $x(0) = -1/4$, and $x'(0) = 0$ we obtain $x = -\frac{1}{4}\cos 4\sqrt{6}\,t$.

5. From $\frac{5}{8}x'' + 40x = 0$, $x(0) = 1/2$, and $x'(0) = 0$ we obtain $x = \frac{1}{2}\cos 8t$.

    (a) $x(\pi/12) = -1/4$, $x(\pi/8) = -1/2$, $x(\pi/6) = -1/4$, $x(\pi/4) = 1/2$, $x(9\pi/32) = \sqrt{2}/4$.

    (b) $x' = -4\sin 8t$ so that $x'(3\pi/16) = 4$ ft/s directed downward.

    (c) If $x = \frac{1}{2}\cos 8t = 0$ then $t = (2n+1)\pi/16$ for $n = 0, 1, 2, \dots$.

7. From $20x'' + 20x = 0$, $x(0) = 0$, and $x'(0) = -10$ we obtain $x = -10\sin t$ and $x' = -10\cos t$.

    (a) The 20 kg mass has the larger amplitude.

    (b) 20 kg: $x'(\pi/4) = -5\sqrt{2}$ m/s, $x'(\pi/2) = 0$ m/s; 50 kg: $x'(\pi/4) = 0$ m/s, $x'(\pi/2) = 10$ m/s

    (c) If $-5\sin 2t = -10\sin t$ then $2\sin t(\cos t - 1) = 0$ so that $t = n\pi$ for $n = 0, 1, 2, \dots$, placing both masses at the equilibrium position. The 50 kg mass is moving upward; the 20 kg mass is moving upward when n is even and downward when n is odd.

9. (a) From $x'' + 16x = 0$, $x(0) = 1/2$, and $x'(0) = 3/2$, we get

$$x(t) = \frac{1}{2}\cos 2t + \frac{3}{4}\sin 2t$$

(b) We use $x(t) = A \sin(\omega t + \phi)$ where $A = \sqrt{c_1^2 + c_2^2} = \sqrt{(1/2)^2 + (3/4)^2} = \sqrt{13}/4$. From this we get

$$\tan \phi = \frac{c_1}{c_2} = \frac{2}{3}$$

$$\phi = \tan^{-1}\left(\frac{2}{3}\right) = 0.588$$

Therefore $x(t) = \frac{\sqrt{13}}{4} \sin(2t + 0.588)$.

(c) We use $x(t) = A \cos(\omega t - \phi)$ where $A = \sqrt{c_1^2 + c_2^2} = \sqrt{(1/2)^2 + (3/4)^2} = \sqrt{13}/4$. This time we get

$$\tan \phi = \frac{c_2}{c_1} = \frac{3}{2}$$

$$\phi = \tan^{-1}\left(\frac{3}{2}\right) = 0.983$$

Therefore $x(t) = \frac{\sqrt{13}}{4} \cos(2t - 0.983)$.

11. From $2x'' + 200x = 0$, $x(0) = -2/3$, and $x'(0) = 5$ we obtain

(a) $x = -\frac{2}{3} \cos 10t + \frac{1}{2} \sin 10t = \frac{5}{6} \sin(10t - 0.927)$.

(b) The amplitude is $5/6$ ft and the period is $2\pi/10 = \pi/5$

(c) $3\pi = \pi k/5$ and $k = 15$ cycles.

(d) If $x = 0$ and the weight is moving downward for the second time, then $10t - 0.927 = 2\pi$ or $t = 0.721$ s.

(e) If $x' = \frac{25}{3} \cos(10t - 0.927) = 0$ then $10t - 0.927 = \pi/2 + n\pi$ or $t = (2n+1)\pi/20 + 0.0927$ for $n = 0, 1, 2, \ldots$.

(f) $x(3) = -0.597$ ft

(g) $x'(3) = -5.814$ ft/s

(h) $x''(3) = 59.702$ ft/s^2

(i) If $x = 0$ then $t = \frac{1}{10}(0.927 + n\pi)$ for $n = 0, 1, 2, \ldots$. The velocity at these times is $x' = \pm 8.33$ ft/s.

(j) If $x = 5/12$ then $t = \frac{1}{10}(\pi/6 + 0.927 + 2n\pi)$ and $t = \frac{1}{10}(5\pi/6 + 0.927 + 2n\pi)$ for $n = 0, 1, 2, \ldots$.

(k) If $x = 5/12$ and $x' < 0$ then $t = \frac{1}{10}(5\pi/6 + 0.927 + 2n\pi)$ for $n = 0,\ 1,\ 2,\ \ldots$.

13. The springs are parallel as in Figure 3.8.5 so the effective spring constant is $k_{\text{eff}} = k_1 + k_2$. From Hooke's Law we find that a mass weighing 20 lb determines the spring constants $k_1 = 40$ and $k_2 = 120$. Then we have

$$k_{\text{eff}} = k_1 + k_2 = 40 + 120 = 160 \text{ lb/ft}$$

$$\frac{20}{32}x'' + 160x = 0$$

$$x'' + 256x = 0$$

$$x(t) = c_1 \cos 16t + c_2 \sin 16t$$

Using the initial condition $x(0) = 0$ we have $c_1 = 0$ and therefore $x(t) = c_2 \sin 16t$. Then $x'(t) = 16c_2 \cos 16t$. Using the condition $x'(0) = 2$ we have $c_2 = 1/8$. Thus

$$x(t) = \frac{1}{8} \sin 16t$$

15. Using $k_1 = 40$ and $k_2 = 120$ we have

$$k_{\text{eff}} = \frac{40 \cdot 120}{40 + 120} = \frac{40 \cdot 120}{160} = 30 \text{ lb/ft}$$

$$\frac{20}{32}x'' + 30x = 0$$

$$x'' + 48x = 0$$

$$x(t) = c_1 \cos 4\sqrt{3}\,t + c_2 \sin 4\sqrt{3}\,t$$

Using the initial condition $x(0) = 0$ we have $c_1 = 0$ and therefore $x(t) = c_2 \sin 4\sqrt{3}\,t$. Then $x'(t) = 4\sqrt{3}\,c_2 \cos 4\sqrt{3}\,t$. Using the condition $x'(0) = 2$ we have $c_2 = \sqrt{3}/6$. Thus

$$x(t) = \frac{\sqrt{3}}{6} \sin 4\sqrt{3}\,t$$

17. For parallel springs the effective spring constant is $k_{\text{eff}} = k_1 + k_2$. When $k = k_1 = k_2$, the effective spring constant is $k_{\text{eff}} = 2k$. Compared to a single spring with constant k, the parallel-spring system is more stiff.

19. For large values of t the differential equation is approximated by $x'' = 0$. The solution of this equation is the linear function $x = c_1 t + c_2$. Thus, for large time, the restoring force will have decayed to the point where the spring is incapable of returning the mass, and the spring will simply keep on stretching.

21. (a) above   **(b)** heading upward

23. (a) below   **(b)** heading upward

25. From $\frac{1}{8}x'' + x' + 2x = 0$, $x(0) = -1$, and $x'(0) = 8$ we obtain $x = 4te^{-4t} - e^{-4t}$ and $x' = 8e^{-4t} - 16te^{-4t}$. If $x = 0$ then $t = 1/4$ second. If $x' = 0$ then $t = 1/2$ second and the extreme displacement is $x = e^{-2}$ feet.

27. (a) From $x'' + 10x' + 16x = 0$, $x(0) = 1$, and $x'(0) = 0$ we obtain $x = \frac{4}{3}e^{-2t} - \frac{1}{3}e^{-8t}$.

(b) From $x'' + 10x' + 16x = 0$, $x(0) = 1$, and $x'(0) = -12$ then $x = -\frac{2}{3}e^{-2t} + \frac{5}{3}e^{-8t}$.

29. (a) From $0.1x'' + 0.4x' + 2x = 0$, $x(0) = -1$, and $x'(0) = 0$ we obtain
$$x = e^{-2t}\left(-\cos 4t - \frac{1}{2}\sin 4t\right).$$

(b) $x = \dfrac{\sqrt{5}}{2}e^{-2t}\sin(4t + 4.25)$

(c) If $x = 0$ (and $t > 0$) then $4t + 4.25 = 2\pi$, 3π, 4π, ... so that the first time heading upward is $t = 1.294$ seconds.

31. From $\frac{5}{16}x'' + \beta x' + 5x = 0$ we find that the roots of the auxiliary equation are
$m = -\frac{8}{5}\beta \pm \frac{4}{5}\sqrt{4\beta^2 - 25}$.

(a) If $4\beta^2 - 25 > 0$ then $\beta > 5/2$.

(b) If $4\beta^2 - 25 = 0$ then $\beta = 5/2$.

(c) If $4\beta^2 - 25 < 0$ then $0 < \beta < 5/2$.

33. If $\frac{1}{2}x'' + \frac{1}{2}x' + 6x = 10\cos 3t$, $x(0) = -2$, and $x'(0) = 0$ then
$$x_c = e^{-t/2}\left(c_1\cos\left(\frac{\sqrt{47}}{2}t\right) + c_2\sin\left(\frac{\sqrt{47}}{2}t\right)\right)$$

and $x_p = \frac{10}{3}(\cos 3t + \sin 3t)$ so that the equation of motion is
$$x = e^{-t/2}\left(-\frac{4}{3}\cos\left(\frac{\sqrt{47}}{2}t\right) - \frac{64}{3\sqrt{47}}\sin\left(\frac{\sqrt{47}}{2}t\right)\right) + \frac{10}{3}(\cos 3t + \sin 3t).$$

35. From $x'' + 8x' + 16x = 8\sin 4t$, $x(0) = 0$, and $x'(0) = 0$ we obtain $x_c = c_1e^{-4t} + c_2te^{-4t}$ and $x_p = -\frac{1}{4}\cos 4t$ so that the equation of motion is
$$x = \frac{1}{4}e^{-4t} + te^{-4t} - \frac{1}{4}\cos 4t.$$

37. From $2x'' + 32x = 68e^{-2t}\cos 4t$, $x(0) = 0$, and $x'(0) = 0$ we obtain $x_c = c_1\cos 4t + c_2\sin 4t$ and $x_p = \frac{1}{2}e^{-2t}\cos 4t - 2e^{-2t}\sin 4t$ so that
$$x = -\frac{1}{2}\cos 4t + \frac{9}{4}\sin 4t + \frac{1}{2}e^{-2t}\cos 4t - 2e^{-2t}\sin 4t.$$

39. (a) By Hooke's law the external force is $F(t) = kh(t)$ so that $mx'' + \beta x' + kx = kh(t)$.

(b) From $\frac{1}{2}x'' + 2x' + 4x = 20\cos t$, $x(0) = 0$, and $x'(0) = 0$ we obtain $x_c = e^{-2t}(c_1\cos 2t + c_2\sin 2t)$ and $x_p = \frac{56}{13}\cos t + \frac{32}{13}\sin t$ so that

$$x = e^{-2t}\left(-\frac{56}{13}\cos 2t - \frac{72}{13}\sin 2t\right) + \frac{56}{13}\cos t + \frac{32}{13}\sin t.$$

41. From $x'' + 4x = -5\sin 2t + 3\cos 2t$, $x(0) = -1$, and $x'(0) = 1$ we obtain $x_c = c_1\cos 2t + c_2\sin 2t$, $x_p = \frac{3}{4}t\sin 2t + \frac{5}{4}t\cos 2t$, and

$$x = -\cos 2t - \frac{1}{8}\sin 2t + \frac{3}{4}t\sin 2t + \frac{5}{4}t\cos 2t.$$

43. (a) From $x'' + \omega^2 x = F_0\cos\gamma t$, $x(0) = 0$, and $x'(0) = 0$ we obtain $x_c = c_1\cos\omega t + c_2\sin\omega t$ and $x_p = (F_0\cos\gamma t)/\left(\omega^2 - \gamma^2\right)$ so that

$$x = -\frac{F_0}{\omega^2 - \gamma^2}\cos\omega t + \frac{F_0}{\omega^2 - \gamma^2}\cos\gamma t.$$

(b) $\displaystyle\lim_{\gamma\to\omega}\frac{F_0}{\omega^2 - \gamma^2}(\cos\gamma t - \cos\omega t) = \lim_{\gamma\to\omega}\frac{-F_0 t\sin\gamma t}{-2\gamma} = \frac{F_0}{2\omega}t\sin\omega t.$

45. (a) From $\cos(u - v) = \cos u\cos v + \sin u\sin v$ and $\cos(u + v) = \cos u\cos v - \sin u\sin v$ we obtain $\sin u\sin v = \frac{1}{2}[\cos(u - v) - \cos(u + v)]$. Letting $u = \frac{1}{2}(\gamma - \omega)t$ and $v = \frac{1}{2}(\gamma + \omega)t$, the result follows.

(b) If $\epsilon = \frac{1}{2}(\gamma - \omega)$ then $\gamma \approx \omega$ so that $x = (F_0/2\epsilon\gamma)\sin\epsilon t\sin\gamma t$.

47. (a) The general solution of the homogeneous equation is

$$\begin{aligned}x_c(t) &= c_1 e^{-\lambda t}\cos\left(\sqrt{\omega^2 - \lambda^2}\,t\right) + c_2 e^{-\lambda t}\sin\left(\sqrt{\omega^2 - \lambda^2}\,t\right)\\ &= Ae^{-\lambda t}\sin\left[\sqrt{\omega^2 - \lambda^2}\,t + \phi\right],\end{aligned}$$

where $A = \sqrt{c_1^2 + c_2^2}$, $\sin\phi = c_1/A$, and $\cos\phi = c_2/A$. Now

$$x_p(t) = \frac{F_0(\omega^2 - \gamma^2)}{(\omega^2 - \gamma^2)^2 + 4\lambda^2\gamma^2}\sin\gamma t + \frac{F_0(-2\lambda\gamma)}{(\omega^2 - \gamma^2)^2 + 4\lambda^2\gamma^2}\cos\gamma t = A\sin(\gamma t + \theta),$$

where

$$\sin\theta = \frac{\dfrac{F_0(-2\lambda\gamma)}{(\omega^2 - \gamma^2)^2 + 4\lambda^2\gamma^2}}{\dfrac{F_0}{\sqrt{\omega^2 - \gamma^2 + 4\lambda^2\gamma^2}}} = \frac{-2\lambda\gamma}{\sqrt{(\omega^2 - \gamma^2)^2 + 4\lambda^2\gamma^2}}$$

and

$$\cos\theta = \frac{\dfrac{F_0(\omega^2 - \gamma^2)}{(\omega^2 - \gamma^2)^2 + 4\lambda^2\gamma^2}}{\dfrac{F_0}{\sqrt{(\omega^2 - \gamma^2)^2 + 4\lambda^2\gamma^2}}} = \frac{\omega^2 - \gamma^2}{\sqrt{(\omega^2 - \gamma^2)^2 + 4\lambda^2\gamma^2}}.$$

(b) If $g'(\gamma) = 0$ then $\gamma\left(\gamma^2 + 2\lambda^2 - \omega^2\right) = 0$ so that $\gamma = 0$ or $\gamma = \sqrt{\omega^2 - 2\lambda^2}$. The first derivative test shows that g has a maximum value at $\gamma = \sqrt{\omega^2 - 2\lambda^2}$. The maximum value of g is

$$g\left(\sqrt{\omega^2 - 2\lambda^2}\right) = F_0 / 2\lambda\sqrt{\omega^2 - \lambda^2}.$$

(c) We identify $\omega^2 = k/m = 4$, $\lambda = \beta/2$, and $\gamma_1 = \sqrt{\omega^2 - 2\lambda^2} = \sqrt{4 - \beta^2/2}$. As $\beta \to 0$, $\gamma_1 \to 2$ and the resonance curve grows without bound at $\gamma_1 = 2$. That is, the system approaches pure resonance.

β	$\gamma1$	g
2.00	1.41	0.58
1.00	1.87	1.03
0.75	1.93	1.36
0.50	1.97	2.02
0.25	1.99	4.01

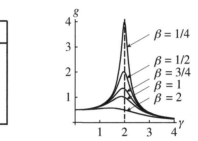

49. Solving $\frac{1}{20}q'' + 2q' + 100q = 0$ we obtain $q(t) = e^{-20t}(c_1 \cos 40t + c_2 \sin 40t)$. The initial conditions $q(0) = 5$ and $q'(0) = 0$ imply $c_1 = 5$ and $c_2 = 5/2$. Thus

$$q(t) = e^{-20t}\left(5\cos 40t + \frac{5}{2}\sin 40t\right) = \frac{5\sqrt{5}}{2}e^{-20t}\sin\left(40t + 1.1071\right)$$

and $q(0.01) \approx 4.5676$ coulombs. The charge is zero for the first time when $40t + 1.1071 = \pi$ or $t \approx 0.0509$ second.

51. Solving $\frac{5}{3}q'' + 10q' + 30q = 300$ we obtain $q(t) = e^{-3t}(c_1 \cos 3t + c_2 \sin 3t) + 10$. The initial conditions $q(0) = q'(0) = 0$ imply $c_1 = c_2 = -10$. Thus

$$q(t) = 10 - 10e^{-3t}(\cos 3t + \sin 3t) \quad \text{and} \quad i(t) = 60e^{-3t}\sin 3t.$$

Solving $i(t) = 0$ we see that the maximum charge occurs when $t = \pi/3$ and $q(\pi/3) \approx 10.432$.

53. Solving $q'' + 2q' + 4q = 0$ we obtain $q_c = e^{-t}\left(\cos\sqrt{3}\,t + \sin\sqrt{3}\,t\right)$. The steady-state charge has the form $q_p = A\cos t + B\sin t$. Substituting into the differential equation we find

$$(3A + 2B)\cos t + (3B - 2A)\sin t = 50\cos t.$$

Thus, $A = 150/13$ and $B = 100/13$. The steady-state charge is

$$q_p(t) = \frac{150}{13}\cos t + \frac{100}{13}\sin t$$

and the steady-state current is

$$i_p(t) = -\frac{150}{13}\sin t + \frac{100}{13}\cos t.$$

55. The differential equation is $\frac{1}{2}q'' + 20q' + 1000q = 100\sin 60t$. To use Example 10 in the text we identify $E_0 = 100$ and $\gamma = 60$. Then

$$X = L\gamma - \frac{1}{c\gamma} = \frac{1}{2}(60) - \frac{1}{0.001(60)} \approx 13.3333,$$

$$Z = \sqrt{X^2 + R^2} = \sqrt{X^2 + 400} \approx 24.0370,$$

and

$$\frac{E_0}{Z} = \frac{100}{Z} \approx 4.1603.$$

From Problem 54, then

$$i_p(t) \approx 4.1603\sin(60t + \phi)$$

where $\sin\phi = -X/Z$ and $\cos\phi = R/Z$. Thus $\tan\phi = -X/R \approx -0.6667$ and ϕ is a fourth quadrant angle. Now $\phi \approx -0.5880$ and

$$i_p(t) = 4.1603\sin(60t - 0.5880).$$

57. Solving $\frac{1}{2}q'' + 10q' + 100q = 150$ we obtain $q(t) = e^{-10t}(c_1\cos 10t + c_2\sin 10t) + 3/2$. The initial conditions $q(0) = 1$ and $q'(0) = 0$ imply $c_1 = c_2 = -1/2$. Thus

$$q(t) = -\frac{1}{2}e^{-10t}(\cos 10t + \sin 10t) + \frac{3}{2}.$$

As $t \to \infty$, $q(t) \to 3/2$.

59. By Problem 54 the amplitude of the steady-state current is E_0/Z, where $Z = \sqrt{X^2 + R^2}$ and $X = L\gamma - 1/C\gamma$. Since E_0 is constant the amplitude will be a maximum when Z is a minimum. Since R is constant, Z will be a minimum when $X = 0$. Solving $L\gamma - 1/C\gamma = 0$ for C we obtain $C = 1/L\gamma^2$.

61. In an LC-series circuit there is no resistor, so the differential equation is

$$L\frac{d^2q}{dt^2} + \frac{1}{C}q = E(t).$$

Then $q(t) = c_1\cos\left(t/\sqrt{LC}\right) + c_2\sin\left(t/\sqrt{LC}\right) + q_p(t)$ where $q_p(t) = A\sin\gamma t + B\cos\gamma t$. Substituting $q_p(t)$ into the differential equation we find

$$\left(\frac{1}{C} - L\gamma^2\right)A\sin\gamma t + \left(\frac{1}{C} - L\gamma^2\right)B\cos\gamma t = E_0\cos\gamma t.$$

Equating coefficients we obtain $A = 0$ and $B = E_0C/(1 - LC\gamma^2)$. Thus, the charge is

$$q(t) = c_1\cos\frac{1}{\sqrt{LC}}t + c_2\sin\frac{1}{\sqrt{LC}}t + \frac{E_0C}{1 - LC\gamma^2}\cos\gamma t.$$

The initial conditions $q(0) = q_0$ and $q'(0) = i_0$ imply $c_1 = q_0 - E_0 C/(1 - LC\gamma^2)$ and $c_2 = i_0 \sqrt{LC}$. The current is $i(t) = q'(t)$ or

$$i(t) = -\frac{c_1}{\sqrt{LC}} \sin \frac{1}{\sqrt{LC}} t + \frac{c_2}{\sqrt{LC}} \cos \frac{1}{\sqrt{LC}} t - \frac{E_0 C \gamma}{1 - LC\gamma^2} \sin \gamma t$$

$$= i_0 \cos \frac{1}{\sqrt{LC}} t - \frac{1}{\sqrt{LC}} \left(q_0 - \frac{E_0 C}{1 - LC\gamma^2} \right) \sin \frac{1}{\sqrt{LC}} t - \frac{E_0 C \gamma}{1 - LC\gamma^2} \sin \gamma t.$$

5.2   Linear Models: Boundary-Value Problems

1. (a) The general solution is

$$y(x) = c_1 + c_2 x + c_3 x^2 + c_4 x^3 + \frac{w_0}{24EI} x^4.$$

The boundary conditions are $y(0) = 0$, $y'(0) = 0$, $y''(L) = 0$, $y'''(L) = 0$. The first two conditions give $c_1 = 0$ and $c_2 = 0$. The conditions at $x = L$ give the system

$$2c_3 + 6c_4 L + \frac{w_0}{2EI} L^2 = 0 \qquad 6c_4 + \frac{w_0}{EI} L = 0.$$

Solving, we obtain $c_3 = w_0 L^2/4EI$ and $c_4 = -w_0 L/6EI$. The deflection is

$$y(x) = \frac{w_0}{24EI} (6L^2 x^2 - 4Lx^3 + x^4).$$

(b)

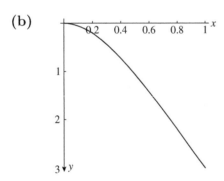

3. (a) The general solution is

$$y(x) = c_1 + c_2 x + c_3 x^2 + c_4 x^3 + \frac{w_0}{24EI} x^4.$$

The boundary conditions are $y(0) = 0$, $y'(0) = 0$, $y(L) = 0$, $y''(L) = 0$. The first two conditions give $c_1 = 0$ and $c_2 = 0$. The conditions at $x = L$ give the system

$$c_3 L^2 + c_4 L^3 + \frac{w_0}{24EI} L^4 = 0$$

$$2c_3 + 6c_4 L + \frac{w_0}{2EI} L^2 = 0.$$

Solving, we obtain $c_3 = w_0 L^2/16EI$ and $c_4 = -5w_0 L/48EI$. The deflection is

$$y(x) = \frac{w_0}{48EI} (3L^2 x^2 - 5Lx^3 + 2x^4).$$

(b)

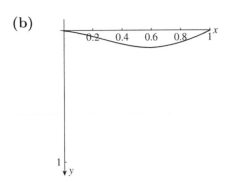

5. (a) The general solution is

$$y(x) = c_1 + c_2 x + c_3 x^2 + c_4 x^3 + \frac{w_0}{120EI} x^5.$$

The boundary conditions are $y(0) = 0$, $y''(0) = 0$, $y(L) = 0$, $y''(L) = 0$. The first two conditions give $c_1 = 0$ and $c_3 = 0$. The conditions at $x = L$ give the system

$$c_2 L + c_4 L^3 + \frac{w_0}{120EI} L^5 = 0$$

$$6c_4 L + \frac{w_0}{6EI} L^3 = 0.$$

Solving, we obtain $c_2 = 7w_0 L^4/360EI$ and $c_4 = -w_0 L^2/36EI$. The deflection is

$$y(x) = \frac{w_0}{360EI}(7L^4 x - 10L^2 x^3 + 3x^5).$$

(b)

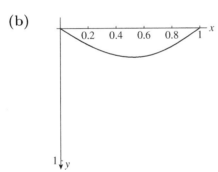

(c) Using a CAS we find the maximum deflection to be 0.234799 when $x = 0.51933$.

7. The general solution of the differential equation is

$$y = c_1 \cosh \sqrt{\frac{P}{EI}}\, x + c_2 \sinh \sqrt{\frac{P}{EI}}\, x + \frac{w_0}{2P} x^2 + \frac{w_0 EI}{P^2}.$$

Setting $y(0) = 0$ we obtain $c_1 = -w_0 EI/P^2$, so that

$$y = -\frac{w_0 EI}{P^2} \cosh \sqrt{\frac{P}{EI}}\, x + c_2 \sinh \sqrt{\frac{P}{EI}}\, x + \frac{w_0}{2P} x^2 + \frac{w_0 EI}{P^2}.$$

Setting $y'(L) = 0$ we find

$$c_2 = \left(\sqrt{\frac{P}{EI}} \frac{w_0 EI}{P^2} \sinh \sqrt{\frac{P}{EI}}\, L - \frac{w_0 L}{P} \right) \Big/ \sqrt{\frac{P}{EI}} \cosh \sqrt{\frac{P}{EI}}\, L.$$

9. This is Example 2 in the text with $L = \pi$. The eigenvalues are $\lambda_n = n^2\pi^2/\pi^2 = n^2$, $n = 1, 2, 3, \ldots$ and the corresponding eigenfunctions are $y_n = \sin(n\pi x/\pi) = \sin nx$, $n = 1, 2, 3, \ldots$.

11. For $\lambda \leq 0$ the only solution of the boundary-value problem is $y = 0$. For $\lambda = \alpha^2 > 0$ we have

$$y = c_1 \cos \alpha x + c_2 \sin \alpha x.$$

Now

$$y'(x) = -c_1\alpha \sin \alpha x + c_2\alpha \cos \alpha x$$

and $y'(0) = 0$ implies $c_2 = 0$, so

$$y(L) = c_1 \cos \alpha L = 0$$

gives

$$\alpha L = \frac{(2n-1)\pi}{2} \quad \text{or} \quad \lambda = \alpha^2 = \frac{(2n-1)^2\pi^2}{4L^2}, \ n = 1, 2, 3, \ldots.$$

The eigenvalues $(2n-1)^2\pi^2/4L^2$ correspond to the eigenfunctions $\cos \dfrac{(2n-1)\pi}{2L}x$ for $n = 1, 2, 3, \ldots$.

13. For $\lambda = -\alpha^2 < 0$ the only solution of the boundary-value problem is $y = 0$. For $\lambda = 0$ we have $y = c_1 x + c_2$. Now $y' = c_1$ and $y'(0) = 0$ implies $c_1 = 0$. Then $y = c_2$ and $y'(\pi) = 0$. Thus, $\lambda = 0$ is an eigenvalue with corresponding eigenfunction $y = 1$. For $\lambda = \alpha^2 > 0$ we have

$$y = c_1 \cos \alpha x + c_2 \sin \alpha x.$$

Now

$$y'(x) = -c_1\alpha \sin \alpha x + c_2\alpha \cos \alpha x$$

and $y'(0) = 0$ implies $c_2 = 0$, so

$$y'(\pi) = -c_1\alpha \sin \alpha\pi = 0$$

gives

$$\alpha\pi = n\pi \quad \text{or} \quad \lambda = \alpha^2 = n^2, \ n = 1, 2, 3, \ldots.$$

The eigenvalues n^2 correspond to the eigenfunctions $\cos nx$ for $n = 0, 1, 2, \ldots$.

15. The auxiliary equation has solutions

$$m = \frac{1}{2}\left(-2 \pm \sqrt{4 - 4(\lambda + 1)}\right) = -1 \pm \alpha.$$

For $\lambda = -\alpha^2 < 0$ we have

$$y = e^{-x}\left(c_1 \cosh \alpha x + c_2 \sinh \alpha x\right).$$

The boundary conditions imply

$$y(0) = c_1 = 0 \qquad\qquad y(5) = c_2 e^{-5} \sinh 5\alpha = 0$$

so $c_1 = c_2 = 0$ and the only solution of the boundary-value problem is $y = 0$.

For $\lambda = 0$ we have

$$y = c_1 e^{-x} + c_2 x e^{-x}$$

and the only solution of the boundary-value problem is $y = 0$.

For $\lambda = \alpha^2 > 0$ we have

$$y = e^{-x}\left(c_1 \cos \alpha x + c_2 \sin \alpha x\right).$$

Now $y(0) = 0$ implies $c_1 = 0$, so

$$y(5) = c_2 e^{-5} \sin 5\alpha = 0$$

gives

$$5\alpha = n\pi \quad \text{or} \quad \lambda = \alpha^2 = \frac{n^2\pi^2}{25}, \quad n = 1, 2, 3, \dots.$$

The eigenvalues $\lambda_n = \dfrac{n^2\pi^2}{25}$ correspond to the eigenfunctions $y_n = e^{-x}\sin\dfrac{n\pi}{5}x$ for $n = 1, 2, 3, \dots$.

17. For $\lambda = \alpha^2 > 0$ a general solution of the given differential equation is

$$y = c_1 \cos\left(\alpha \ln x\right) + c_2 \sin\left(\alpha \ln x\right).$$

Since $\ln 1 = 0$, the boundary condition $y(1) = 0$ implies $c_1 = 0$. Therefore

$$y = c_2 \sin\left(\alpha \ln x\right).$$

Using $\ln e^{\pi} = \pi$ we find that $y\left(e^{\pi}\right) = 0$ implies

$$c_2 \sin \alpha\pi = 0$$

or $\alpha\pi = n\pi$, $n = 1, 2, 3, \dots$. The eigenvalues and eigenfunctions are, in turn,

$$\lambda_n = \alpha^2 = n^2, \quad n = 1,\, 2,\, 3,\, \dots \quad \text{and} \quad y_n = \sin\left(n \ln x\right).$$

For $\lambda \le 0$ the only solution of the boundary-value problem is $y = 0$.

19. For $\lambda = 0$ the general solution is $y = c_1 + c_2 \ln x$. Now $y' = c_2/x$, so $y'(1) = c_2 = 0$ implies $c_2 = 0$. Then $y = c_1$ and $y'(e^2) = 0$ is satisfied for any c_1.

For $\lambda = -\alpha^2 < 0$, $y = c_1 x^{-\alpha} + c_2 x^{\alpha}$. The boundary conditions give $c_1 = c_2 e^{4\alpha}$ and $c_1 = 0$, so that $c_2 = 0$ and $y(x) = 0$.

For $\lambda = \alpha^2 > 0$, $y = c_1 \cos(\alpha \ln x) + c_2 \sin(\alpha \ln x)$. From $y'(1) = 0$ we obtain $c_2 = 0$ and $y = c_1 \cos(\alpha \ln x)$. Now $y' = -c_1(\alpha/x)\sin(\alpha \ln x)$, so $y'(e^2) = -c_1(\alpha/e^2)\sin 2\alpha = 0$ implies $\sin 2\alpha = 0$ or $\alpha = n\pi/2$ and $\lambda_n = \alpha^2 = n^2\pi^2/4$ for $n = 0, 1, 2, \ldots$. The corresponding eigenfunctions are

$$y_n = \cos\left(\frac{n\pi}{2}\ln x\right).$$

Note $n = 0$ in y_n gives $y = 1$.

21. For $\lambda = \alpha^4$, $\alpha > 0$, the general solution of the boundary-value problem

$$y^{(4)} - \lambda y = 0, \quad y(0) = 0, \ y''(0) = 0, \ y(1) = 0, \ y''(1) = 0$$

is

$$y = c_1 \cos \alpha x + c_2 \sin \alpha x + c_3 \cosh \alpha x + c_4 \sinh \alpha x.$$

The boundary conditions $y(0) = 0$, $y''(0) = 0$ give $c_1 + c_3 = 0$ and $-c_1\alpha^2 + c_3\alpha^2 = 0$, from which we conclude $c_1 = c_3 = 0$. Thus, $y = c_2 \sin \alpha x + c_4 \sinh \alpha x$. The boundary conditions $y(1) = 0$, $y''(1) = 0$ then give

$$c_2 \sin \alpha + c_4 \sinh \alpha = 0$$
$$-c_2\alpha^2 \sin \alpha + c_4\alpha^2 \sinh \alpha = 0.$$

In order to have nonzero solutions of this system, we must have the determinant of the coefficients equal zero, that is,

$$\begin{vmatrix} \sin \alpha & \sinh \alpha \\ -\alpha^2 \sin \alpha & \alpha^2 \sinh \alpha \end{vmatrix} = 0 \quad \text{or} \quad 2\alpha^2 \sinh \alpha \sin \alpha = 0$$

But since $\alpha > 0$, the only way that this is satisfied is to have $\sin \alpha = 0$ or $\alpha = n\pi$. The system is then satisfied by choosing $c_2 \neq 0$, $c_4 = 0$, and $\alpha = n\pi$. The eigenvalues and corresponding eigenfunctions are then

$$\lambda_n = \alpha^4 = (n\pi)^4, \ n = 1, 2, 3, \ldots \quad \text{and} \quad y_n = \sin n\pi x.$$

23. If restraints are put on the column at $x = L/4$, $x = L/2$, and $x = 3L/4$, then the critical load will be P_4.

25. If $\lambda = \alpha^2 = P/EI$, then the solution of the differential equation is

$$y = c_1 \cos \alpha x + c_2 \sin \alpha x + c_3 x + c_4.$$

The conditions $y(0) = 0$, $y''(0) = 0$ yield, in turn, $c_1 + c_4 = 0$ and $c_1 = 0$. With $c_1 = 0$ and $c_4 = 0$ the solution is $y = c_2 \sin \alpha x + c_3 x$. The conditions $y(L) = 0$, $y''(L) = 0$, then yield

$$c_2 \sin \alpha L + c_3 L = 0 \qquad \text{and} \qquad -c_2 \alpha^2 \sin \alpha L = 0.$$

Hence, nontrivial solutions of the problem exist only if $\sin \alpha L = 0$. From this point on, the analysis is the same as in Example 3 in the text.

27. The general solution is

$$y = c_1 \cos \sqrt{\frac{\rho}{T}} \, \omega x + c_2 \sin \sqrt{\frac{\rho}{T}} \, \omega x.$$

From $y(0) = 0$ we obtain $c_1 = 0$. Setting $y(L) = 0$ we find $\sqrt{\rho/T} \, \omega L = n\pi$, $n = 1, 2, 3, \ldots$. Thus, critical speeds are $\omega_n = n\pi \sqrt{T}/L\sqrt{\rho}$, $n = 1, 2, 3, \ldots$. The corresponding deflection curves are

$$y(x) = c_2 \sin \frac{n\pi}{L} x, \qquad n = 1, 2, 3, \ldots,$$

where $c_2 \neq 0$.

29. The auxiliary equation is $m^2 + m = m(m+1) = 0$ so that $u(r) = c_1 r^{-1} + c_2$. The boundary conditions $u(a) = u_0$ and $u(b) = u_1$ yield the system $c_1 a^{-1} + c_2 = u_0$, $c_1 b^{-1} + c_2 = u_1$. Solving gives

$$c_1 = \left(\frac{u_0 - u_1}{b - a} \right) ab \quad \text{and} \quad c_2 = \frac{u_1 b - u_0 a}{b - a}.$$

Thus

$$u(r) = \left(\frac{u_0 - u_1}{b - a} \right) \frac{ab}{r} + \frac{u_1 b - u_0 a}{b - a}.$$

31. (a) This is similar to Problem 21 with 1 replaced by L. Thus the eigenvalues and eigenfunctions are, in turn,

$$\lambda_n = \alpha_n^4 = \frac{n^4 \pi^4}{L^4}, \qquad n = 1, 2, 3, \ldots \qquad \text{and} \qquad y_n(x) = \sin\left(\frac{n\pi x}{L} \right)$$

(b)

$$\lambda_n = \alpha_n^4 = \rho \omega_n^2 / EI = n^4 \pi^4 / L^4$$

$$\omega_n^2 = \frac{n^4 \pi^4}{L^4} \frac{EI}{\rho}.$$

Therefore the critical speeds are

$$\omega_n = \frac{n^2 \pi^2}{L^2} \sqrt{\frac{EI}{\rho}}, \qquad n = 1, 2, 3, \ldots .$$

Then the fundamental critical speed is

$$\omega_1 = \frac{\pi^2}{L^2}\sqrt{\frac{EI}{\rho}}.$$

Noe that $\omega_n = n^2\omega_1$.

33. The solution of the initial-value problem

$$x'' + \omega^2 x = 0, \quad x(0) = 0, \ x'(0) = v_0, \ \omega^2 = 10/m$$

is $x(t) = (v_0/\omega)\sin\omega t$. To satisfy the additional boundary condition $x(1) = 0$ we require that $\omega = n\pi$, $n = 1, 2, 3, \ldots$. The eigenvalues $\lambda = \omega^2 = n^2\pi^2$ and eigenfunctions of the problem are then $x(t) = (v_0/n\pi)\sin n\pi t$. Using $\omega^2 = 10/m$ we find that the *only* masses that can pass through the equilibrium position at $t = 1$ are $m_n = 10/n^2\pi^2$. Note for $n = 1$, the heaviest mass $m_1 = 10/\pi^2$ will *not* pass through the equilibrium position on the interval $0 < t < 1$ (the period of $x(t) = (v_0/\pi)\sin\pi t$ is $T = 2$, so on $0 \le t \le 1$ its graph passes through $x = 0$ only at $t = 0$ and $t = 1$). Whereas for $n > 1$, masses of lighter weight will pass through the equilibrium position $n - 1$ times prior to passing through at $t = 1$. For example, if $n = 2$, the period of $x(t) = (v_0/2\pi)\sin 2\pi t$ is $2\pi/2\pi = 1$, the mass will pass through $x = 0$ only *once* ($t = \frac{1}{2}$) prior to $t = 1$; if $n = 3$, the period of $x(t) = (v_0/3\pi)\sin 3\pi t$ is $\frac{2}{3}$, the mass will pass through $x = 0$ *twice* ($t = \frac{1}{3}$ and $t = \frac{2}{3}$) prior to $t = 1$; and so on.

35. (a) The general solution of the differential equation is $y = c_1\cos 4x + c_2\sin 4x$. From $y_0 = y(0) = c_1$ we see that $y = y_0\cos 4x + c_2\sin 4x$. From $y_1 = y(\pi/2) = y_0$ we see that any solution must satisfy $y_0 = y_1$. We also see that when $y_0 = y_1$, $y = y_0\cos 4x + c_2\sin 4x$ is a solution of the boundary-value problem for any choice of c_2. Thus, the boundary-value problem does not have a unique solution for any choice of y_0 and y_1.

(b) Whenever $y_0 = y_1$ there are infinitely many solutions.

(c) When $y_0 \ne y_1$ there are no solutions.

(d) The boundary-value problem has the trivial solution when $y_0 = y_1 = 0$. The solution is not unique.

37. (a) A solution curve has the same y-coordinate at both ends of the interval $[-\pi, \pi]$ and the tangent lines at the endpoints of the interval are parallel.

(b) For $\lambda = 0$ the solution of $y'' = 0$ is $y = c_1 x + c_2$. From the first boundary condition we have

$$y(-\pi) = -c_1\pi + c_2 = y(\pi) = c_1\pi + c_2$$

or $2c_1\pi = 0$. Thus, $c_1 = 0$ and $y = c_2$. This constant solution is seen to satisfy the boundary-value problem.

For $\lambda = -\alpha^2 < 0$ we have $y = c_1 \cosh \alpha x + c_2 \sinh \alpha x$. In this case the first boundary condition gives

$$y(-\pi) = c_1 \cosh(-\alpha\pi) + c_2 \sinh(-\alpha\pi)$$
$$= c_1 \cosh \alpha\pi - c_2 \sinh \alpha\pi$$
$$= y(\pi) = c_1 \cosh \alpha\pi + c_2 \sinh \alpha\pi$$

or $2c_2 \sinh \alpha\pi = 0$. Thus $c_2 = 0$ and $y = c_1 \cosh \alpha x$. The second boundary condition implies in a similar fashion that $c_1 = 0$. Thus, for $\lambda < 0$, the only solution of the boundary-value problem is $y = 0$.

For $\lambda = \alpha^2 > 0$ we have $y = c_1 \cos \alpha x + c_2 \sin \alpha x$. The first boundary condition implies

$$y(-\pi) = c_1 \cos(-\alpha\pi) + c_2 \sin(-\alpha\pi)$$
$$= c_1 \cos \alpha\pi - c_2 \sin \alpha\pi$$
$$= y(\pi) = c_1 \cos \alpha\pi + c_2 \sin \alpha\pi$$

or $2c_2 \sin \alpha\pi = 0$. Similarly, the second boundary condition implies $2c_1\alpha \sin \alpha\pi = 0$. If $c_1 = c_2 = 0$ the solution is $y = 0$. However, if $c_1 \neq 0$ or $c_2 \neq 0$, then $\sin \alpha\pi = 0$, which implies that α must be an integer, n. Therefore, for c_1 and c_2 not both 0, $y = c_1 \cos nx + c_2 \sin nx$ is a nontrivial solution of the boundary-value problem. Since $\cos(-nx) = \cos nx$ and $\sin(-nx) = -\sin nx$, we may assume without loss of generality that the eigenvalues are $\lambda_n = \alpha^2 = n^2$, for n a positive integer. The corresponding eigenfunctions are $y_n = \cos nx$ and $y_n = \sin nx$.

(c)

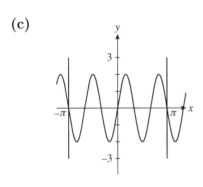

$y = 2\sin 3x$

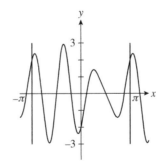

$y = \sin 4x - 2\cos 3x$

39. We see from the graph that $\tan x = -x$ has infinitely many roots. Since $\lambda_n = \alpha_n^2$, there are no new eigenvalues when $\alpha_n < 0$. For $\lambda = 0$, the differential equation $y'' = 0$ has general solution $y = c_1 x + c_2$. The boundary conditions imply $c_1 = c_2 = 0$, so $y = 0$.

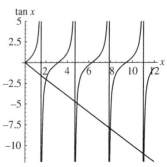

41. In the case when $\lambda = -\alpha^2 < 0$, the solution of the differential equation is $y = c_1 \cosh \alpha x + c_2 \sinh \alpha x$. The condition $y(0) = 0$ gives $c_1 = 0$. The condition $y(1) - \frac{1}{2}y'(1) = 0$ applied to $y = c_2 \sinh \alpha x$ gives $c_2(\sinh \alpha - \frac{1}{2}\alpha \cosh \alpha) = 0$ or $\tanh \alpha = \frac{1}{2}\alpha$. As can be seen from the figure, the graphs of $y = \tanh x$ and $y = \frac{1}{2}x$ intersect at a single point with approximate x-coordinate

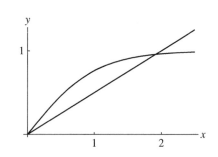

$\alpha_1 = 1.915$. Thus, there is a single negative eigenvalue $\lambda_1 = -\alpha_1^2 \approx -3.667$ and the corresponding eigenfuntion is $y_1 = \sinh 1.915x$. For $\lambda = 0$ the only solution of the boundary-value problem is $y = 0$. For $\lambda = \alpha^2 > 0$ the solution of the differential equation is $y = c_1 \cos \alpha x + c_2 \sin \alpha x$. The condition $y(0) = 0$ gives $c_1 = 0$, so $y = c_2 \sin \alpha x$. The condition $y(1) - \frac{1}{2}y'(1) = 0$ gives $c_2(\sin \alpha - \frac{1}{2}\alpha \cos \alpha) = 0$, so the eigenvalues are $\lambda_n = \alpha_n^2$ when α_n, $n = 2, 3, 4, \ldots$, are the positive roots of $\tan \alpha = \frac{1}{2}\alpha$. Using a CAS we find that the first three values of α are $\alpha_2 = 4.27487$, $\alpha_3 = 7.59655$, and $\alpha_4 = 10.8127$. The first three eigenvalues are then $\lambda_2 = \alpha_2^2 = 18.2738$, $\lambda_3 = \alpha_3^2 = 57.7075$, and $\lambda_4 = \alpha_4^2 = 116.9139$ with corresponding eigenfunctions $y_2 = \sin 4.27487x$, $y_3 = \sin 7.59655x$, and $y_4 = \sin 10.8127x$.

5.3   Nonlinear Models

1. The period corresponding to $x(0) = 1$, $x'(0) = 1$ is approximately 5.6. The period corresponding to $x(0) = 1/2$, $x'(0) = -1$ is approximately 6.2.

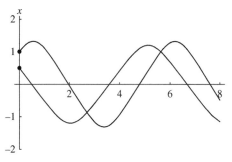

3. The period corresponding to $x(0) = 1$, $x'(0) = 1$ is approximately 5.8. The second initial-value problem does not have a periodic solution.

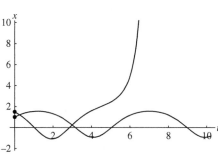

5. From the graph we see that $|x_1| \approx 1.2$.

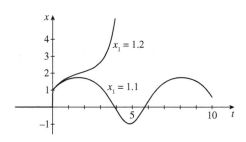

7. Since

$$xe^{0.01x} = x[1 + 0.01x + \frac{1}{2!}(0.01x)^2 + \cdots] \approx x$$

for small values of x, a linearization is $\dfrac{d^2x}{dt^2} + x = 0$.

9. This is a damped hard spring, so x will approach 0 as t approaches ∞.

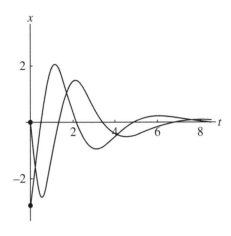

11.

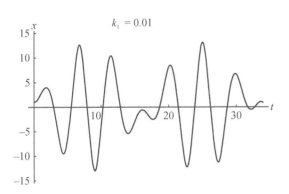

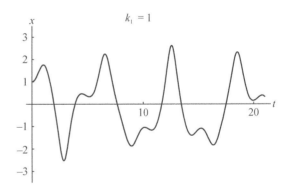

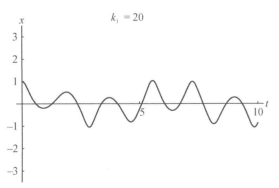

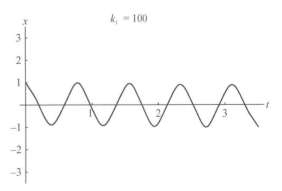

When k_1 is very small the effect of the nonlinearity is greatly diminished, and the system is close to pure resonance.

13. For $\lambda^2 - \omega^2 > 0$ we choose $\lambda = 2$ and $\omega = 1$ with $\theta(0) = 1$ and $\theta'(0) = 2$. For $\lambda^2 - \omega^2 < 0$ we choose $\lambda = 1/3$ and $\omega = 1$ with $\theta(0) = -2$ and $\theta'(0) = 4$. In both cases the motion corresponds to the overdamped and underdamped cases for spring/mass systems.

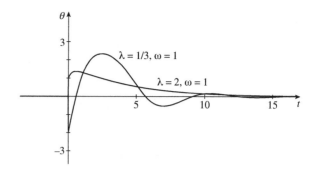

15. (a) Intuitively, one might expect that only half of a 10-pound chain could be lifted by a 5-pound vertical force.

(b) Since $x = 0$ when $t = 0$, and $v = dx/dt = \sqrt{160 - 64x/3}$, we have $v(0) = \sqrt{160} \approx 12.65$ ft/s.

(c) Since x should always be positive, we solve $x(t) = 0$, getting $t = 0$ and $t = \frac{3}{2}\sqrt{5/2} \approx 2.3717$. Since the graph of $x(t)$ is a parabola, the maximum value occurs at $t_m = \frac{3}{4}\sqrt{5/2}$. (This can also be obtained by solving $x'(t) = 0$.) At this time the height of the chain is $x(t_m) \approx 7.5$ ft. This is higher than predicted because of the momentum generated by the force. When the chain is 5 feet high it still has a positive velocity of about 7.3 ft/s, which keeps it going higher for a while.

(d) As discussed in the solution to part **(c)** of this problem, the chain has momentum generated by the force applied to it that will cause it to go higher than expected. It will then fall back to below the expected maximum height, again due to momentum. This, in turn, will cause it to next go higher than expected, and so on.

17. (a) Let (x, y) be the coordinates of S_2 on the curve C. The slope at (x, y) is then

$$\frac{dy}{dx} = \frac{v_1 t - y}{0 - x} = \frac{y - v_1 t}{x} \quad \text{or} \quad xy' - y = -v_1 t.$$

(b) Differentiating with respect to x and using $r = v_1/v_2$ gives

$$xy'' + y' - y' = -v_1 \frac{dt}{dx}$$

$$xy'' = -v_1 \frac{dt}{ds} \frac{ds}{dx}$$

$$xy'' = -v_1 \frac{1}{v_2}\left(-\sqrt{1 + (y')^2}\right)$$

$$xy'' = r\sqrt{1 + (y')^2}.$$

Letting $u = y'$ and separating variables, we obtain

$$x \frac{du}{dx} = r\sqrt{1 + u^2}$$

$$\frac{du}{\sqrt{1 + u^2}} = \frac{r}{x}\, dx$$

$$\sinh^{-1} u = r \ln x + \ln c = \ln(cx^r)$$

$$u = \sinh(\ln cx^r)$$

$$\frac{dy}{dx} = \frac{1}{2}\left(cx^r - \frac{1}{cx^r}\right).$$

At $t = 0$, $dy/dx = 0$ and $x = a$, so $0 = ca^r - 1/ca^r$. Thus $c = 1/a^r$ and

$$\frac{dy}{dx} = \frac{1}{2}\left[\left(\frac{x}{a}\right)^r - \left(\frac{a}{x}\right)^r\right] = \frac{1}{2}\left[\left(\frac{x}{a}\right)^r - \left(\frac{x}{a}\right)^{-r}\right].$$

If $r > 1$ or $r < 1$, integrating gives

$$y = \frac{a}{2}\left[\frac{1}{1+r}\left(\frac{x}{a}\right)^{1+r} - \frac{1}{1-r}\left(\frac{x}{a}\right)^{1-r}\right] + c_1.$$

When $t = 0$, $y = 0$ and $x = a$, so $0 = (a/2)[1/(1+r) - 1/(1-r)] + c_1$. Thus $c_1 = ar/(1-r^2)$ and

$$y = \frac{a}{2}\left[\frac{1}{1+r}\left(\frac{x}{a}\right)^{1+r} - \frac{1}{1-r}\left(\frac{x}{a}\right)^{1-r}\right] + \frac{ar}{1-r^2}.$$

If $r = 1$, then integration gives

$$y = \frac{1}{2}\left[\frac{x^2}{2a} - \frac{1}{a}\ln x\right] + c_2.$$

When $t = 0$, $y = 0$ and $x = a$, so $0 = (1/2)[a/2 - (1/a)\ln a] + c_2$. Thus $c_2 = -(1/2)[a/2 - (1/a)\ln a]$ and

$$y = \frac{1}{2}\left[\frac{x^2}{2a} - \frac{1}{a}\ln x\right] - \frac{1}{2}\left[\frac{a}{2} - \frac{1}{a}\ln a\right] = \frac{1}{2}\left[\frac{1}{2a}(x^2 - a^2) + \frac{1}{a}\ln\frac{a}{x}\right].$$

(c) To see if the paths ever intersect we first note that if $r > 1$, then $v_1 > v_2$ and $y \to \infty$ as $x \to 0^+$. In other words, S_2 always lags behind S_1. Next, if $r < 1$, then $v_1 < v_2$ and $y = ar/(1 - r^2)$ when $x = 0$. In other words, when the submarine's speed is greater than the ship's, their paths will intersect at the point $(0, ar/(1 - r^2))$. Finally, if $r = 1$, then $y \to \infty$ as $x \to 0^+$, meaning S_2 will never catch up with S_1.

19. (a) The auxiliary equation is $m^2 + g/l = 0$, so the general solution of the differential equation is

$$\theta(t) = c_1 \cos\sqrt{\frac{g}{l}}\, t + c_2 \sin\sqrt{\frac{g}{l}}\, t.$$

The initial condtion $\theta(0) = 0$ implies $c_1 = 0$ and $\theta'(0) = \omega_0$ implies $c_2 = \omega_0\sqrt{l/g}$. Thus,

$$\theta(t) = \omega_0\sqrt{\frac{l}{g}}\, \sin\sqrt{\frac{g}{l}}\, t.$$

(b) At $\theta_{\max}$, $\sin\sqrt{g/l}\,t = 1$, so

$$\theta_{\max} = \omega_0\sqrt{\frac{l}{g}} = \frac{m_b}{m_w + m_b}\frac{v_b}{l}\sqrt{\frac{l}{g}} = \frac{m_b}{m_w + m_b}\frac{v_b}{\sqrt{lg}}$$

and

$$v_b = \frac{m_w + m_b}{m_b}\sqrt{lg}\,\theta_{\max}.$$

(c) We have $\cos\theta_{\max} = (l - h)/l = 1 - h/l$. Then

$$\cos\theta_{\max} \approx 1 - \frac{1}{2}\theta_{\max}^2 = 1 - \frac{h}{l}$$

and

$$\theta_{\max}^2 = \frac{2h}{l} \qquad \text{or} \qquad \theta_{\max} = \sqrt{\frac{2h}{l}}.$$

Thus

$$v_b = \frac{m_w + m_b}{m_b}\sqrt{lg}\sqrt{\frac{2h}{l}} = \frac{m_w + m_b}{m_b}\sqrt{2gh}.$$

(d) When $m_b = 5$ g, $m_w = 1$ kg, and $h = 6$ cm, we have

$$v_b = \frac{1005}{5}\sqrt{2\,(980)\,(6)} \approx 21,797 \text{ cm/s}.$$

21. Since $(dx/dt)^2$ is always positive, it is necessary to use $|dx/dt|\,(dx/dt)$ in order to account for the fact that the motion is oscillatory and the velocity (or its square) should be negative when the spring is contracting.

23. (a) Write the differential equation as

$$\frac{d^2\theta}{dt^2} + \omega^2\sin\theta = 0,$$

where $\omega^2 = g/l$. To determine the differences between the Earth and the Moon we take $l = 3$, $\theta(0) = 1$, and $\theta'(0) = 2$. Using $g = 32$ on the Earth and $g = 32 \times 0.165$ on the Moon we obtain the graphs shown in the figure.

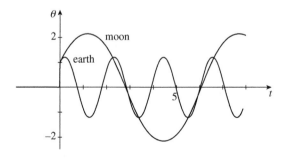

(b) Comparing the apparent periods of the graphs, we see that the pendulum oscillates faster on the Earth than on the Moon and the amplitude is greater on the Moon than on the Earth.

25. (a) The general solution of

$$\frac{d^2\theta}{dt^2} + \theta = 0$$

is $\theta(t) = c_1 \cos t + c_2 \sin t$. From $\theta(0) = \pi/12$ and $\theta'(0) = -1/3$ we find

$$\theta(t) = \left(\frac{\pi}{12}\right)\cos t - \left(\frac{1}{3}\right)\sin t.$$

Setting $\theta(t) = 0$ we have $\tan t = \pi/4$ which implies $t_1 = \tan^{-1}(\pi/4) \approx 0.66577$.

(b) We set $\theta(t) = \theta(0) + \theta'(0)t + \frac{1}{2}\theta''(0)t^2 + \frac{1}{6}\theta'''(0)t^3 + \cdots$ and use $\theta''(t) = -\sin\theta(t)$ together with $\theta(0) = \pi/12$ and $\theta'(0) = -1/3$. Then

$$\theta''(0) = -\sin\left(\frac{\pi}{12}\right) = -\sqrt{2}\,\frac{(\sqrt{3}-1)}{4}$$

and

$$\theta'''(0) = -\cos\theta(0)\cdot\theta'(0) = -\cos\left(\frac{\pi}{12}\right)\left(-\frac{1}{3}\right) = \sqrt{2}\,\frac{\sqrt{3}+1}{12}.$$

Thus

$$\theta(t) = \frac{\pi}{12} - \frac{1}{3}t - \frac{\sqrt{2}\,(\sqrt{3}-1)}{8}t^2 + \frac{\sqrt{2}\,(\sqrt{3}+1)}{72}t^3 + \cdots.$$

(c) Setting $\pi/12 - t/3 = 0$ we obtain $t_1 = \pi/4 \approx 0.785398$.

(d) Setting

$$\frac{\pi}{12} - \frac{1}{3}t - \frac{\sqrt{2}\,(\sqrt{3}-1)}{8}t^2 = 0$$

and using the positive root we obtain $t_1 \approx 0.63088$.

(e) Setting

$$\frac{\pi}{12} - \frac{1}{3}t - \frac{\sqrt{2}\,(\sqrt{3}-1)}{8}t^2 + \frac{\sqrt{2}\,(\sqrt{3}+1)}{72}t^3 = 0$$

we find with the help of a CAS that $t_1 \approx 0.661973$ is the first positive root.

(f) From the output we see that $y(t)$ is an interpolating function on the interval $0 \le t \le 5$, whose graph is shown. The positive root of $y(t) = 0$ near $t = 1$ is $t_1 = 0.666404$.

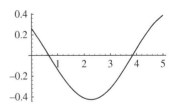

(f) To find the next two positive roots we change the interval used in **NDSolve** and **Plot** from $\{t, 0, 5\}$ to $\{t, 0, 10\}$. We see from the graph that the second and third positive roots are near 4 and 7, respectively. Replacing $\{t, 1\}$ in **FindRoot** with $\{t, 4\}$ and then $\{t, 7\}$ we obtain $t_2 = 3.84411$ and $t_3 = 7.0218$.

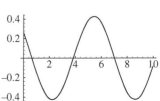

Chapter 5 in Review

1. 8 ft, since $k = 4$

3. 5/4 m, since $x = -\cos 4t + \frac{3}{4}\sin 4t$

5. False; since an external force may exist

7. overdamped

9. $y = 0$ because $\lambda = 8$ is not an eigenvalue

11. The period of a spring/mass system is given by $T = 2\pi/\omega$ where $\omega^2 = k/m = kg/W$, where k is the spring constant, W is the weight of the mass attached to the spring, and g is the acceleration due to gravity. Thus, the period of oscillation is $T = (2\pi/\sqrt{kg})\sqrt{W}$. If the weight of the original mass is W, then $(2\pi/\sqrt{kg})\sqrt{W} = 3$ and $(2\pi/\sqrt{kg})\sqrt{W - 8} = 2$. Dividing, we get $\sqrt{W}/\sqrt{W - 8} = 3/2$ or $W = \frac{9}{4}(W - 8)$. Solving for W we find that the weight of the original mass was 14.4 pounds.

13. We assume that the spring is initially compressed by 4 inches and that the positive direction on the x-axis is in the direction of elongation of the spring. Then, from $\frac{1}{4}x'' + \frac{3}{2}x' + 2x = 0$, $x(0) = -1/3$, and $x'(0) = 0$ we obtain $x = -\frac{2}{3}e^{-2t} + \frac{1}{3}e^{-4t}$.

15. From $mx'' + 4x' + 2x = 0$ we see that nonoscillatory motion results if $16 - 8m \geq 0$ or $0 < m \leq 2$.

17. Writing $\frac{1}{8}x'' + \frac{8}{3}x = \cos\gamma t + \sin\gamma t$ in the form $x'' + \frac{64}{3}x = 8\cos\gamma t + 8\sin\gamma t$ we identify $\omega^2 = \frac{64}{3}$. The system is in a state of pure resonance when $\gamma = \omega = \sqrt{64/3} = 8/\sqrt{3}$.

19. From $\frac{1}{8}x'' + x' + 3x = e^{-t}$, $x(0) = 2$, and $x'(0) = 0$ we obtain

$$x_c = e^{-4t}\left(c_1\cos 2\sqrt{2}\,t + c_2\sin 2\sqrt{2}\,t\right), \quad x_p = \frac{8}{17}e^{-t}, \quad \text{and}$$

$$x = e^{-4t}\left(\frac{26}{17}\cos 2\sqrt{2}\,t + \frac{28\sqrt{2}}{17}\sin 2\sqrt{2}\,t\right) + \frac{8}{17}e^{-t}.$$

21. From $q'' + 10^4 q = 100\sin 50t$, $q(0) = 0$, and $q'(0) = 0$ we obtain $q_c = c_1\cos 100t + c_2\sin 100t$, $q_p = \frac{1}{75}\sin 50t$, and

(a) $q(t) = -\frac{1}{150}\sin 100t + \frac{1}{75}\sin 50t$,

(b) $i(t) = -\frac{2}{3}\cos 100t + \frac{2}{3}\cos 50t$, and

(c) $q(t) = 0$ when $\sin 50t(1 - \cos 50t) = 0$ or $t = n\pi/50$ for $n = 0, 1, 2, \ldots$.

23. For $\lambda = \alpha^2 > 0$ the general solution is $y = c_1 \cos \alpha x + c_2 \sin \alpha x$. Now

$$y(0) = c_1 \qquad \text{and} \qquad y(2\pi) = c_1 \cos 2\pi\alpha + c_2 \sin 2\pi\alpha,$$

so the condition $y(0) = y(2\pi)$ implies

$$c_1 = c_1 \cos 2\pi\alpha + c_2 \sin 2\pi\alpha$$

which is true when $\alpha = \sqrt{\lambda} = n$ or $\lambda = n^2$ for $n = 1, 2, 3, \ldots$. Since

$$y' = -\alpha c_1 \sin \alpha x + \alpha c_2 \cos \alpha x = -n c_1 \sin nx + n c_2 \cos nx,$$

we see that $y'(0) = nc_2 = y'(2\pi)$ for $n = 1, 2, 3, \ldots$ Thus, the eigenvalues are n^2 for $n = 1, 2, 3, \ldots$, with corresponding eigenfunctions $\cos nx$ and $\sin nx$. When $\lambda = 0$, the general solution is $y = c_1 x + c_2$ and the corresponding eigenfunction is $y = 1$.

For $\lambda = -\alpha^2 < 0$ the general solution is $y = c_1 \cosh \alpha x + c_2 \sinh \alpha x$. In this case $y(0) = c_1$ and $y(2\pi) = c_1 \cosh 2\pi\alpha + c_2 \sinh 2\pi\alpha$, so $y(0) = y(2\pi)$ can only be valid for $\alpha = 0$. Thus, there are no eigenvalues corresponding to $\lambda < 0$.

25. Unlike the derivation given in Section 3.8 in the text, the weight mg of the mass m does not appear in the net force since the spring is not stretched by the weight of the mass when it is in the equilibrium position (i.e., there is no $mg - ks$ term in the net force). The only force acting on the mass when it is in motion is the restoring force of the spring. By Newton's second law,

$$m\frac{d^2 x}{dt^2} = -kx \qquad \text{or} \qquad \frac{d^2 x}{dt^2} + \frac{k}{m}x = 0.$$

27. The force of kinetic friction opposing the motion of the mass in μN, where μ is the coefficient of sliding friction and N is the normal component of the weight. Since friction is a force opposite to the direction of motion and since N is pointed directly downward (it is simply the weight of the mass), Newton's second law gives, for motion to the right ($x' > 0$),

$$m\frac{d^2 x}{dt^2} = -kx - \mu mg,$$

and for motion to the left ($x' < 0$),

$$m\frac{d^2 x}{dt^2} = -kx + \mu mg.$$

Traditionally, these two equations are written as one expression

$$m\frac{d^2 x}{dt^2} + f_k \,\text{sgn}(x') + kx = 0,$$

where $f_k = \mu mg$ and

$$\text{sgn}(x') = \begin{cases} 1, & x' > 0 \\ -1, & x' < 0 \end{cases}$$

29. The given initial-value problem is

$$\frac{d^2\theta}{dt^2} + \frac{g}{l}\sin\theta = 0, \qquad \theta(0) = \pi/6, \quad \theta'(0) = 0$$

Differentiating with respect to time t the differential equation yields

$$\frac{d^2\theta}{dt^2} = -\frac{g}{l}\sin\theta, \quad \frac{d^3\theta}{dt^3} = -\frac{g}{l}\cos\theta\,\frac{d\theta}{dt}, \quad \text{and} \quad \frac{d^4\theta}{dt^4} = -\frac{g}{l}\cos\theta\,\frac{d^2\theta}{dt^2} + \frac{g}{l}\sin\theta\left(\frac{d\theta}{dt}\right)^2, \ldots$$

From the differential equation, the initial conditions, and the foregoing derivatives we see at $t = 0$ that $\sin(\theta(0)) = \sin\pi/6 = 1/2$ and $\cos(\theta(0)) = \cos\pi/6 = \sqrt{3}/2$ and so:

$$\theta(0) = \frac{\pi}{6}, \quad \theta'(0) = 0, \quad \theta''(0) = -\frac{g}{2l}, \quad \theta'''(0) = 0, \quad \theta^{(4)}(0) = \frac{\sqrt{3}\,g^2}{4l^2}, \ldots$$

The Maclaurin expansion

$$\theta(t) = \theta(0) + \theta'(0)t + \frac{\theta''(0)}{2!}t^2 + \frac{\theta'''(0)}{3!}t^3 + \frac{\theta^{(4)}(0)}{4!}t^4 + \cdots$$

is the

$$\theta(t) = \frac{\pi}{6} - \frac{g}{4l}t^2 + \frac{\sqrt{3}\,g^2}{96l^2}t^4 + \cdots .$$

31. (a) The linear initial-value problem is

$$\frac{d^2\theta}{dt^2} + \frac{g}{l}\theta_1 = 0, \qquad \theta_1(0) = \theta_0,\ \theta_1'(0) = 0.$$

Applying the initial conditions to $\theta_1(t) = c_1\cos\sqrt{g/l}\,t + c_2\sin\sqrt{g/l}\,t$ gives $c_1 = \theta_0$ and $c_2 = 0$. Therefore the solution is

$$\theta_1(t) = \theta_0\cos\sqrt{g/l}\,t.$$

The string hits the nail when $\theta_1(t_1) = 0$ or $\cos\sqrt{g/l}\,t_1 = 0$. Thus

$$\sqrt{\frac{g}{l}}\,t_1 = \frac{\pi}{2}$$

$$t_1 = \frac{\pi}{2}\sqrt{\frac{l}{g}} \text{ seconds}$$

The interval over which the solution $\theta_1(t)$ is defined is $\left[0, \frac{\pi}{2}\sqrt{\frac{l}{g}}\right]$.

(b) Now $\theta_2(t_1) = \theta_1(t_1) = 0$. Also, the initial angular speed when the string hits the nail satisfies $\theta_2'(t_1) = \theta_1'(t_1)$. Using t_1 found in part (a)

$$\theta_1'(t_1) = -\theta_0\sqrt{\frac{g}{l}}\sin\left(\sqrt{\frac{g}{l}}\,\frac{\pi}{2}\sqrt{\frac{l}{g}}\right) = -\theta_0\sqrt{\frac{g}{l}}\sin\frac{\pi}{2} = -\theta_0\sqrt{\frac{g}{l}}$$

Since the length of the string is $l/4$ the initial-value problem for $\theta_2(t)$ is

$$\frac{d^2}{dt^2} + \frac{g}{l/4}\theta_2 = 0, \quad \theta_2(T) = 0, \quad \theta_2'(T) = -\theta_0\sqrt{\frac{g}{l}}$$

Applying the initial condition $\theta_2(t_1) = 0$ to $\theta_2(t) = c_3\cos 2\sqrt{g/l}\,t + c_2\sin 2\sqrt{g/l}\,t$ gives $c_3 = 0$. Therefore

$$\theta_2(t) = c_4\sin 2\sqrt{\frac{g}{l}}\,t$$

$$\theta_2'(t) = c_4 2\sqrt{\frac{g}{l}}\cos 2\sqrt{\frac{g}{l}}\,t$$

and so

$$\theta_2'(t_1) = c_4 2\sqrt{\frac{g}{l}}\cos\overbrace{\left(2\sqrt{\frac{g}{l}}\,\frac{\pi}{2}\sqrt{\frac{l}{g}}\right)}^{\pi} = -2c_4\sqrt{\frac{g}{l}} = -\theta_0\sqrt{\frac{g}{l}}$$

So $c_4 = \theta_0/2$ and

$$\theta_2(t) = \frac{1}{2}\theta_0\sin 2\sqrt{\frac{g}{l}}\,t.$$

At the time t_2 we have $\theta_2(t) = 0$ for the second time:

$$\theta(t_2) = \frac{1}{2}\theta_0\sin 2\sqrt{\frac{g}{l}}\,t_2 = 0$$

$$2\sqrt{\frac{g}{l}}\,t_2 = 2\pi$$

$$t_2 = \pi\sqrt{\frac{l}{g}}$$

The interval over which $\theta_2(t)$ is defined is $\left[\dfrac{\pi}{2}\sqrt{\dfrac{l}{g}}, \pi\sqrt{\dfrac{l}{g}}\right]$.

Chapter 6

Series Solutions of Linear Equations

6.1 | Review of Power Series

1. $\displaystyle\lim_{n\to\infty}\frac{a_{n+1}}{a_n} = \lim_{n\to\infty}\left|\frac{x^{n+1}/(n+1)}{x^n/n}\right| = \lim_{n\to\infty}\frac{n}{n+1}|x| = |x|$

The series is absolutely convergent on $(-1,1)$. At $x = -1$, the series $\displaystyle\sum_{n=1}^{\infty}\frac{1}{n}$ is the harmonic series which diverges. At $x = 1$, the series $\displaystyle\sum_{n=1}^{\infty}\frac{(-1)^n}{n}$ converges by the alternating series test. Thus, the given series converges on $(-1,1]$, and the radius of converegence is 1.

3. $\displaystyle\lim_{n\to\infty}\frac{a_{n+1}}{a_n} = \lim_{n\to\infty}\left|\frac{2^{n+1}x^{n+1}/(n+1)}{2^n x^n/n}\right| = \lim_{n\to\infty}\frac{2n}{n+1}|x| = 2|x|$

The series is absolutely convergent for $2|x| < 1$ or $|x| < \frac{1}{2}$. The radius of convergence is $R = \frac{1}{2}$. At $x = -\frac{1}{2}$, the series $\displaystyle\sum_{n=1}^{\infty}\frac{(-1)^n}{n}$ converges by the alternating series test. At $x = \frac{1}{2}$, the series $\displaystyle\sum_{n=1}^{\infty}\frac{1}{n}$ is the harmonic series which diverges. Thus, the given series converges on $[-\frac{1}{2}, \frac{1}{2})$, and the radius of convergence is $\frac{1}{2}$.

5. $\displaystyle\lim_{k\to\infty}\frac{a_{k+1}}{a_k} = \lim_{k\to\infty}\left|\frac{(x-5)^{k+1}/10^{k+1}}{(x-5)^k/10^k}\right| = \lim_{k\to\infty}\frac{1}{10}|x-5| = \frac{1}{10}|x-5|$

The series is absolutely convergent for $\frac{1}{10}|x-5| < 1$, $|x-5| < 10$, or on $(-5,15)$. At $x = -5$, the series $\displaystyle\sum_{k=1}^{\infty}(-1)^k\frac{(-10)^n}{10^n} = \sum_{k=1}^{\infty}1$ diverges by the nth term test. At $x = 15$, the series $\displaystyle\sum_{k=1}^{\infty}(-1)^k\frac{10^k}{10^k} = \sum_{k=1}^{\infty}(-1)^k$ diverges by the nth term test. Thus, the series converges on $(-5,15)$, and the radius of convergence is 10.

7. $\displaystyle\lim_{k\to\infty}\frac{a_{k+1}}{a_k}=\lim_{k\to\infty}\left|\frac{(3x-1)^{n+1}/\left[(n+1)^2+(n+1)\right]}{(3x-1)^n/(n^2+n)}\right|=\lim_{k\to\infty}\frac{n^2+n}{n^2+3n+2}|3x-1|=|3x-1|$

The series is absolutely convergent for $|3x-1|<1$ or on $(0,2/3)$. At $x=0$, the series $\displaystyle\sum_{k=1}^{\infty}\frac{(-1)^k}{k^2+k}$ converges by the alternating series test. At $x=2/3$, the series $\displaystyle\sum_{k=1}^{\infty}\frac{1}{k^2+k}$ converges by comparison with the p-series $\displaystyle\sum_{k=1}^{\infty}\frac{1}{k^2}$. Thus, the series converges on $[0,2/3]$, and the radius of convergence is $1/3$.

9. Write the series as $\displaystyle\sum_{k=1}^{\infty}\left(\frac{32}{75}\right)^k x^k$. Then

$$\lim_{k\to\infty}\frac{a_{k+1}}{a_k}=\lim_{k\to\infty}\left|\frac{(32/75)^{k+1}x^{k+1}}{(32/75)^k x^k}\right|=\lim_{k\to\infty}\frac{32}{75}|x|=\frac{32}{75}|x|$$

The series is absolutely convergent for $\frac{32}{75}|x|<1$, or on $(-75/32,75/32)$. At $x=-75/32$, the series $\displaystyle\sum_{k=1}^{\infty}(-1)^k$ diverges by the nth term test. At $x=75/32$, the series $\displaystyle\sum_{k=1}^{\infty}1$ diverges by nth term test. Thus, the series converges on $(-75/32,75/32)$, and the radius of convergence is $75/32$.

11. We replace x by $-x/2$ in the Maclaurin series of e^x.

$$e^{-x/2}=\sum_{n=0}^{\infty}\frac{1}{n!}\left(\frac{-x}{2}\right)^n=\sum_{n=0}^{\infty}\frac{(-1)^n}{n!2^n}x^n$$

13. We factor out a $\dfrac{1}{2}$ and replace x by $-\dfrac{x}{2}$ in the Maclaurin series of $\dfrac{1}{1-x}$.

$$\frac{1}{2+x}=\frac{1}{2}\cdot\frac{1}{1-(-x/2)}=\frac{1}{2}\left[1+\left(-\frac{x}{2}\right)+\left(-\frac{x}{2}\right)^2+\cdots\right]=\frac{1}{2}\left[1-\frac{x}{2}+\frac{x^2}{2^2}-\cdots\right]$$

$$=\frac{1}{2}-\frac{x}{2^2}+\frac{x^2}{2^3}-\cdots=\sum_{n=0}^{\infty}\frac{(-1)^n}{2^{n+1}}x^n$$

15. We replace x by $-x$ in the Maclaurin series of $\ln(1+x)$.

$$\ln(1-x)=-x-\frac{(-x)^2}{2}+\frac{(-x)^3}{3}+\cdots=-x-\frac{x^2}{2}-\frac{x^3}{3}-\cdots=\sum_{n=1}^{\infty}\frac{-1}{n}x^n$$

17. By periodicity $\sin x=\sin\left[(x-2\pi)+2\pi\right]=\sin(x-2\pi)$, so

$$\sin x=\sin(x-2\pi)=\sum_{n=0}^{\infty}\frac{(-1)^n}{(2n+1)!}(x-2\pi)^{2n+1}$$

19. $\displaystyle\sin x\cos x=\left(x-\frac{x^3}{6}+\frac{x^5}{120}-\frac{x^7}{5040}+\cdots\right)\left(1-\frac{x^2}{2}+\frac{x^4}{24}-\frac{x^6}{720}+\cdots\right)$

$$=x-\frac{2x^3}{3}+\frac{2x^5}{15}-\frac{4x^7}{315}+\cdots$$

21. $\sec x = \dfrac{1}{\cos x} = \dfrac{1}{1 - \dfrac{x^2}{2} + \dfrac{x^4}{4!} - \dfrac{x^6}{6!} + \cdots} = 1 + \dfrac{x^2}{2!} + \dfrac{5x^4}{4!} + \dfrac{61x^6}{6!} + \cdots$

Since $\cos(\pi/2) = \cos(-\pi/2) = 0$, the series converges on $(-\pi/2, \pi/2)$.

23. Let $k = n + 2$ so that $n = k - 2$ and

$$\sum_{n=1}^{\infty} nc_n x^{n+2} = \sum_{k=3}^{\infty} (k-2)c_{k-2}x^k.$$

25. In the first summation let $k = n - 1$ so that $n = k + 1$ and

$$\sum_{n=1}^{\infty} nc_n x^{n-1} - \sum_{n=0}^{\infty} c_n x^n = \sum_{k=0}^{\infty} (k+1)c_{k+1}x^k - \sum_{k=0}^{\infty} c_k x^k = \sum_{n=0}^{\infty} [(k+1)c_{k+1} - c_k]x^k.$$

27. In the first summation let $k = n - 1$ and in the second summation let $k = n + 1$. Then $n = k + 1$ in the first summation, $n = k - 1$ in the second summation, and

$$\sum_{n=1}^{\infty} 2nc_n x^{n-1} + \sum_{n=0}^{\infty} 6c_n x^{n+1} = 2 \cdot 1 \cdot c_1 x^0 + \sum_{n=2}^{\infty} 2nc_n x^{n-1} + \sum_{n=0}^{\infty} 6c_n x^{n+1}$$

$$= 2c_1 + \sum_{k=1}^{\infty} 2(k+1)c_{k+1}x^k + \sum_{k=1}^{\infty} 6c_{k-1}x^k$$

$$= 2c_1 + \sum_{k=1}^{\infty} [2(k+1)c_{k+1} + 6c_{k-1}]x^k$$

29. In the first summation let $k = n - 2$ and in the second and third summations let $k = n$. Then $n = k + 2$ in the first summation, $n = k$ in the second and third summations, and

$$\sum_{n=2}^{\infty} n(n-1)c_n x^{n-2} - 2\sum_{n=1}^{\infty} nc_n x^n + \sum_{n=0}^{\infty} c_n x^n$$

$$= \sum_{k=0}^{\infty} (k+2)(k+1)c_{k+2}x^k - \sum_{k=1}^{\infty} 2kc_k x^k + \sum_{k=0}^{\infty} c_k x^k$$

$$= 2c_2 + 6c_3 x + \sum_{k=2}^{\infty} (k+2)(k+1)c_{k+2}x^k + \sum_{k=2}^{\infty} c_{k-2}x^k$$

$$= c_0 + 2c_2 + \sum_{k=1}^{\infty} [(k+2)(k+1)c_{k+2} - (2k-1)c_k]x^k = 0$$

31. Since $y' = \displaystyle\sum_{n=1}^{\infty} \dfrac{(-1)^n 2n}{n!} x^{2n-1}$, we have

$$y' + 2xy = \sum_{n=1}^{\infty} \dfrac{(-1)^n 2n}{n!} x^{2n-1} + 2x \sum_{n=0}^{\infty} \dfrac{(-1)^n}{n!} x^{2n} = 2 \left[\underbrace{\sum_{n=1}^{\infty} \dfrac{(-1)^n}{(n-1)!} x^{2n-1}}_{k=n} + \underbrace{\sum_{n=0}^{\infty} \dfrac{(-1)^n}{n!} x^{2n+1}}_{k=n+1} \right]$$

$$= 2 \left[\sum_{k=1}^{\infty} \dfrac{(-1)^k}{(k-1)!} x^{2k-1} + \sum_{k=1}^{\infty} \dfrac{(-1)^{k-1}}{(k-1)!} x^{2k-1} \right] = 2 \sum_{k=1}^{\infty} \left[\dfrac{(-1)^k}{(k-1)!} + \dfrac{(-1)^{k-1}}{(k-1)!} \right] x^{2k-1}$$

$$= 2 \sum_{k=1}^{\infty} \left[\dfrac{(-1)^k}{(k-1)!} - \dfrac{(-1)^k}{(k-1)!} \right] x^{2k-1} = 0$$

33. In this problem we must take special care with starting values for the indices of summation. Normally when a power series is given in summation notation, successive derivatives of the power series of the unknown function start with an index that is one higher than the preceding one. In this case, the power series starts with $n = 1$ to avoid division by zero. From the first derivative on this is no longer necessary and the index of summation starts again with $n = 1$ for y'. To justify this to yourself you could simply write out the first few term of the power series for y.

Since $y' = \displaystyle\sum_{n=1}^{\infty} (-1)^{n+1} x^{n-1}$ and $y'' = \displaystyle\sum_{n=2}^{\infty} (-1)^{n+1}(n-1) x^{n-2}$

$$(x+1)y'' + y' = (x+1) \sum_{n=2}^{\infty} (-1)^{n+1}(n-1) x^{n-2} + \sum_{n=1}^{\infty} (-1)^{n+1} x^{n-1}$$

$$= \sum_{n=2}^{\infty} (-1)^{n+1}(n-1) x^{n-1} + \sum_{n=2}^{\infty} (-1)^{n+1}(n-1) x^{n-2} + \sum_{n=1}^{\infty} (-1)^{n+1} x^{n-1}$$

$$= -x^0 + x^0 + \underbrace{\sum_{n=2}^{\infty} (-1)^{n+1}(n-1) x^{n-1}}_{k=n-1} + \underbrace{\sum_{n=3}^{\infty} (-1)^{n+1}(n-1) x^{n-2}}_{k=n-2} + \underbrace{\sum_{n=2}^{\infty} (-1)^{n+1} x^{n-1}}_{k=n-1}$$

$$= \sum_{k=1}^{\infty} (-1)^{k+2} k x^k + \sum_{k=1}^{\infty} (-1)^{k+3}(k+1) x^k + \sum_{k=1}^{\infty} (-1)^{k+2} x^k$$

$$= \sum_{k=1}^{\infty} \left[(-1)^{k+2} k - (-1)^{k+2} k - (-1)^{k+2} + (-1)^{k+2} \right] x^k = 0$$

In Problems 35–37 we start with the assumption that $y = \sum\limits_{n=0}^{\infty} c_n x^n$, substitute into the differential equation, and finally find some values of c_n. The solution is then written in terms of elementary functions. (One of the points of power series solutions of differential equations however is that it wont always be possible to express the power series in terms of elementary functions.)

35. Substituting into the differential equation we have

$$y' - 5y = \sum_{n=1}^{\infty} nc_n x^{n-1} - \sum_{n=0}^{\infty} 5c_n x^n = \sum_{k=0}^{\infty} (k+1)c_{k+1}x^k - \sum_{k=0}^{\infty} 5c_k x^k$$

$$= \sum_{k=0}^{\infty} [(k+1)c_{k+1} - 5c_k]\, x^k = 0.$$

Thus $c_{k+1} = \dfrac{5}{k+1}\, c_k$, for $k = 01, 1, 2, \ldots$, and

$$c_1 = \frac{5}{1} c_0 = 5c_0$$

$$c_2 = \frac{5}{2} c_1 = \frac{5^2}{2} c_0$$

$$c_3 = \frac{5}{3} c_2 = \frac{5^3}{3 \cdot 2} c_0$$

$$c_4 = \frac{5}{4} c_3 = \frac{5^4}{4 \cdot 3 \cdot 2} c_0$$

$$\vdots$$

Hence,

$$y = c_0 + 5c_0 x + \frac{5^2}{2} c_0 x^2 + \frac{5^3}{3 \cdot 2} c_0 x^3 + \frac{5^4}{4 \cdot 3 \cdot 2} c_0 x^4 + \cdots$$

and

$$y = c_0 \sum_{k=0}^{\infty} \frac{1}{k!} (5x)^k = c_0 e^{5x}.$$

37. Substituting into the differential equation we have

$$y' - xy = \sum_{n=1}^{\infty} nc_n x^{n-1} - \sum_{n=0}^{\infty} c_n x^{n+1} = \sum_{k=0}^{\infty} (k+1)c_{k+1}x^k - \sum_{k=1}^{\infty} c_{k-1}x^k$$

$$= c_1 + \sum_{k=1}^{\infty} (k+1)c_{k+1}x^k - \sum_{k=1}^{\infty} c_{k+1}x^k$$

$$= c_1 + \sum_{k=1}^{\infty} [(k+1)c_{k+1} - c_{k-1}]\, x^k = 0.$$

Thus $c_1 = 0$ and $c_{k+1} = -\dfrac{1}{4(k+1)}\, c_k$, for $k = 0, 1, 2, \ldots$, and

$$c_2 = \frac{1}{2}\, c_0$$

$$c_3 = \frac{1}{3}\, c_1 = 0$$

$$c_4 = \frac{1}{4}\, c_2 = \frac{1}{4}\left(\frac{1}{2}\, c_0\right) = \frac{1}{2^2 2!}\, c_0$$

$$c_5 = \frac{1}{5}\, c_3 = 0$$

$$c_6 = \frac{1}{6}\, c_4 = \frac{1}{6}\left(\frac{1}{2^2 2!}\, c_0\right) = \frac{1}{2^3 3!}\, c_0$$

$$c_7 = \frac{1}{7}\, c_5 = 0$$

$$c_8 = \frac{1}{8}\, c_6 = \frac{1}{8}\left(\frac{1}{2^3 3!}\, c_0\right) = \frac{1}{2^4 4!}\, c_0$$

$$\vdots$$

Hence,

$$y = c_0 + \frac{1}{2}\, c_0 x^2 + \frac{1}{2^2 2!}\, c_0 x^4 + \frac{1}{2^3 3!}\, c_0 x^6 + \frac{1}{2^4 4!}\, c_0 x^8 + \cdots$$

$$= c_0\left[1 + \left(\frac{x^2}{2}\right) + \frac{1}{2!}\left(\frac{x^2}{2}\right)^2 + \frac{1}{3!}\left(\frac{x^2}{2}\right)^3 + \frac{1}{4!}\left(\frac{x^2}{2}\right)^4 + \cdots\right],$$

and

$$y = c_0 \sum_{k=0}^{\infty} \frac{1}{k!}\left(\frac{x^2}{2}\right)^k = c_0 e^{x^2/2}.$$

39. From the double-angle formula

$$\sin 2x = 2\sin x \cos x \qquad \text{and} \qquad \sin x \cos x = \frac{1}{2}\sin 2x.$$

Therefore we replace x by $2x$ in the Maclaurin series for $\sin x$. This gives

$$\sin x \cos x = \frac{1}{2}\sin 2x = \frac{1}{2}\sum_{n=0}^{\infty} \frac{(-1)^n}{(2n+1)!}(2x)^{2n+1} = \sum_{n=0}^{\infty} \frac{(-4)^n}{(2n+1)!}\, x^{2n+1}$$

$$= x - \frac{2}{3}\, x^3 + \frac{2}{15}\, x^5 - \cdots .$$

6.2 | Solutions About Ordinary Points

1. The singular points of $(x^2 - 25)y'' + 2xy' + y = 0$ are -5 and 5. The distance from 0 to either of these points is 5. The distance from 1 to the closest of these points is 4.

In Problems 3–5 we use

$$y = \sum_{n=0}^{\infty} c_n x^n, \qquad y = \sum_{n=1}^{\infty} n c_n x^{n-1}, \qquad and \quad y'' = \sum_{n=2}^{\infty} n(n-1)c_n x^{n-2}.$$

3. We have

$$y'' + y = \underbrace{\sum_{n=2}^{\infty} n(n-1)c_n x^{n-2}}_{k=n-2} + \underbrace{\sum_{n=0}^{\infty} c_n x^n}_{k=n} = \sum_{k=0}^{\infty} (k+2)(k+1)c_{k+2} x^k + \sum_{k=0}^{\infty} c_k x^k$$

$$= \sum_{k=0}^{\infty} \left[(k+2)(k+1)c_{k+2} + c_k \right] x^k = 0.$$

Thus $c_{k+2} = -\dfrac{c_k}{(k+2)(k+1)}$, for $k = 0, 1, 2, \ldots$, and for $k = 0, 2, 4, 6, \ldots$, we get

$$c_2 = -\frac{c_0}{2!}$$

$$c_4 = \frac{c_0}{4!}$$

$$c_6 = -\frac{c_0}{6!}$$

$$\vdots$$

For $k = 1, 3, 5, 7, \ldots$ we get

$$c_3 = -\frac{c_1}{3!}$$

$$c_5 = \frac{c_1}{5!}$$

$$c_7 = -\frac{c_1}{7!}$$

$$\vdots$$

Hence,

$$y_1(x) = c_0 \left[1 - \frac{1}{2!}x^2 + \frac{1}{4!}x^4 - \frac{1}{6!}x^6 + \cdots \right]$$

and

$$y_2(x) = c_1 \left[x - \frac{1}{3!} x^3 + \frac{1}{5!} x^5 - \frac{1}{7!} x^7 + \cdots \right].$$

The solution $y_1(x)$ is recognized as $y_1(x) = c_0 \cos x$, and the solution $y_2(x)$ is recognized as $y_2(x) = c_1 \sin x$.

5. We have

$$y'' - y' = \underbrace{\sum_{n=2}^{\infty} n(n-1)c_n x^{n-2}}_{k=n-2} - \underbrace{\sum_{n=0}^{\infty} nc_n x^{n-1}}_{k=n-1} = \sum_{k=0}^{\infty} (k+2)(k+1)c_{k+2} x^k - \sum_{k=0}^{\infty} (k+1)c_{k+1} x^k$$

$$= \sum_{k=0}^{\infty} \left[(k+2)(k+1)c_{k+2} - (k+1)c_{k+1} \right] x^k = 0.$$

Thus $c_{k+2} = \dfrac{(k+1)c_{k+1}}{(k+2)(k+1)} = \dfrac{c_{k+1}}{k+2}$, for $k = 0, 1, 2, \ldots$, so

$$c_2 = \frac{c_1}{2!}$$

$$c_3 = \frac{c_2}{3} = \frac{c_1}{3!}$$

$$c_7 = \frac{c_3}{4} = \frac{c_1}{4!}$$

$$\vdots$$

Hence, the solution of the differential equation is

$$y(x) = y_1(x) + y_2(x) = c_0 + c_1 \left[x + \frac{1}{2!} x^2 + \frac{1}{3!} x^3 + \frac{1}{4!} x^4 + \cdots \right]$$

$$= c_0 + c_1 \left[-1 + 1 + x + \frac{1}{2!} x^2 + \frac{1}{3!} x^3 + \frac{1}{4!} x^4 + \cdots \right]$$

$$= c_0 - c_1 + c_1 \sum_{k=0}^{\infty} \frac{1}{k!} x^k.$$

The solutions $y_1(x)$ and $y_2(x)$ are recognized as

$$y_1(x) = c_0 \qquad \text{and} \qquad y_2(x) = -c_1 + c_1 e^x$$

7. Substituting $y = \sum\limits_{n=0}^{\infty} c_n x^n$ into the differential equation we have

$$y'' + xy = \underbrace{\sum_{n=2}^{\infty} n(n-1)c_n x^{n-2}}_{k=n-2} + \underbrace{\sum_{n=0}^{\infty} c_n x^{n+1}}_{k=n+1} = \sum_{k=0}^{\infty}(k+2)(k+1)c_{k+2}x^k + \sum_{k=1}^{\infty} c_{k-1}x^k$$

$$= 2c_2 + \sum_{k=1}^{\infty}[(k+2)(k+1)c_{k+2} + c_{k-1}]x^k = 0.$$

Thus

$$c_2 = 0$$

$$(k+2)(k+1)c_{k+2} + c_{k-1} = 0$$

and

$$c_{k+2} = -\frac{c_{k-1}}{(k+2)(k+1)}, \quad k = 1, 2, 3, \ldots .$$

Choosing $c_0 = 1$ and $c_1 = 0$ we find

$$c_3 = -\frac{1}{6} \qquad\qquad c_4 = c_5 = 0 \qquad\qquad c_6 = \frac{1}{180}$$

and so on. For $c_0 = 0$ and $c_1 = 1$ we obtain

$$c_3 = 0 \qquad c_4 = -\frac{1}{12} \qquad c_5 = c_6 = 0 \qquad c_7 = \frac{1}{504}$$

and so on. Thus, two solutions are

$$y_1 = 1 - \frac{1}{6}x^3 + \frac{1}{180}x^6 - \cdots \qquad \text{and} \qquad y_2 = x - \frac{1}{12}x^4 + \frac{1}{504}x^7 - \cdots .$$

9. Substituting $y = \sum\limits_{n=0}^{\infty} c_n x^n$ into the differential equation we have

$$y'' - 2xy' + y = \underbrace{\sum_{n=2}^{\infty} n(n-1)c_n x^{n-2}}_{k=n-2} - 2\underbrace{\sum_{n=1}^{\infty} nc_n x^n}_{k=n} + \underbrace{\sum_{n=0}^{\infty} c_n x^n}_{k=n}$$

$$= \sum_{k=0}^{\infty}(k+2)(k+1)c_{k+2}x^k - 2\sum_{k=1}^{\infty} kc_k x^k + \sum_{k=0}^{\infty} c_k x^k$$

$$= 2c_2 + c_0 + \sum_{k=1}^{\infty}[(k+2)(k+1)c_{k+2} - (2k-1)c_k]x^k = 0.$$

Thus

$$2c_2 + c_0 = 0$$

$$(k+2)(k+1)c_{k+2} - (2k-1)c_k = 0$$

and

$$c_2 = -\frac{1}{2}c_0$$

$$c_{k+2} = \frac{2k-1}{(k+2)(k+1)}\,c_k, \quad k = 1, 2, 3, \dots .$$

Choosing $c_0 = 1$ and $c_1 = 0$ we find

$$c_2 = -\frac{1}{2} \qquad c_3 = c_5 = c_7 = \cdots = 0 \qquad c_4 = -\frac{1}{8} \qquad c_6 = -\frac{7}{240}$$

and so on. For $c_0 = 0$ and $c_1 = 1$ we obtain

$$c_2 = c_4 = c_6 = \cdots = 0 \qquad c_3 = \frac{1}{6} \qquad c_5 = \frac{1}{24} \qquad c_7 = 1112$$

and so on. Thus, two solutions are

$$y_1 = 1 - \frac{1}{2}x^2 - \frac{1}{8}x^4 - \frac{7}{240}x^6 - \cdots \qquad \text{and} \qquad y_2 = x + \frac{1}{6}x^3 + \frac{1}{24}x^5 + \frac{1}{112}x^7 + \cdots .$$

11. Substituting $y = \displaystyle\sum_{n=0}^{\infty} c_n x^n$ into the differential equation we have

$$y'' + x^2 y' + xy = \underbrace{\sum_{n=2}^{\infty} n(n-1)c_n x^{n-2}}_{k=n-2} + \underbrace{\sum_{n=1}^{\infty} n c_n x^{n+1}}_{k=n+1} + \underbrace{\sum_{n=0}^{\infty} c_n x^{n+1}}_{k=n+1}$$

$$= \sum_{k=0}^{\infty} (k+2)(k+1)c_{k+2} x^k + \sum_{k=2}^{\infty} (k-1)c_{k-1} x^k + \sum_{k=1}^{\infty} c_{k-1} x^k$$

$$= 2c_2 + (6c_3 + c_0)x + \sum_{k=2}^{\infty} [(k+2)(k+1)c_{k+2} + k c_{k-1}]x^k = 0.$$

Thus

$$c_2 = 0$$

$$6c_3 + c_0 = 0$$

$$(k+2)(k+1)c_{k+2} + k c_{k-1} = 0$$

and

$$c_2 = 0$$

$$c_3 = -\frac{1}{6}c_0$$

$$c_{k+2} = -\frac{k}{(k+2)(k+1)}\,c_{k-1}, \quad k = 2, 3, 4, \dots .$$

Choosing $c_0 = 1$ and $c_1 = 0$ we find

$$c_3 = -\frac{1}{6} \qquad\qquad c_4 = c_5 = 0 \qquad\qquad c_6 = \frac{1}{45}$$

and so on. For $c_0 = 0$ and $c_1 = 1$ we obtain

$$c_3 = 0 \qquad c_4 = -\frac{1}{6} \qquad c_5 = c_6 = 0 \qquad c_7 = \frac{5}{252}$$

and so on. Thus, two solutions are

$$y_1 = 1 - \frac{1}{6}x^3 + \frac{1}{45}x^6 - \cdots \qquad \text{and} \qquad y_2 = x - \frac{1}{6}x^4 + \frac{5}{252}x^7 - \cdots .$$

13. Substituting $y = \displaystyle\sum_{n=0}^{\infty} c_n x^n$ into the differential equation we have

$$(x-1)y'' + y' = \underbrace{\sum_{n=2}^{\infty} n(n-1)c_n x^{n-1}}_{k=n-1} - \underbrace{\sum_{n=2}^{\infty} n(n-1)c_n x^{n-2}}_{k=n-2} + \underbrace{\sum_{n=1}^{\infty} n c_n x^{n-1}}_{k=n-1}$$

$$= \sum_{k=1}^{\infty} (k+1)k c_{k+1} x^k - \sum_{k=0}^{\infty} (k+2)(k+1)c_{k+2} x^k + \sum_{k=0}^{\infty} (k+1)c_{k+1} x^k$$

$$= -2c_2 + c_1 + \sum_{k=1}^{\infty} [(k+1)k c_{k+1} - (k+2)(k+1)c_{k+2} + (k+1)c_{k+1}]x^k = 0.$$

Thus

$$-2c_2 + c_1 = 0$$
$$(k+1)^2 c_{k+1} - (k+2)(k+1)c_{k+2} = 0$$

and

$$c_2 = \frac{1}{2}c_1$$

$$c_{k+2} = \frac{k+1}{k+2}c_{k+1}, \quad k = 1, 2, 3, \ldots .$$

Choosing $c_0 = 1$ and $c_1 = 0$ we find $c_2 = c_3 = c_4 = \cdots = 0$. For $c_0 = 0$ and $c_1 = 1$ we obtain

$$c_2 = \frac{1}{2}, \qquad\qquad c_3 = \frac{1}{3}, \qquad\qquad c_4 = \frac{1}{4},$$

and so on. Thus, two solutions are

$$y_1 = 1 \qquad \text{and} \qquad y_2 = x + \frac{1}{2}x^2 + \frac{1}{3}x^3 + \frac{1}{4}x^4 + \cdots .$$

15. Substituting $y = \sum_{n=0}^{\infty} c_n x^n$ into the differential equation we have

$$y'' - (x+1)y' - y = \sum_{n=2}^{\infty} n(n-1)c_n x^{n-2} - \underbrace{\sum_{n=1}^{\infty} nc_n x^n}_{k=n} - \underbrace{\sum_{n=1}^{\infty} nc_n x^{n-1}}_{k=n-1} - \underbrace{\sum_{n=0}^{\infty} c_n x^n}_{k=n}$$

$$\underbrace{\qquad}_{k=n-2}$$

$$= \sum_{k=0}^{\infty} (k+2)(k+1)c_{k+2} x^k - \sum_{k=1}^{\infty} kc_k x^k - \sum_{k=0}^{\infty} (k+1)c_{k+1} x^k - \sum_{k=0}^{\infty} c_k x^k$$

$$= 2c_2 - c_1 - c_0 + \sum_{k=1}^{\infty} [(k+2)(k+1)c_{k+2} - (k+1)c_{k+1} - (k+1)c_k] x^k = 0.$$

Thus

$$2c_2 - c_1 - c_0 = 0$$

$$(k+2)(k+1)c_{k+2} - (k+1)(c_{k+1} + c_k) = 0$$

and

$$c_2 = \frac{c_1 + c_0}{2}$$

$$c_{k+2} = \frac{c_{k+1} + c_k}{k+2}, \quad k = 1, 2, 3, \ldots .$$

Choosing $c_0 = 1$ and $c_1 = 0$ we find

$$c_2 = \frac{1}{2}, \qquad c_3 = \frac{1}{6}, \qquad c_4 = \frac{1}{6},$$

and so on. For $c_0 = 0$ and $c_1 = 1$ we obtain

$$c_2 = \frac{1}{2}, \qquad c_3 = \frac{1}{2}, \qquad c_4 = \frac{1}{4},$$

and so on. Thus, two solutions are

$$y_1 = 1 + \frac{1}{2}x^2 + \frac{1}{6}x^3 + \frac{1}{6}x^4 + \cdots \qquad \text{and} \qquad y_2 = x + \frac{1}{2}x^2 + \frac{1}{2}x^3 + \frac{1}{4}x^4 + \cdots .$$

17. Substituting $y = \sum_{n=0}^{\infty} c_n x^n$ into the differential equation we have

$$\left(x^2 + 2\right) y'' + 3xy' - y = \underbrace{\sum_{n=2}^{\infty} n(n-1)c_n x^n}_{k=n} + 2\underbrace{\sum_{n=2}^{\infty} n(n-1)c_n x^{n-2}}_{k=n-2} + 3\underbrace{\sum_{n=1}^{\infty} nc_n x^n}_{k=n} - \underbrace{\sum_{n=0}^{\infty} c_n x^n}_{k=n}$$

$$= \sum_{k=2}^{\infty} k(k-1)c_k x^k + 2\sum_{k=0}^{\infty} (k+2)(k+1)c_{k+2} x^k + 3\sum_{k=1}^{\infty} kc_k x^k - \sum_{k=0}^{\infty} c_k x^k$$

$$= (4c_2 - c_0) + (12c_3 + 2c_1)x + \sum_{k=2}^{\infty} \left[2(k+2)(k+1)c_{k+2} + \left(k^2 + 2k - 1\right)c_k\right] x^k = 0.$$

Thus

$$4c_2 - c_0 = 0$$

$$12c_3 + 2c_1 = 0$$

$$2(k+2)(k+1)c_{k+2} + \left(k^2 + 2k - 1\right)c_k = 0$$

and

$$c_2 = \frac{1}{4}c_0 \qquad c_3 = -\frac{1}{6}c_1 \qquad c_{k+2} = -\frac{k^2 + 2k - 1}{2(k+2)(k+1)}\,c_k, \quad k = 2, 3, 4, \dots .$$

Choosing $c_0 = 1$ and $c_1 = 0$ we find

$$c_2 = \frac{1}{4} \qquad\qquad c_3 = c_5 = c_7 = \cdots = 0 \qquad\qquad c_4 = -\frac{7}{96}$$

and so on. For $c_0 = 0$ and $c_1 = 1$ we obtain

$$c_2 = c_4 = c_6 = \cdots = 0 \qquad\qquad c_3 = -\frac{1}{6} \qquad\qquad c_5 = \frac{7}{120}$$

and so on. Thus, two solutions are

$$y_1 = 1 + \frac{1}{4}x^2 - \frac{7}{96}x^4 + \cdots \qquad \text{and} \qquad y_2 = x - \frac{1}{6}x^3 + \frac{7}{120}x^5 - \cdots .$$

19. Substituting $y = \displaystyle\sum_{n=0}^{\infty} c_n x^n$ into the differential equation we have

$$(x-1)y'' - xy' + y = \underbrace{\sum_{n=2}^{\infty} n(n-1)c_n x^{n-1}}_{k=n-1} - \underbrace{\sum_{n=2}^{\infty} n(n-1)c_n x^{n-2}}_{k=n-2} - \underbrace{\sum_{n=1}^{\infty} nc_n x^n}_{k=n} + \underbrace{\sum_{n=0}^{\infty} c_n x^n}_{k=n}$$

$$= \sum_{k=1}^{\infty}(k+1)kc_{k+1}x^k - \sum_{k=0}^{\infty}(k+2)(k+1)c_{k+2}x^k - \sum_{k=1}^{\infty} kc_k x^k + \sum_{k=0}^{\infty} c_k x^k$$

$$= -2c_2 + c_0 + \sum_{k=1}^{\infty}[-(k+2)(k+1)c_{k+2} + (k+1)kc_{k+1} - (k-1)c_k]x^k = 0.$$

Thus

$$-2c_2 + c_0 = 0$$

$$-(k+2)(k+1)c_{k+2} + (k-1)kc_{k+1} - (k-1)c_k = 0$$

and

$$c_2 = \frac{1}{2}c_0$$

$$c_{k+2} = \frac{kc_{k+1}}{k+2} - \frac{(k-1)c_k}{(k+2)(k+1)}, \quad k = 1, 2, 3, \dots .$$

Choosing $c_0 = 1$ and $c_1 = 0$ we find

$$c_2 = \frac{1}{2}, \qquad\qquad c_3 = \frac{1}{6}, \qquad\qquad c_4 = 0,$$

and so on. For $c_0 = 0$ and $c_1 = 1$ we obtain $c_2 = c_3 = c_4 = \cdots = 0$. Thus,

$$y = C_1 \left(1 + \frac{1}{2}x^2 + \frac{1}{6}x^3 + \cdots \right) + C_2 x$$

and

$$y' = C_1 \left(x + \frac{1}{2}x^2 + \cdots \right) + C_2.$$

The initial conditions imply $C_1 = -2$ and $C_2 = 6$, so

$$y = -2 \left(1 + \frac{1}{2}x^2 + \frac{1}{6}x^3 + \cdots \right) + 6x = 8x - 2e^x.$$

21. Substituting $y = \displaystyle\sum_{n=0}^{\infty} c_n x^n$ into the differential equation we have

$$y'' - 2xy' + 8y = \underbrace{\sum_{n=2}^{\infty} n(n-1)c_n x^{n-2}}_{k=n-2} - 2\underbrace{\sum_{n=1}^{\infty} n c_n x^n}_{k=n} + 8\underbrace{\sum_{n=0}^{\infty} c_n x^n}_{k=n}$$

$$= \sum_{k=0}^{\infty} (k+2)(k+1)c_{k+2} x^k - 2\sum_{k=1}^{\infty} k c_k x^k + 8\sum_{k=0}^{\infty} c_k x^k$$

$$= 2c_2 + 8c_0 + \sum_{k=1}^{\infty} [(k+2)(k+1)c_{k+2} + (8-2k)c_k]x^k = 0.$$

Thus

$$2c_2 + 8c_0 = 0$$

$$(k+2)(k+1)c_{k+2} + (8-2k)c_k = 0$$

and

$$c_2 = -4c_0 \qquad\qquad c_{k+2} = \frac{2(k-4)}{(k+2)(k+1)} c_k, \quad k = 1, 2, 3, \dots .$$

Choosing $c_0 = 1$ and $c_1 = 0$ we find

$$c_2 = -4 \qquad c_3 = c_5 = c_7 = \cdots = 0 \qquad c_4 = \frac{4}{3} \qquad c_6 = c_8 = c_{10} = \cdots = 0.$$

For $c_0 = 0$ and $c_1 = 1$ we obtain

$$c_2 = c_4 = c_6 = \cdots = 0 \qquad\qquad c_3 = -1 \qquad\qquad c_5 = \frac{1}{10}$$

and so on. Thus,

$$y = C_1\left(1 - 4x^2 + \frac{4}{3}x^4\right) + C_2\left(x - x^3 + \frac{1}{10}x^5 + \cdots\right)$$

and

$$y' = C_1\left(-8x + \frac{16}{3}x^3\right) + C_2\left(1 - 3x^2 + \frac{1}{2}x^4 + \cdots\right).$$

The initial conditions imply $C_1 = 3$ and $C_2 = 0$, so

$$y = 3\left(1 - 4x^2 + \frac{4}{3}x^4\right) = 3 - 12x^2 + 4x^4.$$

23. Substituting $y = \displaystyle\sum_{n=0}^{\infty} c_n x^n$ into the differential equation we have

$$y'' + (\sin x)y = \sum_{n=2}^{\infty} n(n-1)c_n x^{n-2} + \left(x - \frac{1}{6}x^3 + \frac{1}{120}x^5 - \cdots\right)\left(c_0 + c_1 x + c_2 x^2 + \cdots\right)$$

$$= \left[2c_2 + 6c_3 x + 12c_4 x^2 + 20c_5 x^3 + \cdots\right] + \left[c_0 x + c_1 x^2 + \left(c_2 - \frac{1}{6}c_0\right)x^3 + \cdots\right]$$

$$= 2c_2 + (6c_3 + c_0)x + (12c_4 + c_1)x^2 + \left(20c_5 + c_2 - \frac{1}{6}c_0\right)x^3 + \cdots = 0.$$

Thus

$$2c_2 = 0 \qquad 6c_3 + c_0 = 0 \qquad 12c_4 + c_1 = 0 \qquad 20c_5 + c_2 - \frac{1}{6}c_0 = 0$$

and

$$c_2 = 0 \qquad c_3 = -\frac{1}{6}c_0 \qquad c_4 = -\frac{1}{12}c_1 \qquad c_5 = -\frac{1}{20}c_2 + \frac{1}{120}c_0.$$

Choosing $c_0 = 1$ and $c_1 = 0$ we find

$$c_2 = 0, \qquad c_3 = -\frac{1}{6}, \qquad c_4 = 0, \qquad c_5 = \frac{1}{120}$$

and so on. For $c_0 = 0$ and $c_1 = 1$ we obtain

$$c_2 = 0, \qquad c_3 = 0, \qquad c_4 = -\frac{1}{12}, \qquad c_5 = 0$$

and so on. Thus, two solutions are

$$y_1 = 1 - \frac{1}{6}x^3 + \frac{1}{120}x^5 + \cdots \qquad \text{and} \qquad y_2 = x - \frac{1}{12}x^4 + \cdots.$$

25. The singular points of $(\cos x)y'' + y' + 5y = 0$ are odd integer multiples of $\pi/2$. The distance from 0 to either $\pm\pi/2$ is $\pi/2$. The singular point closest to 1 is $\pi/2$. The distance from 1 to the closest singular point is then $\pi/2 - 1$.

27. We identify $P(x) = 0$ and $Q(x) = \sin x/x$. The Taylor series representation for $\sin x/x$ is $1 - x^2/3! + x^4/5! - \cdots$, for $|x| < \infty$. Thus, $Q(x)$ is analytic at $x = 0$ and $x = 0$ is an ordinary point of the differential equation.

29. (a) Substituting $y = \displaystyle\sum_{n=0}^{\infty} c_n x^n$ into the differential equation we have

$$y'' + xy' + y = \sum_{n=2}^{\infty} n(n-1)c_n x^{n-2} + \underbrace{\sum_{n=1}^{\infty} nc_n x^n}_{k=n} + \underbrace{\sum_{n=0}^{\infty} c_n x^n}_{k=n}$$

$$\underbrace{}_{k=n-2}$$

$$= \sum_{k=0}^{\infty} (k+2)(k+1)c_{k+2}x^k + \sum_{k=1}^{\infty} kc_k x^k + \sum_{k=0}^{\infty} c_k x^k$$

$$= (2c_2 + c_0) + \sum_{k=1}^{\infty} \left[(k+2)(k+1)c_{k+2} + (k+1)c_k \right] x^k = 0.$$

Thus

$$2c_2 + c_0 = 0$$

$$(k+2)(k+1)c_{k+2} + (k+1)c_k = 0$$

and

$$c_2 = -\frac{1}{2}c_0 \qquad\qquad c_{k+2} = -\frac{1}{k+2}c_k, \quad k = 1, 2, 3, \ldots .$$

Choosing $c_0 = 1$ and $c_1 = 0$ we find

$$c_2 = -\frac{1}{2}$$

$$c_3 = c_5 = c_7 = \cdots = 0$$

$$c_4 = -\frac{1}{4}\left(-\frac{1}{2}\right) = \frac{1}{2^2 \cdot 2}$$

$$c_6 = -\frac{1}{6}\left(\frac{1}{2^2 \cdot 2}\right) = -\frac{1}{2^3 \cdot 3!}$$

and so on. For $c_0 = 0$ and $c_1 = 1$ we obtain

$$c_2 = c_4 = c_6 = \cdots = 0$$

$$c_3 = -\frac{1}{3} = -\frac{2}{3!}$$

$$c_5 = -\frac{1}{5}\left(-\frac{1}{3}\right) = \frac{1}{5 \cdot 3} = \frac{4 \cdot 2}{5!}$$

$$c_7 = -\frac{1}{7}\left(\frac{4 \cdot 2}{5!}\right) = -\frac{6 \cdot 4 \cdot 2}{7!}$$

and so on. Thus, two solutions are

$$y_1 = \sum_{k=0}^{\infty} \frac{(-1)^k}{2^k \cdot k!} x^{2k} \qquad \text{and} \qquad y_2 = \sum_{k=0}^{\infty} \frac{(-1)^k 2^k k!}{(2k+1)!} x^{2k+1}.$$

(b) For y_1, $S_3 = S_2$ and $S_5 = S_4$, so we plot S_2, S_4, S_6, S_8, and S_{10}.

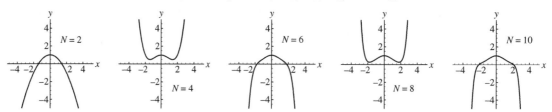

For y_2, $S_3 = S_4$ and $S_5 = S_6$, so we plot S_2, S_4, S_6, S_8, and S_{10}.

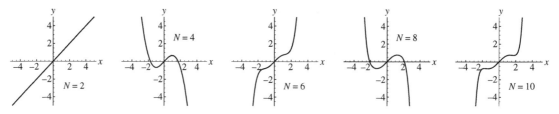

(c)

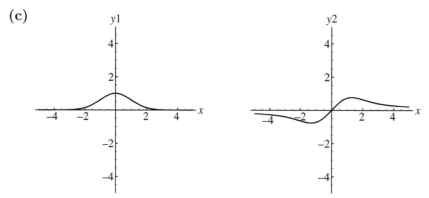

The graphs of y_1 and y_2 obtained from a numerical solver are shown. We see that the partial sum representations indicate the even and odd natures of the solution, but don't really give a very accurate representation of the true solution. Increasing N to about 20 gives a much more accurate representation on $[-4, 4]$.

(d) From $e^x = \displaystyle\sum_{k=0}^{\infty} x^k/k!$ we see that $e^{-x^2/2} = \displaystyle\sum_{k=0}^{\infty} (-x^2/2)^k/k! = \displaystyle\sum_{k=0}^{\infty} (-1)^k x^{2k}/2^k k!$. From (5) of Section 3.2 we have

$$y_2 = y_1 \int \frac{e^{-\int x\,dx}}{y_1^2}\,dx = e^{-x^2/2} \int \frac{e^{-x^2/2}}{(e^{-x^2/2})^2}\,dx = e^{-x^2/2} \int \frac{e^{-x^2/2}}{e^{-x^2}}\,dx = e^{-x^2/2} \int e^{x^2/2}\,dx$$

$$= \sum_{k=0}^{\infty} \frac{(-1)^k}{2^k k!} x^{2k} \int \sum_{k=0}^{\infty} \frac{1}{2^k k!} x^{2k}\,dx = \left(\sum_{k=0}^{\infty} \frac{(-1)^k}{2^k k!} x^{2k} \right) \left(\sum_{k=0}^{\infty} \int \frac{1}{2^k k!} x^{2k}\,dx \right)$$

$$= \left(\sum_{k=0}^{\infty} \frac{(-1)^k}{2^k k!} x^{2k} \right) \left(\sum_{k=0}^{\infty} \frac{1}{(2k+1)2^k k!} x^{2k+1} \right)$$

$$= \left(1 - \frac{1}{2}x^2 + \frac{1}{2^2 \cdot 2}x^4 - \frac{1}{2^3 \cdot 3!}x^6 + \cdots\right)\left(x + \frac{1}{3 \cdot 2}x^3 + \frac{1}{5 \cdot 2^2 \cdot 2}x^5 + \frac{1}{7 \cdot 2^3 \cdot 3!}x^7 + \cdots\right)$$

$$= x - \frac{2}{3!}x^3 + \frac{4 \cdot 2}{5!}x^5 - \frac{6 \cdot 4 \cdot 2}{7!}x^7 + \cdots = \sum_{k=0}^{\infty} \frac{(-1)^k 2^k k!}{(2k+1)!}x^{2k+1}.$$

6.3 │ Solutions About Singular Points

1. Irregular singular point: $x = 0$

3. Irregular singular point: $x = 3$; regular singular point: $x = -3$

5. Regular singular points: $x = 0, \pm 2i$

7. Regular singular points: $x = -3, 2$

9. Irregular singular point: $x = 0$; regular singular points: $x = 2, \pm 5$

11. Writing the differential equation in the form

$$y'' + \frac{5}{x-1}y' + \frac{x}{x+1}y = 0$$

we see that $x_0 = 1$ and $x_0 = -1$ are regular singular points. For $x_0 = 1$ the differential equation can be put in the form

$$(x-1)^2 y'' + 5(x-1)y' + \frac{x(x-1)^2}{x+1}y = 0.$$

In this case $p(x) = 5$ and $q(x) = x(x-1)^2/(x+1)$. For $x_0 = -1$ the differential equation can be put in the form

$$(x+1)^2 y'' + 5(x+1)\frac{x+1}{x-1}y' + x(x+1)y = 0.$$

In this case $p(x) = (x+1)/(x-1)$ and $q(x) = x(x+1)$.

13. We identify $P(x) = 5/3x + 1$ and $Q(x) = -1/3x^2$, so that $p(x) = xP(x) = \frac{5}{3} + x$ and $q(x) = x^2 Q(x) = -\frac{1}{3}$. Then $a_0 = \frac{5}{3}$, $b_0 = -\frac{1}{3}$, and the indicial equation is

$$r(r-1) + \frac{5}{3}r - \frac{1}{3} = r^2 + \frac{2}{3}r - \frac{1}{3} = \frac{1}{3}(3r^2 + 2r - 1) = \frac{1}{3}(3r-1)(r+1) = 0.$$

The indicial roots are $\frac{1}{3}$ and -1. Since these do not differ by an integer we expect to find two series solutions using the method of Frobenius.

15. Substituting $y = \displaystyle\sum_{n=0}^{\infty} c_n x^{n+r}$ into the differential equation and collecting terms, we obtain

$$2xy'' - y' + 2y = \left(2r^2 - 3r\right) c_0 x^{r-1} + \sum_{k=1}^{\infty} \left[2(k+r-1)(k+r)c_k - (k+r)c_k + 2c_{k-1}\right] x^{k+r-1}$$

$$= 0,$$

which implies

$$2r^2 - 3r = r(2r - 3) = 0$$

and

$$(k+r)(2k + 2r - 3)c_k + 2c_{k-1} = 0.$$

The indicial roots are $r = 0$ and $r = 3/2$. For $r = 0$ the recurrence relation is

$$c_k = -\frac{2c_{k-1}}{k(2k-3)}, \quad k = 1, 2, 3, \ldots,$$

and

$$c_1 = 2c_0, \qquad\qquad c_2 = -2c_0, \qquad\qquad c_3 = \frac{4}{9}c_0,$$

and so on. For $r = 3/2$ the recurrence relation is

$$c_k = -\frac{2c_{k-1}}{(2k+3)k}, \quad k = 1, 2, 3, \ldots,$$

and

$$c_1 = -\frac{2}{5}c_0, \qquad\qquad c_2 = \frac{2}{35}c_0, \qquad\qquad c_3 = -\frac{4}{945}c_0,$$

and so on. The general solution on $(0, \infty)$ is

$$y = C_1 \left(1 + 2x - 2x^2 + \frac{4}{9}x^3 + \cdots\right) + C_2 x^{3/2} \left(1 - \frac{2}{5}x + \frac{2}{35}x^2 - \frac{4}{945}x^3 + \cdots\right).$$

17. Substituting $y = \displaystyle\sum_{n=0}^{\infty} c_n x^{n+r}$ into the differential equation and collecting terms, we obtain

$$4xy'' + \frac{1}{2}y' + y = \left(4r^2 - \frac{7}{2}r\right) c_0 x^{r-1} + \sum_{k=1}^{\infty} \left[4(k+r)(k+r-1)c_k + \frac{1}{2}(k+r)c_k + c_{k-1}\right] x^{k+r-1}$$

$$= 0,$$

which implies

$$4r^2 - \frac{7}{2}r = r\left(4r - \frac{7}{2}\right) = 0$$

and

$$\frac{1}{2}(k+r)(8k + 8r - 7)c_k + c_{k-1} = 0.$$

The indicial roots are $r = 0$ and $r = 7/8$. For $r = 0$ the recurrence relation is

$$c_k = -\frac{2c_{k-1}}{k(8k-7)}, \quad k = 1, 2, 3, \ldots,$$

and

$$c_1 = -2c_0, \qquad c_2 = \frac{2}{9}c_0, \qquad c_3 = -\frac{4}{459}c_0,$$

and so on. For $r = 7/8$ the recurrence relation is

$$c_k = -\frac{2c_{k-1}}{(8k+7)k}, \quad k = 1, 2, 3, \ldots,$$

and

$$c_1 = -\frac{2}{15}c_0, \qquad c_2 = 2345c_0, \qquad c_3 = -\frac{4}{32,085}c_0,$$

and so on. The general solution on $(0, \infty)$ is

$$y = C_1\left(1 - 2x + \frac{2}{9}x^2 - \frac{4}{459}x^3 + \cdots\right) + C_2 x^{7/8}\left(1 - \frac{2}{15}x + \frac{2}{345}x^2 - \frac{4}{32,085}x^3 + \cdots\right).$$

19. Substituting $y = \displaystyle\sum_{n=0}^{\infty} c_n x^{n+r}$ into the differential equation and collecting terms, we obtain

$$3xy'' + (2 - x)y' - y = \left(3r^2 - r\right)c_0 x^{r-1}$$

$$+ \sum_{k=1}^{\infty}[3(k+r-1)(k+r)c_k + 2(k+r)c_k - (k+r)c_{k-1}]x^{k+r-1}$$

$$= 0,$$

which implies

$$3r^2 - r = r(3r - 1) = 0$$

and

$$(k+r)(3k+3r-1)c_k - (k+r)c_{k-1} = 0.$$

The indicial roots are $r = 0$ and $r = 1/3$. For $r = 0$ the recurrence relation is

$$c_k = \frac{c_{k-1}}{3k-1}, \quad k = 1, 2, 3, \ldots,$$

and

$$c_1 = \frac{1}{2}c_0, \qquad c_2 = \frac{1}{10}c_0, \qquad c_3 = \frac{1}{80}c_0,$$

and so on. For $r = 1/3$ the recurrence relation is

$$c_k = \frac{c_{k-1}}{3k}, \quad k = 1, 2, 3, \ldots,$$

and

$$c_1 = \frac{1}{3}c_0, \qquad c_2 = \frac{1}{18}c_0, \qquad c_3 = \frac{1}{162}c_0,$$

and so on. The general solution on $(0, \infty)$ is

$$y = C_1 \left(1 + \frac{1}{2}x + \frac{1}{10}x^2 + \frac{1}{80}x^3 + \cdots \right) + C_2 x^{1/3} \left(1 + \frac{1}{3}x + \frac{1}{18}x^2 + \frac{1}{162}x^3 + \cdots \right).$$

21. Substituting $y = \sum_{n=0}^{\infty} c_n x^{n+r}$ into the differential equation and collecting terms, we obtain

$$2xy'' - (3 + 2x)y' + y = \left(2r^2 - 5r\right)c_0 x^{r-1} + \sum_{k=1}^{\infty} [2(k+r)(k+r-1)c_k$$
$$- 3(k+r)c_k - 2(k+r-1)c_{k-1} + c_{k-1}]x^{k+r-1}$$
$$= 0,$$

which implies

$$2r^2 - 5r = r(2r - 5) = 0$$

and

$$(k+r)(2k+2r-5)c_k - (2k+2r-3)c_{k-1} = 0.$$

The indicial roots are $r = 0$ and $r = 5/2$. For $r = 0$ the recurrence relation is

$$c_k = \frac{(2k-3)c_{k-1}}{k(2k-5)}, \quad k = 1, 2, 3, \ldots,$$

and

$$c_1 = \frac{1}{3}c_0, \qquad c_2 = -\frac{1}{6}c_0, \qquad c_3 = -\frac{1}{6}c_0,$$

and so on. For $r = 5/2$ the recurrence relation is

$$c_k = \frac{2(k+1)c_{k-1}}{k(2k+5)}, \quad k = 1, 2, 3, \ldots,$$

and

$$c_1 = \frac{4}{7}c_0, \qquad c_2 = \frac{4}{21}c_0, \qquad c_3 = \frac{32}{693}c_0,$$

and so on. The general solution on $(0, \infty)$ is

$$y = C_1 \left(1 + \frac{1}{3}x - \frac{1}{6}x^2 - \frac{1}{6}x^3 + \cdots \right) + C_2 x^{5/2} \left(1 + \frac{4}{7}x + \frac{4}{21}x^2 + \frac{32}{693}x^3 + \cdots \right).$$

23. Substituting $y = \sum_{n=0}^{\infty} c_n x^{n+r}$ into the differential equation and collecting terms, we obtain

$$9x^2 y'' + 9x^2 y' + 2y = \left(9r^2 - 9r + 2\right)c_0 x^r$$

$$+ \sum_{k=1}^{\infty}[9(k+r)(k+r-1)c_k + 2c_k + 9(k+r-1)c_{k-1}]x^{k+r}$$

$$= 0,$$

which implies

$$9r^2 - 9r + 2 = (3r-1)(3r-2) = 0$$

and

$$[9(k+r)(k+r-1) + 2]c_k + 9(k+r-1)c_{k-1} = 0.$$

The indicial roots are $r = 1/3$ and $r = 2/3$. For $r = 1/3$ the recurrence relation is

$$c_k = -\frac{(3k-2)c_{k-1}}{k(3k-1)}, \quad k = 1, 2, 3, \ldots,$$

and

$$c_1 = -\frac{1}{2}c_0, \qquad c_2 = \frac{1}{5}c_0, \qquad c_3 = -\frac{7}{120}c_0,$$

and so on. For $r = 2/3$ the recurrence relation is

$$c_k = -\frac{(3k-1)c_{k-1}}{k(3k+1)}, \quad k = 1, 2, 3, \ldots,$$

and

$$c_1 = -\frac{1}{2}c_0, \qquad c_2 = \frac{5}{28}c_0, \qquad c_3 = -\frac{1}{21}c_0,$$

and so on. The general solution on $(0, \infty)$ is

$$y = C_1 x^{1/3}\left(1 - \frac{1}{2}x + \frac{1}{5}x^2 - \frac{7}{120}x^3 + \cdots\right) + C_2 x^{2/3}\left(1 - \frac{1}{2}x + \frac{5}{28}x^2 - \frac{1}{21}x^3 + \cdots\right).$$

25. Substituting $y = \sum_{n=0}^{\infty} c_n x^{n+r}$ into the differential equation and collecting terms, we obtain

$$xy'' + 2y' - xy = \left(r^2 + r\right)c_0 x^{r-1} + \left(r^2 + 3r + 2\right)c_1 x^r$$

$$+ \sum_{k=2}^{\infty}[(k+r)(k+r-1)c_k + 2(k+r)c_k - c_{k-2}]x^{k+r-1}$$

$$= 0,$$

which implies

$$r^2 + r = r(r+1) = 0,$$
$$\left(r^2 + 3r + 2\right)c_1 = 0,$$

and
$$(k + r)(k + r + 1)c_k - c_{k-2} = 0.$$

The indicial roots are $r_1 = 0$ and $r_2 = -1$, so $c_1 = 0$. For $r_1 = 0$ the recurrence relation is

$$c_k = \frac{c_{k-2}}{k(k+1)}, \quad k = 2, 3, 4, \ldots,$$

and

$$c_2 = \frac{1}{3!}c_0 \qquad c_3 = c_5 = c_7 = \cdots = 0 \qquad c_4 = \frac{1}{5!}c_0 \qquad c_{2n} = \frac{1}{(2n+1)!}c_0.$$

For $r_2 = -1$ the recurrence relation is

$$c_k = \frac{c_{k-2}}{k(k-1)}, \quad k = 2, 3, 4, \ldots,$$

and

$$c_2 = \frac{1}{2!}c_0 \qquad c_3 = c_5 = c_7 = \cdots = 0 \qquad c_4 = \frac{1}{4!}c_0 \qquad c_{2n} = \frac{1}{(2n)!}c_0.$$

The general solution on $(0, \infty)$ is

$$y = C_1 \sum_{n=0}^{\infty} \frac{1}{(2n+1)!}x^{2n} + C_2 x^{-1} \sum_{n=0}^{\infty} \frac{1}{(2n)!}x^{2n}$$

$$= \frac{1}{x}\left[C_1 \sum_{n=0}^{\infty} \frac{1}{(2n+1)!}x^{2n+1} + C_2 \sum_{n=0}^{\infty} \frac{1}{(2n)!}x^{2n} \right]$$

$$= \frac{1}{x}\left[C_1 \sinh x + C_2 \cosh x \right].$$

27. Substituting $y = \sum_{n=0}^{\infty} c_n x^{n+r}$ into the differential equation and collecting terms, we obtain

$$xy'' - xy' + y = \left(r^2 - r\right)c_0 x^{r-1} + \sum_{k=0}^{\infty}[(k+r+1)(k+r)c_{k+1} - (k+r)c_k + c_k]x^{k+r} = 0$$

which implies

$$r^2 - r = r(r-1) = 0$$

and

$$(k+r+1)(k+r)c_{k+1} - (k+r-1)c_k = 0.$$

The indicial roots are $r_1 = 1$ and $r_2 = 0$. For $r_1 = 1$ the recurrence relation is

$$c_{k+1} = \frac{kc_k}{(k+2)(k+1)}, \quad k = 0, 1, 2, \ldots,$$

and one solution is $y_1 = c_0 x$. A second solution is

$$y_2 = x \int \frac{e^{-\int -1\, dx}}{x^2}\, dx = x \int \frac{e^x}{x^2}\, dx = x \int \frac{1}{x^2} \left(1 + x + \frac{1}{2}x^2 + \frac{1}{3!}x^3 + \cdots \right) dx$$

$$= x \int \left(\frac{1}{x^2} + \frac{1}{x} + \frac{1}{2} + \frac{1}{3!}x + \frac{1}{4!}x^2 + \cdots \right) dx = x \left[-\frac{1}{x} + \ln x + \frac{1}{2}x + \frac{1}{12}x^2 + \frac{1}{72}x^3 + \cdots \right]$$

$$= x \ln x - 1 + \frac{1}{2}x^2 + \frac{1}{12}x^3 + \frac{1}{72}x^4 + \cdots.$$

The general solution on $(0, \infty)$ is

$$y = C_1 x + C_2 y_2(x).$$

29. Substituting $y = \sum_{n=0}^{\infty} c_n x^{n+r}$ into the differential equation and collecting terms, we obtain

$$xy'' + (1-x)y' - y = r^2 c_0 x^{r-1} + \sum_{k=1}^{\infty} [(k+r)(k+r-1)c_k + (k+r)c_k - (k+r)c_{k-1}]x^{k+r-1} = 0,$$

which implies $r^2 = 0$ and

$$(k+r)^2 c_k - (k+r)c_{k-1} = 0.$$

The indicial roots are $r_1 = r_2 = 0$ and the recurrence relation is

$$c_k = \frac{c_{k-1}}{k}, \quad k = 1, 2, 3, \dots.$$

One solution is

$$y_1 = c_0 \left(1 + x + \frac{1}{2}x^2 + \frac{1}{3!}x^3 + \cdots \right) = c_0 e^x.$$

A second solution is

$$y_2 = y_1 \int \frac{e^{-\int (1/x - 1)\, dx}}{e^{2x}}\, dx = e^x \int \frac{e^x/x}{e^{2x}}\, dx = e^x \int \frac{1}{x}e^{-x}\, dx$$

$$= e^x \int \frac{1}{x}\left(1 - x + \frac{1}{2}x^2 - \frac{1}{3!}x^3 + \cdots \right) dx = e^x \int \left(\frac{1}{x} - 1 + \frac{1}{2}x - \frac{1}{3!}x^2 + \cdots \right) dx$$

$$= e^x \left[\ln x - x + \frac{1}{2 \cdot 2}x^2 - \frac{1}{3 \cdot 3!}x^3 + \cdots \right] = e^x \ln x - e^x \sum_{n=1}^{\infty} \frac{(-1)^{n+1}}{n \cdot n!}x^n.$$

The general solution on $(0, \infty)$ is

$$y = C_1 e^x + C_2 e^x \left(\ln x - \sum_{n=1}^{\infty} \frac{(-1)^{n+1}}{n \cdot n!}x^n \right).$$

31. Substituting $y = \sum_{n=0}^{\infty} c_n x^{n+r}$ into the differential equation and collecting terms, we obtain

$$xy'' + (x-6)y' - 3y = (r^2 - 7r)c_0 x^{r-1} + \sum_{k=1}^{\infty} [(k+r)(k+r-1)c_k + (k+r-1)c_{k-1}$$

$$-6(k+r)c_k - 3c_{k-1}]x^{k+r-1} = 0,$$

which implies

$$r^2 - 7r = r(r-7) = 0$$

and

$$(k+r)(k+r-7)c_k + (k+r-4)c_{k-1} = 0.$$

The indicial roots are $r_1 = 7$ and $r_2 = 0$. For $r_1 = 7$ the recurrence relation is

$$(k+7)kc_k + (k+3)c_{k-1} = 0, \quad k = 1, 2, 3, \dots,$$

or

$$c_k = -\frac{k+3}{k(k+7)}c_{k-1}, \quad k = 1, 2, 3, \dots.$$

Taking $c_0 \neq 0$ we obtain

$$c_1 = -\frac{1}{2}c_0 \qquad c_2 = \frac{5}{18}c_0 \qquad c_3 = -\frac{1}{6}c_0,$$

and so on. Thus, the indicial root $r_1 = 7$ yields a single solution. Now, for $r_2 = 0$ the recurrence relation is

$$k(k-7)c_k + (k-4)c_{k-1} = 0, \quad k = 1, 2, 3, \dots.$$

Then

$$-6c_1 - 3c_0 = 0 \qquad\qquad -10c_2 - 2c_1 = 0 \qquad\qquad -12c_3 - c_2 = 0$$

and

$$-12c_4 + 0c_3 = 0 \qquad -10c_5 + c_4 = 0 \qquad -c_6 + 2c_5 = 0 \qquad 0c_7 + 3c_6 = 0$$

$$c_4 = 0 \qquad\qquad c_5 = 0 \qquad\qquad c_6 = 0 \qquad c_7 \text{ is arbitrary}$$

and

$$c_k = -\frac{k-4}{k(k-7)}c_{k-1}, \quad k = 8, 9, 10, \dots.$$

Taking $c_0 \neq 0$ and $c_7 = 0$ we obtain

$$c_1 = -\frac{1}{2}c_0 \qquad c_2 = \frac{1}{10}c_0 \qquad c_3 = -\frac{1}{120}c_0 \qquad c_4 = c_5 = c_6 = \dots = 0.$$

Taking $c_0 = 0$ and $c_7 \neq 0$ we obtain

$$c_1 = c_2 = c_3 = c_4 = c_5 = c_6 = 0 \qquad c_8 = -\frac{1}{2}c_7 \qquad c_9 = \frac{5}{36}c_7 \qquad c_{10} = -\frac{1}{36}c_7,$$

and so on. In this case we obtain the two solutions

$$y_1 = 1 - \frac{1}{2}x + \frac{1}{10}x^2 - \frac{1}{120}x^3 \qquad \text{and} \qquad y_2 = x^7 - \frac{1}{2}x^8 + \frac{5}{36}x^9 - \frac{1}{36}x^{10} + \cdots.$$

33. (a) From $t = 1/x$ we have $dt/dx = -1/x^2 = -t^2$. Then

$$\frac{dy}{dx} = \frac{dy}{dt}\frac{dt}{dx} = -t^2 \frac{dy}{dt}$$

and

$$\frac{d^2y}{dx^2} = \frac{d}{dx}\left(\frac{dy}{dx}\right) = \frac{d}{dx}\left(-t^2\frac{dy}{dt}\right) = -t^2\frac{d^2y}{dt^2}\frac{dt}{dx} - \frac{dy}{dt}\left(2t\frac{dt}{dx}\right) = t^4\frac{d^2y}{dt^2} + 2t^3\frac{dy}{dt}.$$

Now

$$x^4\frac{d^2y}{dx^2} + \lambda y = \frac{1}{t^4}\left(t^4\frac{d^2y}{dt^2} + 2t^3\frac{dy}{dt}\right) + \lambda y = \frac{d^2y}{dt^2} + \frac{2}{t}\frac{dy}{dt} + \lambda y = 0$$

becomes

$$t\frac{d^2y}{dt^2} + 2\frac{dy}{dt} + \lambda t y = 0.$$

(b) Substituting $y = \sum_{n=0}^{\infty} c_n t^{n+r}$ into the differential equation and collecting terms, we obtain

$$t\frac{d^2y}{dt^2} + 2\frac{dy}{dt} + \lambda t y = (r^2 + r)c_0 t^{r-1} + (r^2 + 3r + 2)c_1 t^r$$

$$+ \sum_{k=2}^{\infty}[(k+r)(k+r-1)c_k + 2(k+r)c_k + \lambda c_{k-2}]t^{k+r-1}$$

$$= 0,$$

which implies

$$r^2 + r = r(r+1) = 0,$$
$$\left(r^2 + 3r + 2\right)c_1 = 0,$$

and

$$(k+r)(k+r+1)c_k + \lambda c_{k-2} = 0.$$

The indicial roots are $r_1 = 0$ and $r_2 = -1$, so $c_1 = 0$. For $r_1 = 0$ the recurrence relation is

$$c_k = -\frac{\lambda c_{k-2}}{k(k+1)}, \quad k = 2, 3, 4, \ldots,$$

and

$$c_2 = -\frac{\lambda}{3!}c_0 \qquad c_3 = c_5 = c_7 = \cdots = 0 \qquad c_4 = \frac{\lambda^2}{5!}c_0 \qquad c_{2n} = (-1)^n\frac{\lambda^n}{(2n+1)!}c_0.$$

For $r_2 = -1$ the recurrence relation is

$$c_k = -\frac{\lambda c_{k-2}}{k(k-1)}, \quad k = 2, 3, 4, \ldots,$$

and

$$c_2 = -\frac{\lambda}{2!}c_0 \qquad c_3 = c_5 = c_7 = \cdots = 0 \qquad c_4 = \frac{\lambda^2}{4!}c_0 \qquad c_{2n} = (-1)^n \frac{\lambda^n}{(2n)!}c_0.$$

The general solution on $(0, \infty)$ is

$$y(t) = c_1 \sum_{n=0}^{\infty} \frac{(-1)^n}{(2n+1)!}(\sqrt{\lambda}\,t)^{2n} + c_2 t^{-1} \sum_{n=0}^{\infty} \frac{(-1)^n}{(2n)!}(\sqrt{\lambda}\,t)^{2n}$$

$$= \frac{1}{t}\left[C_1 \sum_{n=0}^{\infty} \frac{(-1)^n}{(2n+1)!}(\sqrt{\lambda}\,t)^{2n+1} + C_2 \sum_{n=0}^{\infty} \frac{(-1)^n}{(2n)!}(\sqrt{\lambda}\,t)^{2n} \right]$$

$$= \frac{1}{t}\left[C_1 \sin\sqrt{\lambda}\,t + C_2 \cos\sqrt{\lambda}\,t \right].$$

(c) Using $t = 1/x$, the solution of the original equation is

$$y(x) = C_1 x \sin\frac{\sqrt{\lambda}}{x} + C_2 x \cos\frac{\sqrt{\lambda}}{x}.$$

35. Express the differential equation in standard form:

$$y''' + P(x)y'' + Q(x)y' + R(x)y = 0.$$

Suppose x_0 is a singular point of the differential equation. Then we say that x_0 is a regular singular point if $(x - x_0)P(x)$, $(x - x_0)^2 Q(x)$, and $(x - x_0)^3 R(x)$ are analytic at $x = x_0$.

37. We write the differential equation in the form $x^2 y'' + (b/a)xy' + (c/a)y = 0$ and identify $a_0 = b/a$ and $b_0 = c/a$ as in (14) in the text. Then the indicial equation is

$$r(r-1) + \frac{b}{a}r + \frac{c}{a} = 0 \qquad \text{or} \qquad ar^2 + (b-a)r + c = 0,$$

which is also the auxiliary equation of $ax^2 y'' + bxy' + cy = 0$.

6.4 | **Special Functions**

1. Since $\nu^2 = 1/9$ the general solution is $y = c_1 J_{1/3}(x) + c_2 J_{-1/3}(x)$.

3. Since $\nu^2 = 25/4$ the general solution is $y = c_1 J_{5/2}(x) + c_2 J_{-5/2}(x)$.

5. Since $\nu^2 = 0$ the general solution is $y = c_1 J_0(x) + c_2 Y_0(x)$.

7. We identify $\alpha = 3$ and $\nu = 2$. Then the general solution is $y = c_1 J_2(3x) + c_2 Y_2(3x)$.

9. We identify $\alpha = 4$ and $\nu = 2/3$. Then the general solution is $y = c_1 I_{2/3}(4x) + c_2 K_{2/3}(4x)$.

11. If $y = x^{-1/2}v(x)$ then

$$y' = x^{-1/2}v'(x) - \frac{1}{2}x^{-3/2}v(x),$$

$$y'' = x^{-1/2}v''(x) - x^{-3/2}v'(x) + \frac{3}{4}x^{-5/2}v(x),$$

and

$$x^2y'' + 2xy' + \alpha^2 x^2 y = x^{3/2}v''(x) + x^{1/2}v'(x) + \left(\alpha^2 x^{3/2} - \frac{1}{4}x^{-1/2}\right)v(x) = 0.$$

Multiplying by $x^{1/2}$ we obtain

$$x^2 v''(x) + xv'(x) + \left(\alpha^2 x^2 - \frac{1}{4}\right)v(x) = 0,$$

whose solution is $v = c_1 J_{1/2}(\alpha x) + c_2 J_{-1/2}(\alpha x)$.
Then $y = c_1 x^{-1/2} J_{1/2}(\alpha x) + c_2 x^{-1/2} J_{-1/2}(\alpha x)$.

13. Write the differential equation in the form $y'' + (2/x)y' + (4/x)y = 0$. This is the form of (18) in the text with $a = -\frac{1}{2}$, $c = \frac{1}{2}$, $b = 4$, and $p = 1$, so, by (19) in the text, the general solution is

$$y = x^{-1/2}[c_1 J_1(4x^{1/2}) + c_2 Y_1(4x^{1/2})].$$

15. Write the differential equation in the form $y'' - (1/x)y' + y = 0$. This is the form of (18) in the text with $a = 1$, $c = 1$, $b = 1$, and $p = 1$, so, by (19) in the text, the general solution is

$$y = x[c_1 J_1(x) + c_2 Y_1(x)].$$

17. Write the differential equation in the form $y'' + (1 - 2/x^2)y = 0$. This is the form of (18) in the text with $a = \frac{1}{2}$, $c = 1$, $b = 1$, and $p = \frac{3}{2}$, so, by (19) in the text, the general solution is

$$y = x^{1/2}[c_1 J_{3/2}(x) + c_2 Y_{3/2}(x)] = x^{1/2}[C_1 J_{3/2}(x) + C_2 J_{-3/2}(x)].$$

19. Write the differential equation in the form $y'' + (3/x)y' + x^2 y = 0$. This is the form of (18) in the text with $a = -1$, $c = 2$, $b = \frac{1}{2}$, and $p = \frac{1}{2}$, so, by (19) in the text, the general solution is

$$y = x^{-1}\left[c_1 J_{1/2}\left(\frac{1}{2}x^2\right) + c_2 Y_{1/2}\left(\frac{1}{2}x^2\right)\right]$$

or

$$y = x^{-1}\left[C_1 J_{1/2}\left(\frac{1}{2}x^2\right) + C_2 J_{-1/2}\left(\frac{1}{2}x^2\right)\right].$$

21. Using the fact that $i^2 = -1$, along with the definition of $J_\nu(x)$ in (7) in the text, we have

$$I_\nu(x) = i^{-\nu} J_\nu(ix) = i^{-\nu} \sum_{n=0}^{\infty} \frac{(-1)^n}{n!\Gamma(1+\nu+n)} \left(\frac{ix}{2}\right)^{2n+\nu}$$

$$= \sum_{n=0}^{\infty} \frac{(-1)^n}{n!\Gamma(1+\nu+n)} i^{2n+\nu-\nu} \left(\frac{x}{2}\right)^{2n+\nu}$$

$$= \sum_{n=0}^{\infty} \frac{(-1)^n}{n!\Gamma(1+\nu+n)} (i^2)^n \left(\frac{x}{2}\right)^{2n+\nu}$$

$$= \sum_{n=0}^{\infty} \frac{(-1)^{2n}}{n!\Gamma(1+\nu+n)} \left(\frac{x}{2}\right)^{2n+\nu}$$

$$= \sum_{n=0}^{\infty} \frac{1}{n!\Gamma(1+\nu+n)} \left(\frac{x}{2}\right)^{2n+\nu},$$

which is a real function.

23. The differential equation has the form of (20) in the text with

$$1 - 2a = 0 \qquad\qquad 2c - 2 = 0 \qquad\qquad b^2c^2 = 1 \qquad\qquad a^2 - p^2c^2 = 0$$

$$a = \frac{1}{2} \qquad\qquad c = 1 \qquad\qquad b = 1 \qquad\qquad p = \frac{1}{2}.$$

Then, by (21) in the text,

$$y = x^{1/2}[c_1 J_{1/2}(x) + c_2 J_{-1/2}(x)] = x^{1/2}\left[c_1 \sqrt{\frac{2}{\pi x}} \sin x + c_2 \sqrt{\frac{2}{\pi x}} \cos x\right] = C_1 \sin x + C_2 \cos x.$$

25. Write the differential equation in the form $y'' + (2/x)y' + (\frac{1}{16}x^2 - 3/4x^2)y = 0$. This is the form of (20) in the text with

$$1 - 2a = 2 \qquad\qquad 2c - 2 = 2 \qquad\qquad b^2c^2 = \frac{1}{16} \qquad\qquad a^2 - p^2c^2 = -\frac{3}{4}$$

$$a = -\frac{1}{2} \qquad\qquad c = 2 \qquad\qquad b = \frac{1}{8} \qquad\qquad p = \frac{1}{2}.$$

Then, by (21) in the text,

$$y = x^{-1/2}\left[c_1 J_{1/2}\left(\frac{1}{8}x^2\right) + c_2 J_{-1/2}\left(\frac{1}{8}x^2\right)\right]$$

$$= x^{-1/2}\left[c_1 \sqrt{\frac{16}{\pi x^2}} \sin\left(\frac{1}{8}x^2\right) + c_2 \sqrt{\frac{16}{\pi x^2}} \cos\left(\frac{1}{8}x^2\right)\right]$$

$$= C_1 x^{-3/2} \sin\left(\frac{1}{8}x^2\right) + C_2 x^{-3/2} \cos\left(\frac{1}{8}x^2\right).$$

27. (a) The recurrence relation follows from

$$-\nu J_\nu(x) + x J_{\nu-1}(x) = -\sum_{n=0}^{\infty} \frac{(-1)^n \nu}{n!\Gamma(1+\nu+n)} \left(\frac{x}{2}\right)^{2n+\nu} + x\sum_{n=0}^{\infty} \frac{(-1)^n}{n!\Gamma(\nu+n)} \left(\frac{x}{2}\right)^{2n+\nu-1}$$

$$= -\sum_{n=0}^{\infty} \frac{(-1)^n \nu}{n!\Gamma(1+\nu+n)} \left(\frac{x}{2}\right)^{2n+\nu} + \sum_{n=0}^{\infty} \frac{(-1)^n (\nu+n)}{n!\Gamma(1+\nu+n)} \cdot 2 \left(\frac{x}{2}\right) \left(\frac{x}{2}\right)^{2n+\nu-1}$$

$$= \sum_{n=0}^{\infty} \frac{(-1)^n (2n+\nu)}{n!\Gamma(1+\nu+n)} \left(\frac{x}{2}\right)^{2n+\nu} = x J_\nu'(x).$$

(b) The formula in part (a) is a linear first-order differential equation in $J_\nu(x)$. An integrating factor for this equation is x^ν, so

$$\frac{d}{dx}\left[x^\nu J_\nu(x)\right] = x^\nu J_{\nu-1}(x).$$

29. Letting $\nu = 1$ in (21) in the text we have

$$x J_0(x) = \frac{d}{dx}[x J_1(x)] \qquad \text{so} \qquad \int_0^x r J_0(r)\,dr = r J_1(r)\Big|_{r=0}^{r=x} = x J_1(x).$$

31. Using $\Gamma\left(1+\dfrac{1}{2}+n\right) = \dfrac{(2n+1)!}{2^{2n+1}n!}\sqrt{\pi}$ we get the following

$$\Gamma\left(1+\frac{1}{2}+n\right) = \frac{(2n+1)!}{2^{2n+1}n!}\sqrt{\pi}$$

$$\left(\frac{1}{2}+n\right)\Gamma\left(\frac{1}{2}+n\right) = \frac{(2n+1)!}{2^{2n+1}n!}\sqrt{\pi}$$

$$\left(1-\frac{1}{2}+n\right)\Gamma\left(1-\frac{1}{2}+n\right) = \frac{(2n+1)!}{2^{2n+1}n!}\sqrt{\pi}$$

$$\Gamma\left(1-\frac{1}{2}+n\right) = \frac{(2n+1)!}{\left(1-\frac{1}{2}+n\right)2^{2n+1}n!}\sqrt{\pi}$$

$$\Gamma\left(1-\frac{1}{2}+n\right) = \frac{(2n+1)!}{\frac{1}{2}(1+2n)\,2^{2n+1}n!}\sqrt{\pi}$$

$$\Gamma\left(1-\frac{1}{2}+n\right) = \frac{(2n)!}{2^{2n}n!}\sqrt{\pi}$$

From the last result we obtain

$$J_{-1/2}(x) = \sum_{n=0}^{\infty} \frac{(-1)^n}{n!\Gamma(1-\frac{1}{2}+n)} \left(\frac{x}{2}\right)^{2n-1/2} = \sum_{n=0}^{\infty} \left(\frac{(-1)^n}{n!\frac{(2n)!}{2^{2n}n!}\sqrt{\pi}}\right) \left(\frac{x}{2}\right)^{2n} \left(\frac{x}{2}\right)^{-1/2}$$

$$= \sqrt{\frac{2}{\pi x}} \cdot \sum_{n=0}^{\infty} \frac{(-1)^n}{(2n)!} x^{2n}$$

The last series is the Maclaurin series for the cosine therefore

$$J_{-1/2}(x) = \sqrt{\frac{2}{\pi x}} \cdot \sum_{n=0}^{\infty} \frac{(-1)^n}{(2n)!} x^{2n} = \sqrt{\frac{2}{\pi x}} \cos x$$

33. (a) To find the spherical Bessel functions $j_1(x)$ and $j_2(x)$ we use the first formula in (30),

$$j_n(x) = \sqrt{\frac{\pi}{2x}} J_{n+1/2}$$

with $n = 1$ and $n = 2$,

$$j_1(x) = \sqrt{\frac{\pi}{2x}} J_{3/2}(x) \quad \text{and} \quad j_2(x) = \sqrt{\frac{\pi}{2x}} J_{5/2}(x).$$

Then from Problem 32 we have

$$J_{3/2}(x) = \sqrt{2\pi x}\left(\frac{\sin x}{x} - \cos x\right) \quad \text{so} \quad j_1(x) = \frac{\sin x}{x^2} - \frac{\cos x}{x}$$

and

$$J_{5/2}(x) = \sqrt{2\pi x}\left(\frac{3\sin x}{x^2} - \frac{3\cos x}{x} - \sin x\right) \quad \text{so} \quad j_2(x) = \left(\frac{3}{x^3} - \frac{1}{x}\right)\sin x - \frac{3\cos x}{x^2}$$

(b) Using a graphing utility to plot the graphs of $j_1(x)$ and $j_2(x)$, we get the red and blue graphcs in the figure to the right.

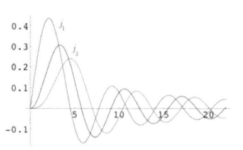

35. Letting

$$s = \frac{2}{\alpha}\sqrt{\frac{k}{m}}\, e^{-\alpha t/2},$$

we have

$$\frac{dx}{dt} = \frac{dx}{ds}\frac{ds}{dt} = \frac{dx}{dt}\left[\frac{2}{\alpha}\sqrt{\frac{k}{m}}\left(-\frac{\alpha}{2}\right)e^{-\alpha t/2}\right] = \frac{dx}{ds}\left(-\sqrt{\frac{k}{m}}\,e^{-\alpha t/2}\right)$$

and

$$\frac{d^2x}{dt^2} = \frac{d}{dt}\left(\frac{dx}{dt}\right) = \frac{dx}{ds}\left(\frac{\alpha}{2}\sqrt{\frac{k}{m}}\,e^{-\alpha t/2}\right) + \frac{d}{dt}\left(\frac{dx}{ds}\right)\left(-\sqrt{\frac{k}{m}}\,e^{-\alpha t/2}\right)$$

$$= \frac{dx}{ds}\left(\frac{\alpha}{2}\sqrt{\frac{k}{m}}\,e^{-\alpha t/2}\right) + \frac{d^2x}{ds^2}\frac{ds}{dt}\left(-\sqrt{\frac{k}{m}}\,e^{-\alpha t/2}\right)$$

$$= \frac{dx}{ds}\left(\frac{\alpha}{2}\sqrt{\frac{k}{m}}\,e^{-\alpha t/2}\right) + \frac{d^2x}{ds^2}\left(\frac{k}{m}\,e^{-\alpha t}\right).$$

Then

$$m\frac{d^2x}{dt^2} + ke^{-\alpha t}x = ke^{-\alpha t}\frac{d^2x}{ds^2} + \frac{m\alpha}{2}\sqrt{\frac{k}{m}}\,e^{-\alpha t/2}\frac{dx}{ds} + ke^{-\alpha t}x = 0.$$

Multiplying by $2^2/\alpha^2 m$ we have

$$\frac{2^2}{\alpha^2}\frac{k}{m}e^{-\alpha t}\frac{d^2x}{ds^2} + \frac{2}{\alpha}\sqrt{\frac{k}{m}}\,e^{-\alpha t/2}\frac{dx}{ds} + \frac{2^2}{\alpha^2}\frac{k}{m}e^{-\alpha t}x = 0$$

or, since $s = (2/\alpha)\sqrt{k/m}\,e^{-\alpha t/2}$,

$$s^2\frac{d^2x}{ds^2} + s\frac{dx}{ds} + s^2x = 0.$$

37. (a) By part (a) of Problem 34, a solution of Airy's equation is $y = x^{1/2}w(\frac{2}{3}\alpha x^{3/2})$, where

$$w(t) = c_1 J_{1/3}(t) + c_2 J_{-1/3}(t)$$

is a solution of Bessel's equation of order $\frac{1}{3}$. Thus, the general solution of Airy's equation for $x > 0$ is

$$y = x^{1/2}w\left(\frac{2}{3}\alpha x^{3/2}\right) = c_1 x^{1/2}J_{1/3}\left(\frac{2}{3}\alpha x^{3/2}\right) + c_2 x^{1/2}J_{-1/3}\left(\frac{2}{3}\alpha x^{3/2}\right).$$

(b) By part (b) of Problem 34, a solution of Airy's equation is $y = x^{1/2}w(\frac{2}{3}\alpha x^{3/2})$, where

$$w(t) = c_1 I_{1/3}(t) + c_2 I_{-1/3}(t)$$

is a solution of modified Bessel's equation of order $\frac{1}{3}$. Thus, the general solution of Airy's equation for $x > 0$ is

$$y = x^{1/2}w\left(\frac{2}{3}\alpha x^{3/2}\right) = c_1 x^{1/2}I_{1/3}\left(\frac{2}{3}\alpha x^{3/2}\right) + c_2 x^{1/2}I_{-1/3}\left(\frac{2}{3}\alpha x^{3/2}\right).$$

39. (a) The differential equation $y'' + (\lambda/x)y = 0$ has the form of (20) in the text with

$$1 - 2a = 0 \qquad 2c - 2 = -1 \qquad b^2 c^2 = \lambda \qquad a^2 - p^2 c^2 = 0$$

$$a = \frac{1}{2} \qquad\qquad c = \frac{1}{2} \qquad\qquad b = 2\sqrt{\lambda} \qquad\qquad p = 1.$$

Then, by (21) in the text,

$$y = x^{1/2}\left[c_1 J_1(2\sqrt{\lambda x}) + c_2 Y_1(2\sqrt{\lambda x})\right].$$

(b) We first note that $y = J_1(t)$ is a solution of Bessel's equation, $t^2 y'' + ty' + (t^2 - 1)y = 0$, with $\nu = 1$. That is,

$$t^2 J_1''(t) + t J_1'(t) + (t^2 - 1)J_1(t) = 0,$$

or, letting $t = 2\sqrt{x}$,

$$4x J_1''\left(2\sqrt{x}\right) + 2\sqrt{x}J_1'\left(2\sqrt{x}\right) + (4x - 1)J_1\left(2\sqrt{x}\right) = 0.$$

Now, if $y = \sqrt{x}J_1(2\sqrt{x})$, we have

$$y' = \sqrt{x}\,J_1'\left(2\sqrt{x}\right)\frac{1}{\sqrt{x}} + \frac{1}{2\sqrt{x}}J_1\left(2\sqrt{x}\right) = J_1'\left(2\sqrt{x}\right) + \frac{1}{2}x^{-1/2}J_1\left(2\sqrt{x}\right)$$

and

$$y'' = x^{-1/2}J_1''\left(2\sqrt{x}\right) + \frac{1}{2x}J_1'\left(2\sqrt{x}\right) - \frac{1}{4}x^{-3/2}J_1\left(2\sqrt{x}\right).$$

Then

$$xy'' + y = \sqrt{x}\,J_1''\,2\sqrt{x} + \frac{1}{2}J_1'\left(2\sqrt{x}\right) - \frac{1}{4}x^{-1/2}J_1\left(2\sqrt{x}\right) + \sqrt{x}\,J\left(2\sqrt{x}\right)$$

$$= \frac{1}{4\sqrt{x}}\left[4xJ_1''\left(2\sqrt{x}\right) + 2\sqrt{x}\,J_1'\left(2\sqrt{x}\right) - J_1\left(2\sqrt{x}\right) + 4xJ\left(2\sqrt{x}\right)\right]$$

$$= 0,$$

and $y = \sqrt{x}\,J_1\left(2\sqrt{x}\right)$ is a solution of Airy's differential equation.

41. (a) We identify $m = 4$, $k = 1$, and $\alpha = 0.1$. Then

$$x(t) = c_1 J_0(10e^{-0.05t}) + c_2 Y_0(10e^{-0.05t})$$

and

$$x'(t) = -0.5c_1 J_0'(10e^{-0.05t}) - 0.5c_2 Y_0'(10e^{-0.05t}).$$

Now $x(0) = 1$ and $x'(0) = -1/2$ imply

$$c_1 J_0(10) + c_2 Y_0(10) = 1$$
$$c_1 J_0'(10) + c_2 Y_0'(10) = 1.$$

Using Cramer's rule we obtain

$$c_1 = \frac{Y_0'(10) - Y_0(10)}{J_0(10)Y_0'(10) - J_0'(10)Y_0(10)} \quad \text{and} \quad c_2 = \frac{J_0(10) - J_0'(10)}{J_0(10)Y_0'(10) - J_0'(10)Y_0(10)}.$$

Using $Y_0' = -Y_1$ and $J_0' = -J_1$ and Table 5.2 we find $c_1 = -4.7860$ and $c_2 = -3.1803$. Thus

$$x(t) = -4.7860 J_0(10e^{-0.05t}) - 3.1803 Y_0(10e^{-0.05t}).$$

(b)

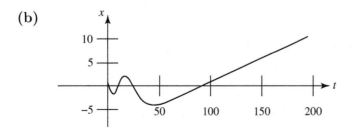

43. (a) Letting $t = L - x$, the boundary-value problem becomes

$$\frac{d^2\theta}{dt^2} + \alpha^2 t\theta = 0, \qquad \theta'(0) = 0, \quad \theta(L) = 0,$$

where $\alpha^2 = \delta g/EI$. This is Airy's differential equation, so by Problem 35 its solution is

$$y = c_1 t^{1/2} J_{1/3}\left(\frac{2}{3}\alpha t^{3/2}\right) + c_2 t^{1/2} J_{-1/3}\left(\frac{2}{3}\alpha t^{3/2}\right) = c_1\theta_1(t) + c_2\theta_2(t).$$

(b) Looking at the series forms of θ_1 and θ_2 we see that $\theta_1'(0) \neq 0$, while $\theta_2'(0) = 0$. Thus, the boundary condition $\theta'(0) = 0$ implies $c_1 = 0$, and so

$$\theta(t) = c_2\sqrt{t}\, J_{-1/3}\left(\frac{2}{3}\alpha t^{3/2}\right).$$

From $\theta(L) = 0$ we have

$$c_2\sqrt{L}\, J_{-1/3}\left(\frac{2}{3}\alpha L^{3/2}\right) = 0,$$

so either $c_2 = 0$, in which case $\theta(t) = 0$, or $J_{-1/3}(\frac{2}{3}\alpha L^{3/2}) = 0$. The column will just start to bend when L is the length corresponding to the smallest positive zero of $J_{-1/3}$.

(c) Using *Mathematica*, the first positive root of $J_{-1/3}(x)$ is $x_1 \approx 1.86635$. Thus $\frac{2}{3}\alpha L^{3/2} = 1.86635$ implies

$$L = \left(\frac{3(1.86635)}{2\alpha}\right)^{2/3} = \left[\frac{9EI}{4\delta g}(1.86635)^2\right]^{1/3}$$

$$= \left[\frac{9(2.6 \times 10^7)\pi(0.05)^4/4}{4(0.28)\pi(0.05)^2}(1.86635)^2\right]^{1/3} \approx 76.9 \text{ in.}$$

45. (a) Since $l' = v$, we integrate to obtain $l(t) = vt + c$. Now $l(0) = l_0$ implies $c = l_0$, so $l(t) = vt + l_0$. Using $\sin\theta \approx \theta$ in $l\, d^2\theta/dt^2 + 2l'\, d\theta/dt + g\sin\theta = 0$ gives

$$(l_0 + vt)\frac{d^2\theta}{dt^2} + 2v\frac{d\theta}{dt} + g\theta = 0.$$

(b) Dividing by v, the differential equation in part (a) becomes

$$\frac{l_0 + vt}{v}\frac{d^2\theta}{dt^2} + 2\frac{d\theta}{dt} + \frac{g}{v}\theta = 0.$$

Letting $x = (l_0 + vt)/v = t + l_0/v$ we have $dx/dt = 1$, so

$$\frac{d\theta}{dt} = \frac{d\theta}{dx}\frac{dx}{dt} = \frac{d\theta}{dx}$$

and

$$\frac{d^2\theta}{dt^2} = \frac{d(d\theta/dt)}{dt} = \frac{d(d\theta/dx)}{dx}\frac{dx}{dt} = \frac{d^2\theta}{dx^2}.$$

Thus, the differential equation becomes

$$x\frac{d^2\theta}{dx^2} + 2\frac{d\theta}{dx} + \frac{g}{v}\theta = 0 \qquad \text{or} \qquad \frac{d^2\theta}{dx^2} + \frac{2}{x}\frac{d\theta}{dx} + \frac{g}{vx}\theta = 0.$$

(c) The differential equation in part (b) has the form of (20) in the text with

$$1 - 2a = 2 \qquad 2c - 2 = -1 \qquad b^2 c^2 = \frac{g}{v} \qquad a^2 - p^2 c^2 = 0$$

$$a = -\frac{1}{2} \qquad c = \frac{1}{2} \qquad b = 2\sqrt{\frac{g}{v}} \qquad p = 1.$$

Then, by (21) in the text,

$$\theta(x) = x^{-1/2}\left[c_1 J_1\left(2\sqrt{\frac{g}{v}}\, x^{1/2}\right) + c_2 Y_1\left(2\sqrt{\frac{g}{v}}\, x^{1/2}\right)\right]$$

or

$$\theta(t) = \sqrt{\frac{v}{l_0 + vt}}\left[c_1 J_1\left(\frac{2}{v}\sqrt{g(l_0 + vt)}\right) + c_2 Y_1\left(\frac{2}{v}\sqrt{g(l_0 + vt)}\right)\right].$$

(d) To simplify calculations, let

$$u = \frac{2}{v}\sqrt{g(l_0 + vt)} = 2\sqrt{\frac{g}{v}}\, x^{1/2},$$

and at $t = 0$ let $u_0 = 2\sqrt{gl_0}/v$. The general solution for $\theta(t)$ can then be written

$$\theta = C_1 u^{-1} J_1(u) + C_2 u^{-1} Y_1(u). \tag{1}$$

Before applying the initial conditions, note that

$$\frac{d\theta}{dt} = \frac{d\theta}{du}\frac{du}{dt}$$

so when $d\theta/dt = 0$ at $t = 0$ we have $d\theta/du = 0$ at $u = u_0$. Also,

$$\frac{d\theta}{du} = C_1 \frac{d}{du}[u^{-1} J_1(u)] + C_2 \frac{d}{du}[u^{-1} Y_1(u)]$$

which, in view of (20) in the text, is the same as

$$\frac{d\theta}{du} = -C_1 u^{-1} J_2(u) - C_2 u^{-1} Y_2(u). \tag{2}$$

Now at $t = 0$, or $u = u_0$, (1) and (2) give the system

$$C_1 u_0^{-1} J_1(u_0) + C_2 u_0^{-1} Y_1(u_0) = \theta_0$$

$$C_1 u_0^{-1} J_2(u_0) + C_2 u_0^{-1} Y_2(u_0) = 0$$

whose solution is easily obtained using Cramer's rule:

$$C_1 = \frac{u_0 \theta_0 Y_2(u_0)}{J_1(u_0)Y_2(u_0) - J_2(u_0)Y_1(u_0)}, \qquad C_2 = \frac{-u_0 \theta_0 J_2(u_0)}{J_1(u_0)Y_2(u_0) - J_2(u_0)Y_1(u_0)}.$$

In view of the given identity these results simplify to

$$C_1 = -\frac{\pi}{2} u_0^2 \theta_0 Y_2(u_0) \qquad \text{and} \qquad C_2 = \frac{\pi}{2} u_0^2 \theta_0 J_2(u_0).$$

The solution is then

$$\theta = \frac{\pi}{2} u_0^2 \theta_0 \left[-Y_2(u_0) \frac{J_1(u)}{u} + J_2(u_0) \frac{Y_1(u)}{u} \right].$$

Returning to $u = (2/v)\sqrt{g(l_0 + vt)}$ and $u_0 = (2/v)\sqrt{gl_0}$, we have

$$\theta(t) = \frac{\pi\sqrt{gl_0}\,\theta_0}{v} \left[-Y_2\left(\frac{2}{v}\sqrt{gl_0}\right) \frac{J_1\left(\frac{2}{v}\sqrt{g(l_0 + vt)}\right)}{\sqrt{l_0 + vt}} \right.$$

$$\left. + J_2\left(\frac{2}{v}\sqrt{gl_0}\right) \frac{Y_1\left(\frac{2}{v}\sqrt{g(l_0 + vt)}\right)}{\sqrt{l_0 + vt}} \right].$$

(e) When $l_0 = 1$ ft, $\theta_0 = \frac{1}{10}$ radian, and $v = \frac{1}{60}$ ft/s, the above function is

$$\theta(t) = -1.69045 \frac{J_1(480\sqrt{2}(1 + t/60))}{\sqrt{1 + t/60}} - 2.79381 \frac{Y_1(480\sqrt{2}\,(1 + t/60))}{\sqrt{1 + t/60}}.$$

The plots of $\theta(t)$ on $[0, 10]$, $[0, 30]$, and $[0, 60]$ are

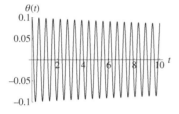 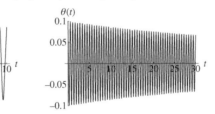 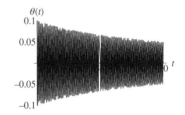

(f) The graphs indicate that $\theta(t)$ decreases as l increases. The graph of $\theta(t)$ on $[0, 300]$ is shown.

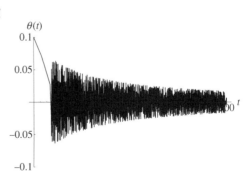

47. The recurrence relation can be written

$$P_{k+1}(x) = \frac{2k+1}{k+1} x P_k(x) - \frac{k}{k+1} P_{k-1}(x), \qquad k = 2,\ 3,\ 4,\ \dots\ .$$

$$k = 1 : \quad P_2(x) = \frac{3}{2}x^2 - \frac{1}{2}$$

$$k = 2 : \quad P_3(x) = \frac{5}{3}x\left(\frac{3}{2}x^2 - \frac{1}{2}\right) - \frac{2}{3}x = \frac{5}{2}x^3 - \frac{3}{2}x$$

$$k = 3 : \quad P_4(x) = \frac{7}{4}x\left(\frac{5}{2}x^3 - \frac{3}{2}x\right) - \frac{3}{4}\left(\frac{3}{2}x^2 - \frac{1}{2}\right) = \frac{35}{8}x^4 - \frac{30}{8}x^2 + \frac{3}{8}$$

$$k = 4 : \quad P_5(x) = \frac{9}{5}x\left(\frac{35}{8}x^4 - \frac{30}{8}x^2 + \frac{3}{8}\right) - \frac{4}{5}\left(\frac{5}{2}x^3 - \frac{3}{2}x\right) = \frac{63}{8}x^5 - \frac{35}{4}x^3 + \frac{15}{8}x$$

$$k = 5 : \quad P_6(x) = \frac{11}{6}x\left(\frac{63}{8}x^5 - \frac{35}{4}x^3 + \frac{15}{8}x\right) - \frac{5}{6}\left(\frac{35}{8}x^4 - \frac{30}{8}x^2 + \frac{3}{8}\right)$$

$$= \frac{231}{16}x^6 - \frac{315}{16}x^4 + \frac{105}{16}x^2 - \frac{5}{16}$$

$$k = 6 : \quad P_7(x) = \frac{13}{7}x\left(\frac{231}{16}x^6 - \frac{315}{16}x^4 + \frac{105}{16}x^2 - \frac{5}{16}\right) - \frac{6}{7}\left(\frac{63}{8}x^5 - \frac{35}{4}x^3 + \frac{15}{8}x\right)$$

$$= \frac{429}{16}x^7 - \frac{693}{16}x^5 + \frac{315}{16}x^3 - \frac{35}{16}x$$

49. The only solutions bounded on $[-1, 1]$ are $y = cP_n(x)$, c a constant and $n = 0, 1, 2, \ldots$. By (iv) of the properties of the Legendre polynomials, $y(0) = 0$ or $P_n(0) = 0$ implies n must be odd. Thus the first three positive eigenvalues correspond to $n = 1, 3,$ and 5 or $\lambda_1 = 1 \cdot 2$, $\lambda_2 = 3 \cdot 4 = 12$, and $\lambda_3 = 5 \cdot 6 = 30$. We can take the eigenfunctions to be $y_1 = P_1(x)$, $y_2 = P_3(x)$, and $y_3 = P_5(x)$.

51.

53. Letting $y = \displaystyle\sum_{k=0}^{\infty} c_k x^k$ we have

$$y' = \sum_{k=1}^{\infty} nc_n x^{n-1} \qquad \text{and} \qquad y'' = \sum_{n=2}^{\infty} n(n+1)c_n x^{n-2}$$

Then, with appropriate substitutions, we have

$$\sum_{k=0}^{\infty} [(k+2)(k+1)c_{k+2} - 2kc_k + 2\alpha c_k]x^k = 0$$

This leads to the recurrence relation

$$c_{k+2} = -\frac{(2\alpha - 2k)}{(k+2)(k+1)}c_k, \qquad \text{for } k = 0, 1, 2, 3, \dots$$

Thus

$$c_2 = -\frac{(2\alpha)}{(2)(1)}c_0 = -\frac{2\alpha}{2!}c_0$$

$$c_3 = -\frac{(2\alpha - 2)}{(3)(2)}c_1 = -\frac{2(\alpha - 1)}{3!}c_1$$

$$c_4 = -\frac{(2\alpha - 4)}{(4)(3)}c_2 = -\frac{2(\alpha - 2)}{4 \cdot 3} \cdot \left(-\frac{2\alpha}{2!}c_0\right) = \frac{2^2\alpha(\alpha - 2)}{4!}c_0$$

$$c_5 = -\frac{(2\alpha - 6)}{(5)(4)}c_3 = -\frac{2(\alpha - 3)}{5 \cdot 4} \cdot \left(-\frac{2(\alpha - 1)}{3!}c_1\right) = \frac{2^2(\alpha - 1)(\alpha - 3)}{5!}c_1$$

$$\vdots$$

Then two solutions are

$$y_1(x) = 1 - \frac{2\alpha}{2!}x^2 + \frac{2^2\alpha(\alpha - 2)}{4!}x^4 - \cdots = 1 + \sum_{k=1}^{\infty} \frac{(-1)^k 2^k \alpha(\alpha - 2)(\alpha - 4)\dots(\alpha - (2k - 2))}{(2k)!}x^{2k}$$

and

$$y_2(x) = x - \frac{2\alpha - 1}{3!}x^3 + \frac{2^2(\alpha - 1)(\alpha - 3)}{5!}x^5 - \cdots$$

$$= 1 + \sum_{k=1}^{\infty} \frac{(-1)^k 2^k (\alpha - 1)(\alpha - 3)\dots(\alpha - (2k - 1))}{(2k + 1)!}x^{2k+1}$$

and the general solution is $y(x) = c_0 y_1(x) + c_1 y_2(x)$.

55. Substitute the assumed solution $y = \sum_{k=0}^{\infty} c_k x^k$ into the equation to get

$$(1 - x^2) \cdot \sum_{k=2}^{\infty} k(k-1)c_k x^{k-2} - x \cdot \sum_{k=1}^{\infty} kc_k x^{k-1} + \alpha^2 \sum_{k=0}^{\infty} c_k x^k = 0$$

From this we get

$$(2c_2 + \alpha^2 c_0) + (6c_3 - c_1 + \alpha^2 c_1)x + \sum_{k=2}^{\infty} [(k+2)(k+1)c_{k+2} - k(k-1)c_k - kc_k + \alpha^2 c_k]x^k = 0$$

Therefore we have

$$c_2 = -\frac{\alpha^2}{2!}c_0, \quad c_3 = \frac{1-\alpha^2}{3!}c_1, \quad \text{and} \quad c_{k+2} = \frac{k^2 - \alpha^2}{(k+2)(k+1)}c_k \text{ for } k = 2, 3, 4, \ldots$$

Therefore we get

$$y = \sum_{k=0}^{\infty} c_k x^k = c_0 + c_1 x + c_2 x^2 + c_3 x^3 + c_4 x^4 + c_5 x^5 + c_6 x^6 + \cdots$$

$$= c_0 + c_1 x - \frac{\alpha^2}{2!}c_0 x^2 + \frac{1-\alpha^2}{3!}c_1 x^3 - \frac{(4-\alpha^2)\alpha^2}{4!}c_0 x^4 + \frac{(9-\alpha^2)(1-\alpha^2)}{5!}c_1 x^5$$

$$- \frac{(16-\alpha^2)(4-\alpha^2)\alpha^2}{6!}c_0 x^6 + \cdots$$

$$= c_0\left[1 - \frac{\alpha^2}{2!}x^2 - \frac{\alpha^2(4-\alpha^2)}{4!}x^4 - \frac{(16-\alpha^2)(4-\alpha^2)\alpha^2}{6!}x^6 + \cdots\right]$$

$$+ c_1\left[x + \frac{1-\alpha^2}{3!}x^3 + \frac{(9-\alpha^2)(1-\alpha^2)}{5!}x^5 + \cdots\right]$$

$$= c_0 y_1(x) + c_1 y_2(x)$$

where we take

$$y_1(x) = 1 - \frac{\alpha^2}{2!}x^2 - \frac{(4-\alpha^2)\alpha^2}{4!}x^4 - \frac{(16-\alpha^2)(4-\alpha^2)\alpha^2}{6!}x^6 + \cdots$$

$$y_2(x) = x + \frac{1-\alpha^2}{3!}x^3 + \frac{(9-\alpha^2)(1-\alpha^2)}{5!}x^5 + \cdots$$

When $\alpha = n$ is a nonnegative even integer $y_1(x)$ terminates at x^n, and when $\alpha = n$ is a positive odd integer $y_2(x)$ terminates at x^n. With $\alpha = n = 5$, $y_2(x)$ yields the fifth degree polynomial solution $y = x - 4x^3 + \frac{16}{5}x^5$.

Chapter 6 in Review

1. False; $J_1(x)$ and $J_{-1}(x)$ are not linearly independent when ν is a positive integer. (In this case $\nu = 1$). The general solution of $x^2 y'' + x y' + (x^2 - 1)y = 0$ is $y = c_1 J_1(x) + c_2 Y_1(x)$.

3. $x = -1$ is the nearest singular point to the ordinary point $x = 0$. Theorem 5.1.1 guarantees the existence of two power series solutions $y = \sum_{n=1}^{\infty} c_n x^n$ of the differential equation that converge at least for $-1 < x < 1$. Since $-\frac{1}{2} \le x \le \frac{1}{2}$ is properly contained in $-1 < x < 1$, both power series must converge for all points contained in $-\frac{1}{2} \le x \le \frac{1}{2}$.

5. The interval of convergence is centered at 4. Since the series converges at -2, it converges at least on the interval $[-2, 10)$. Since it diverges at 13, it converges at most on the interval $[-5, 13)$. Thus, at -7 it does not converge, at 0 and 7 it does converge, and at 10 and 11 it might converge.

7. The differential equation $(x^3 - x^2)y'' + y' + y = 0$ has a regular singular point at $x = 1$ and an irregular singular point at $x = 0$.

9. Substituting $y = \sum_{n=0}^{\infty} c_n x^{n+r}$ into the differential equation we obtain

$$2xy'' + y' + y = (2r^2 - r) c_0 x^{r-1} + \sum_{k=1}^{\infty} [2(k+r)(k+r-1)c_k + (k+r)c_k + c_{k-1}]x^{k+r-1} = 0$$

which implies

$$2r^2 - r = r(2r - 1) = 0$$

and

$$(k+r)(2k + 2r - 1)c_k + c_{k-1} = 0.$$

The indicial roots are $r = 0$ and $r = 1/2$. For $r = 0$ the recurrence relation is

$$c_k = -\frac{c_{k-1}}{k(2k-1)}, \quad k = 1, 2, 3, \ldots,$$

so

$$c_1 = -c_0, \qquad c_2 = \frac{1}{6}c_0, \qquad c_3 = -\frac{1}{90}c_0.$$

For $r = 1/2$ the recurrence relation is

$$c_k = -\frac{c_{k-1}}{k(2k+1)}, \quad k = 1, 2, 3, \ldots,$$

so

$$c_1 = -\frac{1}{3}c_0, \qquad c_2 = \frac{1}{30}c_0, \qquad c_3 = -\frac{1}{630}c_0.$$

Two linearly independent solutions are

$$y_1 = 1 - x + \frac{1}{6}x^2 - \frac{1}{90}x^3 + \cdots$$

and

$$y_2 = x^{1/2}\left(1 - \frac{1}{3}x + \frac{1}{30}x^2 - \frac{1}{630}x^3 + \cdots\right).$$

11. Substituting $y = \sum_{n=0}^{\infty} c_n x^n$ into the differential equation we obtain

$$(x-1)y'' + 3y = (-2c_2 + 3c_0) + \sum_{k=1}^{\infty} [(k+1)kc_{k+1} - (k+2)(k+1)c_{k+2} + 3c_k]x^k = 0$$

which implies $c_2 = 3c_0/2$ and

$$c_{k+2} = \frac{(k+1)kc_{k+1} + 3c_k}{(k+2)(k+1)}, \quad k = 1, 2, 3, \ldots.$$

Choosing $c_0 = 1$ and $c_1 = 0$ we find

$$c_2 = \frac{3}{2}, \qquad\qquad c_3 = \frac{1}{2}, \qquad\qquad c_4 = \frac{5}{8}$$

and so on. For $c_0 = 0$ and $c_1 = 1$ we obtain

$$c_2 = 0, \qquad\qquad c_3 = \frac{1}{2}, \qquad\qquad c_4 = \frac{1}{4}$$

and so on. Thus, two solutions are

$$y_1 = 1 + \frac{3}{2}x^2 + \frac{1}{2}x^3 + \frac{5}{8}x^4 + \cdots$$

and

$$y_2 = x + \frac{1}{2}x^3 + \frac{1}{4}x^4 + \cdots .$$

13. Substituting $y = \displaystyle\sum_{n=0}^{\infty} c_n x^{n+r}$ into the differential equation, we obtain

$$xy'' - (x+2)y' + 2y = (r^2 - 3r)c_0 x^{r-1} + \sum_{k=1}^{\infty} [(k+r)(k+r-3)c_k - (k+r-3)c_{k-1}]x^{k+r-1} = 0,$$

which implies

$$r^2 - 3r = r(r-3) = 0$$

and

$$(k+r)(k+r-3)c_k - (k+r-3)c_{k-1} = 0.$$

The indicial roots are $r_1 = 3$ and $r_2 = 0$. For $r_2 = 0$ the recurrence relation is

$$k(k-3)c_k - (k-3)c_{k-1} = 0, \qquad k = 1, 2, 3, \ldots .$$

Then

$$c_1 - c_0 = 0 \qquad\qquad 2c_2 - c_1 = 0 \qquad\qquad 0c_3 - 0c_2 = 0$$

Therefore c_3 is arbitrary and

$$c_k = \frac{1}{k}c_{k-1}, \qquad k = 4, 5, 6, \ldots .$$

Taking $c_0 \neq 0$ and $c_3 = 0$ we obtain

$$c_1 = c_0 \qquad\qquad c_2 = \frac{1}{2}c_0 \qquad\qquad c_3 = c_4 = c_5 = \cdots = 0.$$

Taking $c_0 = 0$ and $c_3 \neq 0$ we obtain

$$c_0 = c_1 = c_2 = 0 \qquad c_4 = \frac{1}{4}c_3 = \frac{6}{4!}c_3 \qquad c_5 = \frac{1}{5 \cdot 4}c_3 = \frac{6}{5!}c_3 \qquad c_6 = \frac{1}{6 \cdot 5 \cdot 4}c_3 = \frac{6}{6!}c_3,$$

and so on. In this case we obtain the two solutions

$$y_1 = 1 + x + \frac{1}{2}x^2$$

and

$$y_2 = x^3 + \frac{6}{4!}x^4 + \frac{6}{5!}x^5 + \frac{6}{6!}x^6 + \cdots = 6e^x - 6\left(1 + x + \frac{1}{2}x^2\right).$$

15. Substituting $y = \sum_{n=0}^{\infty} c_n x^n$ into the differential equation we have

$$y'' + xy' + 2y = \underbrace{\sum_{n=2}^{\infty} n(n-1)c_n x^{n-2}}_{k=n-2} + \underbrace{\sum_{n=1}^{\infty} nc_n x^n}_{k=n} + 2\underbrace{\sum_{n=0}^{\infty} c_n x^n}_{k=n}$$

$$= \sum_{k=0}^{\infty}(k+2)(k+1)c_{k+2}x^k + \sum_{k=1}^{\infty} kc_k x^k + 2\sum_{k=0}^{\infty} c_k x^k$$

$$= 2c_2 + 2c_0 + \sum_{k=1}^{\infty}[(k+2)(k+1)c_{k+2} + (k+2)c_k]x^k = 0.$$

Thus

$$2c_2 + 2c_0 = 0$$

$$(k+2)(k+1)c_{k+2} + (k+2)c_k = 0$$

and

$$c_2 = -c_0$$

$$c_{k+2} = -\frac{1}{k+1}c_k, \quad k = 1, 2, 3, \ldots.$$

Choosing $c_0 = 1$ and $c_1 = 0$ we find

$$c_2 = -1 \qquad c_3 = c_5 = c_7 = \cdots = 0 \qquad c_4 = \frac{1}{3} \qquad c_6 = -\frac{1}{15}$$

and so on. For $c_0 = 0$ and $c_1 = 1$ we obtain

$$c_2 = c_4 = c_6 = \cdots = 0 \qquad c_3 = -\frac{1}{2} \qquad c_5 = \frac{1}{8} \qquad c_7 = -\frac{1}{48}$$

and so on. Thus, the general solution is

$$y = C_0\left(1 - x^2 + \frac{1}{3}x^4 - \frac{1}{15}x^6 + \cdots\right) + C_1\left(x - \frac{1}{2}x^3 + \frac{1}{8}x^5 - \frac{1}{48}x^7 + \cdots\right)$$

and

$$y' = C_0\left(-2x + \frac{4}{3}x^3 - \frac{2}{5}x^5 + \cdots\right) + C_1\left(1 - \frac{3}{2}x^2 + \frac{5}{8}x^4 - \frac{7}{48}x^6 + \cdots\right).$$

Setting $y(0) = 3$ and $y'(0) = -2$ we find $c_0 = 3$ and $c_1 = -2$. Therefore, the solution of the initial-value problem is

$$y = 3 - 2x - 3x^2 + x^3 + x^4 - \frac{1}{4}x^5 - \frac{1}{5}x^6 + \frac{1}{24}x^7 + \cdots .$$

17. The singular point of $(1 - 2\sin x)y'' + xy = 0$ closest to $x = 0$ is $\pi/6$. Hence a lower bound is $\pi/6$.

19. Writing the differential equation in the form

$$y'' + \left(\frac{1 - \cos x}{x}\right)y' + xy = 0,$$

and noting that

$$\frac{1 - \cos x}{x} = \frac{x}{2} - \frac{x^3}{24} + \frac{x^5}{720} - \cdots$$

is analytic at $x = 0$, we conclude that $x = 0$ is an ordinary point of the differential equation.

21. Substituting $y = \sum_{n=0}^{\infty} c_n x^n$ into the differential equation we have

$$y'' + x^2 y' + 2xy = \underbrace{\sum_{n=2}^{\infty} n(n-1)c_n x^{n-2}}_{k=n-2} + \underbrace{\sum_{n=1}^{\infty} n c_n x^{n+1}}_{k=n+1} + 2\underbrace{\sum_{n=0}^{\infty} c_n x^{n+1}}_{k=n+1}$$

$$= \sum_{k=0}^{\infty}(k+2)(k+1)c_{k+2}x^k + \sum_{k=2}^{\infty}(k-1)c_{k-1}x^k + 2\sum_{k=1}^{\infty} c_{k-1}x^k$$

$$= 2c_2 + (6c_3 + 2c_0)x + \sum_{k=2}^{\infty}[(k+2)(k+1)c_{k+2} + (k+1)c_{k-1}]x^k = 5 - 2x + 10x^3.$$

Thus, equating coefficients of like powers of x gives

$$2c_2 = 5 \qquad 6c_3 + 2c_0 = -2 \qquad 12c_4 + 3c_1 = 0 \qquad 20c_5 + 4c_2 = 10$$

and

$$(k+2)(k+1)c_{k+2} + (k+1)c_{k-1} = 0, \quad k = 4, 5, 6, \ldots,$$

Therefore

$$c_2 = \frac{5}{2} \qquad c_3 = -\frac{1}{3}c_0 - \frac{1}{3} \qquad c_4 = -\frac{1}{4}c_1 \qquad c_5 = \frac{1}{2} - \frac{1}{5}c_2 = \frac{1}{2} - \frac{1}{5}\left(\frac{5}{2}\right) = 0$$

and

$$c_{k+2} = -\frac{1}{k+2}c_{k-1}.$$

Using the recurrence relation, we find

$$c_6 = -\frac{1}{6}c_3 = \frac{1}{3\cdot 6}(c_0 + 1) = \frac{1}{3^2\cdot 2!}c_0 + \frac{1}{3^2\cdot 2!}$$

$$c_7 = -\frac{1}{7}c_4 = \frac{1}{4\cdot 7}c_1$$

$$c_8 = c_{11} = c_{14} = \cdots = 0$$

$$c_9 = -\frac{1}{9}c_6 = -\frac{1}{3^3\cdot 3!}c_0 - \frac{1}{3^3\cdot 3!}$$

$$c_{10} = -\frac{1}{10}c_7 = -\frac{1}{4\cdot 7\cdot 10}c_1$$

$$c_{12} = -\frac{1}{12}c_9 = \frac{1}{3^4\cdot 4!}c_0 + \frac{1}{3^4\cdot 4!}$$

$$c_{13} = -\frac{1}{13}c_0 = \frac{1}{4\cdot 7\cdot 10\cdot 13}c_1$$

and so on. Thus

$$y = c_0\left[1 - \frac{1}{3}x^3 + \frac{1}{3^2\cdot 2!}x^6 - \frac{1}{3^3\cdot 3!}x^9 + \frac{1}{3^4\cdot 4!}x^{12} - \cdots\right]$$

$$+ c_1\left[x - \frac{1}{4}x^4 + \frac{1}{4\cdot 7}x^7 - \frac{1}{4\cdot 7\cdot 10}x^{10} + \frac{1}{4\cdot 7\cdot 10\cdot 13}x^{13} - \cdots\right]$$

$$+ \left[\frac{5}{2}x^2 - \frac{1}{3}x^3 + \frac{1}{3^2\cdot 2!}x^6 - \frac{1}{3^3\cdot 3!}x^9 + \frac{1}{3^4\cdot 4!}x^{12} - \cdots\right].$$

23. (a) From (10) of Section 6.4, with $n = \frac{3}{2}$, we have

$$Y_{3/2}(x) = \frac{-J_{-3/2}(x)}{-1} = J_{-3/2}(x).$$

Then from the solutions of Problems 28 and 32 in Section 6.4 we have

$$J_{-3/2}(x) = -\sqrt{\frac{2}{\pi x}}\left(\frac{\cos x}{x} + \sin x\right) \quad \text{so} \quad Y_{3/2}(x) = -\sqrt{\frac{2}{\pi}}x\left(\frac{\cos x}{x} + \sin x\right)$$

(b) From (15) of Section 6.4 in the text

$$I_{1/2}(x) = i^{-1/2}J_{1/2}(ix) \quad \text{and} \quad I_{-1/2}(x) = i^{1/2}J_{-1/2}(ix)$$

so

$$I_{1/2}(x) = \sqrt{\frac{2}{\pi x}}\sum_{n=0}^{\infty}\frac{1}{(2n+1)!}x^{2n+1} = \sqrt{\frac{2}{\pi x}}\sinh x$$

and

$$I_{-1/2}(x) = \sqrt{\frac{2}{\pi x}}\sum_{n=0}^{\infty}\frac{1}{(2n)!}x^{2n} = \sqrt{\frac{2}{\pi x}}\cosh x.$$

(c) Equation (16) of Section 6.4 in the text and part (b) imply

$$K_{1/2}(x) = \frac{\pi}{2} \frac{I_{-1/2}(x) - I_{1/2}(x)}{\sin\frac{\pi}{2}} = \frac{\pi}{2}\left[\sqrt{\frac{2}{\pi x}}\cosh x - \sqrt{\frac{2}{\pi x}}\sinh x\right]$$

$$= \sqrt{\frac{\pi}{2x}}\left[\frac{e^x + e^{-x}}{2} - \frac{e^x - e^{-x}}{2}\right] = \sqrt{\frac{\pi}{2x}}\,e^{-x}.$$

25. (a) By the binomial theorem we have

$$\left[1 + \left(t^2 - 2xt\right)\right]^{-1/2}$$

$$= 1 - \frac{1}{2}\left(t^2 - 2xt\right) + \frac{(-1/2)(-3/2)}{2!}\left(t^2 - 2xt\right)^2 + \frac{(-1/2)(-3/2)(-5/2)}{3!}\left(t^2 - 2xt\right)^3 + \cdots$$

$$= 1 - \frac{1}{2}(t^2 - 2xt) + \frac{3}{8}(t^2 - 2xt)^2 - \frac{5}{16}(t^2 - 2xt)^3 + \cdots$$

$$= 1 + xt + \frac{1}{2}(3x^2 - 1)t^2 + \frac{1}{2}(5x^3 - 3x)t^3 + \cdots = \sum_{n=0}^{\infty} P_n(x)t^n.$$

(b) Letting $x = 1$ in $\left(1 - 2xt + t^2\right)^{-1/2}$, we have

$$\left(1 - 2t + t^2\right)^{-1/2} = (1-t)^{-1} = \frac{1}{1-t} = 1 + t + t^2 + t^3 + \cdots \qquad (|t| < 1)$$

$$= \sum_{n=0}^{\infty} t^n.$$

From part (a) we have

$$\sum_{n=0}^{\infty} P_n(1)t^n = \left(1 - 2t + t^2\right)^{-1/2} = \sum_{n=0}^{\infty} t^n.$$

Equating the coefficients of corresponding terms in the two series, we see that $P_n(1) = 1$. SImilarly, letting $x = -1$ we have

$$\left(1 + 2t + t^2\right)^{-1/2} = (1+t)^{-1} = \frac{1}{1+t} = 1 - t + t^2 - 3t^3 + \cdots \qquad (|t| < 1)$$

$$= \sum_{n=0}^{\infty} (-1)^n t^n = \sum_{n=0}^{\infty} P_n(-1)t^n,$$

so that $P_n(-1) = (-1)^n$.

27. Using the Product rule gives

$$\frac{d}{dr}\left(r\frac{dT}{dr}\right) = a^2 r\,(T - T_m)$$

$$r\frac{d^2T}{dr^2} + \frac{dT}{dr} - a^2 r\,(T - T_m) = 0$$

$$r^2\frac{d^2T}{dr^2} + r\frac{dT}{dr} - a^2 r^2\,(T - T_m) = 0$$

The given boundary-value problem is

$$r^2 \frac{d^2 T}{dr^2} + r \frac{dT}{dr} - a^2 r^2 (T - 70) = 0, \quad T(1) = 160, \quad \left. \frac{dT}{dr} \right|_{r=3} = 0$$

If we change the dependent variable by means of the substitution $w(r) = T(r) - 70$ the differential equation becomes

$$r^2 \frac{d^2 w}{dr^2} + r \frac{dw}{dr} - a^2 r^2 w = 0.$$

This differential equation is the parametric form of the modified Bessel's equation of order $\nu = 0$ so:

$$w(r) = c_1 I_0(ar) + c_2 K_0(ar)$$
$$T(r) = 70 + w(r)$$
$$T(r) = 70 + c_1 I_0(ar) + c_2 K_0(ar)$$

The derivatives $\frac{d}{dx} I_0(x) = I_1(x)$ and $\frac{d}{dx} K_0(x) = -K_1(x)$ and so the Chain rule gives

$$\frac{d}{dr} I_0 \overbrace{(ar)}^{u} = \frac{dI_0}{du} \frac{du}{dr} = I_0'(ar) \frac{d}{dr} ar = a I_1(ar) \quad \text{and} \quad \frac{d}{dr} K_0(ar) = -a K_1(ar)$$

Therefore

$$T'(r) = c_1 \frac{d}{dr} I_0(ar) + c_2 \frac{d}{dr} K_0(ar) \quad \text{or} \quad T'(r) = c_1 a I_1(ar) - c_2 a K_1(ar)$$

The boundary conditions $T(1) = 160$ and $T'(3) = 0$ then give the linear system of equations

$$c_1 I_0(a) + c_2 K_0(a) = 90$$
$$c_1 I_1(3a) - c_2 K_1(3a) = 0.$$

By Cramer's Rule, the solution of the system is

$$c_1 = \frac{\begin{vmatrix} 90 & K_0(a) \\ 0 & -K_1(3a) \end{vmatrix}}{\begin{vmatrix} I_0(a) & K_0(a) \\ I_1(3a) & -K_1(3a) \end{vmatrix}} = \frac{90 K_1(3a)}{I_0(a) K_1(3a) + I_1(3a) K_0(a)},$$

$$c_2 = \frac{\begin{vmatrix} I_0(a) & 90 \\ I_1(3a) & 0 \end{vmatrix}}{\begin{vmatrix} I_0(a) & K_0(a) \\ I_1(3a) & -K_1(3a) \end{vmatrix}} = \frac{90 I_1(3a)}{I_0(a) K_1(3a) + I_1(3a) K_0(a)}.$$

Hence the solution of the boundary-value problem is

$$T(r) = 70 + c_1 I_0(ar) + c_2 K_0(ar)$$

$$= 70 + \frac{90K_1(3a)I_0(ar)}{I_0(a)K_1(3a) + I_1(3a)K_0(a)} + \frac{90I_1(3a)K_0(ar)}{I_0(a)K_1(3a) + I_1(3a)K_0(a)}$$

$$= 70 + 90\frac{K_1(3a)I_0(ar) + I_1(3a)K_0(ar)}{K_1(3a)I_0(a) + I_1(3a)K_0(a)} \,.$$

Chapter 7

The Laplace Transform

1. $\mathscr{L}\{f(t)\} = \int_0^1 (-e^{-st}) \, dt + \int_1^\infty e^{-st} \, dt = \left. \frac{1}{s} e^{-st} \right|_0^1 - \left. \frac{1}{s} e^{-st} \right|_1^\infty$

$\quad = \frac{1}{s} e^{-s} - \frac{1}{s} - \left(0 - \frac{1}{s} e^{-s} \right) = \frac{2}{s} e^{-s} - \frac{1}{s}, \quad s > 0$

3. $\mathscr{L}\{f(t)\} = \int_0^1 t e^{-st} \, dt + \int_1^\infty e^{-st} \, dt = \left. \left(-\frac{1}{s} t e^{-st} - \frac{1}{s^2} e^{-st} \right) \right|_0^1 - \left. \frac{1}{s} e^{-st} \right|_1^\infty$

$\quad = \left(-\frac{1}{s} e^{-s} - \frac{1}{s^2} e^{-s} \right) - \left(0 - \frac{1}{s^2} \right) - \frac{1}{s} (0 - e^{-s}) = \frac{1}{s^2} (1 - e^{-s}), \quad s > 0$

5. $\mathscr{L}\{f(t)\} = \int_0^\pi (\sin t) e^{-st} \, dt = \left. \left(-\frac{s}{s^2 + 1} e^{-st} \sin t - \frac{1}{s^2 + 1} e^{-st} \cos t \right) \right|_0^\pi$

$\quad = \left(0 + \frac{1}{s^2 + 1} e^{-\pi s} \right) - \left(0 - \frac{1}{s^2 + 1} \right) = \frac{1}{s^2 + 1} (e^{-\pi s} + 1), \quad s > 0$

7. $f(t) = \begin{cases} 0, & 0 < t < 1 \\ t, & t > 1 \end{cases}$

$\quad \mathscr{L}\{f(t)\} = \int_1^\infty t e^{-st} \, dt = \left. \left(-\frac{1}{s} t e^{-st} - \frac{1}{s^2} e^{-st} \right) \right|_1^\infty = \frac{1}{s} e^{-s} + \frac{1}{s^2} e^{-s}, \quad s > 0$

9. $f(t) = \begin{cases} 1 - t, & 0 < t < 1 \\ 0, & t > 1 \end{cases}$

$\quad \mathscr{L}\{f(t)\} = \int_0^1 (1 - t) e^{-st} \, dt + \int_1^\infty 0 e^{-st} \, dt = \int_0^1 (1 - t) e^{-st} \, dt$

$\quad = \left. \left(-\frac{1}{s} (1 - t) e^{-st} + \frac{1}{s^2} e^{-st} \right) \right|_0^1 = \frac{1}{s^2} e^{-s} + \frac{1}{s} - \frac{1}{s^2}, \quad s > 0$

214   CHAPTER 7   THE LAPLACE TRANSFORM

11. $\mathscr{L}\{f(t)\} = \displaystyle\int_0^\infty e^{t+7}e^{-st}\,dt = e^7\int_0^\infty e^{(1-s)t}\,dt$

$$= \frac{e^7}{1-s}e^{(1-s)t}\Big|_0^\infty = 0 - \frac{e^7}{1-s} = \frac{e^7}{s-1}, \quad s > 1$$

13. $\mathscr{L}\{f(t)\} = \displaystyle\int_0^\infty te^{4t}e^{-st}\,dt = \int_0^\infty te^{(4-s)t}\,dt$

$$= \left(\frac{1}{4-s}te^{(4-s)t} - \frac{1}{(4-s)^2}e^{(4-s)t}\right)\Big|_0^\infty = \frac{1}{(4-s)^2}, \quad s > 4$$

15. $\mathscr{L}\{f(t)\} = \displaystyle\int_0^\infty e^{-t}(\sin t)e^{-st}\,dt = \int_0^\infty (\sin t)e^{-(s+1)t}\,dt$

$$= \left(\frac{-(s+1)}{(s+1)^2+1}e^{-(s+1)t}\sin t - \frac{1}{(s+1)^2+1}e^{-(s+1)t}\cos t\right)\Big|_0^\infty$$

$$= \frac{1}{(s+1)^2+1} = \frac{1}{s^2+2s+2}, \quad s > -1$$

17. $\mathscr{L}\{f(t)\} = \displaystyle\int_0^\infty t(\cos t)e^{-st}\,dt$

$$= \left[\left(-\frac{st}{s^2+1} - \frac{s^2-1}{(s^2+1)^2}\right)(\cos t)e^{-st} + \left(\frac{t}{s^2+1} + \frac{2s}{(s^2+1)^2}\right)(\sin t)e^{-st}\right]_0^\infty$$

$$= \frac{s^2-1}{(s^2+1)^2}, \quad s > 0$$

19. $\mathscr{L}\{2t^4\} = 2\dfrac{4!}{s^5} = \dfrac{48}{s^5}$

21. $\mathscr{L}\{4t-10\} = \dfrac{4}{s^2} - \dfrac{10}{s}$

23. $\mathscr{L}\{t^2+6t-3\} = \dfrac{2}{s^3} + \dfrac{6}{s^2} - \dfrac{3}{s}$

25. $\mathscr{L}\{t^3+3t^2+3t+1\} = \dfrac{6}{s^4} + \dfrac{6}{s^3} + \dfrac{3}{s^2} + \dfrac{1}{s}$

27. $\mathscr{L}\{1+e^{4t}\} = \dfrac{1}{s} + \dfrac{1}{s-4}$

29. $\mathscr{L}\{1+2e^{2t}+e^{4t}\} = \dfrac{1}{s} + \dfrac{2}{s-2} + \dfrac{1}{s-4}$

31. $\mathscr{L}\{4t^2-5\sin 3t\} = \dfrac{8}{s^3} - \dfrac{15}{s^2+9}$

33. $\mathscr{L}\{\sinh kt\} = \dfrac{1}{2}\mathscr{L}\{e^{kt}-e^{-kt}\} = \dfrac{1}{2}\left[\dfrac{1}{s-k} - \dfrac{1}{s+k}\right] = \dfrac{k}{s^2-k^2}$

35. $\mathscr{L}\{e^t\sinh t\} = \mathscr{L}\left\{e^t\dfrac{e^t-e^{-t}}{2}\right\} = \mathscr{L}\left\{\dfrac{1}{2}e^{2t} - \dfrac{1}{2}\right\} = \dfrac{1}{2(s-2)} - \dfrac{1}{2s}$

37. $\mathscr{L}\{\sin 2t\cos 2t\} = \mathscr{L}\left\{\dfrac{1}{2}\sin 4t\right\} = \dfrac{2}{s^2+16}$

39. From the addition formula for the sine function, $\sin(4t+5) = \sin 4t \cos 5 + \cos 4t \sin 5$ so

$$\mathscr{L}\{\sin(4t+5)\} = (\cos 5)\,\mathscr{L}\{\sin 4t\} + (\sin 5)\,\mathscr{L}\{\cos 4t\}$$

$$= (\cos 5)\,\frac{4}{s^2+16} + (\sin 5)\,\frac{s}{s^2+16} = \frac{4\cos 5 + (\sin 5)s}{s^2+16}.$$

41. Use integration by parts for $\alpha > 0$ to get

$$\Gamma(\alpha+1) = \int_0^\infty t^\alpha e^{-t}\,dt = -t^\alpha e^{-t}\Big|_0^\infty + \alpha \int_0^\infty t^{\alpha-1} e^{-t}\,dt = \alpha\Gamma(\alpha)$$

43. $\mathscr{L}\left\{t^{-1/2}\right\} = \dfrac{1}{s^{-1/2+1}}\Gamma\left(-\dfrac{1}{2}+1\right) = \dfrac{\Gamma\left(\frac{1}{2}\right)}{s^{1/2}} = \sqrt{\dfrac{\pi}{s}}$

45. $\mathscr{L}\left\{t^{3/2}\right\} = \dfrac{1}{s^{3/2+1}}\Gamma\left(\dfrac{3}{2}+1\right) = \dfrac{\left(\frac{3}{2}\right)\Gamma\left(\frac{3}{2}\right)}{s^{5/2}} = \dfrac{3}{2s^{5/2}}\Gamma\left(\dfrac{1}{2}+1\right) = \dfrac{3}{4s^{5/2}}\Gamma\left(\dfrac{1}{2}\right) = \dfrac{3\sqrt{\pi}}{4s^{5/2}}$

47. The relation will be valid when s is greater than the maximum of c_1 and c_2.

49. Assuming that (c) of Theorem 7.1.1 is applicable with a complex exponent, we have

$$\mathscr{L}\{e^{(a+ib)t}\} = \frac{1}{s-(a+ib)} = \frac{1}{(s-a)-ib}\,\frac{(s-a)+ib}{(s-a)+ib} = \frac{s-a+ib}{(s-a)^2+b^2}.$$

By Euler's formula, $e^{i\theta} = \cos\theta + i\sin\theta$, so

$$\mathscr{L}\{e^{(a+ib)t}\} = \mathscr{L}\{e^{at}e^{ibt}\} = \mathscr{L}\{e^{at}(\cos bt + i\sin bt)\}$$

$$= \mathscr{L}\{e^{at}\cos bt\} + i\mathscr{L}\{e^{at}\sin bt\}$$

$$= \frac{s-a}{(s-a)^2+b^2} + i\,\frac{b}{(s-a)^2+b^2}.$$

Equating real and imaginary parts we get

$$\mathscr{L}\{e^{at}\cos bt\} = \frac{s-a}{(s-a)^2+b^2} \quad\text{and}\quad \mathscr{L}\{e^{at}\sin bt\} = \frac{b}{(s-a)^2+b^2}.$$

51. As written, the function is not defined at $x = 5$. As should be written, $f(x) = 1/(t-5)$, $t \geq 5$, the function f is not bounded on, say, the interval $[4,6]$ because it has an infinite discontiunity at $x = 5$.

53. Applying the definition of the Laplace transform gives us

$$\mathscr{L}\left\{2te^{t^2}\cos e^{t^2}\right\} = \int_0^\infty e^{-st}\,2te^{t^2}\cos e^{t^2}\,dt$$

$$= e^{-st}\sin e^{t^2}\Big|_0^\infty + s\int_0^\infty e^{-st}\sin e^{t^2}\,dt$$

$$= -\sin 1 + s\int_0^\infty e^{-st}\sin e^{t^2}\,dt$$

The last integral exists for $s > 0$ since it's piecewise continuous on $(0, \infty)$ and of exponential order. Alternatively, for $s > 0$

$$\int_0^\infty \left| e^{-st} \sin e^{t^2} \right| dt \le 1 \cdot \int_0^\infty e^{-st}\, dt = \frac{1}{s}$$

The absolute convergence of the integral $\int_0^\infty e^{-st} \sin e^{t^2}\, dt$ implies the convergence of the integral.

55. $\mathscr{L}\left\{ e^{at} \right\} = \dfrac{1}{a} \cdot \dfrac{1}{(s/a - 1)} = \dfrac{1}{s - a}$

57. $\mathscr{L}\left\{ 1 - \cos kt \right\} = \dfrac{1}{k} \cdot \dfrac{1}{(s/k)\left[(s/k)^2 + 1\right]} = \dfrac{1}{k} \cdot \dfrac{k^3}{s\,(s^2 + k^2)} = \dfrac{k^2}{s\,(s^2 + k^2)}$

7.2   The Inverse Transform and Transforms of Derivatives

1. $\mathscr{L}^{-1}\left\{ \dfrac{1}{s^3} \right\} = \dfrac{1}{2}\mathscr{L}^{-1}\left\{ \dfrac{2}{s^3} \right\} = \dfrac{1}{2}t^2$

3. $\mathscr{L}^{-1}\left\{ \dfrac{1}{s^2} - \dfrac{48}{s^5} \right\} = \mathscr{L}^{-1}\left\{ \dfrac{1}{s^2} - \dfrac{48}{24} \cdot \dfrac{4!}{s^5} \right\} = t - 2t^4$

5. $\mathscr{L}^{-1}\left\{ \dfrac{(s+1)^3}{s^4} \right\} = \mathscr{L}^{-1}\left\{ \dfrac{1}{s} + 3 \cdot \dfrac{1}{s^2} + \dfrac{3}{2} \cdot \dfrac{2}{s^3} + \dfrac{1}{6} \cdot \dfrac{3!}{s^4} \right\} = 1 + 3t + \dfrac{3}{2}t^2 + \dfrac{1}{6}t^3$

7. $\mathscr{L}^{-1}\left\{ \dfrac{1}{s^2} - \dfrac{1}{s} + \dfrac{1}{s - 2} \right\} = t - 1 + e^{2t}$

9. $\mathscr{L}^{-1}\left\{ \dfrac{1}{4s + 1} \right\} = \dfrac{1}{4}\mathscr{L}^{-1}\left\{ \dfrac{1}{s + 1/4} \right\} = \dfrac{1}{4}e^{-t/4}$

11. $\mathscr{L}^{-1}\left\{ \dfrac{5}{s^2 + 49} \right\} = \mathscr{L}^{-1}\left\{ \dfrac{5}{7} \cdot \dfrac{7}{s^2 + 49} \right\} = \dfrac{5}{7}\sin 7t$

13. $\mathscr{L}^{-1}\left\{ \dfrac{4s}{4s^2 + 1} \right\} = \mathscr{L}^{-1}\left\{ \dfrac{s}{s^2 + 1/4} \right\} = \cos\dfrac{1}{2}t$

15. $\mathscr{L}^{-1}\left\{ \dfrac{2s - 6}{s^2 + 9} \right\} = \mathscr{L}^{-1}\left\{ 2 \cdot \dfrac{s}{s^2 + 9} - 2 \cdot \dfrac{3}{s^2 + 9} \right\} = 2\cos 3t - 2\sin 3t$

17. $\mathscr{L}^{-1}\left\{ \dfrac{1}{s^2 + 3s} \right\} = \mathscr{L}^{-1}\left\{ \dfrac{1}{3} \cdot \dfrac{1}{s} - \dfrac{1}{3} \cdot \dfrac{1}{s + 3} \right\} = \dfrac{1}{3} - \dfrac{1}{3}e^{-3t}$

19. $\mathscr{L}^{-1}\left\{ \dfrac{s}{s^2 + 2s - 3} \right\} = \mathscr{L}^{-1}\left\{ \dfrac{1}{4} \cdot \dfrac{1}{s - 1} + \dfrac{3}{4} \cdot \dfrac{1}{s + 3} \right\} = \dfrac{1}{4}e^{t} + \dfrac{3}{4}e^{-3t}$

21. $\mathscr{L}^{-1}\left\{ \dfrac{0.9s}{(s - 0.1)(s + 0.2)} \right\} = \mathscr{L}^{-1}\left\{ (0.3) \cdot \dfrac{1}{s - 0.1} + (0.6) \cdot \dfrac{1}{s + 0.2} \right\} = 0.3e^{0.1t} + 0.6e^{-0.2t}$

23. $\mathscr{L}^{-1}\left\{ \dfrac{s}{(s - 2)(s - 3)(s - 6)} \right\} = \mathscr{L}^{-1}\left\{ \dfrac{1}{2} \cdot \dfrac{1}{s - 2} - \dfrac{1}{s - 3} + \dfrac{1}{2} \cdot \dfrac{1}{s - 6} \right\} = \dfrac{1}{2}e^{2t} - e^{3t} + \dfrac{1}{2}e^{6t}$

25. $\mathcal{L}^{-1}\left\{\dfrac{1}{s^3+5s}\right\} = \mathcal{L}^{-1}\left\{\dfrac{1}{s(s^2+5)}\right\} = \mathcal{L}^{-1}\left\{\dfrac{1}{5}\dfrac{1}{s} - \dfrac{1}{5}\dfrac{s}{s^2+5}\right\} = \dfrac{1}{5} - \dfrac{1}{5}\cos\sqrt{5}\,t$

27. $\mathcal{L}^{-1}\left\{\dfrac{2s-4}{(s^2+s)(s^2+1)}\right\} = \mathcal{L}^{-1}\left\{\dfrac{2s-4}{s(s+1)(s^2+1)}\right\}$

$$= \mathcal{L}^{-1}\left\{-\dfrac{4}{s} + \dfrac{3}{s+1} + \dfrac{s}{s^2+1} + \dfrac{3}{s^2+1}\right\}$$

$$= -4 + 3e^{-t} + \cos t + 3\sin t$$

29. $\mathcal{L}^{-1}\left\{\dfrac{1}{(s^2+1)(s^2+4)}\right\} = \mathcal{L}^{-1}\left\{\dfrac{1}{3}\cdot\dfrac{1}{s^2+1} - \dfrac{1}{3}\cdot\dfrac{1}{s^2+4}\right\}$

$$= \mathcal{L}^{-1}\left\{\dfrac{1}{3}\cdot\dfrac{1}{s^2+1} - \dfrac{1}{6}\cdot\dfrac{2}{s^2+4}\right\} = \dfrac{1}{3}\sin t - \dfrac{1}{6}\sin 2t$$

31. Since $f(t) = e^{at}\sinh bt = e^{at}\frac{1}{2}\left(e^{bt}-e^{-bt}\right) = \frac{1}{2}e^{(1+b)t} - \frac{1}{2}e^{(a-b)t}$, then by linearity and part (c) of Theorem 7.1.1

$$\mathcal{L}\left\{\dfrac{1}{2}e^{(a+b)t} - \dfrac{1}{2}e^{(a-b)t}\right\} = \dfrac{1}{2}\dfrac{1}{s-(a+b)} - \dfrac{1}{2}\dfrac{1}{s-(a-b)} = \dfrac{b}{((s-a)-b)((s-a)+b)}$$

$$= \dfrac{b}{(s-a)^2-b^2}$$

$$\mathcal{L}^{-1}\left\{\dfrac{1}{(s-a)^2-b^2}\right\} = \dfrac{1}{b}e^{at}\sinh bt$$

33. By linearity and part (d) of Theorem 7.1.1:

$$\mathcal{L}\{a\sin bt - b\sin at\} = a\dfrac{b}{s^2+b^2} - b\dfrac{a}{s^2+a^2} = \dfrac{a^3b-ab^3}{(s^2+a^2)(s^2+b^2)}$$

$$= \dfrac{ab(a^2-b^2)}{(s^2+a^2)(s^2+b^2)}$$

$$\mathcal{L}^{-1}\left\{\dfrac{1}{(s^2+a^2)(s^2+b^2)}\right\} = \dfrac{a\sin bt - b\sin at}{ab(a^2-b^2)}$$

35. The Laplace transform of the differential equation is

$$s\mathcal{L}\{y\} - y(0) - \mathcal{L}\{y\} = \dfrac{1}{s}.$$

Solving for $\mathcal{L}\{y\}$ we obtain

$$\mathcal{L}\{y\} = -\dfrac{1}{s} + \dfrac{1}{s-1}.$$

Thus

$$y = -1 + e^t.$$

37. The Laplace transform of the differential equation is

$$s\mathscr{L}\{y\} - y(0) + 6\mathscr{L}\{y\} = \frac{1}{s-4}.$$

Solving for $\mathscr{L}\{y\}$ we obtain

$$\mathscr{L}\{y\} = \frac{1}{(s-4)(s+6)} + \frac{2}{s+6} = \frac{1}{10} \cdot \frac{1}{s-4} + \frac{19}{10} \cdot \frac{1}{s+6}.$$

Thus

$$y = \frac{1}{10}e^{4t} + \frac{19}{10}e^{-6t}.$$

39. The Laplace transform of the differential equation is

$$s^2\mathscr{L}\{y\} - sy(0) - y'(0) + 5\left[s\mathscr{L}\{y\} - y(0)\right] + 4\mathscr{L}\{y\} = 0.$$

Solving for $\mathscr{L}\{y\}$ we obtain

$$\mathscr{L}\{y\} = \frac{s+5}{s^2+5s+4} = \frac{4}{3}\frac{1}{s+1} - \frac{1}{3}\frac{1}{s+4}.$$

Thus

$$y = \frac{4}{3}e^{-t} - \frac{1}{3}e^{-4t}.$$

41. The Laplace transform of the differential equation is

$$s^2\mathscr{L}\{y\} - sy(0) - y'(0) + \mathscr{L}\{y\} = \frac{2}{s^2+2}.$$

Solving for $\mathscr{L}\{y\}$ we obtain

$$\mathscr{L}\{y\} = \frac{2}{(s^2+1)(s^2+2)} + \frac{10s}{s^2+1} = \frac{10s}{s^2+1} + \frac{2}{s^2+1} - \frac{2}{s^2+2}.$$

Thus

$$y = 10\cos t + 2\sin t - \sqrt{2}\sin\sqrt{2}\,t.$$

43. The Laplace transform of the differential equation is

$$2\left[s^3\mathscr{L}\{y\} - s^2(0) - sy'(0) - y''(0)\right] + 3\left[s^2\mathscr{L}\{y\} - sy(0) - y'(0)\right] - 3[s\mathscr{L}\{y\} - y(0)] - 2\mathscr{L}\{y\}$$

$$= \frac{1}{s+1}.$$

Solving for $\mathscr{L}\{y\}$ we obtain

$$\mathscr{L}\{y\} = \frac{2s+3}{(s+1)(s-1)(2s+1)(s+2)} = \frac{1}{2}\frac{1}{s+1} + \frac{5}{18}\frac{1}{s-1} - \frac{8}{9}\frac{1}{s+1/2} + \frac{1}{9}\frac{1}{s+2}.$$

Thus

$$y = \frac{1}{2}e^{-t} + \frac{5}{18}e^{t} - \frac{8}{9}e^{-t/2} + \frac{1}{9}e^{-2t}.$$

45. The Laplace transform of the differential equation is

$$s\mathscr{L}\{y\} + \mathscr{L}\{y\} = \frac{s+3}{s^2+6s+13}.$$

Solving for $\mathscr{L}\{y\}$ we obtain

$$\mathscr{L}\{y\} = \frac{s+3}{(s+1)(s^2+6s+13)} = \frac{1}{4}\cdot\frac{1}{s+1} - \frac{1}{4}\cdot\frac{s+1}{s^2+6s+13}$$

$$= \frac{1}{4}\cdot\frac{1}{s+1} - \frac{1}{4}\left(\frac{s+3}{(s+3)^2+4} - \frac{2}{(s+3)^2+4}\right).$$

Thus

$$y = \frac{1}{4}e^{-t} - \frac{1}{4}e^{-3t}\cos 2t + \frac{1}{4}e^{-3t}\sin 2t.$$

47. The Laplace transform of the differential equation in the initial-value problem is

$$\left(s^2+4\right)Y(s) = \frac{10s}{s^2+25}$$

$$Y(s) = \frac{10s}{\left(s^2+4\right)\left(s^2+25\right)}$$

with the identifications $a^2 = 4$ and $b^2 = 25$ or $a^2 - b^2 = -21$ we have from Problem 34:

$$\mathscr{L}^{-1}\left\{\frac{s}{(s^2+a^2)(s^2+b^2)}\right\} = \frac{\cos bt - \cos at}{a^2 - b^2}$$

$$y(t) = 10\mathscr{L}^{-1}\left\{\frac{s}{(s^2+4)(s^2+25)}\right\} = 10\cdot\frac{\cos 5t - \cos 2t}{-21}$$

$$y(t) = \frac{10}{21}\cos 2t - \frac{10}{21}\cos 5t$$

49. (a) Differentiating $f(t) = te^{at}$ we get $f'(t) = ate^{at} + e^{at}$ so $\mathscr{L}\{ate^{at} + e^{at}\} = s\mathscr{L}\{te^{at}\}$, where we have used $f(0) = 0$. Writing the equation as

$$a\mathscr{L}\{te^{at}\} + \mathscr{L}\left\{e^{at}\right\} = s\mathscr{L}\left\{te^{at}\right\}$$

and solving for $\mathscr{L}\{te^{at}\}$ we get

$$\mathscr{L}\left\{te^{at}\right\} = \frac{1}{s-a}\mathscr{L}\left\{e^{at}\right\} = \frac{1}{(s-a)^2}.$$

(b) Starting with $f(t) = t\sin kt$ we have

$$f'(t) = kt\cos kt + \sin kt$$
$$f''(t) = -k^2 t\sin kt + 2k\cos kt.$$

Then

$$\mathscr{L}\left\{-k^2 t\sin t + 2k\cos kt\right\} = s^2\mathscr{L}\{t\sin kt\}$$

where we have used $f(0) = 0$ and $f'(0) = 0$. Writing the above equation as

$$-k^2 \mathscr{L}\{t \sin kt\} + 2k\mathscr{L}\{\cos kt\} = s^2 \mathscr{L}\{t \sin kt\}$$

and solving for $\mathscr{L}\{t \sin kt\}$ gives

$$\mathscr{L}\{t \sin kt\} = \frac{2k}{s^2+k^2}\mathscr{L}\{\cos kt\} = \frac{2k}{s^2+k^2}\frac{s}{s^2+k^2} = \frac{2ks}{(s^2+k^2)^2}.$$

51. For $y'' - 4y' = 6e^{3t} - 3e^{-t}$ the transfer function is $W(s) = 1/(s^2 - 4s)$. The zero-input response is

$$y_0(t) = \mathscr{L}^{-1}\left\{\frac{s-5}{s^2-4s}\right\} = \mathscr{L}^{-1}\left\{\frac{5}{4}\cdot\frac{1}{s} - \frac{1}{4}\cdot\frac{1}{s-4}\right\} = \frac{5}{4} - \frac{1}{4}e^{4t},$$

and the zero-state response is

$$y_1(t) = \mathscr{L}^{-1}\left\{\frac{6}{(s-3)(s^2-4s)} - \frac{3}{(s+1)(s^2-4s)}\right\}$$

$$= \mathscr{L}^{-1}\left\{\frac{27}{20}\cdot\frac{1}{s-4} - \frac{2}{s-3} + \frac{5}{4}\cdot\frac{1}{s} - \frac{3}{5}\cdot\frac{1}{s+1}\right\}$$

$$= \frac{27}{20}e^{4t} - 2e^{3t} + \frac{5}{4} - \frac{3}{5}e^{-t}.$$

7.3   Operational Properties I

1. $\mathscr{L}\{te^{10t}\} = \dfrac{1}{(s-10)^2}$

3. $\mathscr{L}\{t^3 e^{-2t}\} = \dfrac{3!}{(s+2)^4}$

5. $\mathscr{L}\{t(e^t + e^{2t})^2\} = \mathscr{L}\{te^{2t} + 2te^{3t} + te^{4t}\} = \dfrac{1}{(s-2)^2} + \dfrac{2}{(s-3)^2} + \dfrac{1}{(s-4)^2}$

7. $\mathscr{L}\{e^t \sin 3t\} = \dfrac{3}{(s-1)^2+9}$

9. $\mathscr{L}\{(1 - e^t + 3e^{-4t})\cos 5t\} = \mathscr{L}\{\cos 5t - e^t \cos 5t + 3e^{-4t}\cos 5t\}$

$$= \frac{s}{s^2+25} - \frac{s-1}{(s-1)^2+25} + \frac{3(s+4)}{(s+4)^2+25}$$

11. $\mathscr{L}^{-1}\left\{\dfrac{1}{(s+2)^3}\right\} = \mathscr{L}^{-1}\left\{\dfrac{1}{2}\dfrac{2}{(s+2)^3}\right\} = \dfrac{1}{2}t^2 e^{-2t}$

13. $\mathscr{L}^{-1}\left\{\dfrac{1}{s^2-6s+10}\right\} = \mathscr{L}^{-1}\left\{\dfrac{1}{(s-3)^2+1^2}\right\} = e^{3t}\sin t$

15. $\mathscr{L}^{-1}\left\{\dfrac{s}{s^2+4s+5}\right\} = \mathscr{L}^{-1}\left\{\dfrac{s+2}{(s+2)^2+1^2} - 2\dfrac{1}{(s+2)^2+1^2}\right\} = e^{-2t}\cos t - 2e^{-2t}\sin t$

17. $\mathscr{L}^{-1}\left\{\dfrac{s}{(s+1)^2}\right\} = \mathscr{L}^{-1}\left\{\dfrac{s+1-1}{(s+1)^2}\right\} = \mathscr{L}^{-1}\left\{\dfrac{1}{s+1} - \dfrac{1}{(s+1)^2}\right\} = e^{-t} - te^{-t}$

19. $\mathscr{L}^{-1}\left\{\dfrac{2s-1}{s^2(s+1)^3}\right\} = \mathscr{L}^{-1}\left\{\dfrac{5}{s} - \dfrac{1}{s^2} - \dfrac{5}{s+1} - \dfrac{4}{(s+1)^2} - \dfrac{3}{2}\dfrac{2}{(s+1)^3}\right\}$

$$= 5 - t - 5e^{-t} - 4te^{-t} - \frac{3}{2}t^2 e^{-t}$$

21. The Laplace transform of the differential equation is

$$s\mathscr{L}\{y\} - y(0) + 4\mathscr{L}\{y\} = \frac{1}{s+4}.$$

Solving for $\mathscr{L}\{y\}$ we obtain

$$\mathscr{L}\{y\} = \frac{1}{(s+4)^2} + \frac{2}{s+4}.$$

Thus

$$y = te^{-4t} + 2e^{-4t}.$$

23. The Laplace transform of the differential equation is

$$s^2\mathscr{L}\{y\} - sy(0) - y'(0) + 2\left[s\mathscr{L}\{y\} - y(0)\right] + \mathscr{L}\{y\} = 0.$$

Solving for $\mathscr{L}\{y\}$ we obtain

$$\mathscr{L}\{y\} = \frac{s+3}{(s+1)^2} = \frac{1}{s+1} + \frac{2}{(s+1)^2}.$$

Thus

$$y = e^{-t} + 2te^{-t}.$$

25. The Laplace transform of the differential equation is

$$s^2\mathscr{L}\{y\} - sy(0) - y'(0) - 6\left[s\mathscr{L}\{y\} - y(0)\right] + 9\mathscr{L}\{y\} = \frac{1}{s^2}.$$

Solving for $\mathscr{L}\{y\}$ we obtain

$$\mathscr{L}\{y\} = \frac{1+s^2}{s^2(s-3)^2} = \frac{2}{27}\frac{1}{s} + \frac{1}{9}\frac{1}{s^2} - \frac{2}{27}\frac{1}{s-3} + \frac{10}{9}\frac{1}{(s-3)^2}.$$

Thus

$$y = \frac{2}{27} + \frac{1}{9}t - \frac{2}{27}e^{3t} + \frac{10}{9}te^{3t}.$$

27. The Laplace transform of the differential equation is

$$s^2\mathscr{L}\{y\} - sy(0) - y'(0) - 6\left[s\mathscr{L}\{y\} - y(0)\right] + 13\mathscr{L}\{y\} = 0.$$

Solving for $\mathscr{L}\{y\}$ we obtain

$$\mathscr{L}\{y\} = -\frac{3}{s^2-6s+13} = -\frac{3}{2}\frac{2}{(s-3)^2+2^2}.$$

Thus

$$y = -\frac{3}{2}e^{3t}\sin 2t.$$

222 CHAPTER 7 THE LAPLACE TRANSFORM

29. The Laplace transform of the differential equation is

$$s^2 \mathscr{L}\{y\} - sy(0) - y'(0) - [s\mathscr{L}\{y\} - y(0)] = \frac{s-1}{(s-1)^2 + 1}.$$

Solving for $\mathscr{L}\{y\}$ we obtain

$$\mathscr{L}\{y\} = \frac{1}{s(s^2 - 2s + 2)} = \frac{1}{2}\frac{1}{s} - \frac{1}{2}\frac{s-1}{(s-1)^2 + 1} + \frac{1}{2}\frac{1}{(s-1)^2 + 1}.$$

Thus

$$y = \frac{1}{2} - \frac{1}{2}e^t \cos t + \frac{1}{2}e^t \sin t.$$

31. Taking the Laplace transform of both sides of the differential equation and letting $c = y(0)$ we obtain

$$\mathscr{L}\{y''\} + \mathscr{L}\{2y'\} + \mathscr{L}\{y\} = 0$$
$$s^2 \mathscr{L}\{y\} - sy(0) - y'(0) + 2s\mathscr{L}\{y\} - 2y(0) + \mathscr{L}\{y\} = 0$$
$$s^2 \mathscr{L}\{y\} - cs - 2 + 2s\mathscr{L}\{y\} - 2c + \mathscr{L}\{y\} = 0$$
$$\left(s^2 + 2s + 1\right)\mathscr{L}\{y\} = cs + 2c + 2$$

$$\mathscr{L}\{y\} = \frac{cs}{(s+1)^2} + \frac{2c+2}{(s+1)^2}$$

$$= c\frac{s+1-1}{(s+1)^2} + \frac{2c+2}{(s+1)^2}$$

$$= \frac{c}{s+1} + \frac{c+2}{(s+1)^2}.$$

Therefore,

$$y(t) = c\mathscr{L}^{-1}\left\{\frac{1}{s+1}\right\} + (c+2)\mathscr{L}^{-1}\left\{\frac{1}{(s+1)^2}\right\} = ce^{-t} + (c+2)te^{-t}.$$

To find c we let $y(1) = 2$. Then $2 = ce^{-1} + (c+2)e^{-1} = 2(c+1)e^{-1}$ and $c = e - 1$. Thus

$$y(t) = (e-1)e^{-t} + (e+1)te^{-t}.$$

33. Recall from Section 3.8 that $mx'' = -kx - \beta x'$. Now $\beta = 7/8$, $m = W/g = 4/32 = 1/8$ slug, and $4 = 2k$ so that $k = 2$ lb/ft. Thus, the differential equation is $x'' + 7x' + 16x = 0$. The initial conditions are $x(0) = -3/2$ and $x'(0) = 0$. The Laplace transform of the differential equation is

$$s^2 \mathscr{L}\{x\} + \frac{3}{2}s + 7s\mathscr{L}\{x\} + \frac{21}{2} + 16\mathscr{L}\{x\} = 0.$$

Solving for $\mathscr{L}\{x\}$ we obtain

$$\mathscr{L}\{x\} = \frac{-3s/2 - 21/2}{s^2 + 7s + 16} = -\frac{3}{2}\frac{s+7/2}{(s+7/2)^2 + (\sqrt{15}/2)^2} - \frac{7\sqrt{15}}{10}\frac{\sqrt{15}/2}{(s+7/2)^2 + (\sqrt{15}/2)^2}.$$

Thus

$$x = -\frac{3}{2}e^{-7t/2}\cos\frac{\sqrt{15}}{2}t - \frac{7\sqrt{15}}{10}e^{-7t/2}\sin\frac{\sqrt{15}}{2}t.$$

© 2018 Cengage. May not be scanned, copied or duplicated, or posted to a publicly accessible website, in whole or in part.

35. The differential equation is

$$\frac{d^2 q}{dt^2} + 2\lambda dqdt + \omega^2 q = \frac{E_0}{L}, \quad q(0) = q'(0) = 0.$$

The Laplace transform of this equation is

$$s^2 \mathscr{L}\{q\} + 2\lambda s \mathscr{L}\{q\} + \omega^2 \mathscr{L}\{q\} = \frac{E_0}{L} \frac{1}{s}$$

or

$$\left(s^2 + 2\lambda s + \omega^2\right) \mathscr{L}\{q\} = \frac{E_0}{L} \frac{1}{s}.$$

Solving for $\mathscr{L}\{q\}$ and using partial fractions we obtain

$$\mathscr{L}\{q\} = \frac{E_0}{L} \left(\frac{1/\omega^2}{s} - \frac{(1/\omega^2)s + 2\lambda/\omega^2}{s^2 + 2\lambda s + \omega^2} \right) = \frac{E_0}{L\omega^2} \left(\frac{1}{s} - \frac{s + 2\lambda}{s^2 + 2\lambda s + \omega^2} \right).$$

For $\lambda > \omega$ we write $s^2 + 2\lambda s + \omega^2 = (s + \lambda)^2 - \left(\lambda^2 - \omega^2\right)$, so (recalling that $\omega^2 = 1/LC$)

$$\mathscr{L}\{q\} = E_0 C \left(\frac{1}{s} - \frac{s + \lambda}{(s + \lambda)^2 - (\lambda^2 - \omega^2)} - \frac{\lambda}{(s + \lambda)^2 - (\lambda^2 - \omega^2)} \right).$$

Thus for $\lambda > \omega$,

$$q(t) = E_0 C \left[1 - e^{-\lambda t} \left(\cosh \sqrt{\lambda^2 - \omega^2}\, t - \frac{\lambda}{\sqrt{\lambda^2 - \omega^2}} \sinh \sqrt{\lambda^2 - \omega^2}\, t \right) \right].$$

For $\lambda < \omega$ we write $s^2 + 2\lambda s + \omega^2 = (s + \lambda)^2 + \left(\omega^2 - \lambda^2\right)$, so

$$\mathscr{L}\{q\} = E_0 C \left(\frac{1}{s} - \frac{s + \lambda}{(s + \lambda)^2 + (\omega^2 - \lambda^2)} - \frac{\lambda}{(s + \lambda)^2 + (\omega^2 - \lambda^2)} \right).$$

Thus for $\lambda < \omega$,

$$q(t) = E_0 C \left[1 - e^{-\lambda t} \left(\cos \sqrt{\omega^2 - \lambda^2}\, t - \frac{\lambda}{\sqrt{\omega^2 - \lambda^2}} \sin \sqrt{\omega^2 - \lambda^2}\, t \right) \right].$$

For $\lambda = \omega$, $s^2 + 2\lambda + \omega^2 = (s + \lambda)^2$ and

$$\mathscr{L}\{q\} = \frac{E_0}{L} \frac{1}{s(s + \lambda)^2} = \frac{E_0}{L} \left(\frac{1/\lambda^2}{s} - \frac{1/\lambda^2}{s + \lambda} - \frac{1/\lambda}{(s + \lambda)^2} \right) = \frac{E_0}{L\lambda^2} \left(\frac{1}{s} - \frac{1}{s + \lambda} - \frac{\lambda}{(s + \lambda)^2} \right).$$

Thus for $\lambda = \omega$,

$$q(t) = E_0 C \left(1 - e^{-\lambda t} - \lambda t e^{-\lambda t} \right).$$

37. $\mathscr{L}\{(t - 1)\mathscr{U}(t - 1)\} = \dfrac{e^{-s}}{s^2}$

39. $\mathcal{L}\{t\mathcal{U}(t-2)\} = \mathcal{L}\{(t-2)\mathcal{U}(t-2)+2\mathcal{U}(t-2)\} = \dfrac{e^{-2s}}{s^2} + \dfrac{2e^{-2s}}{s}$

Alternatively, (16) of this section could be used:

$$\mathcal{L}\{t\mathcal{U}(t-2)\} = e^{-2s}\mathcal{L}\{t+2\} = e^{-2s}\left(\dfrac{1}{s^2}+\dfrac{2}{s}\right).$$

41. $\mathcal{L}\{\cos 2t\,\mathcal{U}(t-\pi)\} = \mathcal{L}\{\cos 2(t-\pi)\mathcal{U}(t-\pi)\} = \dfrac{se^{-\pi s}}{s^2+4}$

Alternatively, (16) of this section could be used:

$$\mathcal{L}\{\cos 2t\,\mathcal{U}(t-\pi)\} = e^{-\pi s}\mathcal{L}\{\cos 2(t+\pi)\} = e^{-\pi s}\mathcal{L}\{\cos 2t\} = e^{-\pi s}\dfrac{s}{s^2+4}.$$

43. $\mathcal{L}^{-1}\left\{\dfrac{e^{-2s}}{s^3}\right\} = \mathcal{L}^{-1}\left\{\dfrac{1}{2}\cdot\dfrac{2}{s^3}e^{-2s}\right\} = \dfrac{1}{2}(t-2)^2\mathcal{U}(t-2)$

45. $\mathcal{L}^{-1}\left\{\dfrac{e^{-\pi s}}{s^2+1}\right\} = \sin(t-\pi)\mathcal{U}(t-\pi) = -\sin t\,\mathcal{U}(t-\pi)$

47. $\mathcal{L}^{-1}\left\{\dfrac{e^{-s}}{s(s+1)}\right\} = \mathcal{L}^{-1}\left\{\dfrac{e^{-s}}{s}-\dfrac{e^{-s}}{s+1}\right\} = \mathcal{U}(t-1) - e^{-(t-1)}\mathcal{U}(t-1)$

49. (c)   **51.** (f)   **53.** (a)

55. $\mathcal{L}\{2-4\mathcal{U}(t-3)\} = \dfrac{2}{s} - \dfrac{4}{s}e^{-3s}$

57. $\mathcal{L}\{t^2\mathcal{U}(t-1)\} = \mathcal{L}\left\{\left[(t-1)^2+2t-1\right]\mathcal{U}(t-1)\right\}$

$$= \mathcal{L}\left\{\left[(t-1)^2+2(t-1)-1\right]\mathcal{U}(t-1)\right\} = \left(\dfrac{2}{s^3}+\dfrac{2}{s^2}+\dfrac{1}{s}\right)e^{-s}$$

Alternatively, by (16) of this section,

$$\mathcal{L}\{t^2\mathcal{U}(t-1)\} = e^{-s}\mathcal{L}\{t^2+2t+1\} = e^{-s}\left(\dfrac{2}{s^3}+\dfrac{2}{s^2}+\dfrac{1}{s}\right).$$

59. $\mathcal{L}\{t-t\mathcal{U}(t-2)\} = \mathcal{L}\{t-(t-2)\mathcal{U}(t-2)-2\mathcal{U}(t-2)\} = \dfrac{1}{s^2}-\dfrac{e^{-2s}}{s^2}-\dfrac{2e^{-2s}}{s}$

61. $\mathcal{L}\{f(t)\} = \mathcal{L}\{\mathcal{U}(t-a)-\mathcal{U}(t-b)\} = \dfrac{e^{-as}}{s}-\dfrac{e^{-bs}}{s}$

63. The Laplace transform of the differential equation is

$$s\mathcal{L}\{y\} - y(0) + \mathcal{L}\{y\} = \dfrac{5}{s}e^{-s}.$$

Solving for $\mathcal{L}\{y\}$ we obtain

$$\mathcal{L}\{y\} = \dfrac{5e^{-s}}{s(s+1)} = 5e^{-s}\left[\dfrac{1}{s}-\dfrac{1}{s+1}\right].$$

Thus

$$y = 5\mathcal{U}(t-1) - 5e^{-(t-1)}\mathcal{U}(t-1).$$

65. The Laplace transform of the differential equation is

$$s\mathscr{L}\{y\} - y(0) + 2\mathscr{L}\{y\} = \frac{1}{s^2} - e^{-s}\frac{s+1}{s^2}.$$

Solving for $\mathscr{L}\{y\}$ we obtain

$$\mathscr{L}\{y\} = \frac{1}{s^2(s+2)} - e^{-s}\frac{s+1}{s^2(s+2)} = -\frac{1}{4}\frac{1}{s} + \frac{1}{2}\frac{1}{s^2} + \frac{1}{4}\frac{1}{s+2} - e^{-s}\left[\frac{1}{4}\frac{1}{s} + \frac{1}{2}\frac{1}{s^2} - \frac{1}{4}\frac{1}{s+2}\right].$$

Thus

$$y = -\frac{1}{4} + \frac{1}{2}t + \frac{1}{4}e^{-2t} - \left[\frac{1}{4} + \frac{1}{2}(t-1) - \frac{1}{4}e^{-2(t-1)}\right]\mathscr{U}(t-1).$$

67. The Laplace transform of the differential equation is

$$s^2\mathscr{L}\{y\} - sy(0) - y'(0) + 4\mathscr{L}\{y\} = e^{-2\pi s}\frac{1}{s^2+1}.$$

Solving for $\mathscr{L}\{y\}$ we obtain

$$\mathscr{L}\{y\} = \frac{s}{s^2+4} + e^{-2\pi s}\left[\frac{1}{3}\frac{1}{s^2+1} - \frac{1}{6}\frac{2}{s^2+4}\right].$$

Thus

$$y = \cos 2t + \left[\frac{1}{3}\sin(t-2\pi) - \frac{1}{6}\sin 2(t-2\pi)\right]\mathscr{U}(t-2\pi).$$

69. The Laplace transform of the differential equation is

$$s^2\mathscr{L}\{y\} - sy(0) - y'(0) + \mathscr{L}\{y\} = \frac{e^{-\pi s}}{s} - \frac{e^{-2\pi s}}{s}.$$

Solving for $\mathscr{L}\{y\}$ we obtain

$$\mathscr{L}\{y\} = e^{-\pi s}\left[\frac{1}{s} - \frac{s}{s^2+1}\right] - e^{-2\pi s}\left[\frac{1}{s} - \frac{s}{s^2+1}\right] + \frac{1}{s^2+1}.$$

Thus

$$y = [1 - \cos(t-\pi)]\mathscr{U}(t-\pi) - [1 - \cos(t-2\pi)]\mathscr{U}(t-2\pi) + \sin t.$$

71. Recall from Section 3.8 that $mx'' = -kx + f(t)$. Now $m = W/g = 32/32 = 1$ slug, and $32 = 2k$ so that $k = 16$ lb/ft. Thus, the differential equation is $x'' + 16x = f(t)$. The initial conditions are $x(0) = 0$, $x'(0) = 0$. Also, since

$$f(t) = \begin{cases} 20t, & 0 \le t < 5 \\ 0, & t \ge 5 \end{cases}$$

and $20t = 20(t-5) + 100$ we can write

$$f(t) = 20t - 20t\mathscr{U}(t-5) = 20t - 20(t-5)\mathscr{U}(t-5) - 100\mathscr{U}(t-5).$$

The Laplace transform of the differential equation is

$$s^2\mathscr{L}\{x\} + 16\mathscr{L}\{x\} = \frac{20}{s^2} - \frac{20}{s^2}e^{-5s} - \frac{100}{s}e^{-5s}.$$

Solving for $\mathscr{L}\{x\}$ we obtain

$$\mathscr{L}\{x\} = \frac{20}{s^2(s^2 + 16)} - \frac{20}{s^2(s^2 + 16)}e^{-5s} - \frac{100}{s(s^2 + 16)}e^{-5s}$$

$$= \left(\frac{5}{4}\cdot\frac{1}{s^2} - \frac{5}{16}\cdot\frac{4}{s^2 + 16}\right)\left(1 - e^{-5s}\right) - \left(\frac{25}{4}\cdot\frac{1}{s} - \frac{25}{4}\cdot\frac{s}{s^2 + 16}\right)e^{-5s}.$$

Thus

$$x(t) = \frac{5}{4}t - \frac{5}{16}\sin 4t - \left[\frac{5}{4}(t - 5) - \frac{5}{16}\sin 4(t - 5)\right]\mathscr{U}(t - 5) - \left[\frac{25}{4} - \frac{25}{4}\cos 4(t - 5)\right]\mathscr{U}(t - 5)$$

$$= \frac{5}{4}t - \frac{5}{16}\sin 4t - \frac{5}{4}t\mathscr{U}(t - 5) + \frac{5}{16}\sin 4(t - 5)\mathscr{U}(t - 5) + \frac{25}{4}\cos 4(t - 5)\mathscr{U}(t - 5).$$

73. The differential equation is

$$2.5\frac{dq}{dt} + 12.5q = 5\mathscr{U}(t - 3).$$

The Laplace transform of this equation is

$$s\mathscr{L}\{q\} + 5\mathscr{L}\{q\} = \frac{2}{s}e^{-3s}.$$

Solving for $\mathscr{L}\{q\}$ we obtain

$$\mathscr{L}\{q\} = \frac{2}{s(s + 5)}e^{-3s} = \left(\frac{2}{5}\cdot\frac{1}{s} - \frac{2}{5}\cdot\frac{1}{s + 5}\right)e^{-3s}.$$

Thus

$$q(t) = \frac{2}{5}\mathscr{U}(t - 3) - \frac{2}{5}e^{-5(t-3)}\mathscr{U}(t - 3).$$

75. (a) The differential equation is

$$\frac{di}{dt} + 10i = \sin t + \cos\left(t - \frac{3\pi}{2}\right)\mathscr{U}\left(t - \frac{3\pi}{2}\right), \quad i(0) = 0.$$

The Laplace transform of this equation is

$$s\mathscr{L}\{i\} + 10\mathscr{L}\{i\} = \frac{1}{s^2 + 1} + \frac{se^{-3\pi s/2}}{s^2 + 1}.$$

Solving for $\mathscr{L}\{i\}$ we obtain

$$\mathscr{L}\{i\} = \frac{1}{(s^2 + 1)(s + 10)} + \frac{s}{(s^2 + 1)(s + 10)}e^{-3\pi s/2}$$

$$= \frac{1}{101}\left(\frac{1}{s + 10} - \frac{s}{s^2 + 1} + \frac{10}{s^2 + 1}\right) + \frac{1}{101}\left(\frac{-10}{s + 10} + \frac{10s}{s^2 + 1} + \frac{1}{s^2 + 1}\right)e^{-3\pi s/2}.$$

Thus

$$i(t) = \frac{1}{101}\left(e^{-10t} - \cos t + 10\sin t\right)$$

$$+ \frac{1}{101}\left[-10e^{-10(t-3\pi/2)} + 10\cos\left(t - \frac{3\pi}{2}\right) + \sin\left(t - \frac{3\pi}{2}\right)\right]\mathscr{U}\left(t - \frac{3\pi}{2}\right).$$

(b)

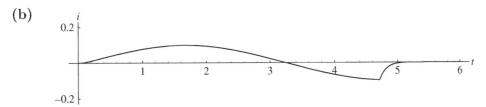

The maximum value of $i(t)$ is approximately 0.1 at $t = 1.7$, the minimum is approximately -0.1 at 4.7.

77. The differential equation is

$$EI\frac{d^4y}{dx^4} = w_0[1 - \mathscr{U}\left(x - \frac{L}{2}\right)].$$

Taking the Laplace transform of both sides and using $y(0) = y'(0) = 0$ we obtain

$$s^4\mathscr{L}\{y\} - sy''(0) - y'''(0) = \frac{w_0}{EI}\frac{1}{s}\left(1 - e^{-Ls/2}\right).$$

Letting $y''(0) = c_1$ and $y'''(0) = c_2$ we have

$$\mathscr{L}\{y\} = \frac{c_1}{s^3} + \frac{c_2}{s^4} + \frac{w_0}{EI}\frac{1}{s^5}\left(1 - e^{-Ls/2}\right)$$

so that

$$y(x) = \frac{1}{2}c_1x^2 + \frac{1}{6}c_2x^3 + \frac{1}{24}\frac{w_0}{EI}\left[x^4 - \left(x - \frac{L}{2}\right)^4\mathscr{U}\left(x - \frac{L}{2}\right)\right].$$

To find c_1 and c_2 we compute

$$y''(x) = c_1 + c_2x + \frac{1}{2}\frac{w_0}{EI}\left[x^2 - \left(x - \frac{L}{2}\right)^2\mathscr{U}\left(x - \frac{L}{2}\right)\right]$$

and

$$y'''(x) = c_2 + \frac{w_0}{EI}\left[x - \left(x - \frac{L}{2}\right)\mathscr{U}\left(x - \frac{L}{2}\right)\right].$$

Then $y''(L) = y'''(L) = 0$ yields the system

$$c_1 + c_2L + \frac{1}{2}\frac{w_0}{EI}\left[L^2 - \left(\frac{L}{2}\right)^2\right] = 0 \qquad c_1 + c_2L + \frac{3}{8}\frac{w_0L^2}{EI} = 0$$

or

$$c_2 + \frac{w_0}{EI}\left(\frac{L}{2}\right) = 0. \qquad c_2 + \frac{1}{2}\frac{w_0L}{EI} = 0.$$

Solving for c_1 and c_2 we obtain $c_1 = w_0 L^2/8EI$ and $c_2 = -w_0 L/2EI$. Thus

$$y(x) = \frac{w_0}{EI}\left[\frac{1}{16}L^2 x^2 - \frac{1}{12}Lx^3 + \frac{1}{24}x^4 - \frac{1}{24}\left(x - \frac{L}{2}\right)^4 \mathscr{U}\left(x - \frac{L}{2}\right)\right].$$

79. The differential equation is

$$EI\frac{d^4 y}{dx^4} = \frac{2w_0}{L}\left[\frac{L}{2} - x + \left(x - \frac{L}{2}\right)\mathscr{U}\left(x - \frac{L}{2}\right)\right].$$

Taking the Laplace transform of both sides and using $y(0) = y'(0) = 0$ we obtain

$$s^4 \mathscr{L}\{y\} - sy''(0) - y'''(0) = \frac{2w_0}{EIL}\left[\frac{L}{2s} - \frac{1}{s^2} + \frac{1}{s^2}e^{-Ls/2}\right].$$

Letting $y''(0) = c_1$ and $y'''(0) = c_2$ we have

$$\mathscr{L}\{y\} = \frac{c_1}{s^3} + \frac{c_2}{s^4} + \frac{2w_0}{EIL}\left[\frac{L}{2s^5} - \frac{1}{s^6} + \frac{1}{s^6}e^{-Ls/2}\right]$$

so that

$$y(x) = \frac{1}{2}c_1 x^2 + \frac{1}{6}c_2 x^3 + \frac{2w_0}{EIL}\left[\frac{L}{48}x^4 - \frac{1}{120}x^5 + \frac{1}{120}\left(x - \frac{L}{2}\right)^5 \mathscr{U}\left(x - \frac{L}{2}\right)\right]$$

$$= \frac{1}{2}c_1 x^2 + \frac{1}{6}c_2 x^3 + \frac{w_0}{60EIL}\left[\frac{5L}{2}x^4 - x^5 + \left(x - \frac{L}{2}\right)^5 \mathscr{U}\left(x - \frac{L}{2}\right)\right].$$

To find c_1 and c_2 we compute

$$y''(x) = c_1 + c_2 x + \frac{w_0}{60EIL}\left[30Lx^2 - 20x^3 + 20\left(x - \frac{L}{2}\right)^3 \mathscr{U}\left(x - \frac{L}{2}\right)\right]$$

and

$$y'''(x) = c_2 + \frac{w_0}{60EIL}\left[60Lx - 60x^2 + 60\left(x - \frac{L}{2}\right)^2 \mathscr{U}\left(x - \frac{L}{2}\right)\right].$$

Then $y''(L) = y'''(L) = 0$ yields the system

$$c_1 + c_2 L + \frac{w_0}{60EIL}\left[30L^3 - 20L^3 + \frac{5}{2}L^3\right] = 0 \qquad c_1 + c_2 L + \frac{5w_0 L^2}{24EI} = 0$$

$$\text{or}$$

$$c_2 + \frac{w_0}{60EIL}[60L^2 - 60L^2 + 15L^2] = 0. \qquad c_2 + \frac{w_0 L}{4EI} = 0.$$

Solving for c_1 and c_2 we obtain $c_1 = w_0 L^2/24EI$ and $c_2 = -w_0 L/4EI$. Thus

$$y(x) = \frac{w_0 L^2}{48EI}x^2 - \frac{w_0 L}{24EI}x^3 + \frac{w_0}{60EIL}\left[\frac{5L}{2}x^4 - x^5 + \left(x - \frac{L}{2}\right)^5 \mathscr{U}\left(x - \frac{L}{2}\right)\right].$$

81. (a) The temperature T of the cake inside the oven is modeled by

$$\frac{dT}{dt} = k(T - T_m)$$

where T_m is the ambient temperature of the oven. For $0 \le t \le 4$, we have

$$T_m = 70 + \frac{300 - 70}{4 - 0} t = 70 + 57.5t.$$

Hence for $t \ge 0$,

$$T_m = \begin{cases} 70 + 57.5t, & 0 \le t < 4 \\ 300, & t \ge 4. \end{cases}$$

In terms of the unit step function,

$$T_m = (70 + 57.5t)[1 - \mathscr{U}(t - 4)] + 300\,\mathscr{U}(t - 4) = 70 + 57.5t + (230 - 57.5t)\mathscr{U}(t - 4).$$

The initial-value problem is then

$$\frac{dT}{dt} = k[T - 70 - 57.5t - (230 - 57.5t)\mathscr{U}(t - 4)], \qquad T(0) = 70.$$

(b) Let $t(s) = \mathscr{L}\{T(t)\}$. Transforming the equation, using $230 - 57.5t = -57.5(t - 4)$ and Theorem 7.3.2, gives

$$st(s) - 70 = k\left(t(s) - \frac{70}{s} - \frac{57.5}{s^2} + \frac{57.5}{s^2}e^{-4s}\right)$$

or

$$t(s) = \frac{70}{s - k} - \frac{70k}{s(s - k)} - \frac{57.5k}{s^2(s - k)} + \frac{57.5k}{s^2(s - k)}e^{-4s}.$$

After using partial functions, the inverse transform is then

$$T(t) = 70 + 57.5\left(\frac{1}{k} + t - \frac{1}{k}e^{kt}\right) - 57.5\left(\frac{1}{k} + t - 4 - \frac{1}{k}e^{k(t-4)}\right)\mathscr{U}(t - 4).$$

Of course, the obvious question is: What is k? If the cake is supposed to bake for, say, 20 minutes, then $T(20) = 300$. That is,

$$300 = 70 + 57.5\left(\frac{1}{k} + 20 - \frac{1}{k}e^{20k}\right) - 57.5\left(\frac{1}{k} + 16 - \frac{1}{k}e^{16k}\right).$$

But this equation has no physically meaningful solution. This should be no surprise since the model predicts the asymptotic behavior $T(t) \to 300$ as t increases. Using $T(20) = 299$ instead, we find, with the help of a CAS, that $k \approx -0.3$.

83. (a) From Theorem 7.3.1 we have $\mathscr{L}\{te^{kti}\} = 1/(s - ki)^2$. Then, using Euler's formula,

$$\mathscr{L}\{te^{kti}\} = \mathscr{L}\{t\cos kt + it\sin kt\} = \mathscr{L}\{t\cos kt\} + i\mathscr{L}\{t\sin kt\}$$

$$= \frac{1}{(s - ki)^2} = \frac{(s + ki)^2}{(s^2 + k^2)^2} = \frac{s^2 - k^2}{(s^2 + k^2)^2} + i\frac{2ks}{(s^2 + k^2)^2}.$$

Equating real and imaginary parts we have

$$\mathscr{L}\{t\cos kt\} = \frac{s^2 - k^2}{(s^2 + k^2)^2} \quad \text{and} \quad \mathscr{L}\{t\sin kt\} = \frac{2ks}{(s^2 + k^2)^2}.$$

(b) The Laplace transform of the differential equation is

$$s^2 \mathscr{L}\{x\} + \omega^2 \mathscr{L}\{x\} = \frac{s}{s^2 + \omega^2}.$$

Solving for $\mathscr{L}\{x\}$ we obtain $\mathscr{L}\{x\} = s/(s^2 + \omega^2)^2$. Thus $x = (1/2\omega)t \sin \omega t$.

7.4   Operational Properties II

1. $\mathscr{L}\{te^{-10t}\} = -\dfrac{d}{ds}\left(\dfrac{1}{s+10}\right) = \dfrac{1}{(s+10)^2}$

3. $\mathscr{L}\{t \cos 2t\} = -\dfrac{d}{ds}\left(\dfrac{s}{s^2+4}\right) = \dfrac{s^2-4}{(s^2+4)^2}$

5. $\mathscr{L}\{t^2 \sinh t\} = \dfrac{d^2}{ds^2}\left(\dfrac{1}{s^2-1}\right) = \dfrac{6s^2+2}{(s^2-1)^3}$

7. $\mathscr{L}\{te^{2t}\sin 6t\} = -\dfrac{d}{ds}\left(\dfrac{6}{(s-2)^2+36}\right) = \dfrac{12(s-2)}{[(s-2)^2+36]^2}$

9. The Laplace transform of the differential equation is

$$s\mathscr{L}\{y\} + \mathscr{L}\{y\} = \frac{2s}{(s^2+1)^2}.$$

Solving for $\mathscr{L}\{y\}$ we obtain

$$\mathscr{L}\{y\} = \frac{2s}{(s+1)(s^2+1)^2} = -\frac{1}{2}\frac{1}{s+1} - \frac{1}{2}\frac{1}{s^2+1} + \frac{1}{2}\frac{s}{s^2+1} + \frac{1}{(s^2+1)^2} + \frac{s}{(s^2+1)^2}.$$

Thus

$$y(t) = -\frac{1}{2}e^{-t} - \frac{1}{2}\sin t + \frac{1}{2}\cos t + \frac{1}{2}(\sin t - t\cos t) + \frac{1}{2}t\sin t$$

$$= -\frac{1}{2}e^{-t} + \frac{1}{2}\cos t - \frac{1}{2}t\cos t + \frac{1}{2}t\sin t.$$

11. The Laplace transform of the differential equation is

$$s^2 \mathscr{L}\{y\} - sy(0) - y'(0) + 9\mathscr{L}\{y\} = \frac{s}{s^2+9}.$$

Letting $y(0) = 2$ and $y'(0) = 5$ and solving for $\mathscr{L}\{y\}$ we obtain

$$\mathscr{L}\{y\} = \frac{2s^3 + 5s^2 + 19s - 45}{(s^2+9)^2} = \frac{2s}{s^2+9} + \frac{5}{s^2+9} + \frac{s}{(s^2+9)^2}.$$

Thus

$$y = 2\cos 3t + \frac{5}{3}\sin 3t + \frac{1}{6}t\sin 3t.$$

13. The Laplace transform of the differential equation is

$$s^2\mathscr{L}\{y\} - sy(0) - y'(0) + 16\mathscr{L}\{y\} = \mathscr{L}\{\cos 4t - \cos 4t\,\mathscr{U}(t-\pi)\}$$

or by (16) of Section 7.3 in the text,

$$(s^2+16)\mathscr{L}\{y\} = 1 + \frac{s}{s^2+16} - e^{-\pi s}\mathscr{L}\{\cos 4(t+\pi)\}$$

$$= 1 + \frac{s}{s^2+16} - e^{-\pi s}\mathscr{L}\{\cos 4t\} = 1 + \frac{s}{s^2+16} - \frac{s}{s^2+16}e^{-\pi s}.$$

Thus

$$\mathscr{L}\{y\} = \frac{1}{s^2+16} + \frac{s}{(s^2+16)^2} - \frac{s}{(s^2+16)^2}e^{-\pi s}$$

and

$$y = \frac{1}{4}\sin 4t + \frac{1}{8}t\sin 4t - \frac{1}{8}(t-\pi)\sin 4(t-\pi)\,\mathscr{U}(t-\pi).$$

15.

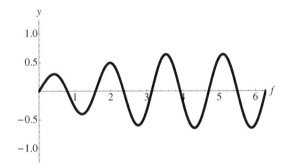

17. From (7) of Section 7.2 in the text along with Theorem 7.4.1,

$$\mathscr{L}\{ty''\} = -\frac{d}{ds}\mathscr{L}\{y''\} = -\frac{d}{ds}[s^2Y(s) - sy(0) - y'(0)] = -s^2\frac{dY}{ds} - 2sY + y(0),$$

so that the transform of the given second-order differential equation is the linear first-order differential equation in $Y(s)$:

$$s^2Y' + 3sY = -\frac{4}{s^3} \qquad \text{or} \qquad Y' + \frac{3}{s}Y = -\frac{4}{s^5}.$$

The solution of the latter equation is $Y(s) = 4/s^4 + c/s^3$, so

$$y(t) = \mathscr{L}^{-1}\{Y(s)\} = \frac{2}{3}t^3 + \frac{c}{2}t^2.$$

19. Identify $f(\tau) = 4\tau$ and $g(t-\tau) = 3(t-\tau)^2$. Therefore,

$$f * g = \int_0^t (4\tau)\,3(t-\tau)^2\,d\tau = 12\int_0^t \left(t^2\tau - 2t\tau^2 + \tau^3\right)d\tau = 12\left(t^2\cdot\frac{1}{2}t^2 - 2t\cdot\frac{1}{3}t^3 + \frac{1}{4}t^4\right) = t^4$$

Then

$$\mathscr{L}\{f * g\} = \mathscr{L}\{t^4\} = \frac{4!}{s^5} = \frac{24}{s^5}$$

21. Identify $f(\tau) = e^{-\tau}$ and $g(t - \tau) = e^{t-\tau}$. Therefore,

$$f * g = \int_0^t e^{-\tau} e^{t-\tau}\, d\tau = e^t \int_0^t e^{-2\tau}\, d\tau = e^t \cdot \left(-\frac{1}{2} e^{-2t} + \frac{1}{2} \right) = -\frac{1}{2} e^{-t} + \frac{1}{2} e^t$$

Then

$$\mathscr{L}\{f * g\} = \mathscr{L}\left\{ -\frac{1}{2} e^{-t} + \frac{1}{2} e^t \right\} = -\frac{1}{2} \frac{1}{s+1} + \frac{1}{2} \frac{1}{s-1} = \frac{1}{s^2 - 1}$$

23. $\mathscr{L}\{1 * t^3\} = \dfrac{1}{s} \dfrac{3!s^4}{=} \dfrac{6}{s^5}$

25. $\mathscr{L}\{e^{-t} * e^t \cos t\} = \dfrac{s-1}{(s+1)\left[(s-1)^2 + 1\right]}$

27. $\mathscr{L}\left\{ \displaystyle\int_0^t e^\tau\, d\tau \right\} = \dfrac{1}{s}\mathscr{L}\{e^t\} = \dfrac{1}{s(s-1)}$

29. $\mathscr{L}\left\{ \displaystyle\int_0^t e^{-\tau} \cos \tau\, d\tau \right\} = \dfrac{1}{s}\mathscr{L}\{e^{-t} \cos t\} = \dfrac{1}{s} \dfrac{s+1}{(s+1)^2 + 1} = \dfrac{s+1}{s\,(s^2 + 2s + 2)}$

31. $\mathscr{L}\left\{ \displaystyle\int_0^t \tau e^{t-\tau}\, d\tau \right\} = \mathscr{L}\{t\}\mathscr{L}\{e^t\} = \dfrac{1}{s^2(s-1)}$

33. $\mathscr{L}\left\{ t \displaystyle\int_0^t \sin \tau\, d\tau \right\} = -\dfrac{d}{ds}\mathscr{L}\left\{ \displaystyle\int_0^t \sin \tau\, d\tau \right\} = -\dfrac{d}{ds}\left(\dfrac{1}{s} \dfrac{1}{s^2 + 1} \right) = \dfrac{3s^2 + 1}{s^2\,(s^2 + 1)^2}$

35. $\mathscr{L}^{-1}\left\{ \dfrac{1}{s(s-1)} \right\} = \mathscr{L}^{-1}\left\{ \dfrac{1/(s-1)}{s} \right\} = \displaystyle\int_0^t e^\tau\, d\tau = e^t - 1$

37. $\mathscr{L}^{-1}\left\{ \dfrac{1}{s^3(s-1)} \right\} = \mathscr{L}^{-1}\left\{ \dfrac{1/s^2(s-1)}{s} \right\} = \displaystyle\int_0^t (e^\tau - \tau - 1)\, d\tau = e^t - \dfrac{1}{2}t^2 - t - 1$

39. (a) The result in (4) in the text is $\mathscr{L}^{-1}\{F(s)G(s)\} = f * g$, so identify

$$F(s) = \frac{2k^3}{(s^2 + k^2)^2} \qquad \text{and} \qquad G(s) = \frac{4s}{s^2 + k^2}.$$

Then

$$f(t) = \sin kt - kt \cos kt \qquad \text{and} \qquad g(t) = 4 \cos kt$$

so

$$\mathscr{L}^{-1}\left\{ \frac{8k^3 s}{(s^2 + k^2)^3} \right\} = \mathscr{L}^{-1}\{F(s)G(s)\} = f * g = 4 \int_0^t f(\tau) g(t - \tau)\, dt$$

$$= 4 \int_0^t (\sin k\tau - k\tau \cos k\tau) \cos k(t - \tau)\, d\tau.$$

Using a CAS to evaluate the integral we get

$$\mathscr{L}^{-1}\left\{ \frac{8k^3 s}{(s^2 + k^2)^3} \right\} = t \sin kt - kt^2 \cos kt.$$

(b) Observe from part (a) that

$$\mathscr{L}\{t(\sin kt - kt \cos kt)\} = \frac{8k^3 s}{(s^2 + k^2)^3},$$

and from Theorem 7.4.1 that $\mathscr{L}\{tf(t)\} = -F'(s)$. We saw in (5) in the text that

$$\mathscr{L}\{\sin kt - kt \cos kt\} = 2k^3/(s^2 + k^2)^2,$$

so

$$\mathscr{L}\{t(\sin kt - kt \cos kt)\} = -\frac{d}{ds}\frac{2k^3}{(s^2 + k^2)^2} = \frac{8k^3 s}{(s^2 + k^2)^3}.$$

41. The Laplace transform of the given equation is

$$\mathscr{L}\{f\} + \mathscr{L}\{t\}\mathscr{L}\{f\} = \mathscr{L}\{t\}.$$

Solving for $\mathscr{L}\{f\}$ we obtain $\mathscr{L}\{f\} = \dfrac{1}{s^2 + 1}$. Thus, $f(t) = \sin t$.

43. The Laplace transform of the given equation is

$$\mathscr{L}\{f\} = \mathscr{L}\{te^t\} + \mathscr{L}\{t\}\mathscr{L}\{f\}.$$

Solving for $\mathscr{L}\{f\}$ we obtain

$$\mathscr{L}\{f\} = \frac{s^2}{(s-1)^3(s+1)} = \frac{1}{8}\frac{1}{s-1} + \frac{3}{4}\frac{1}{(s-1)^2} + \frac{1}{4}\frac{2}{(s-1)^3} - \frac{1}{8}\frac{1}{s+1}.$$

Thus

$$f(t) = \frac{1}{8}e^t + \frac{3}{4}te^t + \frac{1}{4}t^2 e^t - \frac{1}{8}e^{-t}$$

45. The Laplace transform of the given equation is

$$\mathscr{L}\{f\} + \mathscr{L}\{1\}\mathscr{L}\{f\} = \mathscr{L}\{1\}.$$

Solving for $\mathscr{L}\{f\}$ we obtain $\mathscr{L}\{f\} = \dfrac{1}{s+1}$. Thus, $f(t) = e^{-t}$.

47. The Laplace transform of the given equation is

$$\mathscr{L}\{f\} = \mathscr{L}\{1\} + \mathscr{L}\{t\} - \mathscr{L}\left\{\frac{8}{3}\int_0^t (t - \tau)^3 f(\tau)\,d\tau\right\}$$

$$= \frac{1}{s} + \frac{1}{s^2} + \frac{8}{3}\mathscr{L}\{t^3\}\mathscr{L}\{f\} = \frac{1}{s} + \frac{1}{s^2} + \frac{16}{s^4}\mathscr{L}\{f\}.$$

Solving for $\mathscr{L}\{f\}$ we obtain

$$\mathscr{L}\{f\} = \frac{s^2(s+1)}{s^4 - 16} = \frac{1}{8}\frac{1}{s+2} + \frac{3}{8}\frac{1}{s-2} + \frac{1}{4}\frac{2}{s^2 + 4} + \frac{1}{2}\frac{s}{s^2 + 4}.$$

Thus

$$f(t) = \frac{1}{8}e^{-2t} + \frac{3}{8}e^{2t} + \frac{1}{4}\sin 2t + \frac{1}{2}\cos 2t.$$

49. The Laplace transform of the given equation is

$$s\mathscr{L}\{y\} - y(0) = \mathscr{L}\{1\} - \mathscr{L}\{\sin t\} - \mathscr{L}\{1\}\mathscr{L}\{y\}.$$

Solving for $\mathscr{L}\{y\}$ we obtain

$$\mathscr{L}\{y\} = \frac{s^2 - s + 1}{(s^2 + 1)^2} = \frac{1}{s^2 + 1} - \frac{1}{2}\frac{2s}{(s^2 + 1)^2}.$$

Thus

$$y = \sin t - \frac{1}{2}t\sin t.$$

51. The differential equation is

$$0.1\frac{di}{dt} + 3i + \frac{1}{0.05}\int_0^t i(\tau)\,d\tau = 100\left[\mathscr{U}(t-1) - \mathscr{U}(t-2)\right]$$

or

$$\frac{di}{dt} + 30i + 200\int_0^t i(\tau)\,d\tau = 1000\left[\mathscr{U}(t-1) - \mathscr{U}(t-2)\right],$$

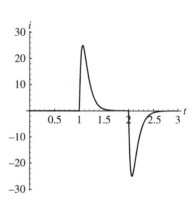

where $i(0) = 0$. The Laplace transform of the differential equation is

$$s\mathscr{L}\{i\} - y(0) + 30\mathscr{L}\{i\} + \frac{200}{s}\mathscr{L}\{i\} = \frac{1000}{s}(e^{-s} - e^{-2s}).$$

Solving for $\mathscr{L}\{i\}$ we obtain

$$\mathscr{L}\{i\} = \frac{1000e^{-s} - 1000e^{-2s}}{s^2 + 30s + 200} = \left(\frac{100}{s+10} - \frac{100}{s+20}\right)(e^{-s} - e^{-2s}).$$

Thus

$$i(t) = 100\left(e^{-10(t-1)} - e^{-20(t-1)}\right)\mathscr{U}(t-1) - 100\left(e^{-10(t-2)} - e^{-20(t-2)}\right)\mathscr{U}(t-2).$$

53. $\mathscr{L}\{f(t)\} = \dfrac{1}{1 - e^{-2as}}\left[\displaystyle\int_0^a e^{-st}\,dt - \int_a^{2a} e^{-st}\,dt\right] = \dfrac{(1 - e^{-as})^2}{s(1 - e^{-2as})} = \dfrac{1 - e^{-as}}{s(1 + e^{-as})}$

55. Using integration by parts,

$$\mathscr{L}\{f(t)\} = \frac{1}{1 - e^{-bs}}\int_0^b \frac{a}{b}te^{-st}\,dt = \frac{a}{s}\left(\frac{1}{bs} - \frac{1}{e^{bs} - 1}\right).$$

57. $\mathscr{L}\{f(t)\} = \dfrac{1}{1 - e^{-\pi s}}\displaystyle\int_0^\pi e^{-st}\sin t\,dt = \dfrac{1}{s^2 + 1}\cdot\dfrac{e^{\pi s/2} + e^{-\pi s/2}}{e^{\pi s/2} - e^{-\pi s/2}} = \dfrac{1}{s^2 + 1}\coth\dfrac{\pi s}{2}$

59. The differential equation is $L\,di/dt + Ri = E(t)$, where $i(0) = 0$. The Laplace transform of the equation is

$$Ls\mathscr{L}\{i\} + R\mathscr{L}\{i\} = \mathscr{L}\{E(t)\}.$$

From Problem 53 we have $\mathscr{L}\{E(t)\} = (1 - e^{-s})/s(1 + e^{-s})$. Thus

$$(Ls + R)\mathscr{L}\{i\} = \frac{1 - e^{-s}}{s(1 + e^{-s})}$$

and

$$\mathscr{L}\{i\} = \frac{1}{L}\frac{1 - e^{-s}}{s(s + R/L)(1 + e^{-s})} = \frac{1}{L}\frac{1 - e^{-s}}{s(s + R/L)}\frac{1}{1 + e^{-s}}$$

$$= \frac{1}{R}\left(\frac{1}{s} - \frac{1}{s + R/L}\right)(1 - e^{-s})(1 - e^{-s} + e^{-2s} - e^{-3s} + e^{-4s} - \cdots)$$

$$= \frac{1}{R}\left(\frac{1}{s} - \frac{1}{s + R/L}\right)(1 - 2e^{-s} + 2e^{-2s} - 2e^{-3s} + 2e^{-4s} - \cdots).$$

Therefore,

$$i(t) = \frac{1}{R}\left(1 - e^{-Rt/L}\right) - \frac{2}{R}\left(1 - e^{-R(t-1)/L}\right)\mathscr{U}(t - 1)$$

$$+ \frac{2}{R}\left(1 - e^{-R(t-2)/L}\right)\mathscr{U}(t - 2) - \frac{2}{R}\left(1 - e^{-R(t-3)/L}\right)\mathscr{U}(t - 3) + \cdots$$

$$= \frac{1}{R}\left(1 - e^{-Rt/L}\right) + \frac{2}{R}\sum_{n=1}^{\infty}\left(1 - e^{-R(t-n)/L}\right)\mathscr{U}(t - n).$$

The graph of $i(t)$ with $L = 1$ and $R = 1$ is shown below.

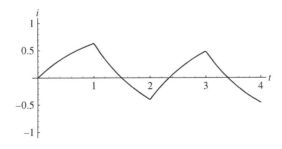

61. The differential equation is $x'' + 2x' + 10x = 20f(t)$, where $f(t)$ is the meander function in Problem 53 with $a = \pi$. Using the initial conditions $x(0) = x'(0) = 0$ and taking the Laplace transform we obtain

$$(s^2 + 2s + 10)\mathscr{L}\{x(t)\} = \frac{20}{s}(1 - e^{-\pi s})\frac{1}{1 + e^{-\pi s}}$$

$$= \frac{20}{s}(1 - e^{-\pi s})(1 - e^{-\pi s} + e^{-2\pi s} - e^{-3\pi s} + \cdots)$$

$$= \frac{20}{s}(1 - 2e^{-\pi s} + 2e^{-2\pi s} - 2e^{-3\pi s} + \cdots)$$

$$= \frac{20}{s} + \frac{40}{s}\sum_{n=1}^{\infty}(-1)^n e^{-n\pi s}.$$

Then

$$\mathscr{L}\{x(t)\} = \frac{20}{s(s^2 + 2s + 10)} + \frac{40}{s(s^2 + 2s + 10)}\sum_{n=1}^{\infty}(-1)^n e^{-n\pi s}$$

$$= \frac{2}{s} - \frac{2s + 4}{s^2 + 2s + 10} + \sum_{n=1}^{\infty}(-1)^n\left[\frac{4}{s} - \frac{4s + 8}{s^2 + 2s + 10}\right]e^{-n\pi s}$$

$$= \frac{2}{s} - \frac{2(s + 1) + 2}{(s + 1)^2 + 9} + 4\sum_{n=1}^{\infty}(-1)^n\left[\frac{1}{s} - \frac{(s + 1) + 1}{(s + 1)^2 + 9}\right]e^{-n\pi s}$$

and

$$x(t) = 2\left(1 - e^{-t}\cos 3t - \frac{1}{3}e^{-t}\sin 3t\right)$$

$$+ 4\sum_{n=1}^{\infty}(-1)^n\left[1 - e^{-(t - n\pi)}\cos 3(t - n\pi) - \frac{1}{3}e^{-(t - n\pi)}\sin 3(t - n\pi)\right]\mathscr{U}(t - n\pi).$$

The graph of $x(t)$ is shown below.

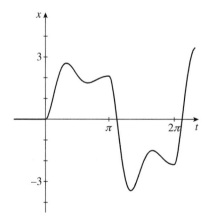

63. $f(t) = -\frac{1}{t}\mathscr{L}^{-1}\left\{\frac{d}{ds}\left[\ln(s - 3) - \ln(s + 1)\right]\right\} = -\frac{1}{t}\mathscr{L}^{-1}\left\{\frac{1}{s - 3} - \frac{1}{s + 1}\right\}$

$$= -\frac{1}{t}\left(e^{3t} - e^{-t}\right)$$

65. (a) Using Theorem 7.4.1, the Laplace transform of the differential equation is

$$-\frac{d}{ds}\left[s^2 Y - sy(0) - y'(0)\right] + sY - y(0) + \frac{d}{ds}\left[sY - y(0)\right] + nY$$

$$= -\frac{d}{ds}\left[s^2 Y\right] + sY + \frac{d}{ds}\left[sY\right] + nY$$

$$= -s^2\left(\frac{dY}{ds}\right) - 2sY + sY + s\left(\frac{dY}{ds}\right) + Y + nY$$

$$= (s - s^2)\left(\frac{dY}{ds}\right) + (1 + n - s)Y = 0.$$

Separating variables, we find

$$\frac{dY}{Y} = \frac{1+n-s}{s^2 - s}\,ds = \left(\frac{n}{s-1} - \frac{1+n}{s}\right)ds$$

$$\ln Y = n\ln(s-1) - (1+n)\ln s + c$$

$$Y = c_1 \frac{(s-1)^n}{s^{1+n}}.$$

Since the differential equation is homogeneous, any constant multiple of a solution will still be a solution, so for convenience we take $c_1 = 1$. The following polynomials are solutions of Laguerre's differential equation:

$$n = 0: \quad L_0(t) = \mathscr{L}^{-1}\left\{\frac{1}{s}\right\} = 1$$

$$n = 1: \quad L_1(t) = \mathscr{L}^{-1}\left\{\frac{s-1}{s^2}\right\} = \mathscr{L}^{-1}\left\{\frac{1}{s} - \frac{1}{s^2}\right\} = 1 - t$$

$$n = 2: \quad L_2(t) = \mathscr{L}^{-1}\left\{\frac{(s-1)^2}{s^3}\right\} = \mathscr{L}^{-1}\left\{\frac{1}{s} - \frac{2}{s^2} + \frac{1}{s^3}\right\} = 1 - 2t + \frac{1}{2}t^2$$

$$n = 3: \quad L_3(t) = \mathscr{L}^{-1}\left\{\frac{(s-1)^3}{s^4}\right\} = \mathscr{L}^{-1}\left\{\frac{1}{s} - \frac{3}{s^2} + \frac{3}{s^3} - \frac{1}{s^4}\right\} = 1 - 3t + \frac{3}{2}t^2 - \frac{1}{6}t^3$$

$$n = 4: \quad L_4(t) = \mathscr{L}^{-1}\left\{\frac{(s-1)^4}{s^5}\right\} = \mathscr{L}^{-1}\left\{\frac{1}{s} - \frac{4}{s^2} + \frac{6}{s^3} - \frac{4}{s^4} + \frac{1}{s^5}\right\}$$

$$= 1 - 4t + 3t^2 - \frac{2}{3}t^3 + \frac{1}{24}t^4.$$

(b) Letting $f(t) = t^n e^{-t}$ we note that $f^{(k)}(0) = 0$ for $k = 0, 1, 2, \ldots, n-1$ and $f^{(n)}(0) = n!$.

Now, by the first translation theorem,

$$\mathscr{L}\left\{\frac{e^t}{n!}\frac{d^n}{dt^n}t^n e^{-t}\right\} = \frac{1}{n!}\mathscr{L}\{e^t f^{(n)}(t)\} = \frac{1}{n!}\mathscr{L}\{f^{(n)}(t)\}\bigg|_{s\to s-1}$$

$$= \frac{1}{n!}\left[s^n\mathscr{L}\{t^n e^{-t}\} - s^{n-1}f(0) - s^{n-2}f'(0) - \cdots - f^{(n-1)}(0)\right]_{s\to s-1}$$

$$= \frac{1}{n!}\left[s^n\mathscr{L}\{t^n e^{-t}\}\right]_{s\to s-1}$$

$$= \frac{1}{n!}\left[s^n\frac{n!}{(s+1)^{n+1}}\right]_{s\to s-1} = \frac{(s-1)^n}{s^{n+1}} = Y,$$

where $Y = \mathscr{L}\{L_n(t)\}$. Thus

$$L_n(t) = \frac{e^t}{n!}\frac{d^n}{dt^n}(t^n e^{-t}), \quad n = 0, 1, 2, \ldots.$$

67. Take the transform of both sides of the equation to get

$$\mathscr{L}\{f(t)\} = \mathscr{L}\{e^t\} + \mathscr{L}\left\{e^t \cdot \int_0^t f(\tau)e^{-\tau}\,d\tau\right\}$$

$$F(s) = \frac{1}{s-1} + \left[\mathscr{L}\{f(t)e^{-t}\}\mathscr{L}\{1\}\right]_{s\to s-1}$$

$$F(s) = \frac{1}{s-1} + \left[\mathscr{L}\{f(t)e^{-t}\}\right]_{s\to s-1} \cdot \left[\mathscr{L}\{1\}\right]_{s\to s-1}$$

$$F(s) = frac1s - 1 + \left[F(s+1)\right]_{s\to s-1} \cdot \left[\frac{1}{s}\right]_{s\to s-1}$$

$$F(s) = \frac{1}{s-1} + F(s) \cdot \frac{1}{s-1}$$

Solve the last equation for $F(s)$ to get $F(s) = 1/(s-2)$ therefore $f(t) = e^{2t}$.

69. We know that

$$\mathscr{L}\{t\sin 4t\| = -\frac{d}{ds}\left(\frac{4}{s^2+16}\right) = \frac{8s}{(s^2+16)^2},$$

and so for $s > 0$,

$$F(s) = \int_0^\infty te^{-st}\sin 4t\,dt = \frac{8s}{(s^2+16)^2}.$$

Therefore

$$F(2) = \int_0^\infty te^{2t}\sin 4t\,dt = \frac{8\cdot 2}{(2^2+16)^2} = \frac{16}{400} = \frac{1}{25}.$$

71. (a) Using the definition of the Laplace transform and integration by parts,

$$\mathscr{L}\{\ln t\} = \int_0^\infty e^{-st}\ln t\,dt = e^{-st}(t\ln t - t)\bigg|_0^\infty + s\int_0^\infty e^{-st}(t\ln t - t)\,dt$$

$$= s\mathscr{L}\{t\ln t\} - s\mathscr{L}\{t\} = s\mathscr{L}\{t\ln t\} - \frac{1}{s}.$$

(b) Letting $Y(s) = \mathscr{L}\{\ln t\}$ we have

$$Y = s\left(-\frac{d}{ds}Y\right) - \frac{1}{s}$$

by Theorem 7.4.1. That is,

$$s\frac{dY}{ds} + Y = -\frac{1}{s}$$

$$\frac{d}{ds}[sY] = -\frac{1}{s}$$

$$sY = -\ln s + c$$

$$Y = \frac{c}{s} - \frac{1}{s}\ln s, s > 0.$$

(c) Now $Y(1) = \mathscr{L}\{\ln t\}\Big|_{s=1} = \int_0^\infty e^{-t}\ln t\,dt = -\gamma.$ Thus

$$-\gamma = Y(1) = c - \ln 1 = c$$

and

$$Y = \mathscr{L}\{\ln t\} = -\frac{\gamma}{s} - \frac{1}{s}\ln s.$$

73. The solution is

$$y(t) = \frac{1}{6}e^t - \frac{1}{6}e^{-t/2}\cos\frac{\sqrt{15}\,t}{2} - \frac{1}{2\sqrt{15}}e^{-t/2}\sin\frac{\sqrt{15}\,t}{2}.$$

7.5 | The Dirac Delta Function

1. The Laplace transform of the differential equation yields

$$\mathscr{L}\{y\} = \frac{1}{s-3}e^{-2s}$$

so that

$$y = e^{3(t-2)}\,\mathscr{U}(t-2).$$

3. The Laplace transform of the differential equation yields

$$\mathscr{L}\{y\} = \frac{1}{s^2+1}\left(1 + e^{-2\pi s}\right)$$

so that

$$y = \sin t + \sin t\,\mathscr{U}(t - 2\pi).$$

5. The Laplace transform of the differential equation yields

$$\mathscr{L}\{y\} = \frac{1}{s^2 + 1}\left(e^{-\pi s/2} + e^{-3\pi s/2}\right)$$

so that

$$y = \sin\left(t - \frac{\pi}{2}\right)\mathscr{U}\left(t - \frac{\pi}{2}\right) + \sin\left(t - \frac{3\pi}{2}\right)\mathscr{U}\left(t - \frac{3\pi}{2}\right)$$

$$= -\cos t\,\mathscr{U}\left(t - \frac{\pi}{2}\right) + \cos t\,\mathscr{U}\left(t - \frac{3\pi}{2}\right).$$

7. The Laplace transform of the differential equation yields

$$\mathscr{L}\{y\} = \frac{1}{s^2 + 2s}(1 + e^{-s}) = \left[\frac{1}{2}\frac{1}{s} - \frac{1}{2}\frac{1}{s+2}\right](1 + e^{-s})$$

so that

$$y = \frac{1}{2} - \frac{1}{2}e^{-2t} + \left[\frac{1}{2} - \frac{1}{2}e^{-2(t-1)}\right]\mathscr{U}(t - 1).$$

9. The Laplace transform of the differential equation yields

$$\mathscr{L}\{y\} = \frac{1}{(s+2)^2 + 1}e^{-2\pi s}$$

so that

$$y = e^{-2(t-2\pi)}\sin t\,\mathscr{U}(t - 2\pi).$$

11. The Laplace transform of the differential equation yields

$$\mathscr{L}\{y\} = \frac{4 + s}{s^2 + 4s + 13} + \frac{e^{-\pi s} + e^{-3\pi s}}{s^2 + 4s + 13}$$

$$= \frac{2}{3}\frac{3}{(s+2)^2 + 3^2} + \frac{s+2}{(s+2)^2 + 3^2} + \frac{1}{3}\frac{3}{(s+2)^2 + 3^2}\left(e^{-\pi s} + e^{-3\pi s}\right)$$

so that

$$y = \frac{2}{3}e^{-2t}\sin 3t + e^{-2t}\cos 3t + \frac{1}{3}e^{-2(t-\pi)}\sin 3(t - \pi)\mathscr{U}(t - \pi)$$

$$+ \frac{1}{3}e^{-2(t-3\pi)}\sin 3(t - 3\pi)\mathscr{U}(t - 3\pi).$$

13. The Laplace transform of the differential equation yields

$$\mathscr{L}\{y''\} + \mathscr{L}\{y\} = \sum_{k=1}^{\infty}\mathscr{L}\{\delta(t - k\pi)\}$$

$$(s^2 + 1)Y(s) = 1 + \sum k = 1^{\infty}e^{-k\pi s}$$

$$Y(s) = \frac{1}{s^2 + 1} + \sum_{k=1}^{\infty}\frac{e^{-k\pi s}}{s^2 + 1}$$

so that

$$y(t) = \sin t + \sum_{k=1}^{\infty} \sin(t - k\pi)\,\mathscr{U}(t - k\pi)$$

$$= \sin t + \sin t \sum_{k=1}^{\infty} (-1)^k\,\mathscr{U}(t - k\pi) \qquad \longleftarrow \qquad \sin(t - k\pi) = (-1)^k \sin t$$

$$= \sin t - \sin t\,\mathscr{U}(t - \pi) + \sin t\,\mathscr{U}(t - 2\pi) - \sin t\,\mathscr{U}(t - 3\pi) + \sin t\,\mathscr{U}(t - 4\pi) - \cdots$$

$$= \begin{cases} \sin t, & 0 \le t < \pi \\ 0, & \pi \le t < 2\pi \\ \sin t, & 2\pi \le t < 3\pi \\ 0, & 3\pi \le t < 4\pi \\ \sin t, & 4\pi \le t < 5\pi \\ 0, & 5\pi \le t < 6\pi \\ \sin t, & 6\pi \le t < 7\pi \\ 0, & 7\pi \le t < 8\pi \\ \vdots \end{cases}$$

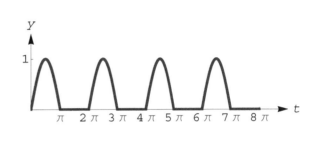

The graph of $y(t)$ given on the right is the half-wave rectification of $\sin t$. Also, see Figure 7.4.11 in Exercises 7.4.

15. Take the transform of the equation $y^{(4)} = \dfrac{w_0}{EI}\,\delta\left(x - \dfrac{L}{2}\right)$ and solve for $Y(s)$ to get

$$s^2 Y(s) - s^3 y(0) - s^2 y'(0) - sy''(0) - y'''(0) = \frac{w_0}{EI}\, e^{(-L/2)s}$$

$$s^4 Y(s) - sy''(0) - y'''(0) = \frac{w_0}{EI}\, e^{(-L/2)s}$$

$$Y(s) = \frac{1}{s^3}\, y''(0) + \frac{1}{s^4}\, y'''(0) + \frac{w_0}{EI}\,\frac{1}{s^4}\, e^{(-L/2)s}$$

The inverse transform is then $y(x) = \dfrac{1}{2}\, y''(0)x^2 + \dfrac{1}{6}\, y'''(0)x^3 + \dfrac{w_0}{6EI}\left(x - \dfrac{L}{2}\right)^3 \mathscr{U}\left(x - \dfrac{L}{2}\right)$.

Using the conditions $y''(L) = y'''(L) = 0$ to find $y''(0)$ and $y'''(0)$ we get

$$y(x) = \frac{1}{2}\left(\frac{w_0 L}{2EI}\right)x^2 - \frac{1}{6}\left(\frac{w_0}{EI}\right)x^3 + \frac{w_0}{6EI}\left(x - \frac{L}{2}\right)^3 \mathscr{U}\left(x - \frac{L}{2}\right)$$

We could also write the solution as

$$y(x) = \begin{cases} \left(\dfrac{w_0 L}{4EI}\right)x^2 - \left(\dfrac{w_0}{6EI}\right)x^3, & 0 \le x < \dfrac{L}{2} \\[4mm] \left(\dfrac{w_0 L}{4EI}\right)x^2 - \left(\dfrac{w_0}{6EI}\right)x^3 + \dfrac{w_0}{6EI}\left(x - \dfrac{L}{2}\right)^3, & \dfrac{L}{2} \le x \le L \end{cases}$$

17. You should disagree. Although formal manipulations of the Laplace transform lead to $y(t) = \frac{1}{3}e^{-t}\sin 3t$ in both cases, this function does not satisfy the initial condition $y'(0) = 0$ of the second initial-value problem.

7.6 Systems of Linear Differential Equations

1. Taking the Laplace transform of the system gives

$$s\mathscr{L}\{x\} = -\mathscr{L}\{x\} + \mathscr{L}\{y\}$$
$$s\mathscr{L}\{y\} - 1 = 2\mathscr{L}\{x\}$$

so that

$$\mathscr{L}\{x\} = \frac{1}{(s-1)(s+2)} = \frac{1}{3}\frac{1}{s-1} - \frac{1}{3}\frac{1}{s+2}$$

and

$$\mathscr{L}\{y\} = \frac{1}{s} + \frac{2}{s(s-1)(s+2)} = \frac{2}{3}\frac{1}{s-1} + \frac{1}{3}\frac{1}{s+2}.$$

Then

$$x = \frac{1}{3}e^t - \frac{1}{3}e^{-2t} \qquad \text{and} \qquad y = \frac{2}{3}e^t + \frac{1}{3}e^{-2t}.$$

3. Taking the Laplace transform of the system gives

$$s\mathscr{L}\{x\} + 1 = \mathscr{L}\{x\} - 2\mathscr{L}\{y\}$$
$$s\mathscr{L}\{y\} - 2 = 5\mathscr{L}\{x\} - \mathscr{L}\{y\}$$

so that

$$\mathscr{L}\{x\} = \frac{-s-5}{s^2+9} = -\frac{s}{s^2+9} - \frac{5}{3}\frac{3}{s^2+9}$$

and

$$x = -\cos 3t - \frac{5}{3}\sin 3t.$$

Then

$$y = \frac{1}{2}x - \frac{1}{2}x' = 2\cos 3t - \frac{7}{3}\sin 3t.$$

5. Taking the Laplace transform of the system gives

$$(2s-2)\mathscr{L}\{x\} + s\mathscr{L}\{y\} = \frac{1}{s}$$

$$(s-3)\mathscr{L}\{x\} + (s-3)\mathscr{L}\{y\} = \frac{2}{s}$$

so that

$$\mathscr{L}\{x\} = \frac{-s-3}{s(s-2)(s-3)} = -\frac{1}{2}\frac{1}{s} + \frac{5}{2}\frac{1}{s-2} - \frac{2}{s-3}$$

and

$$\mathscr{L}\{y\} = \frac{3s-1}{s(s-2)(s-3)} = -\frac{1}{6}\frac{1}{s} - \frac{5}{2}\frac{1}{s-2} + \frac{8}{3}\frac{1}{s-3}.$$

Then

$$x = -\frac{1}{2} + \frac{5}{2}e^{2t} - 2e^{3t} \quad \text{and} \quad y = -\frac{1}{6} - \frac{5}{2}e^{2t} + \frac{8}{3}e^{3t}.$$

7. Taking the Laplace transform of the system gives

$$(s^2 + 1)\mathscr{L}\{x\} - \mathscr{L}\{y\} = -2$$
$$-\mathscr{L}\{x\} + (s^2 + 1)\mathscr{L}\{y\} = 1$$

so that

$$\mathscr{L}\{x\} = \frac{-2s^2 - 1}{s^4 + 2s^2} = -\frac{1}{2}\frac{1}{s^2} - \frac{3}{2}\frac{1}{s^2 + 2}$$

and

$$x = -\frac{1}{2}t - \frac{3}{2\sqrt{2}}\sin\sqrt{2}\,t.$$

Then

$$y = x'' + x = -\frac{1}{2}t + \frac{3}{2\sqrt{2}}\sin\sqrt{2}\,t.$$

9. Adding the equations and then subtracting them gives

$$\frac{d^2x}{dt^2} = \frac{1}{2}t^2 + 2t$$

$$\frac{d^2y}{dt^2} = \frac{1}{2}t^2 - 2t.$$

Taking the Laplace transform of the system gives

$$\mathscr{L}\{x\} = 8\frac{1}{s} + \frac{1}{24}\frac{4!}{s^5} + \frac{1}{3}\frac{3!}{s^4}$$

and

$$\mathscr{L}\{y\} = \frac{1}{24}\frac{4!}{s^5} - \frac{1}{3}\frac{3!}{s^4}$$

so that

$$x = 8 + \frac{1}{24}t^4 + \frac{1}{3}t^3 \quad \text{and} \quad y = \frac{1}{24}t^4 - \frac{1}{3}t^3.$$

11. Taking the Laplace transform of the system gives

$$s^2\mathscr{L}\{x\} + 3(s+1)\mathscr{L}\{y\} = 2$$
$$s^2\mathscr{L}\{x\} + 3\mathscr{L}\{y\} = \frac{1}{(s+1)^2}$$

so that

$$\mathscr{L}\{x\} = -\frac{2s+1}{s^3(s+1)} = \frac{1}{s} + \frac{1}{s^2} + \frac{1}{2}\frac{2}{s^3} - \frac{1}{s+1}.$$

Then

$$x = 1 + t + \frac{1}{2}t^2 - e^{-t}$$

and

$$y = \frac{1}{3}te^{-t} - \frac{1}{3}x'' = \frac{1}{3}te^{-t} + \frac{1}{3}e^{-t} - \frac{1}{3}.$$

13. The system is

$$x_1'' = -3x_1 + 2(x_2 - x_1)$$
$$x_2'' = -2(x_2 - x_1)$$

$$x_1(0) = 0 \qquad x_1'(0) = 1 \qquad x_2(0) = 1 \qquad x_2'(0) = 0.$$

Taking the Laplace transform of the system gives

$$(s^2 + 5)\mathscr{L}\{x_1\} - 2\mathscr{L}\{x_2\} = 1$$
$$-2\mathscr{L}\{x_1\} + (s^2 + 2)\mathscr{L}\{x_2\} = s$$

so that

$$\mathscr{L}\{x_1\} = \frac{s^2 + 2s + 2}{s^4 + 7s^2 + 6} = \frac{2}{5}\frac{s}{s^2 + 1} + \frac{1}{5}\frac{1}{s^2 + 1} - \frac{2}{5}\frac{s}{s^2 + 6} + \frac{4}{5\sqrt{6}}\frac{\sqrt{6}}{s^2 + 6}$$

and

$$\mathscr{L}\{x_2\} = \frac{s^3 + 5s + 2}{(s^2 + 1)(s^2 + 6)} = \frac{4}{5}\frac{s}{s^2 + 1} + \frac{2}{5}\frac{1}{s^2 + 1} + \frac{1}{5}\frac{s}{s^2 + 6} - \frac{2}{5\sqrt{6}}\frac{\sqrt{6}}{s^2 + 6}.$$

Then

$$x_1 = \frac{2}{5}\cos t + \frac{1}{5}\sin t - \frac{2}{5}\cos\sqrt{6}\,t + \frac{4}{5\sqrt{6}}\sin\sqrt{6}\,t$$

and

$$x_2 = \frac{4}{5}\cos t + \frac{2}{5}\sin t + \frac{1}{5}\cos\sqrt{6}\,t - \frac{2}{5\sqrt{6}}\sin\sqrt{6}\,t.$$

15. (a) By Kirchhoff's first law we have $i_1 = i_2 + i_3$. By Kirchhoff's second law, on each loop we have $E(t) = Ri_1 + L_1 i_2'$ and $E(t) = Ri_1 + L_2 i_3'$ or $L_1 i_2' + Ri_2 + Ri_3 = E(t)$ and $L_2 i_3' + Ri_2 + Ri_3 = E(t)$.

(b) Taking the Laplace transform of the system

$$0.01 i_2' + 5i_2 + 5i_3 = 100$$
$$0.0125 i_3' + 5i_2 + 5i_3 = 100$$

gives

$$(s + 500)\mathscr{L}\{i_2\} + 500\mathscr{L}\{i_3\} = \frac{10{,}000}{s}$$

$$400\mathscr{L}\{i_2\} + (s + 400)\mathscr{L}\{i_3\} = \frac{8{,}000}{s}$$

so that

$$\mathscr{L}\{i_3\} = \frac{8{,}000}{s^2 + 900s} = \frac{80}{9}\frac{1}{s} - \frac{80}{9}\frac{1}{s + 900}.$$

Then

$$i_3 = \frac{80}{9} - \frac{80}{9}e^{-900t} \qquad \text{and} \qquad i_2 = 20 - 0.0025 i_3' - i_3 = \frac{100}{9} - \frac{100}{9}e^{-900t}.$$

(c) $i_1 = i_2 + i_3 = 20 - 20e^{-900t}$

17. Taking the Laplace transform of the system

$$i_2' + 11i_2 + 6i_3 = 50 \sin t$$

$$i_3' + 6i_2 + 6i_3 = 50 \sin t$$

gives

$$(s + 11)\mathscr{L}\{i_2\} + 6\mathscr{L}\{i_3\} = \frac{50}{s^2 + 1}$$

$$6\mathscr{L}\{i_2\} + (s + 6)\mathscr{L}\{i_3\} = \frac{50}{s^2 + 1}$$

so that

$$\mathscr{L}\{i_2\} = \frac{50s}{(s + 2)(s + 15)(s^2 + 1)} = -\frac{20}{13}\frac{1}{s + 2} + \frac{375}{1469}\frac{1}{s + 15} + \frac{145}{113}\frac{s}{s^2 + 1} + \frac{85}{113}\frac{1}{s^2 + 1}.$$

Then

$$i_2 = -\frac{20}{13}e^{-2t} + \frac{375}{1469}e^{-15t} + \frac{145}{113}\cos t + \frac{85}{113}\sin t$$

and

$$i_3 = \frac{25}{3}\sin t - \frac{1}{6}i_2' - \frac{11}{6}i_2 = \frac{30}{13}e^{-2t} + \frac{250}{1469}e^{-15t} - \frac{280}{113}\cos t + \frac{810}{113}\sin t.$$

19. Taking the Laplace transform of the system

$$2i_1' + 50i_2 = 60$$

$$0.005i_2' + i_2 - i_1 = 0$$

gives

$$2s\mathscr{L}\{i_1\} + 50\mathscr{L}\{i_2\} = \frac{60}{s}$$

$$-200\mathscr{L}\{i_1\} + (s + 200)\mathscr{L}\{i_2\} = 0$$

so that

$$\mathscr{L}\{i_2\} = \frac{6{,}000}{s(s^2 + 200s + 5{,}000)}$$

$$= \frac{6}{5}\frac{1}{s} - \frac{6}{5}\frac{s + 100}{(s + 100)^2 - (50\sqrt{2})^2} - \frac{6\sqrt{2}}{5}\frac{50\sqrt{2}}{(s + 100)^2 - (50\sqrt{2})^2}.$$

Then

$$i_2 = \frac{6}{5} - \frac{6}{5}e^{-100t}\cosh 50\sqrt{2}\,t - \frac{6\sqrt{2}}{5}e^{-100t}\sinh 50\sqrt{2}\,t$$

and

$$i_1 = 0.005i_2' + i_2 = \frac{6}{5} - \frac{6}{5}e^{-100t}\cosh 50\sqrt{2}\,t - \frac{9\sqrt{2}}{10}e^{-100t}\sinh 50\sqrt{2}\,t.$$

21. (a) Taking the Laplace transform of the system

$$4\theta_1'' + \theta_2'' + 8\theta_1 = 0$$
$$\theta_1'' + \theta_2'' + 2\theta_2 = 0$$

gives

$$4\left(s^2 + 2\right)\mathscr{L}\{\theta_1\} + s^2\mathscr{L}\{\theta_2\} = 3s$$
$$s^2\mathscr{L}\{\theta_1\} + \left(s^2 + 2\right)\mathscr{L}\{\theta_2\} = 0$$

so that

$$\left(3s^2 + 4\right)\left(s^2 + 4\right)\mathscr{L}\{\theta_2\} = -3s^3$$

or

$$\mathscr{L}\{\theta_2\} = \frac{1}{2}\frac{s}{s^2 + 4/3} - \frac{3}{2}\frac{s}{s^2 + 4}.$$

Then

$$\theta_2 = \frac{1}{2}\cos\frac{2}{\sqrt{3}}t - \frac{3}{2}\cos 2t \qquad \text{and} \qquad \theta_1'' = -\theta_2'' - 2\theta_2$$

so that

$$\theta_1 = \frac{1}{4}\cos\frac{2}{\sqrt{3}}t + \frac{3}{4}\cos 2t.$$

(b)

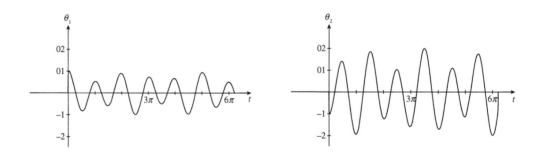

Mass m_2 has extreme displacements of greater magnitude. Mass m_1 first passes through its equilibrium position at about $t = 0.87$, and mass m_2 first passes through its equilibrium position at about $t = 0.66$. The motion of the pendulums is not periodic since $\cos\left(2t/\sqrt{3}\right)$ has period $\sqrt{3}\,\pi$, $\cos 2t$ has period π, and the ratio of these periods is $\sqrt{3}$, which is not a rational number.

(c)

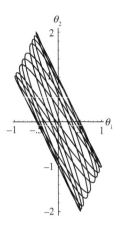

The Lissajous curve is plotted for $0 \le t \le 30$.

(d)

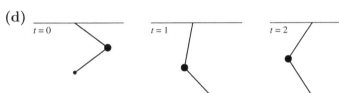

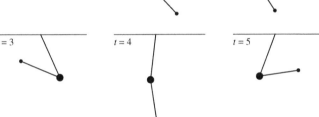

t	θ_1	θ_2
1	−0.2111	0.8263
2	−0.6585	0.6438
3	0.4830	−1.9145
4	−0.1325	0.1715
5	−0.4111	1.6951
6	0.8327	−0.8662
7	0.0458	−0.3186
8	−0.9639	0.9452
9	0.3534	−1.2741
10	0.4370	−0.3502

   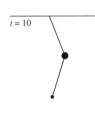

(e) Using a CAS to solve $\theta_1(t) = \theta_2(t)$ we see that $\theta_1 = \theta_2$ (so that the double pendulum is straight out) when t is about 0.75 seconds.

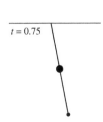

(f) To make a movie of the pendulum it is necessary to locate the mass in the plane as a function of time. Suppose that the upper arm is attached to the origin and that the equilibrium position lies along the negative y-axis. Then mass m_1 is at $(x, (t), y_1(t))$ and mass m_2 is at $(x_2(t), y_2(t))$, where

$$x_1(t) = 16 \sin \theta_1(t) \qquad \text{and} \qquad y_1(t) = -16 \cos \theta_1(t)$$

and

$$x_2(t) = x_1(t) + 16\sin\theta_2(t) \qquad \text{and} \qquad y_2(t) = y_1(t) - 16\cos\theta_2(t).$$

A reasonable movie can be constructed by letting t range from 0 to 10 in increments of 0.1 seconds.

Chapter 7 in Review

1. $\mathscr{L}\{f(t)\} = \int_0^1 te^{-st}\,dt + \int_1^\infty (2-t)e^{-st}\,dt = \dfrac{1}{s^2} - \dfrac{2}{s^2}e^{-s}$

3. False; consider $f(t) = t^{-1/2}$.

5. True, since $\lim_{s\to\infty} F(s) = 1 \neq 0$. (See Theorem 7.2.3 in the text.)

7. $\mathscr{L}\{e^{-7t}\} = \dfrac{1}{s+7}$

9. $\mathscr{L}\{\sin 2t\} = \dfrac{2}{s^2+4}$

11. $\mathscr{L}\{t\sin 2t\} = -\dfrac{d}{ds}\left[\dfrac{2}{s^2+4}\right] = \dfrac{4s}{(s^2+4)^2}$

13. $\mathscr{L}^{-1}\left\{\dfrac{20}{s^6}\right\} = \mathscr{L}^{-1}\left\{\dfrac{1}{6}\dfrac{5!}{s^6}\right\} = \dfrac{1}{6}t^5$

15. $\mathscr{L}^{-1}\left\{\dfrac{1}{(s-5)^3}\right\} = \dfrac{1}{2}\mathscr{L}^{-1}\left\{\dfrac{2}{(s-5)^3}\right\} = \dfrac{1}{2}t^2 e^{5t}$

17. $\mathscr{L}^{-1}\left\{\dfrac{s}{s^2-10s+29}\right\} = \mathscr{L}^{-1}\left\{\dfrac{s-5}{(s-5)^2+2^2} + \dfrac{5}{2}\dfrac{2}{(s-5)^2+2^2}\right\} = e^{5t}\cos 2t + \dfrac{5}{2}e^{5t}\sin 2t$

19. $\mathscr{L}^{-1}\left\{\dfrac{s+\pi}{s^2+\pi^2}e^{-s}\right\} = \mathscr{L}^{-1}\left\{\dfrac{s}{s^2+\pi^2}e^{-s} + \dfrac{\pi}{s^2+\pi^2}e^{-s}\right\}$
$$= \cos\pi(t-1)\mathscr{U}(t-1) + \sin\pi(t-1)\mathscr{U}(t-1)$$

21. $\mathscr{L}\{e^{-5t}\}$ exists for $s > -5$.

23. $\mathscr{L}\{e^{at}f(t-k)\mathscr{U}(t-k)\} = e^{-ks}\mathscr{L}\{e^{a(t+k)}f(t)\} = e^{-ks}e^{ak}\mathscr{L}\{e^{at}f(t)\} = e^{-k(s-a)}F(s-a)$

25. $f(t)\mathscr{U}(t-t_0)$

27. $f(t-t_0)\mathscr{U}(t-t_0)$

29. $f(t) = t - [(t-1)+1]\mathscr{U}(t-1) + \mathscr{U}(t-1) - \mathscr{U}(t-4) = t - (t-1)\mathscr{U}(t-1) - \mathscr{U}(t-4)$
$$\mathscr{L}\{f(t)\} = \dfrac{1}{s^2} - \dfrac{1}{s^2}e^{-s} - \dfrac{1}{s}e^{-4s}$$
$$\mathscr{L}\{e^t f(t)\} = \dfrac{1}{(s-1)^2} - \dfrac{1}{(s-1)^2}e^{-(s-1)} - \dfrac{1}{s-1}e^{-4(s-1)}$$

31. $f(t) = 2 - 2\mathcal{U}(t-2) + [(t-2) + 2]\mathcal{U}(t-2) = 2 + (t-2)\mathcal{U}(t-2)$

$$\mathcal{L}\{f(t)\} = \frac{2}{s} + \frac{1}{s^2}e^{-2s}$$

$$\mathcal{L}\{e^t f(t)\} = \frac{2}{s-1} + \frac{1}{(s-1)^2}e^{-2(s-1)}$$

33. The graph of

$$f(t) = -1 + 2\sum_{k=1}^{\infty}(-1)^{k+1}\,\mathcal{U}(t-k) = -1 + 2\mathcal{U}(t-1) - 2\mathcal{U}(t-2) + 2\mathcal{U}(t-3) - \cdots$$

is

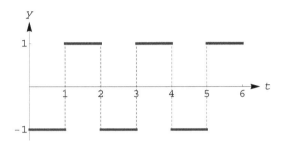

One way of proceeding to find the Laplace transform is to take the transform term-by-term of the series:

$$\mathcal{L}\{f(t)\} = -\frac{1}{s} + \frac{2}{s}e^{-s} - \frac{2}{s}e^{-2s} + \frac{2}{s}e^{-3s} - \cdots \qquad \longleftarrow \qquad \text{geometric series}$$

For $s > 0$,

$$\mathcal{L}\{f(t)\} = -\frac{1}{s} + \frac{2}{s}\left[e^{-s} - e^{-2s} + e^{-3s} - \cdots\right] = -\frac{1}{s} + \frac{2}{s}\cdot\frac{e^{-s}}{1+e^{-s}}$$

$$= \frac{e^{-s} - 1}{s\left(1 + e^{-s}\right)}$$

Alternatively, since f is a periodic functions it can also be defined by

$$f(t) = \begin{cases} -1, & 0 \le t < 1 \\ 1, & 1 \le t < 2, \end{cases} \qquad \text{where } f(t+2) = f(t).$$

By Theorem 7.4.3 with $p = 2$ we get

$$\mathcal{L}\{f(t)\} = \frac{1}{1-e^{-2s}}\left(\int_0^1 (-1)\,e^{-st}\,dt + \int_1^2 (1)\,e^{-st}\,dt\right)$$

$$= \frac{1}{1-e^{-2s}}\left(\frac{1}{s}e^{-s} - \frac{1}{s} - \frac{1}{s}e^{-2s} + \frac{1}{s}e^{-s}\right) = \frac{-1}{s\left(1-e^{-2s}\right)}\left(1 - 2e^{-s} + e^{-2s}\right)$$

$$= \frac{-\left(1-e^{-s}\right)^2}{s\left(1-e^{-2s}\right)}$$

Using $1 - e^{-2s} = (1 + e^{-s})(1 - e^{-s})$ and algebra the last expression is the same as

$$\mathscr{L}\{f(t)\} = \frac{e^{-s} - 1}{s(1 + e^{-s})}.$$

35. Taking the Laplace transform of the differential equation we obtain

$$\mathscr{L}\{y\} = \frac{5}{(s-1)^2} + \frac{1}{2}\frac{2}{(s-1)^3}$$

so that

$$y = 5te^t + \frac{1}{2}t^2 e^t.$$

37. Taking the Laplace transform of the given differential equation we obtain

$$\mathscr{L}\{y\} = \frac{s^3 + 6s^2 + 1}{s^2(s+1)(s+5)} - \frac{1}{s^2(s+1)(s+5)}e^{-2s} - \frac{2}{s(s+1)(s+5)}e^{-2s}$$

$$= -\frac{6}{25}\cdot\frac{1}{s} + \frac{1}{5}\cdot\frac{1}{s^2} + \frac{3}{2}\cdot\frac{1}{s+1} - \frac{13}{50}\cdot\frac{1}{s+5}$$

$$- \left(-\frac{6}{25}\cdot\frac{1}{s} + \frac{1}{5}\cdot\frac{1}{s^2} + \frac{1}{4}\cdot\frac{1}{s+1} - \frac{1}{100}\cdot\frac{1}{s+5}\right)e^{-2s}$$

$$- \left(\frac{2}{5}\cdot\frac{1}{s} - \frac{1}{2}\cdot\frac{1}{s+1} + \frac{1}{10}\cdot\frac{1}{s+5}\right)e^{-2s}$$

so that

$$y = -\frac{6}{25} + \frac{1}{5}t + \frac{3}{2}e^{-t} - \frac{13}{50}e^{-5t} - \frac{4}{25}\mathscr{U}(t-2) - \frac{1}{5}(t-2)\mathscr{U}(t-2)$$

$$+ \frac{1}{4}e^{-(t-2)}\mathscr{U}(t-2) - \frac{9}{100}e^{-5(t-2)}\mathscr{U}(t-2).$$

39. The function in Figure 7.R.10 is

$$f(t) = \begin{cases} 0, & 0 \le t < 1 \\ t - 1, & 1 \le t < 2 \\ 3 - t, & 2 \le t < 3 \\ 0, & t \ge 3 \end{cases}$$

or

$$f(t) = (t-1)\mathscr{U}(t-1) - 2(t-2)\mathscr{U}(t-2) + (t-3)\mathscr{U}(t-3).$$

The transform of the differential equation is

$$sY(s) - 1 + 2Y(s) = \frac{e^{-s}}{s^2} - \frac{2e^{-2s}}{s^2} + \frac{e^{-3s}}{s^2}$$

so

$$Y(s) = \frac{1}{s+2} + \frac{1}{s^2(s+2)}e^{-s} - \frac{2}{s^2(s+2)}e^{-2s} + \frac{1}{s^2(s+2)}e^{-3s},$$

and

$$y(t) = e^{-2t} + \left[-\frac{1}{4} + \frac{1}{2}(t-1) + \frac{1}{4}e^{-2(t-1)} \right] \mathscr{U}(t-1)$$

$$- 2\left[-\frac{1}{4} + \frac{1}{2}(t-2) + \frac{1}{4}e^{-2(t-2)} \right] \mathscr{U}(t-2)$$

$$+ \left[-\frac{1}{4} + \frac{1}{2}(t-3) + \frac{1}{4}e^{-2(t-3)} \right] \mathscr{U}(t-3).$$

41. Taking the Laplace transform of the integral equation we obtain

$$\mathscr{L}\{y\} = \frac{1}{s} + \frac{1}{s^2} + \frac{1}{2}\frac{2}{s^3}$$

so that

$$y(t) = 1 + t + \frac{1}{2}t^2.$$

43. Taking the Laplace transform of the system gives

$$s\mathscr{L}\{x\} + \mathscr{L}\{y\} = \frac{1}{s^2} + 1$$

$$4\mathscr{L}\{x\} + s\mathscr{L}\{y\} = 2$$

so that

$$\mathscr{L}\{x\} = \frac{s^2 - 2s + 1}{s(s-2)(s+2)} = -\frac{1}{4}\frac{1}{s} + \frac{1}{8}\frac{1}{s-2} + \frac{9}{8}\frac{1}{s+2}.$$

Then

$$x = -\frac{1}{4} + \frac{1}{8}e^{2t} + \frac{9}{8}e^{-2t} \quad \text{and} \quad y = -x' + t = \frac{9}{4}e^{-2t} - \frac{1}{4}e^{2t} + t.$$

45. The integral equation is

$$10i + 2\int_0^t i(\tau)\,d\tau = 2t^2 + 2t.$$

Taking the Laplace transform we obtain

$$\mathscr{L}\{i\} = \left(\frac{4}{s^3} + \frac{2}{s^2}\right)\frac{s}{10s+2} = \frac{s+2}{s^2(5s+2)} = -\frac{9}{s} + \frac{2}{s^2} + \frac{45}{5s+1} = -\frac{9}{s} + \frac{2}{s^2} + \frac{9}{s+1/5}.$$

Thus

$$i(t) = -9 + 2t + 9e^{-t/5}.$$

47. Taking the Laplace transform of the given differential equation we obtain

$$\mathscr{L}\{y\} = \frac{2w_0}{EIL}\left(\frac{L}{48}\cdot\frac{4!}{s^5} - \frac{1}{120}\cdot\frac{5!}{s^6} + \frac{1}{120}\cdot\frac{5!}{s^6}e^{-sL/2}\right) + \frac{c_1}{2}\cdot\frac{2!}{s^3} + \frac{c_2}{6}\cdot\frac{3!}{s^4}$$

so that

$$y = \frac{2w_0}{EIL}\left[\frac{L}{48}x^4 - \frac{1}{120}x^5 + \frac{1}{120}\left(x - \frac{L}{2}\right)^5\mathscr{U}\left(x - \frac{L}{2}\right) + \frac{c_1}{2}x^2 + \frac{c_2}{6}x^3\right]$$

where $y''(0) = c_1$ and $y'''(0) = c_2$. Using $y''(L) = 0$ and $y'''(L) = 0$ we find

$$c_1 = w_0 L^2/24EI, \qquad c_2 = -w_0 L/4EI.$$

Hence

$$y = \frac{w_0}{12EIL}\left[-\frac{1}{5}x^5 + \frac{L}{2}x^4 - \frac{L^2}{2}x^3 + \frac{L^3}{4}x^2 + \frac{1}{5}\left(x - \frac{L}{2}\right)^5 \mathscr{U}\left(x - \frac{L}{2}\right)\right].$$

49. (a) With $\omega^2 = g/l$ and $K = k/m$ the system of differential equations is

$$\theta_1'' + \omega^2\theta_1 = -K(\theta_1 - \theta_2)$$
$$\theta_2'' + \omega^2\theta_2 = K(\theta_1 - \theta_2).$$

Denoting the Laplace transform of $\theta(t)$ by $\Theta(s)$ we have that the Laplace transform of the system is

$$(s^2 + \omega^2)\Theta_1(s) = -K\Theta_1(s) + K\Theta_2(s) + s\theta_0$$
$$(s^2 + \omega^2)\Theta_2(s) = K\Theta_1(s) - K\Theta_2(s) + s\psi_0.$$

If we add the two equations, we get

$$\Theta_1(s) + \Theta_2(s) = (\theta_0 + \psi_0)\frac{s}{s^2 + \omega^2}$$

which implies

$$\theta_1(t) + \theta_2(t) = (\theta_0 + \psi_0)\cos\omega t.$$

This enables us to solve for first, say, $\theta_1(t)$ and then find $\theta_2(t)$ from

$$\theta_2(t) = -\theta_1(t) + (\theta_0 + \psi_0)\cos\omega t.$$

Now solving

$$(s^2 + \omega^2 + K)\Theta_1(s) - K\Theta_2(s) = s\theta_0$$
$$-k\Theta_1(s) + (s^2 + \omega^2 + K)\Theta_2(s) = s\psi_0$$

gives

$$[(s^2 + \omega^2 + K)^2 - K^2]\Theta_1(s) = s(s^2 + \omega^2 + K)\theta_0 + Ks\psi_0.$$

Factoring the difference of two squares and using partial fractions we get

$$\Theta_1(s) = \frac{s(s^2 + \omega^2 + K)\theta_0 + Ks\psi_0}{(s^2 + \omega^2)(s^2 + \omega^2 + 2K)} = \frac{\theta_0 + \psi_0}{2}\frac{s}{s^2 + \omega^2} + \frac{\theta_0 - \psi_0}{2}\frac{s}{s^2 + \omega^2 + 2K},$$

so

$$\theta_1(t) = \frac{\theta_0 + \psi_0}{2}\cos\omega t + \frac{\theta_0 - \psi_0}{2}\cos\sqrt{\omega^2 + 2K}\,t.$$

Then from $\theta_2(t) = -\theta_1(t) + (\theta_0 + \psi_0)\cos\omega t$ we get

$$\theta_2(t) = \frac{\theta_0 + \psi_0}{2}\cos\omega t - \frac{\theta_0 - \psi_0}{2}\cos\sqrt{\omega^2 + 2K}\,t.$$

(b) With the initial conditions $\theta_1(0) = \theta_0$, $\theta_1'(0) = 0$, $\theta_2(0) = \theta_0$, $\theta_2'(0) = 0$ we have

$$\theta_1(t) = \theta_0 \cos \omega t, \qquad \theta_2(t) = \theta_0 \cos \omega t.$$

Physically this means that both pendulums swing in the same direction as if they were free since the spring exerts no influence on the motion ($\theta_1(t)$ and $\theta_2(t)$ are free of K). With the initial conditions $\theta_1(0) = \theta_0$, $\theta_1'(0) = 0$, $\theta_2(0) = -\theta_0$, $\theta_2'(0) = 0$ we have

$$\theta_1(t) = \theta_0 \cos \sqrt{\omega^2 + 2K}\, t, \qquad \theta_2(t) = -\theta_0 \cos \sqrt{\omega^2 + 2K}\, t.$$

Physically this means that both pendulums swing in the opposite directions, stretching and compressing the spring. The amplitude of both displacements is $|\theta_0|$. Moreover, $\theta_1(t) = \theta_0$ and $\theta_2(t) = -\theta_0$ at precisely the same times. At these times the spring is stretched to its maximum.

51. (a) Rewriting the system as

$$\frac{d^2}{dt^2} = 0$$

$$\frac{d^2 y}{dt^2} = -g.$$

Then, taking the Laplace of each equation, we have

$$s^2 X(s) - sx(0) - x'(0) = 0$$

$$s^2 Y(s) - sy(0) - y'(0) = \frac{g}{s}.$$

Using $x(0) = 0$, $x'(0) = v_0 \cos t$, $y(0) = 0$, $y'(0) = v_0 \sin \theta$ where $v_0 = |\mathbf{v_0}|$, we have

$$\begin{cases} s^2 X(s) = v_0 \cos \theta \\ s^2 Y(s) = v_0 \sin \theta - \dfrac{g}{s} \end{cases} \quad \text{or} \quad \begin{cases} X(s) = (v_0 \cos \theta)\,\dfrac{1}{s^2} \\ Y(s) = (v_0 \sin \theta)\,\dfrac{1}{s^2} - \dfrac{g}{s^3} \end{cases}$$

Then

$$\begin{cases} x(t) = (v_0 \cos \theta)\, t \\ y(t) = (v_0 \sin \theta)\, t - \dfrac{1}{2} g t^2 \end{cases}$$

(b) Substituting $t = \dfrac{x}{v_0 \cos \theta}$ into the equation for $y(t)$ yields

$$y(x) = -\frac{1}{2} g \left(\frac{x}{v_0 \cos \theta} \right)^2 + (v_0 \sin \theta) \left(\frac{x}{v_0 \cos \theta} \right) = \left(-\frac{1}{2} \left(\frac{g}{v_0^2 \cos^2 \theta} \right) x + \tan \theta \right) x$$

The projectile hits the ground when $y = 0$. This occurs when $x = 0$, which is the initial condition, or

$$0 = \frac{-g}{2v_0^2 \cos^2 \theta} x + \tan \theta$$

$$x = \frac{(\tan \theta) \, 2v_0^2 \cos^2 \theta}{g} = \frac{v_0^2 \sin 2\theta}{g}$$

which is the horizontal range.

(c) When $0 < \theta < \pi/2$ the complementary angle of θ is $\pi/2 - \theta$. Substituting this into the result of part (b) we have

$$R\left(\frac{\pi}{2} - \theta\right) = \frac{v_0^2 \sin 2\left(\frac{\pi}{2} - \theta\right)}{g} = \frac{v_0^2 \sin (\pi - 2\theta)}{g} = \frac{v_0^2 \left(\sin \pi \cos 2\theta - \cos \pi \sin 2\theta\right)}{g}$$

$$= \frac{v_0^2 \sin 2\theta}{g} = R(\theta).$$

(d) When $g = 32$, $\theta = 38°$ and $v_0 = 300$ w ehave from part (b) that

$$R = \frac{300^2}{32} \sin 76° \approx 2729 \text{ ft.}$$

Solving

$$x(t) = (300 \cos 38°)t = 2729$$

for t, we see that the projectile hits the ground after about 11.54 sec.

(e) For $\theta = 38°$ the curve is shown in blue, while for $\theta = 52°$ the curve is shown in red.

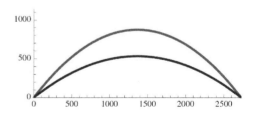

Chapter 8

Systems of Linear First-Order Differential Equations

8.1 Preliminary Theory – Linear Systems

1. Let $\mathbf{X} = \begin{pmatrix} x \\ y \end{pmatrix}$. Then $\mathbf{X}' = \begin{pmatrix} 3 & -5 \\ 4 & 8 \end{pmatrix} \mathbf{X}$.

3. Let $\mathbf{X} = \begin{pmatrix} x \\ y \\ z \end{pmatrix}$. Then $\mathbf{X}' = \begin{pmatrix} -3 & 4 & -9 \\ 6 & -1 & 0 \\ 10 & 4 & 3 \end{pmatrix} \mathbf{X}$.

5. Let $\mathbf{X} = \begin{pmatrix} x \\ y \\ z \end{pmatrix}$. Then $\mathbf{X}' = \begin{pmatrix} 1 & -1 & 1 \\ 2 & 1 & -1 \\ 1 & 1 & 1 \end{pmatrix} \mathbf{X} + \begin{pmatrix} 0 \\ -3t^2 \\ t^2 \end{pmatrix} + \begin{pmatrix} t \\ 0 \\ -t \end{pmatrix} + \begin{pmatrix} -1 \\ 0 \\ 2 \end{pmatrix}$.

7. $\dfrac{dx}{dt} = 4x + 2y + e^t; \quad \dfrac{dy}{dt} = -x + 3y - e^t$

9. $\dfrac{dx}{dt} = x - y + 2z + e^{-t} - 3t; \quad \dfrac{dy}{dt} = 3x - 4y + z + 2e^{-t} + t; \quad \dfrac{dz}{dt} = -2x + 5y + 6z + 2e^{-t} - t$

11. Since

$$\mathbf{X}' = \begin{pmatrix} -5 \\ -10 \end{pmatrix} e^{-5t} \quad \text{and} \quad \begin{pmatrix} 3 & -4 \\ 4 & -7 \end{pmatrix} \mathbf{X} = \begin{pmatrix} -5 \\ -10 \end{pmatrix} e^{-5t}$$

we see that

$$\mathbf{X}' = \begin{pmatrix} 3 & -4 \\ 4 & -7 \end{pmatrix} \mathbf{X}.$$

13. Since

$$\mathbf{X}' = \begin{pmatrix} \frac{3}{2} \\ -3 \end{pmatrix} e^{-3t/2} \quad \text{and} \quad \begin{pmatrix} -1 & \frac{1}{4} \\ 1 & -1 \end{pmatrix} \mathbf{X} = \begin{pmatrix} \frac{3}{2} \\ -3 \end{pmatrix} e^{-3t/2}$$

we see that

$$\mathbf{X}' = \begin{pmatrix} -1 & 1/4 \\ 1 & -1 \end{pmatrix} \mathbf{X}.$$

15. Since

$$\mathbf{X}' = \begin{pmatrix} 0 \\ 0 \\ 0 \end{pmatrix} \quad \text{and} \quad \begin{pmatrix} 1 & 2 & 1 \\ 6 & -1 & 0 \\ -1 & -2 & -1 \end{pmatrix} \mathbf{X} = \begin{pmatrix} 0 \\ 0 \\ 0 \end{pmatrix}$$

we see that

$$\mathbf{X}' = \begin{pmatrix} 1 & 2 & 1 \\ 6 & -1 & 0 \\ -1 & -2 & -1 \end{pmatrix} \mathbf{X}.$$

17. Yes, since $W(\mathbf{X}_1, \mathbf{X}_2) = -2e^{-8t} \neq 0$ the set $\mathbf{X}_1$, $\mathbf{X}_2$ is linearly independent on the interval $(-\infty, \infty)$.

19. No, since $W(\mathbf{X}_1, \mathbf{X}_2, \mathbf{X}_3) = 0$ the set $\mathbf{X}_1$, $\mathbf{X}_2$, $\mathbf{X}_3$ is linearly dependent on the interval $(-\infty, \infty)$.

21. Since

$$\mathbf{X}'_p = \begin{pmatrix} 2 \\ -1 \end{pmatrix} \quad \text{and} \quad \begin{pmatrix} 1 & 4 \\ 3 & 2 \end{pmatrix} \mathbf{X}_p + \begin{pmatrix} 2 \\ -4 \end{pmatrix} t + \begin{pmatrix} -7 \\ -18 \end{pmatrix} = \begin{pmatrix} 2 \\ -1 \end{pmatrix}$$

we see that

$$\mathbf{X}'_p = \begin{pmatrix} 1 & 4 \\ 3 & 2 \end{pmatrix} \mathbf{X}_p + \begin{pmatrix} 2 \\ -4 \end{pmatrix} t + \begin{pmatrix} -7 \\ -18 \end{pmatrix}.$$

23. Since

$$\mathbf{X}'_p = \begin{pmatrix} 2 \\ 0 \end{pmatrix} e^t + \begin{pmatrix} 1 \\ -1 \end{pmatrix} te^t \quad \text{and} \quad \begin{pmatrix} 2 & 1 \\ 3 & 4 \end{pmatrix} \mathbf{X}_p - \begin{pmatrix} 1 \\ 7 \end{pmatrix} e^t = \begin{pmatrix} 2 \\ 0 \end{pmatrix} e^t + \begin{pmatrix} 1 \\ -1 \end{pmatrix} te^t$$

we see that

$$\mathbf{X}'_p = \begin{pmatrix} 2 & 1 \\ 3 & 4 \end{pmatrix} \mathbf{X}_p - \begin{pmatrix} 1 \\ 7 \end{pmatrix} e^t.$$

25. Let

$$\mathbf{X}_1 = \begin{pmatrix} 6 \\ -1 \\ -5 \end{pmatrix} e^{-t}, \quad \mathbf{X}_2 = \begin{pmatrix} -3 \\ 1 \\ 1 \end{pmatrix} e^{-2t}, \quad \mathbf{X}_3 = \begin{pmatrix} 2 \\ 1 \\ 1 \end{pmatrix} e^{3t}, \quad \text{and} \quad \mathbf{A} = \begin{pmatrix} 0 & 6 & 0 \\ 1 & 0 & 1 \\ 1 & 1 & 0 \end{pmatrix}.$$

Then

$$\mathbf{X}'_1 = \begin{pmatrix} -6 \\ 1 \\ 5 \end{pmatrix} e^{-t} = \mathbf{A}\mathbf{X}_1, \quad \mathbf{X}'_2 = \begin{pmatrix} 6 \\ -2 \\ -2 \end{pmatrix} e^{-2t} = \mathbf{A}\mathbf{X}_2, \quad \mathbf{X}'_3 = \begin{pmatrix} 6 \\ 3 \\ 3 \end{pmatrix} e^{3t} = \mathbf{A}\mathbf{X}_3,$$

and $W(\mathbf{X}_1, \mathbf{X}_2, \mathbf{X}_3) = 20 \neq 0$ so that $\mathbf{X}_1$, $\mathbf{X}_2$, and $\mathbf{X}_3$ form a fundamental set for $\mathbf{X}' = \mathbf{A}\mathbf{X}$ on the interval $(-\infty, \infty)$.

8.2 │ Homogeneous Linear Systems

1. The system is

$$\mathbf{X}' = \begin{pmatrix} 1 & 2 \\ 4 & 3 \end{pmatrix} \mathbf{X}$$

and $\det(\mathbf{A} - \lambda\mathbf{I}) = (\lambda - 5)(\lambda + 1) = 0$. For $\lambda_1 = 5$ we obtain

$$\left(\begin{array}{cc|c} -4 & 2 & 0 \\ 4 & -2 & 0 \end{array} \right) \longrightarrow \left(\begin{array}{cc|c} 1 & -\frac{1}{2} & 0 \\ 0 & 0 & 0 \end{array} \right) \quad \text{so that} \quad \mathbf{K}_1 = \begin{pmatrix} 1 \\ 2 \end{pmatrix}.$$

For $\lambda_2 = -1$ we obtain

$$\left(\begin{array}{cc|c} 2 & 2 & 0 \\ 4 & 4 & 0 \end{array} \right) \longrightarrow \left(\begin{array}{cc|c} 1 & 1 & 0 \\ 0 & 0 & 0 \end{array} \right) \quad \text{so that} \quad \mathbf{K}_2 = \begin{pmatrix} -1 \\ 1 \end{pmatrix}.$$

Then

$$\mathbf{X} = c_1 \begin{pmatrix} 1 \\ 2 \end{pmatrix} e^{5t} + c_2 \begin{pmatrix} -1 \\ 1 \end{pmatrix} e^{-t}.$$

3. The system is

$$\mathbf{X}' = \begin{pmatrix} -4 & 2 \\ -\frac{5}{2} & 2 \end{pmatrix} \mathbf{X}$$

and $\det(\mathbf{A} - \lambda\mathbf{I}) = (\lambda - 1)(\lambda + 3) = 0$. For $\lambda_1 = 1$ we obtain

$$\left(\begin{array}{cc|c} -5 & 2 & 0 \\ -\frac{5}{2} & 1 & 0 \end{array} \right) \longrightarrow \left(\begin{array}{cc|c} -5 & 2 & 0 \\ 0 & 0 & 0 \end{array} \right) \quad \text{so that} \quad \mathbf{K}_1 = \begin{pmatrix} 2 \\ 5 \end{pmatrix}.$$

For $\lambda_2 = -3$ we obtain

$$\left(\begin{array}{cc|c} -1 & 2 & 0 \\ -\frac{5}{2} & 5 & 0 \end{array} \right) \longrightarrow \left(\begin{array}{cc|c} -1 & 2 & 0 \\ 0 & 0 & 0 \end{array} \right) \quad \text{so that} \quad \mathbf{K}_2 = \begin{pmatrix} 2 \\ 1 \end{pmatrix}.$$

Then

$$\mathbf{X} = c_1 \begin{pmatrix} 2 \\ 5 \end{pmatrix} e^{t} + c_2 \begin{pmatrix} 2 \\ 1 \end{pmatrix} e^{-3t}.$$

5. The system is

$$\mathbf{X}' = \begin{pmatrix} 10 & -5 \\ 8 & -12 \end{pmatrix} \mathbf{X}$$

and $\det(\mathbf{A} - \lambda\mathbf{I}) = (\lambda - 8)(\lambda + 10) = 0$. For $\lambda_1 = 8$ we obtain

$$\left(\begin{array}{cc|c} 2 & -5 & 0 \\ 8 & -20 & 0 \end{array} \right) \longrightarrow \left(\begin{array}{cc|c} 1 & -\frac{5}{2} & 0 \\ 0 & 0 & 0 \end{array} \right) \quad \text{so that} \quad \mathbf{K}_1 = \begin{pmatrix} 5 \\ 2 \end{pmatrix}.$$

For $\lambda_2 = -10$ we obtain

$$\begin{pmatrix} 20 & -5 & \bigm| & 0 \\ 8 & -2 & \bigm| & 0 \end{pmatrix} \longrightarrow \begin{pmatrix} 1 & -\frac{1}{4} & \bigm| & 0 \\ 0 & 0 & \bigm| & 0 \end{pmatrix} \quad \text{so that} \quad \mathbf{K}_2 = \begin{pmatrix} 1 \\ 4 \end{pmatrix}.$$

Then

$$\mathbf{X} = c_1 \begin{pmatrix} 5 \\ 2 \end{pmatrix} e^{8t} + c_2 \begin{pmatrix} 1 \\ 4 \end{pmatrix} e^{-10t}.$$

7. The system is

$$\mathbf{X}' = \begin{pmatrix} 1 & 1 & -1 \\ 0 & 2 & 0 \\ 0 & 1 & -1 \end{pmatrix} \mathbf{X}$$

and $\det(\mathbf{A} - \lambda \mathbf{I}) = (\lambda - 1)(2 - \lambda)(\lambda + 1) = 0$. For $\lambda_1 = 1$, $\lambda_2 = 2$, and $\lambda_3 = -1$ we obtain

$$\mathbf{K}_1 = \begin{pmatrix} 1 \\ 0 \\ 0 \end{pmatrix}, \quad \mathbf{K}_2 = \begin{pmatrix} 2 \\ 3 \\ 1 \end{pmatrix}, \quad \text{and} \quad \mathbf{K}_3 = \begin{pmatrix} 1 \\ 0 \\ 2 \end{pmatrix},$$

so that

$$\mathbf{X} = c_1 \begin{pmatrix} 1 \\ 0 \\ 0 \end{pmatrix} e^t + c_2 \begin{pmatrix} 2 \\ 3 \\ 1 \end{pmatrix} e^{2t} + c_3 \begin{pmatrix} 1 \\ 0 \\ 2 \end{pmatrix} e^{-t}.$$

9. We have $\det(\mathbf{A} - \lambda \mathbf{I}) = -(\lambda + 1)(\lambda - 3)(\lambda + 2) = 0$. For $\lambda_1 = -1$, $\lambda_2 = 3$, and $\lambda_3 = -2$ we obtain

$$\mathbf{K}_1 = \begin{pmatrix} -1 \\ 0 \\ 1 \end{pmatrix}, \quad \mathbf{K}_2 = \begin{pmatrix} 1 \\ 4 \\ 3 \end{pmatrix}, \quad \text{and} \quad \mathbf{K}_3 = \begin{pmatrix} 1 \\ -1 \\ 3 \end{pmatrix},$$

so that

$$\mathbf{X} = c_1 \begin{pmatrix} -1 \\ 0 \\ 1 \end{pmatrix} e^{-t} + c_2 \begin{pmatrix} 1 \\ 4 \\ 3 \end{pmatrix} e^{3t} + c_3 \begin{pmatrix} 1 \\ -1 \\ 3 \end{pmatrix} e^{-2t}.$$

11. We have $\det(\mathbf{A} - \lambda \mathbf{I}) = -(\lambda + 1)(\lambda + 1/2)(\lambda + 3/2) = 0$. For $\lambda_1 = -1$, $\lambda_2 = -1/2$, and $\lambda_3 = -3/2$ we obtain

$$\mathbf{K}_1 = \begin{pmatrix} 4 \\ 0 \\ -1 \end{pmatrix}, \quad \mathbf{K}_2 = \begin{pmatrix} -12 \\ 6 \\ 5 \end{pmatrix}, \quad \text{and} \quad \mathbf{K}_3 = \begin{pmatrix} 4 \\ 2 \\ -1 \end{pmatrix},$$

so that

$$\mathbf{X} = c_1 \begin{pmatrix} 4 \\ 0 \\ -1 \end{pmatrix} e^{-t} + c_2 \begin{pmatrix} -12 \\ 6 \\ 5 \end{pmatrix} e^{-t/2} + c_3 \begin{pmatrix} 4 \\ 2 \\ -1 \end{pmatrix} e^{-3t/2}.$$

13. We have $\det(\mathbf{A} - \lambda\mathbf{I}) = (\lambda + 1/2)(\lambda - 1/2) = 0$. For $\lambda_1 = -1/2$ and $\lambda_2 = 1/2$ we obtain

$$\mathbf{K}_1 = \begin{pmatrix} 0 \\ 1 \end{pmatrix} \quad \text{and} \quad \mathbf{K}_2 = \begin{pmatrix} 1 \\ 1 \end{pmatrix},$$

so that

$$\mathbf{X} = c_1 \begin{pmatrix} 0 \\ 1 \end{pmatrix} e^{-t/2} + c_2 \begin{pmatrix} 1 \\ 1 \end{pmatrix} e^{t/2}.$$

If

$$\mathbf{X}(0) = \begin{pmatrix} 3 \\ 5 \end{pmatrix}$$

then $c_1 = 2$ and $c_2 = 3$. The solutions is

$$\mathbf{X} = 2 \begin{pmatrix} 0 \\ 1 \end{pmatrix} e^{-t/2} + 3 \begin{pmatrix} 1 \\ 1 \end{pmatrix} e^{t/2}.$$

15. (a) From the discussion in Section 2.9 we get the system

$$\begin{cases} \dfrac{dx_1}{dt} = -\dfrac{3}{100}x_1 + \dfrac{1}{100}x_2 \\[2mm] \dfrac{dx_2}{dt} = \dfrac{1}{50}x_1 - \dfrac{1}{50}x_2 \end{cases} \qquad \text{Thus} \qquad \mathbf{X}' = \begin{pmatrix} -\frac{3}{100} & \frac{1}{100} \\ \frac{1}{50} & -\frac{1}{50} \end{pmatrix} \mathbf{X}$$

(b) The eigenvalues and eigenvectors of the coefficient matrix are found by solving $\det(\mathbf{A} - \lambda\mathbf{I}) = 0$ to get

$$\lambda_1 = -\frac{1}{25} \quad \text{so} \quad \mathbf{K}_1 = \begin{pmatrix} -1 \\ 1 \end{pmatrix} \quad \text{and} \quad \lambda_2 = -\frac{1}{100} \quad \text{so} \quad \mathbf{K}_2 = \begin{pmatrix} 1 \\ 2 \end{pmatrix}$$

The general solution is then

$$\mathbf{X}(t) = c_1 \mathbf{K}_1 e^{\lambda_1 t} + c_2 \mathbf{K}_2 e^{\lambda_2 t} = c_1 \begin{pmatrix} -1 \\ 1 \end{pmatrix} e^{-t/25} + c_2 \begin{pmatrix} 1 \\ 2 \end{pmatrix} e^{-t/100}$$

Using the initial conditions, we get

$$\mathbf{X}(0) = c_1 \begin{pmatrix} -1 \\ 1 \end{pmatrix} + c_2 \begin{pmatrix} 1 \\ 2 \end{pmatrix} = \begin{pmatrix} -c_1 + c_2 \\ c_1 + c_2 \end{pmatrix} = \begin{pmatrix} 20 \\ 5 \end{pmatrix} \quad \text{so} \quad c_1 = -\frac{35}{3} \quad \text{and} \quad c_2 = \frac{25}{3}$$

Thus

$$\mathbf{X}(t) = -\frac{35}{3} \begin{pmatrix} -1 \\ 1 \end{pmatrix} e^{-t/25} + \frac{25}{3} \begin{pmatrix} 1 \\ 2 \end{pmatrix} e^{-t/100}$$

Hence

$$x_1(t) = \frac{35}{3}e^{-t/25} + \frac{25}{3}e^{-t/100} \quad \text{and} \quad x_2(t) = -\frac{35}{3}e^{-t/25} + \frac{50}{3}e^{-t/100}$$

(c)

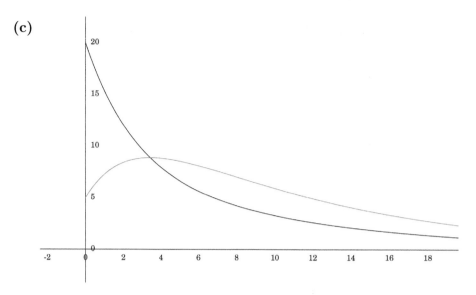

(d) Set $x_1(t) = x_2(t)$ and, using a scientific calculator, we find that $t \approx 34.3277$ min.

17. $\mathbf{X} = c_1 \begin{pmatrix} 0.382175 \\ 0.851161 \\ 0.359815 \end{pmatrix} e^{8.58979t} + c_2 \begin{pmatrix} 0.405188 \\ -0.676043 \\ 0.615458 \end{pmatrix} e^{2.25684t} + c_3 \begin{pmatrix} -0.923562 \\ -0.132174 \\ 0.35995 \end{pmatrix} e^{-0.0466321t}$

19. (a)

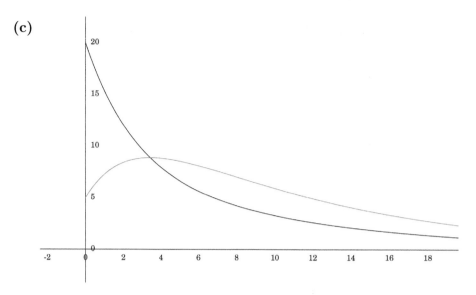

(b) Letting $c_1 = 1$ and $c_2 = 0$ we get $x = 5e^{8t}$, $y = 2e^{8t}$. Eliminating the parameter we find $y = \frac{2}{5}x$, $x > 0$. When $c_1 = -1$ and $c_2 = 0$ we find $y = \frac{2}{5}x$, $x < 0$. Letting $c_1 = 0$ and $c_2 = 1$ we get $x = e^{-10t}$, $y = 4e^{-10t}$. Eliminating the parameter we find $y = 4x$, $x > 0$. Letting $c_1 = 0$ and $c_2 = -1$ we find $y = 4x$, $x < 0$.

(c) The eigenvectors $\mathbf{K}_1 = (5, 2)$ and $\mathbf{K}_2 = (1, 4)$ are shown in the figure in part (**A**).

21. We have $\det(\mathbf{A} - \lambda\mathbf{I}) = \lambda^2 = 0$. For $\lambda_1 = 0$ we obtain

$$\mathbf{K} = \begin{pmatrix} 1 \\ 3 \end{pmatrix}.$$

A solution of $(\mathbf{A} - \lambda_1\mathbf{I})\mathbf{P} = \mathbf{K}$ is

$$\mathbf{P} = \begin{pmatrix} 1 \\ 2 \end{pmatrix}$$

so that

$$\mathbf{X} = c_1 \begin{pmatrix} 1 \\ 3 \end{pmatrix} + c_2 \left[\begin{pmatrix} 1 \\ 3 \end{pmatrix} t + \begin{pmatrix} 1 \\ 2 \end{pmatrix} \right].$$

23. We have $\det(\mathbf{A} - \lambda\mathbf{I}) = (\lambda - 2)^2 = 0$. For $\lambda_1 = 2$ we obtain

$$\mathbf{K} = \begin{pmatrix} 1 \\ 1 \end{pmatrix}.$$

A solution of $(\mathbf{A} - \lambda_1\mathbf{I})\mathbf{P} = \mathbf{K}$ is

$$\mathbf{P} = \begin{pmatrix} -\frac{1}{3} \\ 0 \end{pmatrix}$$

so that

$$\mathbf{X} = c_1 \begin{pmatrix} 1 \\ 1 \end{pmatrix} e^{2t} + c_2 \left[\begin{pmatrix} 1 \\ 1 \end{pmatrix} te^{2t} + \begin{pmatrix} -\frac{1}{3} \\ 0 \end{pmatrix} e^{2t} \right].$$

25. We have $\det(\mathbf{A} - \lambda\mathbf{I}) = (1 - \lambda)(\lambda - 2)^2 = 0$. For $\lambda_1 = 1$ we obtain

$$\mathbf{K}_1 = \begin{pmatrix} 1 \\ 1 \\ 1 \end{pmatrix}.$$

For $\lambda_2 = 2$ we obtain

$$\mathbf{K}_2 = \begin{pmatrix} 1 \\ 0 \\ 1 \end{pmatrix} \quad \text{and} \quad \mathbf{K}_3 = \begin{pmatrix} 1 \\ 1 \\ 0 \end{pmatrix}.$$

Then

$$\mathbf{X} = c_1 \begin{pmatrix} 1 \\ 1 \\ 1 \end{pmatrix} e^t + c_2 \begin{pmatrix} 1 \\ 0 \\ 1 \end{pmatrix} e^{2t} + c_3 \begin{pmatrix} 1 \\ 1 \\ 0 \end{pmatrix} e^{2t}.$$

27. We have $\det(\mathbf{A} - \lambda\mathbf{I}) = -\lambda(5 - \lambda)^2 = 0$. For $\lambda_1 = 0$ we obtain

$$\mathbf{K}_1 = \begin{pmatrix} -4 \\ -5 \\ 2 \end{pmatrix}.$$

For $\lambda_2 = 5$ we obtain

$$\mathbf{K} = \begin{pmatrix} -2 \\ 0 \\ 1 \end{pmatrix}.$$

A solution of $(\mathbf{A} - \lambda_1\mathbf{I})\mathbf{P} = \mathbf{K}$ is

$$\mathbf{P} = \begin{pmatrix} \frac{5}{2} \\ \frac{1}{2} \\ 0 \end{pmatrix}$$

so that

$$\mathbf{X} = c_1 \begin{pmatrix} -4 \\ -5 \\ 2 \end{pmatrix} + c_2 \begin{pmatrix} -2 \\ 0 \\ 1 \end{pmatrix} e^{5t} + c_3 \left[\begin{pmatrix} -2 \\ 0 \\ 1 \end{pmatrix} t e^{5t} + \begin{pmatrix} \frac{5}{2} \\ \frac{1}{2} \\ 0 \end{pmatrix} e^{5t} \right].$$

29. We have $\det(\mathbf{A} - \lambda \mathbf{I}) = -(\lambda - 1)^3 = 0$. For $\lambda_1 = 1$ we obtain

$$\mathbf{K} = \begin{pmatrix} 0 \\ 1 \\ 1 \end{pmatrix}.$$

Solutions of $(\mathbf{A} - \lambda_1 \mathbf{I})\mathbf{P} = \mathbf{K}$ and $(\mathbf{A} - \lambda_1 \mathbf{I})\mathbf{Q} = \mathbf{P}$ are

$$\mathbf{P} = \begin{pmatrix} 0 \\ 1 \\ 0 \end{pmatrix} \quad \text{and} \quad \mathbf{Q} = \begin{pmatrix} \frac{1}{2} \\ 0 \\ 0 \end{pmatrix}$$

so that

$$\mathbf{X} = c_1 \begin{pmatrix} 0 \\ 1 \\ 1 \end{pmatrix} e^{t} + c_2 \left[\begin{pmatrix} 0 \\ 1 \\ 1 \end{pmatrix} t e^{t} + \begin{pmatrix} 0 \\ 1 \\ 0 \end{pmatrix} e^{t} \right] + c_3 \left[\begin{pmatrix} 0 \\ 1 \\ 1 \end{pmatrix} \frac{t^2}{2} e^{t} + \begin{pmatrix} 0 \\ 1 \\ 0 \end{pmatrix} t e^{t} + \begin{pmatrix} \frac{1}{2} \\ 0 \\ 0 \end{pmatrix} e^{t} \right].$$

31. We have $\det(\mathbf{A} - \lambda \mathbf{I}) = (\lambda - 4)^2 = 0$. For $\lambda_1 = 4$ we obtain

$$\mathbf{K} = \begin{pmatrix} 2 \\ 1 \end{pmatrix}.$$

A solution of $(\mathbf{A} - \lambda_1 \mathbf{I})\mathbf{P} = \mathbf{K}$ is

$$\mathbf{P} = \begin{pmatrix} 1 \\ 1 \end{pmatrix}$$

so that

$$\mathbf{X} = c_1 \begin{pmatrix} 2 \\ 1 \end{pmatrix} e^{4t} + c_2 \left[\begin{pmatrix} 2 \\ 1 \end{pmatrix} t e^{4t} + \begin{pmatrix} 1 \\ 1 \end{pmatrix} e^{4t} \right].$$

If

$$\mathbf{X}(0) = \begin{pmatrix} -1 \\ 6 \end{pmatrix}$$

then $c_1 = -7$ and $c_2 = 13$. The solutions is

$$\mathbf{X} = -7 \begin{pmatrix} 2 \\ 1 \end{pmatrix} e^{4t} + 13 \begin{pmatrix} 2t + 1 \\ t + 1 \end{pmatrix} e^{4t}.$$

33. In this case $\det(\mathbf{A} - \lambda\mathbf{I}) = (2 - \lambda)^5$, and $\lambda_1 = 2$ is an eigenvalue of multiplicity 5. Linearly independent eigenvectors are

$$\mathbf{K}_1 = \begin{pmatrix} 1 \\ 0 \\ 0 \\ 0 \\ 0 \end{pmatrix}, \qquad \mathbf{K}_2 = \begin{pmatrix} 0 \\ 0 \\ 1 \\ 0 \\ 0 \end{pmatrix}, \qquad \text{and} \qquad \mathbf{K}_3 = \begin{pmatrix} 0 \\ 0 \\ 0 \\ 1 \\ 0 \end{pmatrix}.$$

In Problem 35 the form of the answer will vary according to the choice of eigenvector. For example, if $\mathbf{K}_1$ is chosen to be $\begin{pmatrix} 1 \\ 2 - i \end{pmatrix}$ the solution has the form

$$\mathbf{X} = c_1 \begin{pmatrix} \cos t \\ 2\cos t + \sin t \end{pmatrix} e^{4t} + c_2 \begin{pmatrix} \sin t \\ 2\sin t - \cos t \end{pmatrix} e^{4t}.$$

35. We have $\det(\mathbf{A} - \lambda\mathbf{I}) = \lambda^2 - 8\lambda + 17 = 0$. For $\lambda_1 = 4 + i$ we obtain

$$\mathbf{K}_1 = \begin{pmatrix} 2 + i \\ 5 \end{pmatrix}$$

so that

$$\mathbf{X}_1 = \begin{pmatrix} 2 + i \\ 5 \end{pmatrix} e^{(4+i)t} = \begin{pmatrix} 2\cos t - \sin t \\ 5\cos t \end{pmatrix} e^{4t} + i \begin{pmatrix} \cos t + 2\sin t \\ 5\sin t \end{pmatrix} e^{4t}.$$

Then

$$\mathbf{X} = c_1 \begin{pmatrix} 2\cos t - \sin t \\ 5\cos t \end{pmatrix} e^{4t} + c_2 \begin{pmatrix} \cos t + 2\sin t \\ 5\sin t \end{pmatrix} e^{4t}.$$

37. We have $\det(\mathbf{A} - \lambda\mathbf{I}) = \lambda^2 - 8\lambda + 17 = 0$. For $\lambda_1 = 4 + i$ we obtain

$$\mathbf{K}_1 = \begin{pmatrix} -1 - i \\ 2 \end{pmatrix}$$

so that

$$\mathbf{X}_1 = \begin{pmatrix} -1 - i \\ 2 \end{pmatrix} e^{(4+i)t} = \begin{pmatrix} \sin t - \cos t \\ 2\cos t \end{pmatrix} e^{4t} + i \begin{pmatrix} -\sin t - \cos t \\ 2\sin t \end{pmatrix} e^{4t}.$$

Then

$$\mathbf{X} = c_1 \begin{pmatrix} \sin t - \cos t \\ 2\cos t \end{pmatrix} e^{4t} + c_2 \begin{pmatrix} -\sin t - \cos t \\ 2\sin t \end{pmatrix} e^{4t}.$$

39. We have $\det(\mathbf{A} - \lambda\mathbf{I}) = \lambda^2 + 9 = 0$. For $\lambda_1 = 3i$ we obtain

$$\mathbf{K}_1 = \begin{pmatrix} 4 + 3i \\ 5 \end{pmatrix}$$

so that

$$\mathbf{X}_1 = \begin{pmatrix} 4 + 3i \\ 5 \end{pmatrix} e^{3it} = \begin{pmatrix} 4\cos 3t - 3\sin 3t \\ 5\cos 3t \end{pmatrix} + i \begin{pmatrix} 4\sin 3t + 3\cos 3t \\ 5\sin 3t \end{pmatrix}.$$

Then

$$\mathbf{X} = c_1 \begin{pmatrix} 4\cos 3t - 3\sin 3t \\ 5\cos 3t \end{pmatrix} + c_2 \begin{pmatrix} 4\sin 3t + 3\cos 3t \\ 5\sin 3t \end{pmatrix}.$$

41. We have $\det(\mathbf{A} - \lambda\mathbf{I}) = -\lambda\left(\lambda^2 + 1\right) = 0$. For $\lambda_1 = 0$ we obtain

$$\mathbf{K}_1 = \begin{pmatrix} 1 \\ 0 \\ 0 \end{pmatrix}.$$

For $\lambda_2 = i$ we obtain

$$\mathbf{K}_2 = \begin{pmatrix} -i \\ i \\ 1 \end{pmatrix}$$

so that

$$\mathbf{X}_2 = \begin{pmatrix} -i \\ i \\ 1 \end{pmatrix} e^{it} = \begin{pmatrix} \sin t \\ -\sin t \\ \cos t \end{pmatrix} + i \begin{pmatrix} -\cos t \\ \cos t \\ \sin t \end{pmatrix}.$$

Then

$$\mathbf{X} = c_1 \begin{pmatrix} 1 \\ 0 \\ 0 \end{pmatrix} + c_2 \begin{pmatrix} \sin t \\ -\sin t \\ \cos t \end{pmatrix} + c_3 \begin{pmatrix} -\cos t \\ \cos t \\ \sin t \end{pmatrix}.$$

43. We have $\det(\mathbf{A} - \lambda\mathbf{I}) = (1 - \lambda)(\lambda^2 - 2\lambda + 2) = 0$. For $\lambda_1 = 1$ we obtain

$$\mathbf{K}_1 = \begin{pmatrix} 0 \\ 2 \\ 1 \end{pmatrix}.$$

For $\lambda_2 = 1 + i$ we obtain

$$\mathbf{K}_2 = \begin{pmatrix} 1 \\ i \\ i \end{pmatrix}$$

so that

$$\mathbf{X}_2 = \begin{pmatrix} 1 \\ i \\ i \end{pmatrix} e^{(1+i)t} = \begin{pmatrix} \cos t \\ -\sin t \\ -\sin t \end{pmatrix} e^t + i \begin{pmatrix} \sin t \\ \cos t \\ \cos t \end{pmatrix} e^t.$$

Then

$$\mathbf{X} = c_1 \begin{pmatrix} 0 \\ 2 \\ 1 \end{pmatrix} e^t + c_2 \begin{pmatrix} \cos t \\ -\sin t \\ -\sin t \end{pmatrix} e^t + c_3 \begin{pmatrix} \sin t \\ \cos t \\ \cos t \end{pmatrix} e^t.$$

45. We have $\det(\mathbf{A} - \lambda\mathbf{I}) = (2 - \lambda)(\lambda^2 + 4\lambda + 13) = 0$. For $\lambda_1 = 2$ we obtain

$$\mathbf{K}_1 = \begin{pmatrix} 28 \\ -5 \\ 25 \end{pmatrix}.$$

For $\lambda_2 = -2 + 3i$ we obtain

$$\mathbf{K}_2 = \begin{pmatrix} 4 + 3i \\ -5 \\ 0 \end{pmatrix}$$

so that

$$\mathbf{X}_2 = \begin{pmatrix} 4 + 3i \\ -5 \\ 0 \end{pmatrix} e^{(-2+3i)t} = \begin{pmatrix} 4\cos 3t - 3\sin 3t \\ -5\cos 3t \\ 0 \end{pmatrix} e^{-2t} + i \begin{pmatrix} 4\sin 3t + 3\cos 3t \\ -5\sin 3t \\ 0 \end{pmatrix} e^{-2t}.$$

Then

$$\mathbf{X} = c_1 \begin{pmatrix} 28 \\ -5 \\ 25 \end{pmatrix} e^{2t} + c_2 \begin{pmatrix} 4\cos 3t - 3\sin 3t \\ -5\cos 3t \\ 0 \end{pmatrix} e^{-2t} + c_3 \begin{pmatrix} 4\sin 3t + 3\cos 3t \\ -5\sin 3t \\ 0 \end{pmatrix} e^{-2t}.$$

47. We have $\det(\mathbf{A} - \lambda\mathbf{I}) = (1 - \lambda)(\lambda^2 + 25) = 0$. For $\lambda_1 = 1$ we obtain

$$\mathbf{K}_1 = \begin{pmatrix} 25 \\ -7 \\ 6 \end{pmatrix}.$$

For $\lambda_2 = 5i$ we obtain

$$\mathbf{K}_2 = \begin{pmatrix} 1 + 5i \\ 1 \\ 1 \end{pmatrix}$$

so that

$$\mathbf{X}_2 = \begin{pmatrix} 1 + 5i \\ 1 \\ 1 \end{pmatrix} e^{5it} = \begin{pmatrix} \cos 5t - 5\sin 5t \\ \cos 5t \\ \cos 5t \end{pmatrix} + i \begin{pmatrix} \sin 5t + 5\cos 5t \\ \sin 5t \\ \sin 5t \end{pmatrix}.$$

Then

$$\mathbf{X} = c_1 \begin{pmatrix} 25 \\ -7 \\ 6 \end{pmatrix} e^{t} + c_2 \begin{pmatrix} \cos 5t - 5\sin 5t \\ \cos 5t \\ \cos 5t \end{pmatrix} + c_3 \begin{pmatrix} \sin 5t + 5\cos 5t \\ \sin 5t \\ \sin 5t \end{pmatrix}.$$

If

$$\mathbf{X}(0) = \begin{pmatrix} 4 \\ 6 \\ -7 \end{pmatrix}$$

then $c_1 = c_2 = -1$ and $c_3 = 6$. The solution is

$$\mathbf{X} = -1 \begin{pmatrix} 25 \\ -7 \\ 6 \end{pmatrix} e^t - \begin{pmatrix} \cos 5t - 5\sin 5t \\ \cos 5t \\ \cos 5t \end{pmatrix} + 6 \begin{pmatrix} \sin 5t + 5\cos 5t \\ \sin 5t \\ \sin 5t \end{pmatrix}.$$

49. (a) The system to solve is

$$\begin{cases} \dfrac{dx_1}{dt} = -\dfrac{1}{20}x_1 + \dfrac{1}{10}x_3 \\[2mm] \dfrac{dx_2}{dt} = \dfrac{1}{20}x_1 - \dfrac{1}{20}x_2 \\[2mm] \dfrac{dx_3}{dt} = \dfrac{1}{20}x_2 - \dfrac{1}{10}x_3 \end{cases} \quad \text{Thus} \quad \mathbf{X}' = \begin{pmatrix} -\frac{1}{20} & 0 & \frac{1}{10} \\ \frac{1}{20} & -\frac{1}{20} & 0 \\ 0 & \frac{1}{20} & -\frac{1}{10} \end{pmatrix} \mathbf{X}$$

(b) The eigenvalues and eigenvectors of the coefficient matrix are found by solving $\det(\mathbf{A} - \lambda\mathbf{I}) = 0$ to get

$$\lambda_1 = 0 \quad \text{so} \quad \mathbf{K}_1 = \begin{pmatrix} 2 \\ 2 \\ 1 \end{pmatrix}, \quad \lambda_2 = -\frac{1}{10} + \frac{1}{20}i \quad \text{so} \quad \mathbf{K}_2 = \begin{pmatrix} -1-i \\ i \\ 1 \end{pmatrix},$$

$$\text{and} \quad \lambda_3 = -\frac{1}{10} - \frac{1}{20}i \quad \text{so} \quad \mathbf{K}_3 = \begin{pmatrix} -1+i \\ -i \\ 1 \end{pmatrix}.$$

The general solution is then

$$\mathbf{X}(t) = c_1 \begin{pmatrix} 2 \\ 2 \\ 1 \end{pmatrix} + c_2 \begin{pmatrix} -\cos\frac{1}{20}t + \sin\frac{1}{20}t \\ -\sin\frac{1}{20}t \\ \cos\frac{1}{20}t \end{pmatrix} e^{-t/10} + c_3 \begin{pmatrix} -\cos\frac{1}{20}t - \sin\frac{1}{20}t \\ \cos\frac{1}{20}t \\ \sin\frac{1}{20}t \end{pmatrix} e^{-t/10}$$

Using the initial conditions, we get

$$\mathbf{X}(0) = c_1 \begin{pmatrix} 2 \\ 2 \\ 1 \end{pmatrix} + c_2 \begin{pmatrix} -1 \\ 0 \\ 1 \end{pmatrix} + c_3 \begin{pmatrix} -1 \\ 1 \\ 0 \end{pmatrix} = \begin{pmatrix} 30 \\ 20 \\ 5 \end{pmatrix}$$

$$c_1 = 11, \quad c_2 = -6, \quad \text{and} \quad c_3 = -2$$

$$\mathbf{X}(t) = 11 \begin{pmatrix} 2 \\ 2 \\ 1 \end{pmatrix} - 6 \begin{pmatrix} -\cos\frac{1}{20}t + \sin\frac{1}{20}t \\ -\sin\frac{1}{20}t \\ \cos\frac{1}{20}t \end{pmatrix} e^{-t/10} - 2 \begin{pmatrix} -\cos\frac{1}{20}t - \sin\frac{1}{20}t \\ \cos\frac{1}{20}t \\ \sin\frac{1}{20}t \end{pmatrix} e^{-t/10}$$

51.

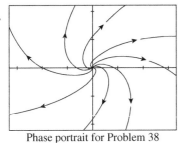

Phase portrait for Problem 38

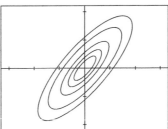

Phase portrait for Problem 39

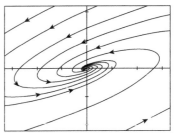

Phase portrait for Problem 40

53. (a) From $\det(\mathbf{A} - \lambda\mathbf{I}) = \lambda(\lambda - 2) = 0$ we get $\lambda_1 = 0$ and $\lambda_2 = 2$. For $\lambda_1 = 0$ we obtain

$$\begin{pmatrix} 1 & 1 & | & 0 \\ 1 & 1 & | & 0 \end{pmatrix} \longrightarrow \begin{pmatrix} 1 & 1 & | & 0 \\ 0 & 0 & | & 0 \end{pmatrix} \quad \text{so that} \quad \mathbf{K}_1 = \begin{pmatrix} -1 \\ 1 \end{pmatrix}.$$

For $\lambda_2 = 2$ we obtain

$$\begin{pmatrix} -1 & 1 & | & 0 \\ 1 & -1 & | & 0 \end{pmatrix} \longrightarrow \begin{pmatrix} -1 & 1 & | & 0 \\ 0 & 0 & | & 0 \end{pmatrix} \quad \text{so that} \quad \mathbf{K}_2 = \begin{pmatrix} 1 \\ 1 \end{pmatrix}.$$

Then

$$\mathbf{X} = c_1 \begin{pmatrix} -1 \\ 1 \end{pmatrix} + c_2 \begin{pmatrix} 1 \\ 1 \end{pmatrix} e^{2t}.$$

The line $y = -x$ is not a trajectory of the system. Trajectories are $x = -c_1 + c_2 e^{2t}$, $y = c_1 + c_2 e^{2t}$ or $y = x + 2c_1$. This is a family of lines perpendicular to the line $y = -x$. All of the constant solutions of the system do, however, lie on the line $y = -x$.

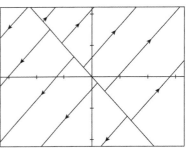

(b) From $\det(\mathbf{A} - \lambda\mathbf{I}) = \lambda^2 = 0$ we get $\lambda_1 = 0$ and

$$\mathbf{K} = \begin{pmatrix} -1 \\ 1 \end{pmatrix}.$$

A solution of $(\mathbf{A} - \lambda_1\mathbf{I})\mathbf{P} = \mathbf{K}$ is

$$\mathbf{P} = \begin{pmatrix} -1 \\ 0 \end{pmatrix}$$

so that

$$\mathbf{X} = c_1 \begin{pmatrix} -1 \\ 1 \end{pmatrix} + c_2 \left[\begin{pmatrix} -1 \\ 1 \end{pmatrix} t + \begin{pmatrix} -1 \\ 0 \end{pmatrix} \right].$$

All trajectories are parallel to $y = -x$, but $y = -x$ is not a trajectory. There are constant solutions of the system, however, that do lie on the line $y = -x$.

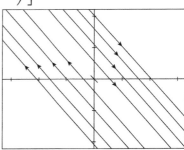

55. From $x = 2\cos 2t - 2\sin 2t$, $y = -\cos 2t$ we find $x + 2y = -2\sin 2t$. Then

$$(x + 2y)^2 = 4\sin^2 2t = 4(1 - \cos^2 2t) = 4 - 4\cos^2 2t = 4 - 4y^2$$

and

$$x^2 + 4xy + 4y^2 = 4 - 4y^2 \quad \text{or} \quad x^2 + 4xy + 8y^2 = 4.$$

This is a rotated conic section and, from the discriminant $b^2 - 4ac = 16 - 32 < 0$, we see that the curve is an ellipse.

8.3 | Nonhomogeneous Linear Systems

1. Solving

$$\det(\mathbf{A} - \lambda\mathbf{I}) = \begin{vmatrix} 2 - \lambda & 3 \\ -1 & -2 - \lambda \end{vmatrix} = \lambda^2 - 1 = (\lambda - 1)(\lambda + 1) = 0$$

we obtain eigenvalues $\lambda_1 = -1$ and $\lambda_2 = 1$. Corresponding eigenvectors are

$$\mathbf{K}_1 = \begin{pmatrix} -1 \\ 1 \end{pmatrix} \quad \text{and} \quad \mathbf{K}_2 = \begin{pmatrix} -3 \\ 1 \end{pmatrix}.$$

Thus

$$\mathbf{X}_c = c_1 \begin{pmatrix} -1 \\ 1 \end{pmatrix} e^{-t} + c_2 \begin{pmatrix} -3 \\ 1 \end{pmatrix} e^t.$$

Substituting

$$\mathbf{X}_p = \begin{pmatrix} a_1 \\ b_1 \end{pmatrix}$$

into the system yields

$$2a_1 + 3b_1 = 7$$
$$-a_1 - 2b_1 = -5,$$

from which we obtain $a_1 = -1$ and $b_1 = 3$. Then

$$\mathbf{X}(t) = c_1 \begin{pmatrix} -1 \\ 1 \end{pmatrix} e^{-t} + c_2 \begin{pmatrix} -3 \\ 1 \end{pmatrix} e^t + \begin{pmatrix} -1 \\ 3 \end{pmatrix}.$$

3. Solving

$$\det(\mathbf{A} - \lambda\mathbf{I}) = \begin{vmatrix} 1 - \lambda & 3 \\ 3 & 1 - \lambda \end{vmatrix} = \lambda^2 - 2\lambda - 8 = (\lambda - 4)(\lambda + 2) = 0$$

we obtain eigenvalues $\lambda_1 = -2$ and $\lambda_2 = 4$. Corresponding eigenvectors are

$$\mathbf{K}_1 = \begin{pmatrix} 1 \\ -1 \end{pmatrix} \quad \text{and} \quad \mathbf{K}_2 = \begin{pmatrix} 1 \\ 1 \end{pmatrix}.$$

Thus

$$\mathbf{X}_c = c_1 \begin{pmatrix} 1 \\ -1 \end{pmatrix} e^{-2t} + c_2 \begin{pmatrix} 1 \\ 1 \end{pmatrix} e^{4t}.$$

Substituting

$$\mathbf{X}_p = \begin{pmatrix} a_3 \\ b_3 \end{pmatrix} t^2 + \begin{pmatrix} a_2 \\ b_2 \end{pmatrix} t + \begin{pmatrix} a_1 \\ b_1 \end{pmatrix}$$

into the system yields

$$a_3 + 3b_3 = 2 \qquad a_2 + 3b_2 = 2a_3 \qquad a_1 + 3b_1 = a_2$$
$$3a_3 + b_3 = 0 \qquad 3a_2 + b_2 + 1 = 2b_3 \qquad 3a_1 + b_1 + 5 = b_2$$

from which we obtain $a_3 = -1/4$, $b_3 = 3/4$, $a_2 = 1/4$, $b_2 = -1/4$, $a_1 = -2$, and $b_1 = 3/4$.
Then

$$\mathbf{X}(t) = c_1 \begin{pmatrix} 1 \\ -1 \end{pmatrix} e^{-2t} + c_2 \begin{pmatrix} 1 \\ 1 \end{pmatrix} e^{4t} + \begin{pmatrix} -\frac{1}{4} \\ \frac{3}{4} \end{pmatrix} t^2 + \begin{pmatrix} \frac{1}{4} \\ -\frac{1}{4} \end{pmatrix} t + \begin{pmatrix} -2 \\ \frac{3}{4} \end{pmatrix}.$$

5. Solving

$$\det(\mathbf{A} - \lambda\mathbf{I}) = \begin{vmatrix} 4-\lambda & \frac{1}{3} \\ 9 & 6-\lambda \end{vmatrix} = \lambda^2 - 10\lambda + 21 = (\lambda-3)(\lambda-7) = 0$$

we obtain the eigenvalues $\lambda_1 = 3$ and $\lambda_2 = 7$. Corresponding eigenvectors are

$$\mathbf{K}_1 = \begin{pmatrix} 1 \\ -3 \end{pmatrix} \quad \text{and} \quad \mathbf{K}_2 = \begin{pmatrix} 1 \\ 9 \end{pmatrix}.$$

Thus

$$\mathbf{X}_c = c_1 \begin{pmatrix} 1 \\ -3 \end{pmatrix} e^{3t} + c_2 \begin{pmatrix} 1 \\ 9 \end{pmatrix} e^{7t}.$$

Substituting

$$\mathbf{X}_p = \begin{pmatrix} a_1 \\ b_1 \end{pmatrix} e^t$$

into the system yields

$$3a_1 + \frac{1}{3}b_1 = 3$$
$$9a_1 + 5b_1 = -10$$

from which we obtain $a_1 = 55/36$ and $b_1 = -19/4$. Then

$$\mathbf{X}(t) = c_1 \begin{pmatrix} 1 \\ -3 \end{pmatrix} e^{3t} + c_2 \begin{pmatrix} 1 \\ 9 \end{pmatrix} e^{7t} + \begin{pmatrix} \frac{55}{36} \\ -\frac{19}{4} \end{pmatrix} e^t.$$

7. Solving

$$\det(\mathbf{A} - \lambda\mathbf{I}) = \begin{vmatrix} 1 - \lambda & 1 & 1 \\ 0 & 2 - \lambda & 3 \\ 0 & 0 & 5 - \lambda \end{vmatrix} = (1 - \lambda)(2 - \lambda)(5 - \lambda) = 0$$

we obtain the eigenvalues $\lambda_1 = 1$, $\lambda_2 = 2$, and $\lambda_3 = 5$. Corresponding eigenvectors are

$$\mathbf{K}_1 = \begin{pmatrix} 1 \\ 0 \\ 0 \end{pmatrix}, \quad \mathbf{K}_2 = \begin{pmatrix} 1 \\ 1 \\ 0 \end{pmatrix} \quad \text{and} \quad \mathbf{K}_3 = \begin{pmatrix} 1 \\ 2 \\ 2 \end{pmatrix}.$$

Thus

$$\mathbf{X}_c = C_1 \begin{pmatrix} 1 \\ 0 \\ 0 \end{pmatrix} e^t + C_2 \begin{pmatrix} 1 \\ 1 \\ 0 \end{pmatrix} e^{2t} + C_3 \begin{pmatrix} 1 \\ 2 \\ 2 \end{pmatrix} e^{5t}.$$

Substituting

$$\mathbf{X}_p = \begin{pmatrix} a_1 \\ b_1 \\ c_1 \end{pmatrix} e^{4t}$$

into the system yields

$$-3a_1 + b_1 + c_1 = -1$$
$$-2b_1 + 3c_1 = 1$$
$$c_1 = -2$$

from which we obtain $c_1 = -2$, $b_1 = -7/2$, and $a_1 = -3/2$. Then

$$\mathbf{X}(t) = C_1 \begin{pmatrix} 1 \\ 0 \\ 0 \end{pmatrix} e^t + C_2 \begin{pmatrix} 1 \\ 1 \\ 0 \end{pmatrix} e^{2t} + C_3 \begin{pmatrix} 1 \\ 2 \\ 2 \end{pmatrix} e^{5t} + \begin{pmatrix} -\frac{3}{2} \\ -\frac{7}{2} \\ -2 \end{pmatrix} e^{4t}.$$

9. First solve the associated homogeneous system

$$\mathbf{X}' = \begin{pmatrix} -1 & -2 \\ 3 & 4 \end{pmatrix} \mathbf{X}$$

The eigenvalues and eigenvectors of the coefficient matrix are found by solving $\det(\mathbf{A} - \lambda\mathbf{I}) = 0$ to get

$$\lambda_1 = 1 \quad \text{so} \quad \mathbf{K}_1 = \begin{pmatrix} 1 \\ -1 \end{pmatrix} \quad \text{and} \quad \lambda_2 = 2 \quad \text{so} \quad \mathbf{K}_2 = \begin{pmatrix} 2 \\ -3 \end{pmatrix}$$

The complementary solution is then

$$\mathbf{X}_c(t) = c_1 \mathbf{K}_1 e^{\lambda_1 t} + c_2 \mathbf{K}_2 e^{\lambda_2 t} = c_1 \begin{pmatrix} 1 \\ -1 \end{pmatrix} e^t + c_2 \begin{pmatrix} 2 \\ -3 \end{pmatrix} e^{2t}$$

Based on the form of $\mathbf{F}(t)$, guess $\mathbf{X}_p = \begin{pmatrix} a_1 \\ b_1 \end{pmatrix}$ and force it into the original system to get

$\mathbf{X}_p = \begin{pmatrix} -9 \\ 6 \end{pmatrix}$. The general solution is then

$$\mathbf{X} = \mathbf{X}_c + \mathbf{X}_p = c_1 \begin{pmatrix} 1 \\ -1 \end{pmatrix} e^t + c_2 \begin{pmatrix} 2 \\ -3 \end{pmatrix} e^{2t} + \begin{pmatrix} -9 \\ 6 \end{pmatrix}$$

Next use the initial condition to solve for c_1 and c_2:

$$\mathbf{X}(0) = c_1 \begin{pmatrix} 1 \\ -1 \end{pmatrix} + c_2 \begin{pmatrix} 2 \\ -3 \end{pmatrix} + \begin{pmatrix} -9 \\ 6 \end{pmatrix} = \begin{pmatrix} -4 \\ 5 \end{pmatrix}$$

$$c_1 = 13 \quad \text{and} \quad c_2 = -4$$

$$\mathbf{X} = 13 \begin{pmatrix} 1 \\ -1 \end{pmatrix} e^t - 4 \begin{pmatrix} 2 \\ -3 \end{pmatrix} e^{2t} + \begin{pmatrix} -9 \\ 6 \end{pmatrix}$$

11. (a) From the discussion in Section 2.9 we get the system

$$\begin{cases} \dfrac{dx_1}{dt} = -\dfrac{3}{100}x_1 + \dfrac{1}{100}x_2 \\[2mm] \dfrac{dx_2}{dt} = \dfrac{1}{50}x_1 - \dfrac{1}{25}x_2 + 1 \end{cases} \quad \text{Thus} \quad \mathbf{X}' = \begin{pmatrix} -\frac{3}{100} & \frac{1}{100} \\ \frac{1}{50} & -\frac{1}{25} \end{pmatrix} \mathbf{X} + \begin{pmatrix} 0 \\ 1 \end{pmatrix}$$

(b) First solve the associated homogeneous system

$$\mathbf{X}' = \begin{pmatrix} -\frac{3}{100} & \frac{1}{100} \\ \frac{1}{50} & -\frac{1}{25} \end{pmatrix} \mathbf{X}$$

The eigenvalues and eigenvectors of the coefficient matrix are found by solving $\det(\mathbf{A} - \lambda\mathbf{I}) = 0$ to get

$$\lambda_1 = -\frac{1}{20} \quad \text{so} \quad \mathbf{K}_1 = \begin{pmatrix} -1 \\ 2 \end{pmatrix} \quad \text{and} \quad \lambda_2 = -\frac{1}{50} \quad \text{so} \quad \mathbf{K}_2 = \begin{pmatrix} 1 \\ 1 \end{pmatrix}$$

The complementary solution is then

$$\mathbf{X}_c(t) = c_1 \mathbf{K}_1 e^{\lambda_1 t} + c_2 \mathbf{K}_2 e^{\lambda_2 t} = c_1 \begin{pmatrix} -1 \\ 2 \end{pmatrix} e^{-t/20} + c_2 \begin{pmatrix} 1 \\ 1 \end{pmatrix} e^{-t/50}$$

Based on the form of $\mathbf{F}(t)$, guess $\mathbf{X}_p = \begin{pmatrix} a_1 \\ b_1 \end{pmatrix}$ and force it into the original system to get

$\mathbf{X}_p = \begin{pmatrix} 10 \\ 30 \end{pmatrix}$. The general solution is then

$$\mathbf{X} = \mathbf{X}_c + \mathbf{X}_p = c_1 \begin{pmatrix} -1 \\ 2 \end{pmatrix} e^{-t/20} + c_2 \begin{pmatrix} 1 \\ 1 \end{pmatrix} e^{-t/50} + \begin{pmatrix} 10 \\ 30 \end{pmatrix}$$

Next use the initial condition to solve for c_1 and c_2:

$$\mathbf{X}(0) = c_1 \begin{pmatrix} -1 \\ 2 \end{pmatrix} + c_2 \left[\begin{pmatrix} 1 \\ 1 \end{pmatrix} \right] + \begin{pmatrix} 10 \\ 30 \end{pmatrix} = \begin{pmatrix} 60 \\ 10 \end{pmatrix}$$

$$c_1 = -\frac{70}{3} \quad \text{and} \quad c_2 = \frac{80}{3}$$

$$\mathbf{X} = -\frac{70}{3} \begin{pmatrix} -1 \\ 2 \end{pmatrix} e^{-t/20} + \frac{80}{3} \begin{pmatrix} 1 \\ 1 \end{pmatrix} e^{-t/50} + \begin{pmatrix} 10 \\ 30 \end{pmatrix}$$

(c) The solution $\mathbf{X}(t) \to \begin{pmatrix} 10 \\ 30 \end{pmatrix}$ as $t \to \infty$. Over a long period of time the total amount of salt in the system of tanks approaches 40 lb.

(d)

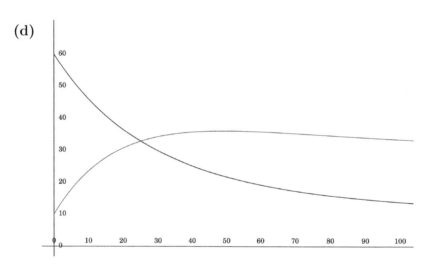

13. From

$$\mathbf{X}' = \begin{pmatrix} 3 & -3 \\ 2 & -2 \end{pmatrix} \mathbf{X} + \begin{pmatrix} 4 \\ -1 \end{pmatrix}$$

we obtain

$$\mathbf{X}_c = c_1 \begin{pmatrix} 1 \\ 1 \end{pmatrix} + c_2 \begin{pmatrix} 3 \\ 2 \end{pmatrix} e^t.$$

Then

$$\mathbf{\Phi} = \begin{pmatrix} 1 & 3e^t \\ 1 & 2e^t \end{pmatrix} \quad \text{and} \quad \mathbf{\Phi}^{-1} = \begin{pmatrix} -2 & 3 \\ e^{-t} & -e^{-t} \end{pmatrix}$$

so that

$$\mathbf{U} = \int \mathbf{\Phi}^{-1} \mathbf{F} \, dt = \int \begin{pmatrix} -11 \\ 5e^{-t} \end{pmatrix} dt = \begin{pmatrix} -11t \\ -5e^{-t} \end{pmatrix}$$

and

$$\mathbf{X}_p = \mathbf{\Phi} \mathbf{U} = \begin{pmatrix} -11 \\ -11 \end{pmatrix} t + \begin{pmatrix} -15 \\ -10 \end{pmatrix}.$$

The solution is

$$\mathbf{X} = \mathbf{X}_c + \mathbf{X}_p = c_1 \begin{pmatrix} 1 \\ 1 \end{pmatrix} + c_2 \begin{pmatrix} 3 \\ 2 \end{pmatrix} e^t - \begin{pmatrix} 11 \\ 11 \end{pmatrix} t - \begin{pmatrix} 15 \\ 10 \end{pmatrix}$$

15. From

$$\mathbf{X}' = \begin{pmatrix} 3 & -5 \\ \frac{3}{4} & -1 \end{pmatrix} \mathbf{X} + \begin{pmatrix} 1 \\ -1 \end{pmatrix} e^{t/2}$$

we obtain

$$\mathbf{X}_c = c_1 \begin{pmatrix} 10 \\ 3 \end{pmatrix} e^{3t/2} + c_2 \begin{pmatrix} 2 \\ 1 \end{pmatrix} e^{t/2}.$$

Then

$$\mathbf{\Phi} = \begin{pmatrix} 10e^{3t/2} & 2e^{t/2} \\ 3e^{3t/2} & e^{t/2} \end{pmatrix} \quad \text{and} \quad \mathbf{\Phi}^{-1} = \begin{pmatrix} \frac{1}{4}e^{-3t/2} & -\frac{1}{2}e^{-3t/2} \\ -\frac{3}{4}e^{-t/2} & \frac{5}{2}e^{-t/2} \end{pmatrix}$$

so that

$$\mathbf{U} = \int \mathbf{\Phi}^{-1}\mathbf{F}\, dt = \int \begin{pmatrix} \frac{3}{4}e^{-t} \\ -\frac{13}{4} \end{pmatrix} dt = \begin{pmatrix} -\frac{3}{4}e^{-t} \\ -\frac{13}{4}t \end{pmatrix}$$

and

$$\mathbf{X}_p = \mathbf{\Phi}\mathbf{U} = \begin{pmatrix} -\frac{13}{2} \\ -\frac{13}{4} \end{pmatrix} te^{t/2} + \begin{pmatrix} -\frac{15}{2} \\ -\frac{9}{4} \end{pmatrix} e^{t/2}.$$

The solution is

$$\mathbf{X} = \mathbf{X}_c + \mathbf{X}_p = c_1 \begin{pmatrix} 10 \\ 3 \end{pmatrix} e^{3t/2} + c_2 \begin{pmatrix} 2 \\ 1 \end{pmatrix} e^{t/2} - \begin{pmatrix} \frac{13}{2} \\ \frac{13}{4} \end{pmatrix} te^{t/2} - \begin{pmatrix} \frac{15}{2} \\ \frac{9}{4} \end{pmatrix} e^{t/2}$$

17. From

$$\mathbf{X}' = \begin{pmatrix} 0 & 2 \\ -1 & 3 \end{pmatrix} \mathbf{X} + \begin{pmatrix} 1 \\ -1 \end{pmatrix} e^t$$

we obtain

$$\mathbf{X}_c = c_1 \begin{pmatrix} 2 \\ 1 \end{pmatrix} e^t + c_2 \begin{pmatrix} 1 \\ 1 \end{pmatrix} e^{2t}.$$

Then

$$\mathbf{\Phi} = \begin{pmatrix} 2e^t & e^{2t} \\ e^t & e^{2t} \end{pmatrix} \quad \text{and} \quad \mathbf{\Phi}^{-1} = \begin{pmatrix} e^{-t} & -e^{-t} \\ -e^{-2t} & 2e^{-2t} \end{pmatrix}$$

so that

$$\mathbf{U} = \int \mathbf{\Phi}^{-1}\mathbf{F}\, dt = \int \begin{pmatrix} 2 \\ -3e^{-t} \end{pmatrix} dt = \begin{pmatrix} 2t \\ 3e^{-t} \end{pmatrix}$$

and

$$\mathbf{X}_p = \mathbf{\Phi}\mathbf{U} = \begin{pmatrix} 4 \\ 2 \end{pmatrix} te^t + \begin{pmatrix} 3 \\ 3 \end{pmatrix} e^t.$$

The solution is

$$\mathbf{X} = \mathbf{X}_c + \mathbf{X}_p = c_1 \begin{pmatrix} 2 \\ 1 \end{pmatrix} e^t + c_2 \begin{pmatrix} 1 \\ 1 \end{pmatrix} e^{2t} + \begin{pmatrix} 4 \\ 2 \end{pmatrix} te^t + \begin{pmatrix} 3 \\ 3 \end{pmatrix} e^t$$

19. From

$$\mathbf{X}' = \begin{pmatrix} 1 & 8 \\ 1 & -1 \end{pmatrix} \mathbf{X} + \begin{pmatrix} 12 \\ 12 \end{pmatrix} t$$

we obtain

$$\mathbf{X}_c = c_1 \begin{pmatrix} 4 \\ 1 \end{pmatrix} e^{3t} + c_2 \begin{pmatrix} -2 \\ 1 \end{pmatrix} e^{-3t}.$$

Then

$$\mathbf{\Phi} = \begin{pmatrix} 4e^{3t} & -2e^{-3t} \\ e^{3t} & e^{-3t} \end{pmatrix} \quad \text{and} \quad \mathbf{\Phi}^{-1} = \begin{pmatrix} \frac{1}{6}e^{-3t} & \frac{1}{3}e^{-3t} \\ -\frac{1}{6}e^{3t} & \frac{2}{3}e^{3t} \end{pmatrix}$$

so that

$$\mathbf{U} = \int \mathbf{\Phi}^{-1}\mathbf{F}\,dt = \int \begin{pmatrix} 6te^{-3t} \\ 6te^{3t} \end{pmatrix} dt = \begin{pmatrix} -2te^{-3t} - \frac{2}{3}e^{-3t} \\ 2te^{3t} - \frac{2}{3}e^{3t} \end{pmatrix}$$

and

$$\mathbf{X}_p = \mathbf{\Phi}\mathbf{U} = \begin{pmatrix} -12 \\ 0 \end{pmatrix} t + \begin{pmatrix} -\frac{4}{3} \\ -\frac{4}{3} \end{pmatrix}.$$

The solution is

$$\mathbf{X} = \mathbf{X}_c + \mathbf{X}_p = c_1 \begin{pmatrix} 4 \\ 1 \end{pmatrix} e^{3t} + c_2 \begin{pmatrix} -2 \\ 1 \end{pmatrix} e^{-3t} + \begin{pmatrix} -12 \\ 0 \end{pmatrix} t - \begin{pmatrix} \frac{4}{3} \\ \frac{4}{3} \end{pmatrix}$$

21. From

$$\mathbf{X}' = \begin{pmatrix} 3 & 2 \\ -2 & -1 \end{pmatrix} \mathbf{X} + \begin{pmatrix} 2 \\ 1 \end{pmatrix} e^{-t}$$

we obtain

$$\mathbf{X}_c = c_1 \begin{pmatrix} 1 \\ -1 \end{pmatrix} e^t + c_2 \left[\begin{pmatrix} 1 \\ -1 \end{pmatrix} te^t + \begin{pmatrix} 0 \\ \frac{1}{2} \end{pmatrix} e^t \right].$$

Then

$$\mathbf{\Phi} = \begin{pmatrix} e^t & te^t \\ -e^t & \frac{1}{2}e^t - te^t \end{pmatrix} \quad \text{and} \quad \mathbf{\Phi}^{-1} = \begin{pmatrix} e^{-t} - 2te^{-t} & -2te^{-t} \\ 2e^{-t} & 2e^{-t} \end{pmatrix}$$

so that

$$\mathbf{U} = \int \mathbf{\Phi}^{-1}\mathbf{F}\,dt = \int \begin{pmatrix} 2e^{-2t} - 6te^{-2t} \\ 6e^{-2t} \end{pmatrix} dt = \begin{pmatrix} \frac{1}{2}e^{-2t} + 3te^{-2t} \\ -3e^{-2t} \end{pmatrix}$$

and

$$\mathbf{X}_p = \mathbf{\Phi}\mathbf{U} = \begin{pmatrix} \frac{1}{2} \\ -2 \end{pmatrix} e^{-t}.$$

The solution is

$$\mathbf{X} = \mathbf{X}_c + \mathbf{X}_p = c_1 \begin{pmatrix} 1 \\ -1 \end{pmatrix} e^t + c_2 \begin{pmatrix} t \\ \frac{1}{2} - t \end{pmatrix} e^t + \begin{pmatrix} \frac{1}{2} \\ -2 \end{pmatrix} e^{-t}$$

23. From

$$\mathbf{X}' = \begin{pmatrix} 0 & -1 \\ 1 & 0 \end{pmatrix} \mathbf{X} + \begin{pmatrix} \sec t \\ 0 \end{pmatrix}$$

we obtain

$$\mathbf{X}_c = c_1 \begin{pmatrix} \cos t \\ \sin t \end{pmatrix} + c_2 \begin{pmatrix} \sin t \\ -\cos t \end{pmatrix}.$$

Then

$$\mathbf{\Phi} = \begin{pmatrix} \cos t & \sin t \\ \sin t & -\cos t \end{pmatrix} \quad \text{and} \quad \mathbf{\Phi}^{-1} = \begin{pmatrix} \cos t & \sin t \\ \sin t & -\cos t \end{pmatrix}$$

so that

$$\mathbf{U} = \int \mathbf{\Phi}^{-1} \mathbf{F}\, dt = \int \begin{pmatrix} 1 \\ \tan t \end{pmatrix} dt = \begin{pmatrix} t \\ -\ln|\cos t| \end{pmatrix}$$

and

$$\mathbf{X}_p = \mathbf{\Phi}\mathbf{U} = \begin{pmatrix} t\cos t - \sin t \ln|\cos t| \\ t \sin t + \cos t \ln|\cos t| \end{pmatrix}.$$

The solution is

$$\mathbf{X} = \mathbf{X}_c + \mathbf{X}_p = c_1 \begin{pmatrix} \cos t \\ \sin t \end{pmatrix} + c_2 \begin{pmatrix} \sin t \\ -\cos t \end{pmatrix} + \begin{pmatrix} t\cos t - \sin t \ln|\cos t| \\ t \sin t + \cos t \ln|\cos t| \end{pmatrix}$$

25. From

$$\mathbf{X}' = \begin{pmatrix} 1 & -1 \\ 1 & 1 \end{pmatrix} \mathbf{X} + \begin{pmatrix} \cos t \\ \sin t \end{pmatrix} e^t$$

we obtain

$$\mathbf{X}_c = c_1 \begin{pmatrix} -\sin t \\ \cos t \end{pmatrix} e^t + c_2 \begin{pmatrix} \cos t \\ \sin t \end{pmatrix} e^t.$$

Then

$$\mathbf{\Phi} = \begin{pmatrix} -\sin t & \cos t \\ \cos t & \sin t \end{pmatrix} e^t \quad \text{and} \quad \mathbf{\Phi}^{-1} = \begin{pmatrix} -\sin t & \cos t \\ \cos t & \sin t \end{pmatrix} e^{-t}$$

so that

$$\mathbf{U} = \int \mathbf{\Phi}^{-1} \mathbf{F}\, dt = \int \begin{pmatrix} 0 \\ 1 \end{pmatrix} dt = \begin{pmatrix} 0 \\ t \end{pmatrix}$$

and

$$\mathbf{X}_p = \mathbf{\Phi}\mathbf{U} = \begin{pmatrix} \cos t \\ \sin t \end{pmatrix} t e^t.$$

The solution is

$$\mathbf{X} = \mathbf{X}_c + \mathbf{X}_p = c_1 \begin{pmatrix} -\sin t \\ \cos t \end{pmatrix} e^t + c_2 \begin{pmatrix} \cos t \\ \sin t \end{pmatrix} e^t + \begin{pmatrix} \cos t \\ \sin t \end{pmatrix} t e^t$$

27. From

$$\mathbf{X}' = \begin{pmatrix} 0 & 1 \\ -1 & 0 \end{pmatrix} \mathbf{X} + \begin{pmatrix} 0 \\ \sec t \tan t \end{pmatrix}$$

we obtain

$$\mathbf{X}_c = c_1 \begin{pmatrix} \cos t \\ -\sin t \end{pmatrix} + c_2 \begin{pmatrix} \sin t \\ \cos t \end{pmatrix}.$$

Then

$$\mathbf{\Phi} = \begin{pmatrix} \cos t & \sin t \\ -\sin t & \cos t \end{pmatrix} t \quad \text{and} \quad \mathbf{\Phi}^{-1} = \begin{pmatrix} \cos t & -\sin t \\ \sin t & \cos t \end{pmatrix}$$

so that

$$\mathbf{U} = \int \mathbf{\Phi}^{-1}\mathbf{F}\, dt = \int \begin{pmatrix} -\tan^2 t \\ \tan t \end{pmatrix} dt = \begin{pmatrix} t - \tan t \\ -\ln|\cos t| \end{pmatrix}$$

and

$$\mathbf{X}_p = \mathbf{\Phi U} = \begin{pmatrix} \cos t \\ -\sin t \end{pmatrix} t + \begin{pmatrix} -\sin t \\ \sin t \tan t \end{pmatrix} - \begin{pmatrix} \sin t \\ \cos t \end{pmatrix} \ln|\cos t|.$$

The solution is

$$\mathbf{X} = \mathbf{X}_c + \mathbf{X}_p = c_1 \begin{pmatrix} \cos t \\ -\sin t \end{pmatrix} + c_2 \begin{pmatrix} \sin t \\ \cos t \end{pmatrix} + \begin{pmatrix} \cos t \\ -\sin t \end{pmatrix} t + \begin{pmatrix} -\sin t \\ \sin t \tan t \end{pmatrix} - \begin{pmatrix} \sin t \\ \cos t \end{pmatrix} \ln|\cos t|$$

29. From

$$\mathbf{X}' = \begin{pmatrix} 1 & 2 \\ -\frac{1}{2} & 1 \end{pmatrix} \mathbf{X} + \begin{pmatrix} \csc t \\ \sec t \end{pmatrix} e^t$$

we obtain

$$\mathbf{X}_c = c_1 \begin{pmatrix} 2\sin t \\ \cos t \end{pmatrix} e^t + c_2 \begin{pmatrix} 2\cos t \\ -\sin t \end{pmatrix} e^t.$$

Then

$$\mathbf{\Phi} = \begin{pmatrix} 2\sin t & 2\cos t \\ \cos t & -\sin t \end{pmatrix} e^t \quad \text{and} \quad \mathbf{\Phi}^{-1} = \begin{pmatrix} \frac{1}{2}\sin t & \cos t \\ \frac{1}{2}\cos t & -\sin t \end{pmatrix} e^{-t}$$

so that

$$\mathbf{U} = \int \mathbf{\Phi}^{-1}\mathbf{F}\, dt = \int \begin{pmatrix} \frac{3}{2} \\ \frac{1}{2}\cot t - \tan t \end{pmatrix} dt = \begin{pmatrix} \frac{3}{2}t \\ \frac{1}{2}\ln|\sin t| + \ln|\cos t| \end{pmatrix}$$

and

$$\mathbf{X}_p = \mathbf{\Phi U} = \begin{pmatrix} 3\sin t \\ \frac{3}{2}\cos t \end{pmatrix} te^t + \begin{pmatrix} \cos t \\ -\frac{1}{2}\sin t \end{pmatrix} e^t \ln|\sin t| + \begin{pmatrix} 2\cos t \\ -\sin t \end{pmatrix} e^t \ln|\cos t|.$$

The solution is

$$\mathbf{X} = \mathbf{X}_c + \mathbf{X}_p = c_1 \begin{pmatrix} 2\sin t \\ \cos t \end{pmatrix} e^t + c_2 \begin{pmatrix} 2\cos t \\ -\sin t \end{pmatrix} e^t + \begin{pmatrix} 3\sin t \\ \frac{3}{2}\cos t \end{pmatrix} te^t$$

$$+ \begin{pmatrix} \cos t \\ -\frac{1}{2}\sin t \end{pmatrix} e^t \ln|\sin t| + \begin{pmatrix} 2\cos t \\ -\sin t \end{pmatrix} e^t \ln|\cos t|$$

31. From

$$\mathbf{X}' = \begin{pmatrix} 1 & 1 & 0 \\ 1 & 1 & 0 \\ 0 & 0 & 3 \end{pmatrix} \mathbf{X} + \begin{pmatrix} e^t \\ e^{2t} \\ te^{3t} \end{pmatrix}$$

we obtain

$$\mathbf{X}_c = c_1 \begin{pmatrix} 1 \\ -1 \\ 0 \end{pmatrix} + c_2 \begin{pmatrix} 1 \\ 1 \\ 0 \end{pmatrix} e^{2t} + c_3 \begin{pmatrix} 0 \\ 0 \\ 1 \end{pmatrix} e^{3t}.$$

Then

$$\mathbf{\Phi} = \begin{pmatrix} 1 & e^{2t} & 0 \\ -1 & e^{2t} & 0 \\ 0 & 0 & e^{3t} \end{pmatrix} \quad \text{and} \quad \mathbf{\Phi}^{-1} = \begin{pmatrix} \frac{1}{2} & -\frac{1}{2} & 0 \\ \frac{1}{2}e^{-2t} & \frac{1}{2}e^{-2t} & 0 \\ 0 & 0 & e^{-3t} \end{pmatrix}$$

so that

$$\mathbf{U} = \int \mathbf{\Phi}^{-1}\mathbf{F}\, dt = \int \begin{pmatrix} \frac{1}{2}e^t - \frac{1}{2}e^{2t} \\ \frac{1}{2}e^{-t} + \frac{1}{2} \\ t \end{pmatrix} dt = \begin{pmatrix} \frac{1}{2}e^t - \frac{1}{4}e^{2t} \\ -\frac{1}{2}e^{-t} + \frac{1}{2}t \\ \frac{1}{2}t^2 \end{pmatrix}$$

and

$$\mathbf{X}_p = \mathbf{\Phi}\mathbf{U} = \begin{pmatrix} -\frac{1}{4}e^{2t} + \frac{1}{2}te^{2t} \\ -e^t + \frac{1}{4}e^{2t} + \frac{1}{2}te^{2t} \\ \frac{1}{2}t^2 e^{3t} \end{pmatrix}.$$

The solution is

$$\mathbf{X} = \mathbf{X}_c + \mathbf{X}_p = c_1 \begin{pmatrix} 1 \\ -1 \\ 0 \end{pmatrix} + c_2 \begin{pmatrix} 1 \\ 1 \\ 0 \end{pmatrix} e^{2t} + c_3 \begin{pmatrix} 0 \\ 0 \\ 1 \end{pmatrix} e^{3t} + \begin{pmatrix} -\frac{1}{4}e^{2t} + \frac{1}{2}te^{2t} \\ -e^t + \frac{1}{4}e^{2t} + \frac{1}{2}te^{2t} \\ \frac{1}{2}t^2 e^{3t} \end{pmatrix}$$

33. From

$$\mathbf{X}' = \begin{pmatrix} 3 & -1 \\ -1 & 3 \end{pmatrix} \mathbf{X} + \begin{pmatrix} 4e^{2t} \\ 4e^{4t} \end{pmatrix}$$

we obtain

$$\mathbf{\Phi} = \begin{pmatrix} -e^{4t} & e^{2t} \\ e^{4t} & e^{2t} \end{pmatrix}, \quad \mathbf{\Phi}^{-1} = \begin{pmatrix} -\frac{1}{2}e^{-4t} & \frac{1}{2}e^{-4t} \\ \frac{1}{2}e^{-2t} & \frac{1}{2}e^{-2t} \end{pmatrix},$$

and

$$\mathbf{X} = \mathbf{\Phi}\mathbf{\Phi}^{-1}(0)\mathbf{X}(0) + \mathbf{\Phi}\int_0^t \mathbf{\Phi}^{-1}\mathbf{F}\, ds = \mathbf{\Phi} \cdot \begin{pmatrix} 0 \\ 1 \end{pmatrix} + \mathbf{\Phi} \cdot \begin{pmatrix} e^{-2t} + 2t - 1 \\ e^{2t} + 2t - 1 \end{pmatrix}$$

$$= \begin{pmatrix} 2 \\ 2 \end{pmatrix} te^{2t} + \begin{pmatrix} -1 \\ 1 \end{pmatrix} e^{2t} + \begin{pmatrix} -2 \\ 2 \end{pmatrix} te^{4t} + \begin{pmatrix} 2 \\ 0 \end{pmatrix} e^{4t}.$$

35. Let $\mathbf{X} = \begin{pmatrix} i_1 \\ i_2 \end{pmatrix}$ so that

$$\mathbf{X}' = \begin{pmatrix} -11 & 3 \\ 3 & -3 \end{pmatrix} \mathbf{X} + \begin{pmatrix} 100\sin t \\ 0 \end{pmatrix}$$

and

$$\mathbf{X}_C = c_1 \begin{pmatrix} 1 \\ 3 \end{pmatrix} e^{-2t} + c_2 \begin{pmatrix} 3 \\ -1 \end{pmatrix} e^{-12t}.$$

Then

$$\boldsymbol{\Phi} = \begin{pmatrix} e^{-2t} & 3e^{-12t} \\ 3e^{-2t} & -e^{-12t} \end{pmatrix}, \quad \boldsymbol{\Phi}^{-1} = \begin{pmatrix} \frac{1}{10}e^{2t} & \frac{3}{0}e^{2t} \\ \frac{3}{10}e^{12t} & -\frac{1}{10}e^{12t} \end{pmatrix},$$

$$\mathbf{U} = \int \boldsymbol{\Phi}^{-1}\mathbf{F}\,dt = \int \begin{pmatrix} 10e^{2t}\sin t \\ 30e^{12t}\sin t \end{pmatrix} dt = \begin{pmatrix} 2e^{2t}(2\sin t - \cos t) \\ \frac{6}{29}e^{12t}(12\sin t - \cos t) \end{pmatrix},$$

and

$$\mathbf{X}_p = \boldsymbol{\Phi}\mathbf{U} = \begin{pmatrix} \frac{332}{29}\sin t - \frac{76}{29}\cos t \\ \frac{276}{29}\sin t - \frac{168}{29}\cos t \end{pmatrix}$$

so that

$$\mathbf{X} = c_1 \begin{pmatrix} 1 \\ 3 \end{pmatrix} e^{-2t} + c_2 \begin{pmatrix} 3 \\ -1 \end{pmatrix} e^{-12t} + \mathbf{X}_p.$$

If $\mathbf{X}(0) = \begin{pmatrix} 0 \\ 0 \end{pmatrix}$ then $c_1 = 2$ and $c_2 = \frac{6}{29}$.

The solutions is

$$\mathbf{X} = 2 \begin{pmatrix} 1 \\ 3 \end{pmatrix} e^{-2t} + \frac{6}{29} \begin{pmatrix} 3 \\ -1 \end{pmatrix} e^{-12t} - \frac{4}{29} \begin{pmatrix} 19 \\ 42 \end{pmatrix} \cos t + \frac{4}{29} \begin{pmatrix} 83 \\ 69 \end{pmatrix} \sin t.$$

37. (a) The eigenvalues are 0, 1, 3, and 4, with corresponding eigenvectors

$$\begin{pmatrix} -6 \\ -4 \\ 1 \\ 2 \end{pmatrix}, \quad \begin{pmatrix} 2 \\ 1 \\ 0 \\ 0 \end{pmatrix}, \quad \begin{pmatrix} 3 \\ 1 \\ 2 \\ 1 \end{pmatrix}, \quad \text{and} \quad \begin{pmatrix} -1 \\ 1 \\ 0 \\ 0 \end{pmatrix}.$$

(b) $\boldsymbol{\Phi} = \begin{pmatrix} -6 & 2e^t & 3e^{3t} & -e^{4t} \\ -4 & e^t & e^{3t} & e^{4t} \\ 1 & 0 & 2e^{3t} & 0 \\ 2 & 0 & e^{3t} & 0 \end{pmatrix}, \quad \boldsymbol{\Phi}^{-1} = \begin{pmatrix} 0 & 0 & -\frac{1}{3} & \frac{2}{3} \\ \frac{1}{3}e^{-t} & \frac{1}{3}e^{-t} & -2e^{-t} & \frac{8}{3}e^{-t} \\ 0 & 0 & \frac{2}{3}e^{-3t} & -\frac{1}{3}e^{-3t} \\ -\frac{1}{3}e^{-4t} & \frac{2}{3}e^{-4t} & 0 & \frac{1}{3}e^{-4t} \end{pmatrix}$

(c) $\Phi^{-1}(t)\mathbf{F}(t) = \begin{pmatrix} \frac{2}{3} - \frac{1}{3}e^{2t} \\ \frac{1}{3}e^{-2t} + \frac{8}{3}e^{-t} - 2e^t + \frac{1}{3}t \\ -\frac{1}{3}e^{-3t} + \frac{2}{3}e^{-t} \\ \frac{2}{3}e^{-5t} + \frac{1}{3}e^{-4t} - \frac{1}{3}te^{-3t} \end{pmatrix},$

$\int \Phi^{-1}(t)\mathbf{F}(t)\,dt = \begin{pmatrix} -\frac{1}{6}e^{2t} + \frac{2}{3}t \\ -\frac{1}{6}e^{-2t} - \frac{8}{3}e^{-t} - 2e^t + \frac{1}{6}t^2 \\ \frac{1}{9}e^{-3t} - \frac{2}{3}e^{-t} \\ -\frac{2}{15}e^{-5t} - \frac{1}{12}e^{-4t} + \frac{1}{27}e^{-3t} + \frac{1}{9}te^{-3t} \end{pmatrix},$

$\mathbf{X}_p(t) = \Phi(t)\int \Phi^{-1}(t)\mathbf{F}(t)\,dt$

$= \begin{pmatrix} -5e^{2t} - \frac{1}{5}e^{-t} - \frac{1}{27}e^t - \frac{1}{9}te^t + \frac{1}{3}t^2e^t - 4t - \frac{59}{12} \\ -2e^{2t} - \frac{3}{10}e^{-t} + \frac{1}{27}e^t + \frac{1}{9}te^t + \frac{1}{6}t^2e^t - \frac{8}{3}t - \frac{95}{36} \\ -\frac{3}{2}e^{2t} + \frac{2}{3}t + \frac{2}{9} \\ -e^{2t} + \frac{4}{3}t - \frac{1}{9} \end{pmatrix},$

$\mathbf{X}_c(t) = \Phi(t)\mathbf{C} = \begin{pmatrix} -6c_1 + 2c_2 e^t + 3c_3 e^{3t} - c_4 e^{4t} \\ -4c_1 + c_2 e^t + c_3 e^{3t} + c_4 e^{4t} \\ c_1 + 2c_3 e^{3t} \\ 2c_1 + c_3 e^{3t} \end{pmatrix},$

$\mathbf{X}(t) = \Phi(t)\mathbf{C} + \Phi(t)\int \Phi^{-1}(t)\mathbf{F}(t)\,dt = \begin{pmatrix} -6c_1 + 2c_2 e^t + 3c_3 e^{3t} - c_4 e^{4t} \\ -4c_1 + c_2 e^t + c_3 e^{3t} + c_4 e^{4t} \\ c_1 + 2c_3 e^{3t} \\ 2c_1 + c_3 e^{3t} \end{pmatrix}$

$+ \begin{pmatrix} -5e^{2t} - \frac{1}{5}e^{-t} - \frac{1}{27}e^t - \frac{1}{9}te^t + \frac{1}{3}t^2e^t - 4t - \frac{59}{12} \\ -2e^{2t} - \frac{3}{10}e^{-t} + \frac{1}{27}e^t + \frac{1}{9}te^t + \frac{1}{6}t^2e^t - \frac{8}{3}t - \frac{95}{36} \\ -\frac{3}{2}e^{2t} + \frac{2}{3}t + \frac{2}{9} \\ -e^{2t} + \frac{4}{3}t - \frac{1}{9} \end{pmatrix}$

(d) $\mathbf{X}(t) = c_1 \begin{pmatrix} -6 \\ -4 \\ 1 \\ 2 \end{pmatrix} + c_2 \begin{pmatrix} 2 \\ 1 \\ 0 \\ 0 \end{pmatrix} e^t + c_3 \begin{pmatrix} 3 \\ 1 \\ 2 \\ 1 \end{pmatrix} e^{3t} + c_4 \begin{pmatrix} -1 \\ 1 \\ 0 \\ 0 \end{pmatrix} e^{4t}$

$+ \begin{pmatrix} -5e^{2t} - \frac{1}{5} e^{-t} - \frac{1}{27} e^t - \frac{1}{9} te^t + \frac{1}{3} t^2 e^t - 4t - \frac{59}{12} \\ -2e^{2t} - \frac{3}{10} e^{-t} + \frac{1}{27} e^t + \frac{1}{9} te^t + \frac{1}{6} t^2 e^t - \frac{8}{3}t - \frac{95}{36} \\ -\frac{3}{2} e^{2t} + \frac{2}{3} t + \frac{2}{9} \\ -e^{2t} + \frac{4}{3} t - \frac{1}{9} \end{pmatrix}$

8.4 ⎸ **Matrix Exponential**

1. For $\mathbf{A} = \begin{pmatrix} 1 & 0 \\ 0 & 2 \end{pmatrix}$ we have

$$\mathbf{A}^2 = \begin{pmatrix} 1 & 0 \\ 0 & 2 \end{pmatrix} \begin{pmatrix} 1 & 0 \\ 0 & 2 \end{pmatrix} = \begin{pmatrix} 1 & 0 \\ 0 & 4 \end{pmatrix},$$

$$\mathbf{A}^3 = \mathbf{A}\mathbf{A}^2 = \begin{pmatrix} 1 & 0 \\ 0 & 2 \end{pmatrix} \begin{pmatrix} 1 & 0 \\ 0 & 4 \end{pmatrix} = \begin{pmatrix} 1 & 0 \\ 0 & 8 \end{pmatrix},$$

$$\mathbf{A}^4 = \mathbf{A}\mathbf{A}^3 = \begin{pmatrix} 1 & 0 \\ 0 & 2 \end{pmatrix} \begin{pmatrix} 1 & 0 \\ 0 & 8 \end{pmatrix} = \begin{pmatrix} 1 & 0 \\ 0 & 16 \end{pmatrix},$$

and so on. In general

$$\mathbf{A}^k = \begin{pmatrix} 1 & 0 \\ 0 & 2^k \end{pmatrix} \quad \text{for} \quad k = 1, 2, 3, \ldots .$$

Thus

$$e^{\mathbf{A}t} = \mathbf{I} + \frac{\mathbf{A}}{1!}t + \frac{\mathbf{A}^2}{2!}t^2 + \frac{\mathbf{A}^3}{3!}t^3 + \cdots$$

$$= \begin{pmatrix} 1 & 0 \\ 0 & 1 \end{pmatrix} + \frac{1}{1!}\begin{pmatrix} 1 & 0 \\ 0 & 2 \end{pmatrix}t + \frac{1}{2!}\begin{pmatrix} 1 & 0 \\ 0 & 4 \end{pmatrix}t^2 + \frac{1}{3!}\begin{pmatrix} 1 & 0 \\ 0 & 8 \end{pmatrix}t^3 + \cdots$$

$$= \begin{pmatrix} 1 + t + \frac{t^2}{2!} + \frac{t^3}{3!} + \cdots & 0 \\ 0 & 1 + t + \frac{(2t)^2}{2!} + \frac{(2t)^3}{3!} + \cdots \end{pmatrix} = \begin{pmatrix} e^t & 0 \\ 0 & e^{2t} \end{pmatrix}$$

and

$$e^{-\mathbf{A}t} = \begin{pmatrix} e^{-t} & 0 \\ 0 & e^{-2t} \end{pmatrix}.$$

3. For

$$\mathbf{A} = \begin{pmatrix} 1 & 1 & 1 \\ 1 & 1 & 1 \\ -2 & -2 & -2 \end{pmatrix}$$

we have

$$\mathbf{A}^2 = \begin{pmatrix} 1 & 1 & 1 \\ 1 & 1 & 1 \\ -2 & -2 & -2 \end{pmatrix} \begin{pmatrix} 1 & 1 & 1 \\ 1 & 1 & 1 \\ -2 & -2 & -2 \end{pmatrix} = \begin{pmatrix} 0 & 0 & 0 \\ 0 & 0 & 0 \\ 0 & 0 & 0 \end{pmatrix}.$$

Thus, $\mathbf{A}^3 = \mathbf{A}^4 = \mathbf{A}^5 = \cdots = \mathbf{0}$ and

$$e^{\mathbf{A}t} = \mathbf{I} + \mathbf{A}t = \begin{pmatrix} 1 & 0 & 0 \\ 0 & 1 & 0 \\ 0 & 0 & 1 \end{pmatrix} + \begin{pmatrix} t & t & t \\ t & t & t \\ -2t & -2t & -2t \end{pmatrix} = \begin{pmatrix} t+1 & t & t \\ t & t+1 & t \\ -2t & -2t & -2t+1 \end{pmatrix}.$$

5. Using the result of Problem 1,

$$\mathbf{X} = \begin{pmatrix} e^t & 0 \\ 0 & e^{2t} \end{pmatrix} \begin{pmatrix} c_1 \\ c_2 \end{pmatrix} = c_1 \begin{pmatrix} e^t \\ 0 \end{pmatrix} + c_2 \begin{pmatrix} 0 \\ e^{2t} \end{pmatrix}.$$

7. Using the result of Problem 3,

$$\mathbf{X} = \begin{pmatrix} t+1 & t & t \\ t & t+1 & t \\ -2t & -2t & -2t+1 \end{pmatrix} \begin{pmatrix} c_1 \\ c_2 \\ c_3 \end{pmatrix} = c_1 \begin{pmatrix} t+1 \\ t \\ -2t \end{pmatrix} + c_2 \begin{pmatrix} t \\ t+1 \\ -2t \end{pmatrix} + c_3 \begin{pmatrix} t \\ t \\ -2t+1 \end{pmatrix}.$$

9. To solve

$$\mathbf{X}' = \begin{pmatrix} 1 & 0 \\ 0 & 2 \end{pmatrix} \mathbf{X} + \begin{pmatrix} 3 \\ -1 \end{pmatrix}$$

we identify $t_0 = 0$, $\mathbf{F}(t) = \begin{pmatrix} 3 \\ -1 \end{pmatrix}$, and use the results of Problem 1 and Equation (6) in the text.

$$\mathbf{X}(t) = e^{\mathbf{A}t}\mathbf{C} + e^{\mathbf{A}t}\int_{t_0}^{t} e^{-\mathbf{A}s}\mathbf{F}(s)\,ds$$

$$= \begin{pmatrix} e^t & 0 \\ 0 & e^{2t} \end{pmatrix}\begin{pmatrix} c_1 \\ c_2 \end{pmatrix} + \begin{pmatrix} e^t & 0 \\ 0 & e^{2t} \end{pmatrix}\int_0^t \begin{pmatrix} e^{-s} & 0 \\ 0 & e^{-2s} \end{pmatrix}\begin{pmatrix} 3 \\ -1 \end{pmatrix}ds$$

$$= \begin{pmatrix} c_1 e^t \\ c_2 e^{2t} \end{pmatrix} + \begin{pmatrix} e^t & 0 \\ 0 & e^{2t} \end{pmatrix}\int_0^t \begin{pmatrix} 3e^{-s} \\ -e^{-2s} \end{pmatrix}ds$$

$$= \begin{pmatrix} c_1 e^t \\ c_2 e^{2t} \end{pmatrix} + \begin{pmatrix} e^t & 0 \\ 0 & e^{2t} \end{pmatrix}\begin{pmatrix} -3e^{-s} \\ \frac{1}{2}e^{-2s} \end{pmatrix}\Big|_0^t$$

$$= \begin{pmatrix} c_1 e^t \\ c_2 e^{2t} \end{pmatrix} + \begin{pmatrix} e^t & 0 \\ 0 & e^{2t} \end{pmatrix}\begin{pmatrix} -3e^{-t} + 3 \\ \frac{1}{2}e^{-2t} - \frac{1}{2} \end{pmatrix}$$

$$= \begin{pmatrix} c_1 e^t \\ c_2 e^{2t} \end{pmatrix} + \begin{pmatrix} -3 + 3e^t \\ \frac{1}{2} - \frac{1}{2}e^{2t} \end{pmatrix} = c_3\begin{pmatrix} 1 \\ 0 \end{pmatrix}e^t + c_4\begin{pmatrix} 0 \\ 1 \end{pmatrix}e^{2t} + \begin{pmatrix} -3 \\ \frac{1}{2} \end{pmatrix}.$$

11. To solve

$$\mathbf{X}' = \begin{pmatrix} 0 & 1 \\ 1 & 0 \end{pmatrix}\mathbf{X} + \begin{pmatrix} 1 \\ 1 \end{pmatrix}$$

we identify $t_0 = 0$, $\mathbf{F}(t) = \begin{pmatrix} 1 \\ 1 \end{pmatrix}$, and use the results of Problem 2 and Equation (6) in the text.

$$\mathbf{X}(t) = e^{\mathbf{A}t}\mathbf{C} + e^{\mathbf{A}t}\int_{t_0}^{t} e^{-\mathbf{A}s}\mathbf{F}(s)\,ds$$

$$= \begin{pmatrix} \cosh t & \sinh t \\ \sinh t & \cosh t \end{pmatrix}\begin{pmatrix} c_1 \\ c_2 \end{pmatrix} + \begin{pmatrix} \cosh t & \sinh t \\ \sinh t & \cosh t \end{pmatrix}\int_0^t \begin{pmatrix} \cosh s & -\sinh s \\ -\sinh s & \cosh s \end{pmatrix}\begin{pmatrix} 1 \\ 1 \end{pmatrix}ds$$

$$= \begin{pmatrix} c_1\cosh t + c_2\sinh t \\ c_1\sinh t + c_2\cosh t \end{pmatrix} + \begin{pmatrix} \cosh t & \sinh t \\ \sinh t & \cosh t \end{pmatrix}\int_0^t \begin{pmatrix} \cosh s - \sinh s \\ -\sinh s + \cosh s \end{pmatrix}ds$$

$$= \begin{pmatrix} c_1\cosh t + c_2\sinh t \\ c_1\sinh t + c_2\cosh t \end{pmatrix} + \begin{pmatrix} \cosh t & \sinh t \\ \sinh t & \cosh t \end{pmatrix}\begin{pmatrix} \sinh s - \cosh s \\ -\cosh s + \sinh s \end{pmatrix}\Big|_0^t$$

$$= \begin{pmatrix} c_1\cosh t + c_2\sinh t \\ c_1\sinh t + c_2\cosh t \end{pmatrix} + \begin{pmatrix} \cosh t & \sinh t \\ \sinh t & \cosh t \end{pmatrix}\begin{pmatrix} \sinh t - \cosh t + 1 \\ -\cosh t + \sinh t + 1 \end{pmatrix}$$

$$= \begin{pmatrix} c_1\cosh t + c_2\sinh t \\ c_1\sinh t + c_2\cosh t \end{pmatrix} + \begin{pmatrix} \sinh^2 t - \cosh^2 t + \cosh t + \sinh t \\ \sinh^2 t - \cosh^2 t + \sinh t + \cosh t \end{pmatrix}$$

$$= c_1\begin{pmatrix} \cosh t \\ \sinh t \end{pmatrix} + c_2\begin{pmatrix} \sinh t \\ \cosh t \end{pmatrix} + \begin{pmatrix} \cosh t \\ \sinh t \end{pmatrix} + \begin{pmatrix} \sinh t \\ \cosh t \end{pmatrix} - \begin{pmatrix} 1 \\ 1 \end{pmatrix}$$

$$= c_3 \begin{pmatrix} \cosh t \\ \sinh t \end{pmatrix} + c_4 \begin{pmatrix} \sinh t \\ \cosh t \end{pmatrix} - \begin{pmatrix} 1 \\ 1 \end{pmatrix}.$$

13. We have

$$\mathbf{X}(0) = c_1 \begin{pmatrix} 1 \\ 0 \\ 0 \end{pmatrix} + c_2 \begin{pmatrix} 0 \\ 1 \\ 0 \end{pmatrix} + c_3 \begin{pmatrix} 0 \\ 0 \\ 1 \end{pmatrix} = \begin{pmatrix} c_1 \\ c_2 \\ c_3 \end{pmatrix} = \begin{pmatrix} 1 \\ -4 \\ 6 \end{pmatrix}.$$

Thus, the solution of the initial-value problem is

$$\mathbf{X} = \begin{pmatrix} t+1 \\ t \\ -2t \end{pmatrix} - 4 \begin{pmatrix} t \\ t+1 \\ -2t \end{pmatrix} + 6 \begin{pmatrix} t \\ t \\ -2t+1 \end{pmatrix}.$$

15. From $s\mathbf{I} - \mathbf{A} = \begin{pmatrix} s-4 & -3 \\ 4 & s+4 \end{pmatrix}$ we find

$$(s\mathbf{I} - \mathbf{A})^{-1} = \begin{pmatrix} \dfrac{3/2}{s-2} - \dfrac{1/2}{s+2} & \dfrac{3/4}{s-2} - \dfrac{3/4}{s+2} \\[3mm] \dfrac{-1}{s-2} + \dfrac{1}{s+2} & \dfrac{-1/2}{s-2} + \dfrac{3/2}{s+2} \end{pmatrix}$$

and

$$e^{\mathbf{A}t} = \begin{pmatrix} \frac{3}{2}e^{2t} - \frac{1}{2}e^{-2t} & \frac{3}{4}e^{2t} - \frac{3}{4}e^{-2t} \\[2mm] -e^{2t} + e^{-2t} & -\frac{1}{2}e^{2t} + \frac{3}{2}e^{-2t} \end{pmatrix}.$$

The general solution of the system is then

$$\mathbf{X} = e^{\mathbf{A}t}\mathbf{C} = \begin{pmatrix} \frac{3}{2}e^{2t} - \frac{1}{2}e^{-2t} & \frac{3}{4}e^{2t} - \frac{3}{4}e^{-2t} \\[2mm] -e^{2t} + e^{-2t} & -\frac{1}{2}e^{2t} + \frac{3}{2}e^{-2t} \end{pmatrix} \begin{pmatrix} c_1 \\ c_2 \end{pmatrix}$$

$$= c_1 \begin{pmatrix} \frac{3}{2} \\ -1 \end{pmatrix} e^{2t} + c_1 \begin{pmatrix} -\frac{1}{2} \\ 1 \end{pmatrix} e^{-2t} + c_2 \begin{pmatrix} \frac{3}{4} \\ -\frac{1}{2} \end{pmatrix} e^{2t} + c_2 \begin{pmatrix} -\frac{3}{4} \\ \frac{3}{2} \end{pmatrix} e^{-2t}$$

$$= \left(\frac{1}{2}c_1 + \frac{1}{4}c_2 \right) \begin{pmatrix} 3 \\ -2 \end{pmatrix} e^{2t} + \left(-\frac{1}{2}c_1 - \frac{3}{4}c_2 \right) \begin{pmatrix} 1 \\ -2 \end{pmatrix} e^{-2t}$$

$$= c_3 \begin{pmatrix} 3 \\ -2 \end{pmatrix} e^{2t} + c_4 \begin{pmatrix} 1 \\ -2 \end{pmatrix} e^{-2t}.$$

17. From $s\mathbf{I} - \mathbf{A} = \begin{pmatrix} s-5 & 9 \\ -1 & s+1 \end{pmatrix}$ we find

$$(s\mathbf{I} - \mathbf{A})^{-1} = \begin{pmatrix} \dfrac{1}{s-2} + \dfrac{3}{(s-2)^2} & -\dfrac{9}{(s-2)^2} \\[4mm] \dfrac{1}{(s-2)^2} & \dfrac{1}{s-2} - \dfrac{3}{(s-2)^2} \end{pmatrix}$$

and

$$e^{\mathbf{A}t} = \begin{pmatrix} e^{2t} + 3te^{2t} & -9te^{2t} \\ te^{2t} & e^{2t} - 3te^{2t} \end{pmatrix}.$$

The general solution of the system is then

$$\mathbf{X} = e^{\mathbf{A}t}\mathbf{C} = \begin{pmatrix} e^{2t} + 3te^{2t} & -9te^{2t} \\ te^{2t} & e^{2t} - 3te^{2t} \end{pmatrix} \begin{pmatrix} c_1 \\ c_2 \end{pmatrix}$$

$$= c_1 \begin{pmatrix} 1 \\ 0 \end{pmatrix} e^{2t} + c_1 \begin{pmatrix} 3 \\ 1 \end{pmatrix} te^{2t} + c_2 \begin{pmatrix} 0 \\ 1 \end{pmatrix} e^{2t} + c_2 \begin{pmatrix} -9 \\ -3 \end{pmatrix} te^{2t}$$

$$= c_1 \begin{pmatrix} 1 + 3t \\ t \end{pmatrix} e^{2t} + c_2 \begin{pmatrix} -9t \\ 1 - 3t \end{pmatrix} e^{2t}.$$

19. Solving

$$\det(\mathbf{A} - \lambda\mathbf{I}) = \begin{vmatrix} 2 - \lambda & 1 \\ -3 & 6 - \lambda \end{vmatrix} = \lambda^2 - 8\lambda + 15 = (\lambda - 3)(\lambda - 5) = 0$$

we find eigenvalues $\lambda_1 = 3$ and $\lambda_2 = 5$. Corresponding eigenvectors are

$$\mathbf{K}_1 = \begin{pmatrix} 1 \\ 1 \end{pmatrix} \quad \text{and} \quad \mathbf{K}_2 = \begin{pmatrix} 1 \\ 3 \end{pmatrix}.$$

Then

$$\mathbf{P} = \begin{pmatrix} 1 & 1 \\ 1 & 3 \end{pmatrix}, \quad \mathbf{P}^{-1} = \begin{pmatrix} 3/2 & -1/2 \\ -1/2 & 1/2 \end{pmatrix}, \quad \text{and} \quad \mathbf{D} = \begin{pmatrix} 3 & 0 \\ 0 & 5 \end{pmatrix},$$

so that

$$\mathbf{PDP}^{-1} = \begin{pmatrix} 2 & 1 \\ -3 & 6 \end{pmatrix}.$$

21. From equation (3) in the text

$$e^{t\mathbf{A}} = e^{t\mathbf{PDP}^{-1}} = \mathbf{I} + t(\mathbf{PDP}^{-1}) + \frac{1}{2!}t^2(\mathbf{PDP}^{-1})^2 + \frac{1}{3!}t^3(\mathbf{PDP}^{-1})^3 + \cdots$$

$$= \mathbf{P}\left[\mathbf{I} + t\mathbf{D} + \frac{1}{2!}(t\mathbf{D})^2 + \frac{1}{3!}(t\mathbf{D})^3 + \cdots\right]\mathbf{P}^{-1} = \mathbf{P}e^{t\mathbf{D}}\mathbf{P}^{-1}.$$

23. From Problems 19, 21, and 22 and Equation (1) in the text

$$\mathbf{X} = e^{\mathbf{A}t}\mathbf{C} = \mathbf{P}e^{\mathbf{D}t}\mathbf{P}^{-1}\mathbf{C} = \begin{pmatrix} 1 & 1 \\ 1 & 3 \end{pmatrix} \begin{pmatrix} e^{3t} & 0 \\ 0 & e^{5t} \end{pmatrix} \begin{pmatrix} \frac{3}{2} & -\frac{1}{2} \\ -\frac{1}{2} & \frac{1}{2} \end{pmatrix} \begin{pmatrix} c_1 \\ c_2 \end{pmatrix}$$

$$= \begin{pmatrix} \frac{3}{2}e^{3t} - \frac{1}{2}e^{5t} & -\frac{1}{2}e^{3t} + \frac{1}{2}e^{5t} \\ \frac{3}{2}e^{3t} - \frac{3}{2}e^{5t} & -\frac{1}{2}e^{3t} + \frac{3}{2}e^{5t} \end{pmatrix} \begin{pmatrix} c_1 \\ c_2 \end{pmatrix}.$$

25. If $\det(s\mathbf{I} - \mathbf{A}) = 0$, then s is an eigenvalue of $\mathbf{A}$. Thus $s\mathbf{I} - \mathbf{A}$ has an inverse if s is not an eigenvalue of $\mathbf{A}$. For the purposes of the discussion in this section, we take s to be larger than the largest eigenvalue of $\mathbf{A}$. Under this condition $s\mathbf{I} - \mathbf{A}$ has an inverse.

27. (a) The following commands can be used in *Mathematica*:

```
A = {{4, 2},{3, 3}};
c = {c1, c2};
m = MatrixExp[A t];
sol = Expand[m.c]
Collect[sol, {c1, c2}]//MatrixForm
```

The output gives

$$x(t) = c_1\left(\frac{2}{5}e^t + \frac{3}{5}e^{6t}\right) + c_2\left(-\frac{2}{5}e^t + \frac{2}{5}e^{6t}\right)$$

$$y(t) = c_1\left(-\frac{3}{5}e^t + \frac{3}{5}e^{6t}\right) + c_2\left(\frac{3}{5}e^t + \frac{2}{5}e^{6t}\right).$$

The eigenvalues are 1 and 6 with corresponding eigenvectors

$$\begin{pmatrix} -2 \\ 3 \end{pmatrix} \quad \text{and} \quad \begin{pmatrix} 1 \\ 1 \end{pmatrix},$$

so the solution of the system is

$$\mathbf{X}(t) = b_1 \begin{pmatrix} -2 \\ 3 \end{pmatrix} e^t + b_2 \begin{pmatrix} 1 \\ 1 \end{pmatrix} e^{6t}$$

or

$$x(t) = -2b_1 e^t + b_2 e^{6t}$$
$$y(t) = 3b_1 e^t + b_2 e^{6t}.$$

If we replace b_1 with $-\frac{1}{5}c_1 + \frac{1}{5}c_2$ and b_2 with $\frac{3}{5}c_1 + \frac{2}{5}c_2$, we obtain the solution found using the matrix exponential.

(b) $x(t) = c_1 e^{-2t}\cos t - (c_1 + c_2)e^{-2t}\sin t$

$y(t) = c_2 e^{-2t}\cos t + (2c_1 + c_2)e^{-2t}\sin t$

Chapter 8 in Review

1. If $\mathbf{X} = k\begin{pmatrix} 4 \\ 5 \end{pmatrix}$, then $\mathbf{X}' = \mathbf{0}$ and

$$k\begin{pmatrix} 1 & 4 \\ 2 & -1 \end{pmatrix}\begin{pmatrix} 4 \\ 5 \end{pmatrix} - \begin{pmatrix} 8 \\ 1 \end{pmatrix} = k\begin{pmatrix} 24 \\ 3 \end{pmatrix} - \begin{pmatrix} 8 \\ 1 \end{pmatrix} = \begin{pmatrix} 0 \\ 0 \end{pmatrix}.$$

We see that $k = \frac{1}{3}$.

3. Since

$$\begin{pmatrix} 4 & 6 & 6 \\ 1 & 3 & 2 \\ -1 & -4 & -3 \end{pmatrix} \begin{pmatrix} 3 \\ 1 \\ -1 \end{pmatrix} = \begin{pmatrix} 12 \\ 4 \\ -4 \end{pmatrix} = 4 \begin{pmatrix} 3 \\ 1 \\ -1 \end{pmatrix},$$

we see that $\lambda = 4$ is an eigenvalue with eigenvector $\mathbf{K}_3$. The corresponding solution is $\mathbf{X}_3 = \mathbf{K}_3 e^{4t}$.

5. We have $\det(\mathbf{A} - \lambda\mathbf{I}) = (\lambda - 1)^2 = 0$ and $\mathbf{K} = \begin{pmatrix} 1 \\ -1 \end{pmatrix}$. A solution to $(\mathbf{A} - \lambda\mathbf{I})\mathbf{P} = \mathbf{K}$ is

$\mathbf{P} = \begin{pmatrix} 0 \\ 1 \end{pmatrix}$ so that

$$\mathbf{X} = c_1 \begin{pmatrix} 1 \\ -1 \end{pmatrix} e^t + c_2 \left[\begin{pmatrix} 1 \\ -1 \end{pmatrix} te^t + \begin{pmatrix} 0 \\ 1 \end{pmatrix} e^t \right].$$

7. We have $\det(\mathbf{A} - \lambda\mathbf{I}) = \lambda^2 - 2\lambda + 5 = 0$. For $\lambda = 1 + 2i$ we obtain $\mathbf{K}_1 = \begin{pmatrix} 1 \\ i \end{pmatrix}$ and

$$\mathbf{X}_1 = \begin{pmatrix} 1 \\ i \end{pmatrix} e^{(1+2i)t} = \begin{pmatrix} \cos 2t \\ -\sin 2t \end{pmatrix} e^t + i \begin{pmatrix} \sin 2t \\ \cos 2t \end{pmatrix} e^t.$$

Then

$$\mathbf{X} = c_1 \begin{pmatrix} \cos 2t \\ -\sin 2t \end{pmatrix} e^t + c_2 \begin{pmatrix} \sin 2t \\ \cos 2t \end{pmatrix} e^t.$$

9. We have $\det(\mathbf{A} - \lambda\mathbf{I}) = -(\lambda - 2)(\lambda - 4)(\lambda + 3) = 0$ so that

$$\mathbf{X} = c_1 \begin{pmatrix} -2 \\ 3 \\ 1 \end{pmatrix} e^{2t} + c_2 \begin{pmatrix} 0 \\ 1 \\ 1 \end{pmatrix} e^{4t} + c_3 \begin{pmatrix} 7 \\ 12 \\ -16 \end{pmatrix} e^{-3t}.$$

11. We have

$$\mathbf{X}_c = c_1 \begin{pmatrix} 1 \\ 0 \end{pmatrix} e^{2t} + c_2 \begin{pmatrix} 4 \\ 1 \end{pmatrix} e^{4t}.$$

Then

$$\mathbf{\Phi} = \begin{pmatrix} e^{2t} & 4e^{4t} \\ 0 & e^{4t} \end{pmatrix}, \quad \mathbf{\Phi}^{-1} = \begin{pmatrix} e^{-2t} & -4e^{-2t} \\ 0 & e^{-4t} \end{pmatrix},$$

and

$$\mathbf{U} = \int \mathbf{\Phi}^{-1}\mathbf{F}\, dt = \int \begin{pmatrix} 2e^{-2t} - 64te^{-2t} \\ 16te^{-4t} \end{pmatrix} dt = \begin{pmatrix} 15e^{-2t} + 32te^{-2t} \\ -e^{-4t} - 4te^{-4t} \end{pmatrix},$$

so that

$$\mathbf{X}_p = \mathbf{\Phi}\mathbf{U} = \begin{pmatrix} 11 + 16t \\ -1 - 4t \end{pmatrix}.$$

The solution is

$$\mathbf{X} = \mathbf{X}_c + \mathbf{X}_p = c_1 \begin{pmatrix} 1 \\ 0 \end{pmatrix} e^{2t} + c_2 \begin{pmatrix} 4 \\ 1 \end{pmatrix} e^{4t} + \begin{pmatrix} 11 + 16t \\ -1 - 4t \end{pmatrix}.$$

13. We have

$$\mathbf{X}_c = c_1 \begin{pmatrix} \cos t + \sin t \\ 2\cos t \end{pmatrix} + c_2 \begin{pmatrix} \sin t - \cos t \\ 2\sin t \end{pmatrix}.$$

Then

$$\mathbf{\Phi} = \begin{pmatrix} \cos t + \sin t & \sin t - \cos t \\ 2\cos t & 2\sin t \end{pmatrix}, \quad \mathbf{\Phi}^{-1} = \begin{pmatrix} \sin t & \frac{1}{2}\cos t - \frac{1}{2}\sin t \\ -\cos t & \frac{1}{2}\cos t + \frac{1}{2}\sin t \end{pmatrix},$$

and

$$\mathbf{U} = \int \mathbf{\Phi}^{-1}\mathbf{F}\,dt = \int \begin{pmatrix} \frac{1}{2}\sin t - \frac{1}{2}\cos t + \frac{1}{2}\csc t \\ -\frac{1}{2}\sin t - \frac{1}{2}\cos t + \frac{1}{2}\csc t \end{pmatrix} dt$$

$$= \begin{pmatrix} -\frac{1}{2}\cos t - \frac{1}{2}\sin t + \frac{1}{2}\ln|\csc t - \cot t| \\ \frac{1}{2}\cos t - \frac{1}{2}\sin t + \frac{1}{2}\ln|\csc t - \cot t| \end{pmatrix},$$

so that

$$\mathbf{X}_p = \mathbf{\Phi}\mathbf{U} = \begin{pmatrix} -1 \\ -1 \end{pmatrix} + \begin{pmatrix} \sin t \\ \sin t + \cos t \end{pmatrix} \ln|\csc t - \cot t|.$$

The solution is

$$\mathbf{X} = \mathbf{X}_c + \mathbf{X}_p$$

$$= c_1 \begin{pmatrix} \cos t + \sin t \\ 2\cos t \end{pmatrix} + c_2 \begin{pmatrix} \sin t - \cos t \\ 2\sin t \end{pmatrix} + \begin{pmatrix} -1 \\ -1 \end{pmatrix} + \begin{pmatrix} \sin t \\ \sin t + \cos t \end{pmatrix} \ln|\csc t - \cot t|.$$

15. (a) Letting

$$\mathbf{K} = \begin{pmatrix} k_1 \\ k_2 \\ k_3 \end{pmatrix}$$

we note that $(\mathbf{A} - 2\mathbf{I})\mathbf{K} = \mathbf{0}$ implies that $3k_1 + 3k_2 + 3k_3 = 0$, so $k_1 = -(k_2 + k_3)$. Choosing $k_2 = 0$, $k_3 = 1$ and then $k_2 = 1$, $k_3 = 0$ we get

$$\mathbf{K}_1 = \begin{pmatrix} -1 \\ 0 \\ 1 \end{pmatrix} \quad \text{and} \quad \mathbf{K}_2 = \begin{pmatrix} -1 \\ 1 \\ 0 \end{pmatrix},$$

respectively. Thus,

$$\mathbf{X}_1 = \begin{pmatrix} -1 \\ 0 \\ 1 \end{pmatrix} e^{2t} \quad \text{and} \quad \mathbf{X}_2 = \begin{pmatrix} -1 \\ 1 \\ 0 \end{pmatrix} e^{2t}$$

are two solutions.

(b) From $\det(\mathbf{A} - \lambda\mathbf{I}) = \lambda^2(3 - \lambda) = 0$ we see that $\lambda_1 = 3$, and $\lambda_2 = 0$ is an eigenvalue of multiplicity two. Letting

$$\mathbf{K} = \begin{pmatrix} k_1 \\ k_2 \\ k_3 \end{pmatrix},$$

as in part $(\mathbf{A})$, we note that $(\mathbf{A} - 0\mathbf{I})\mathbf{K} = \mathbf{A}\mathbf{K} = \mathbf{0}$ implies that $k_1 + k_2 + k_3 = 0$, so $k_1 = -(k_2 + k_3)$. Choosing $k_2 = 0$, $k_3 = 1$, and then $k_2 = 1$, $k_3 = 0$ we get

$$\mathbf{K}_2 = \begin{pmatrix} -1 \\ 0 \\ 1 \end{pmatrix} \quad \text{and} \quad \mathbf{K}_3 = \begin{pmatrix} -1 \\ 1 \\ 0 \end{pmatrix},$$

respectively. Since an eigenvector corresponding to $\lambda_1 = 3$ is

$$\mathbf{K}_1 = \begin{pmatrix} 1 \\ 1 \\ 1 \end{pmatrix},$$

the general solution of the system is

$$\mathbf{X} = c_1 \begin{pmatrix} 1 \\ 1 \\ 1 \end{pmatrix} e^{3t} + c_2 \begin{pmatrix} -1 \\ 0 \\ 1 \end{pmatrix} + c_3 \begin{pmatrix} -1 \\ 1 \\ 0 \end{pmatrix}.$$

Chapter 9

Numerical Solutions of Ordinary Differential Equations

9.1 | Euler Methods and Error Analysis

1.

h = 0.1		h = 0.05	
x_n	y_n	x_n	y_n
1.00	5.0000	1.00	5.0000
1.10	3.9900	1.05	4.4475
1.20	3.2546	1.10	3.9763
1.30	2.7236	1.15	3.5751
1.40	2.3451	1.20	3.2342
1.50	2.0801	1.25	2.9452
		1.30	2.7009
		1.35	2.4952
		1.40	2.3226
		1.45	2.1786
		1.50	2.0592

3.

h = 0.1		h = 0.05	
x_n	y_n	x_n	y_n
0.00	0.0000	0.00	0.0000
0.10	0.1005	0.05	0.0501
0.20	0.2030	0.10	0.1004
0.30	0.3098	0.15	0.1512
0.40	0.4234	0.20	0.2028
0.50	0.5470	0.25	0.2554
		0.30	0.3095
		0.35	0.3652
		0.40	0.4230
		0.45	0.4832
		0.50	0.5465

5.

h = 0.1		h = 0.05	
x_n	y_n	x_n	y_n
0.00	0.0000	0.00	0.0000
0.10	0.0952	0.05	0.0488
0.20	0.1822	0.10	0.0953
0.30	0.2622	0.15	0.1397
0.40	0.3363	0.20	0.1823
0.50	0.4053	0.25	0.2231
		0.30	0.2623
		0.35	0.3001
		0.40	0.3364
		0.45	0.3715
		0.50	0.4054

7.

h = 0.1		h = 0.05	
x_n	y_n	x_n	y_n
0.00	0.5000	0.00	0.5000
0.10	0.5215	0.05	0.5116
0.20	0.5362	0.10	0.5214
0.30	0.5449	0.15	0.5294
0.40	0.5490	0.20	0.5359
0.50	0.5503	0.25	0.5408
		0.30	0.5444
		0.35	0.5469
		0.40	0.5484
		0.45	0.5492
		0.50	0.5495

9.

$h = 0.1$

x_n	y_n
1.00	1.0000
1.10	1.0095
1.20	1.0404
1.30	1.0967
1.40	1.1866
1.50	1.3260

$h = 0.05$

x_n	y_n
1.00	1.0000
1.05	1.0024
1.10	1.0100
1.15	1.0228
1.20	1.0414
1.25	1.0663
1.30	1.0984
1.35	1.1389
1.40	1.1895
1.45	1.2526
1.50	1.3315

11. To obtain the analytic solution use the substitution $u = x + y - 1$. The resulting differential equation in $u(x)$ will be separable.

$h = 0.1$

x_n	y_n	Actual Value
0.00	2.0000	2.0000
1.10	2.1220	2.1230
0.20	2.3049	2.3085
0.30	2.5858	2.5958
0.40	3.0378	3.0650
0.50	3.8254	3.9082

$h = 0.05$

x_n	y_n	Actual Value
0.00	2.0000	2.0000
0.05	2.0553	2.0554
0.10	2.1228	2.1230
0.15	2.2056	2.2061
0.20	2.3075	2.3085
0.25	2.4342	2.4358
0.30	2.5931	2.5958
0.35	2.7953	2.7997
0.40	3.0574	3.0650
0.45	3.4057	3.4189
0.50	3.8840	3.9082

13. (a) Using Euler's method we obtain $y(0.1) \approx y_1 = 1.2$.

(b) Using $y'' = 4e^{2x}$ we see that the local truncation error is

$$y''(c)\frac{h^2}{2} = 4e^{2c}\frac{(0.1)^2}{2} = 0.02e^{2c}.$$

Since e^{2x} is an increasing function, $e^{2c} \le e^{2(0.1)} = e^{0.2}$ for $0 \le c \le 0.1$. Thus an upper bound for the local truncation error is $0.02e^{0.2} = 0.0244$.

(c) Since $y(0.1) = e^{0.2} = 1.2214$, the actual error is $y(0.1) - y_1 = 0.0214$, which is less than 0.0244.

(d) Using Euler's method with $h = 0.05$ we obtain $y(0.1) \approx y_2 = 1.21$.

(e) The error in (d) is $1.2214 - 1.21 = 0.0114$. With global truncation error $O(h)$, when the step size is halved we expect the error for $h = 0.05$ to be one-half the error when $h = 0.1$. Comparing 0.0114 with 0.0214 we see that this is the case.

15. (a) Using Euler's method we obtain $y(0.1) \approx y_1 = 0.8$.

(b) Using $y'' = 5e^{-2x}$ we see that the local truncation error is

$$5e^{-2c} \frac{(0.1)^2}{2} = 0.025e^{-2c}.$$

Since e^{-2x} is a decreasing function, $e^{-2c} \leq e^0 = 1$ for $0 \leq c \leq 0.1$. Thus an upper bound for the local truncation error is $0.025(1) = 0.025$.

(c) Since $y(0.1) = 0.8234$, the actual error is $y(0.1) - y_1 = 0.0234$, which is less than 0.025.

(d) Using Euler's method with $h = 0.05$ we obtain $y(0.1) \approx y_2 = 0.8125$.

(e) The error in (d) is $0.8234 - 0.8125 = 0.0109$. With global truncation error $O(h)$, when the step size is halved we expect the error for $h = 0.05$ to be one-half the error when $h = 0.1$. Comparing 0.0109 with 0.0234 we see that this is the case.

17. (a) Using $y'' = 38e^{-3(x-1)}$ we see that the local truncation error is

$$y''(c) \frac{h^2}{2} = 38e^{-3(c-1)} \frac{h^2}{2} = 19h^2 e^{-3(c-1)}.$$

(b) Since $e^{-3(x-1)}$ is a decreasing function for $1 \leq x \leq 1.5$, $e^{-3(c-1)} \leq e^{-3(1-1)} = 1$ for $1 \leq c \leq 1.5$ and

$$y''(c) \frac{h^2}{2} \leq 19(0.1)^2(1) = 0.19.$$

(c) Using Euler's method with $h = 0.1$ we obtain $y(1.5) \approx 1.8207$. With $h = 0.05$ we obtain $y(1.5) \approx 1.9424$.

(d) Since $y(1.5) = 2.0532$, the error for $h = 0.1$ is $E_{0.1} = 0.2325$, while the error for $h = 0.05$ is $E_{0.05} = 0.1109$. With global truncation error $O(h)$ we expect $E_{0.1}/E_{0.05} \approx 2$. We actually have $E_{0.1}/E_{0.05} = 2.10$.

19. (a) Using $y'' = -1/(x+1)^2$ we see that the local truncation error is

$$\left| y''(c) \frac{h^2}{2} \right| = \frac{1}{(c+1)^2} \frac{h^2}{2}.$$

(b) Since $1/(x+1)^2$ is a decreasing function for $0 \leq x \leq 0.5$, $1/(c+1)^2 \leq 1/(0+1)^2 = 1$ for $0 \leq c \leq 0.5$ and

$$\left| y''(c) \frac{h^2}{2} \right| \leq (1) \frac{(0.1)^2}{2} = 0.005.$$

(c) Using Euler's method with $h = 0.1$ we obtain $y(0.5) \approx 0.4198$. With $h = 0.05$ we obtain $y(0.5) \approx 0.4124$.

(d) Since $y(0.5) = 0.4055$, the error for $h = 0.1$ is $E_{0.1} = 0.0143$, while the error for $h = 0.05$ is $E_{0.05} = 0.0069$. With global truncation error $O(h)$ we expect $E_{0.1}/E_{0.05} \approx 2$. We actually have $E_{0.1}/E_{0.05} = 2.06$.

21. Because y_{n+1}^* depends on y_n and is used to determine y_{n+1}, all of the y_n^* cannot be computed at one time independently of the corresponding y_n values. For example, the computation of y_4^* involves the value of y_3.

9.2   Runge–Kutta Methods

1.

x_n	y_n	Actual Value
0.00	2.0000	2.0000
1.10	2.1230	2.1230
0.20	2.3085	2.3085
0.30	2.5958	2.5958
0.40	3.0649	3.0650
0.50	3.9078	3.9082

3.

x_n	y_n
1.00	5.0000
1.10	3.9724
1.20	3.2284
1.30	2.6945
1.40	2.3163
1.50	2.0533

5.

x_n	y_n
0.00	0.0000
0.10	0.1003
0.20	0.2027
0.30	0.3093
0.40	0.4228
0.50	0.5463

7.

x_n	y_n
0.00	0.0000
0.10	0.0953
0.20	0.1823
0.30	0.2624
0.40	0.3365
0.50	0.4055

9.

x_n	y_n
0.00	0.5000
0.10	0.5213
0.20	0.5358
0.30	0.5443
0.40	0.5482
0.50	0.5493

11.

x_n	y_n
1.00	1.0000
1.10	1.0101
1.20	1.0417
1.30	1.0989
1.40	1.1905
1.50	1.3333

13. (a) Write the equation in the form

$$\frac{dv}{dt} = 32 - 0.025v^2 = f(t, v).$$

(b)

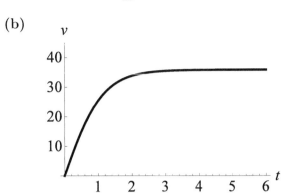

t_n	y_n
0.0	0.0000
1.0	25.2570
2.0	32.9390
3.0	34.9772
4.0	35.5503
5.0	35.7128

(c) Separating variables and using partial fractions we have

$$\frac{1}{2\sqrt{32}} \left(\frac{1}{\sqrt{32} - \sqrt{0.025}\, v} + \frac{1}{\sqrt{32} + \sqrt{0.025}\, v} \right) dv = dt$$

and

$$\frac{1}{2\sqrt{32}\sqrt{0.025}} \left(\ln \left| \sqrt{32} + \sqrt{0.025}\, v \right| - \ln \left| \sqrt{32} - \sqrt{0.025}\, v \right| \right) = t + c.$$

Since $v(0) = 0$ we find $c = 0$. Solving for v we obtain

$$v(t) = \frac{16\sqrt{5}(e^{\sqrt{3.2}\,t} - 1)}{e^{\sqrt{3.2}\,t} + 1}$$

and $v(5) \approx 35.7678$. Alternatively, the solution can be expressed as

$$v(t) = \sqrt{\frac{mg}{k}} \tanh \sqrt{\frac{kg}{m}}\, t.$$

15. (a)

x_n	$h = 0.05$	$h = 0.1$
1.00	1.0000	1.0000
1.05	1.1112	
1.10	1.2511	1.2511
1.15	1.4348	
1.20	1.6934	1.6934
1.25	2.1047	
1.30	2.9560	2.9425
1.35	7.8981	
1.40	1.0608×10^{15}	903.0282

(b)

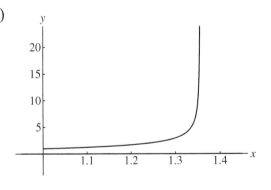

17. (a) Using the RK4 method we obtain $y(0.1) \approx y_1 = 0.823416667$.

(b) Using $y^{(5)}(x) = -40e^{-2x}$ we see that the local truncation error is

$$40e^{-2c}\, \frac{(0.1)^5}{120} = 0.000003333.$$

Since e^{-2x} is a decreasing function, $e^{-2c} \leq e^0 = 1$ for $0 \leq c \leq 0.1$. Thus an upper bound for the local truncation error is $0.000003333(1) = 0.000003333$.

(c) Since $y(0.1) = 0.823413441$, the actual error is $|y(0.1) - y_1| = 0.000003225$, which is less than 0.000003333.

(d) Using the RK4 method with $h = 0.05$ we obtain $y(0.1) \approx y_2 = 0.823413627$.

(e) The error in **(d)** is $|0.823413441 - 0.823413627| = 0.000000185$. With global truncation error $O(h^4)$, when the step size is halved we expect the error for $h = 0.05$ to be one-sixteenth the error when $h = 0.1$. Comparing 0.000000185 with 0.000003225 we see that this is the case.

19. (a) Using $y^{(5)} = 24/(x+1)^5$ we see that the local truncation error is

$$y^{(5)}(c)\frac{h^5}{120} = \frac{1}{(c+1)^5}\frac{h^5}{5}.$$

(b) Since $1/(x+1)^5$ is a decreasing function for $0 \leq x \leq 0.5$, $1/(c+1)^5 \leq 1/(0+1)^5 = 1$ for $0 \leq c \leq 0.5$ and

$$y^{(5)}(c)\frac{h^5}{5} \leq (1)\frac{(0.1)^5}{5} = 0.000002.$$

(c) Using the RK4 method with $h = 0.1$ we obtain $y(0.5) \approx 0.405465168$. With $h = 0.05$ we obtain $y(0.5) \approx 0.405465111$.

21. (a) For $y' + y = 10\sin 3x$ an integrating factor is e^x so that

$$\frac{d}{dx}[e^x y] = 10e^x \sin 3x$$

$$e^x y = e^x \sin 3x - 3e^x \cos 3x + c$$

$$y = \sin 3x - 3\cos 3x + ce^{-x}.$$

When $x = 0$, $y = 0$, so $0 = -3 + c$ and $c = 3$. The solution is

$$y = \sin 3x - 3\cos 3x + 3e^{-x}.$$

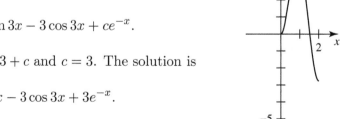

Using Newton's method we find that $x = 1.53235$ is the only positive root in $[0, 2]$.

(b) Using the RK4 method with $h = 0.1$ we obtain the table of values shown. These values are used to obtain an interpolating function in *Mathematica*. The graph of the interpolating function is shown. Using *Mathematica*'s root finding capability we see that the only positive root in $[0, 2]$ is $x = 1.53236$.

x_n	y_n
0.0	0.0000
0.1	0.1440
0.2	0.5448
0.3	1.1409
0.4	1.8559
0.5	2.6049
0.6	3.3019
0.7	3.8675
0.8	4.2356
0.9	4.3593
1.0	4.2147

x_n	y_n
1.0	4.2147
1.1	3.8033
1.2	3.1513
1.3	2.3076
1.4	1.3390
1.5	0.3243
1.6	−0.6530
1.7	−1.5117
1.8	−2.1809
1.9	−2.6061
2.0	−2.7539

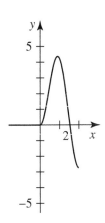

9.3   Multistep Methods

In the tables in this section "ABM" stands for Adams-Bashforth-Moulton.

1. Writing the differential equation in the form $y' - y = x - 1$ we see that an integrating factor is $e^{-\int dx} = e^{-x}$, so that

$$\frac{d}{dx}\left[e^{-x}y\right] = (x-1)e^{-x}$$

and

$$y = e^x(-xe^{-x} + c) = -x + ce^x.$$

From $y(0) = 1$ we find $c = 1$, so the solution of the initial-value problem is $y = -x + e^x$. Actual values of the analytic solution above are compared with the approximated values in the table.

x_n	y_n	Actual	
0.0	1.00000000	1.00000000	init. cond.
0.2	1.02140000	1.02140276	RK4
0.4	1.09181796	1.09182470	RK4
0.6	1.22210646	1.22211880	RK4
0.8	1.42552788	1.42554093	ABM

3. The first predictor is $y_4^* = 0.73318477$.

x_n	y_n	
0.0	1.00000000	init. cond.
0.2	0.73280000	RK4
0.4	0.64608032	RK4
0.6	0.65851653	RK4
0.8	0.72319464	ABM

5. The first predictor for $h = 0.2$ is $y_4^* = 1.02343488$.

x_n	$h = 0.2$		$h = 0.1$	
0.0	0.00000000	init. cond.	0.00000000	init. cond.
0.1			0.10033459	RK4
0.2	0.20270741	RK4	0.20270988	RK4
0.3			0.30933604	RK4
0.4	0.42278899	RK4	0.42279808	ABM
0.5			0.54631491	ABM
0.6	0.68413340	RK4	0.68416105	ABM
0.7			0.84233188	ABM
0.8	1.02969040	ABM	1.02971420	ABM
0.9			1.26028800	ABM
1.0	1.55685960	ABM	1.55762558	ABM

7. The first predictor for $h = 0.2$ is $y_4^* = 0.13618654$.

x_n	$h = 0.2$		$h = 0.1$	
0.0	0.00000000	init. cond.	0.00000000	init. cond.
0.1			0.00033209	RK4
0.2	0.00262739	RK4	0.00262486	RK4
0.3			0.00868768	RK4
0.4	0.02005764	RK4	0.02004821	ABM
0.5			0.03787884	ABM
0.6	0.06296284	RK4	0.06294717	ABM
0.7			0.09563116	ABM
0.8	0.13598600	ABM	0.13596515	ABM
0.9			0.18370712	ABM
1.0	0.23854783	ABM	0.23841344	ABM

9.4   Higher-Order Equations and Systems

1. The substitution $y' = u$ leads to the iteration formulas

$$y_{n+1} = y_n + hu_n, \qquad u_{n+1} = u_n + h(4u_n - 4y_n).$$

The initial conditions are $y_0 = -2$ and $u_0 = 1$. Then

$$y_1 = y_0 + 0.1u_0 = -2 + 0.1(1) = -1.9$$
$$u_1 = u_0 + 0.1(4u_0 - 4y_0) = 1 + 0.1(4 + 8) = 2.2$$
$$y_2 = y_1 + 0.1u_1 = -1.9 + 0.1(2.2) = -1.68.$$

The general solution of the differential equation is $y = c_1 e^{2x} + c_2 x e^{2x}$. From the initial conditions we find $c_1 = -2$ and $c_2 = 5$. Thus $y = -2e^{2x} + 5xe^{2x}$ and $y(0.2) \approx 1.4918$.

3. The substitution $y' = u$ leads to the system

$$y' = u, \qquad u' = 4u - 4y.$$

Using formula (4), we obtain the table shown.

x_n	$h = 0.2$ y_n	$h = 0.2$ u_n	$h = 0.1$ y_n	$h = 0.1$ u_n
0.0	−2.0000	1.0000	−2.0000	1.0000
0.1			−1.8321	2.4427
0.2	−1.4928	4.4731	−1.4919	4.4753

5. The substitution $y' = u$ leads to the system

$$y' = u, \qquad u' = 2u - 2y + e^t \cos t.$$

Using formula (4), we obtain the table shown.

x_n	$h = 0.2$ y_n	$h = 0.2$ u_n	$h = 0.1$ y_n	$h = 0.1$ u_n
0.0	1.0000	2.0000	1.0000	2.0000
0.1			1.2155	2.3150
0.2	1.4640	2.6594	1.4640	2.6594

7.

t_n	$h = 0.2$ x_n	$h = 0.2$ y_n	$h = 0.1$ x_n	$h = 0.1$ y_n
0.0	6.0000	2.0000	6.0000	2.0000
0.1			7.0731	2.6524
0.2	8.3055	3.4199	8.3055	3.4199

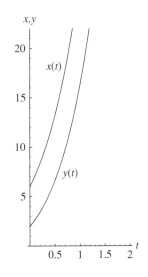

9.

t_n	$h = 0.2$ x_n	$h = 0.2$ y_n	$h = 0.1$ x_n	$h = 0.1$ y_n
0.0	−3.0000	5.0000	−3.0000	5.0000
0.1			−3.4790	4.6707
0.2	−3.9123	4.2857	−3.9123	4.2857

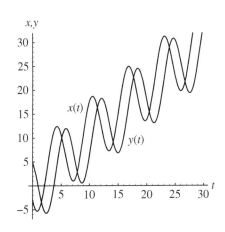

11. Solving for x' and y' we obtain the system

$$x' = -2x + y + 5t$$
$$y' = 2x + y - 2t$$

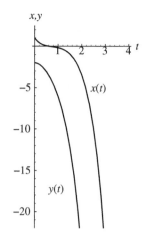

t_n	$h = 0.2$ x_n	$h = 0.2$ y_n	$h = 0.1$ x_n	$h = 0.1$ y_n
0.0	1.0000	−2.0000	1.0000	−2.0000
0.1			0.6594	−2.0476
0.2	0.4179	−2.1824	0.4173	−2.1821

9.5   Second-Order Boundary-Value Problems

1. We identify $P(x) = 0$, $Q(x) = 9$, $f(x) = 0$, and $h = (2-0)/4 = 0.5$. Then the finite difference equation is

$$y_{i+1} + 0.25y_i + y_{i-1} = 0.$$

The solution of the corresponding linear system gives

x	0.0	0.5	1.0	1.5	2.0
y	4.0000	−5.6774	−2.5807	−6.3226	1.0000

3. We identify $P(x) = 2$, $Q(x) = 1$, $f(x) = 5x$, and $h = (1-0)/5 = 0.2$. Then the finite difference equation is

$$1.2y_{i+1} - 1.96y_i + 0.8y_{i-1} = 0.04(5x_i).$$

The solution of the corresponding linear system gives

x	0.0	0.2	0.4	0.6	0.8	1.0
y	0.0000	−0.2259	−0.3356	−0.3308	−0.2167	0.0000

5. We identify $P(x) = -4$, $Q(x) = 4$, $f(x) = (1+x)e^{2x}$, and $h = (1-0)/6 = 0.1667$. Then the finite difference equation is

$$0.6667y_{i+1} - 1.8889y_i + 1.3333y_{i-1} = 0.2778(1 + x_i)e^{2x_i}.$$

The solution of the corresponding linear system gives

x	0.0000	0.1667	0.3333	0.5000	0.6667	0.8333	1.0000
y	3.0000	3.3751	3.6306	3.6448	3.2355	2.1411	0.0000

7. We identify $P(x) = 3/x$, $Q(x) = 3/x^2$, $f(x) = 0$, and $h = (2-1)/8 = 0.125$. Then the finite difference equation is

$$\left(1 + \frac{0.1875}{x_i}\right)y_{i+1} + \left(-2 + \frac{0.0469}{x_i^2}\right)y_i + \left(1 - \frac{0.1875}{x_i}\right)y_{i-1} = 0.$$

The solution of the corresponding linear system gives

x	1.000	1.125	1.250	1.375	1.500	1.625	1.750	1.875	2.000
y	5.0000	3.8842	2.9640	2.2064	1.5826	1.0681	0.6430	0.2913	0.0000

9. We identify $P(x) = 1 - x$, $Q(x) = x$, $f(x) = x$, and $h = (1 - 0)/10 = 0.1$. Then the finite difference equation is

$$[1 + 0.05(1 - x_i)]y_{i+1} + [-2 + 0.01x_i]y_i + [1 - 0.05(1 - x_i)]y_{i-1} = 0.01x_i.$$

The solution of the corresponding linear system gives

x	0.0	0.1	0.2	0.3	0.4	0.5	0.6
y	0.0000	0.2660	0.5097	0.7357	0.9471	1.1465	1.3353

0.7	0.8	0.9	1.0
1.5149	1.6855	1.8474	2.0000

11. We identify $P(x) = 0$, $Q(x) = -4$, $f(x) = 0$, and $h = (1 - 0)/8 = 0.125$. Then the finite difference equation is

$$y_{i+1} - 2.0625y_i + y_{i-1} = 0.$$

The solution of the corresponding linear system gives

x	0.000	0.125	0.250	0.375	0.500	0.625	0.750	0.875	1.000
y	0.0000	0.3492	0.7202	1.1363	1.6233	2.2118	2.9386	3.8490	5.0000

13. (a) The difference equation

$$\left(1 + \frac{h}{2}P_i\right)y_{i+1} + (-2 + h^2Q_i)y_i + \left(1 - \frac{h}{2}P_i\right)y_{i-1} = h^2 f_i$$

is the same as equation (8) in the text. The equations are the same because the derivation was based only on the differential equation, not the boundary conditions. If we allow i to range from 0 to $n - 1$ we obtain n equations in the $n + 1$ unknowns $y_{-1}, y_0, y_1, \ldots, y_{n-1}$. Since y_n is one of the given boundary conditions, it is not an unknown.

(b) Identifying $y_0 = y(0)$, $y_{-1} = y(0 - h)$, and $y_1 = y(0 + h)$ we have from equation (5) in the text

$$\frac{1}{2h}[y_1 - y_{-1}] = y'(0) = 1 \quad \text{or} \quad y_1 - y_{-1} = 2h.$$

The difference equation corresponding to $i = 0$,

$$\left(1 + \frac{h}{2}P_0\right)y_1 + (-2 + h^2Q_0)y_0 + \left(1 - \frac{h}{2}P_0\right)y_{-1} = h^2 f_0$$

becomes, with $y_{-1} = y_1 - 2h$,

$$\left(1 + \frac{h}{2}P_0\right)y_1 + (-2 + h^2Q_0)y_0 + \left(1 - \frac{h}{2}P_0\right)(y_1 - 2h) = h^2 f_0$$

or

$$2y_1 + (-2 + h^2 Q_0)y_0 = h^2 f_0 + 2h - P_0.$$

Alternatively, we may simply add the equation $y_1 - y_{-1} = 2h$ to the list of n difference equations obtaining $n + 1$ equations in the $n + 1$ unknowns $y_{-1}, y_0, y_1, \ldots, y_{n-1}$.

(c) Using $n = 5$ we obtain

x	0.0	0.2	0.4	0.6	0.8	1.0
y	−2.2755	−2.0755	−1.8589	−1.6126	−1.3275	−1.0000

Chapter 9 in Review

1.

x_n	Euler $h = 0.1$	Euler $h = 0.05$	Imp. Euler $h = 0.1$	Imp. Euler $h = 0.05$	RK4 $h = 0.1$	RK4 $h = 0.05$
1.00	2.0000	2.0000	2.0000	2.0000	2.0000	2.0000
1.05		2.0693		2.0735		2.0736
1.10	2.1386	2.1469	2.1549	2.1554	2.1556	2.1556
1.15		2.2328		2.2459		2.2462
1.20	2.3097	2.3272	2.3439	2.3450	2.3454	2.3454
1.25		2.4299		2.4527		2.4532
1.30	2.5136	2.5409	2.5672	2.5689	2.5695	2.5695
1.35		2.6604		2.6937		2.6944
1.40	2.7504	2.7883	2.8246	2.8269	2.8278	2.8278
1.45		2.9245		2.9686		2.9696
1.50	3.0201	3.0690	3.1157	3.1187	3.1197	3.1197

3.

x_n	Euler $h = 0.1$	Euler $h = 0.05$	Imp. Euler $h = 0.1$	Imp. Euler $h = 0.05$	RK4 $h = 0.1$	RK4 $h = 0.05$
0.50	0.5000	0.5000	0.5000	0.5000	0.5000	0.5000
0.55		0.5500		0.5512		0.5512
0.60	0.6000	0.6024	0.6048	0.6049	0.6049	0.6049
0.65		0.6573		0.6609		0.6610
0.70	0.7090	0.7144	0.7191	0.7193	0.7194	0.7194
0.75		0.7739		0.7800		0.7801
0.80	0.8283	0.8356	0.8427	0.8430	0.8431	0.8431
0.85		0.8996		0.9082		0.9083
0.90	0.9559	0.9657	0.9752	0.9755	0.9757	0.9757
0.95		1.0340		1.0451		1.0452
1.00	1.0921	1.1044	1.1163	1.1168	1.1169	1.1169

5. Using

$$y_{n+1} = y_n + hu_n, \qquad\qquad\qquad y_0 = 3$$
$$u_{n+1} = u_n + h(2x_n + 1)y_n, \qquad\qquad u_0 = 1$$

we obtain (when $h = 0.2$) $y_1 = y(0.2) = y_0 + hu_0 = 3 + (0.2)1 = 3.2$. When $h = 0.1$ we have

$$y_1 = y_0 + 0.1u_0 = 3 + (0.1)1 = 3.1$$
$$u_1 = u_0 + 0.1(2x_0 + 1)y_0 = 1 + 0.1(1)3 = 1.3$$
$$y_2 = y_1 + 0.1u_1 = 3.1 + 0.1(1.3) = 3.23.$$

7. Using $x_0 = 1$, $y_0 = 2$, and $h = 0.1$ we have

$$x_1 = x_0 + h(x_0 + y_0) = 1 + 0.1(1 + 2) = 1.3$$
$$y_1 = y_0 + h(x_0 - y_0) = 2 + 0.1(1 - 2) = 1.9$$

and

$$x_2 = x_1 + h(x_1 + y_1) = 1.3 + 0.1(1.3 + 1.9) = 1.62$$
$$y_2 = y_1 + h(x_1 - y_1) = 1.9 + 0.1(1.3 - 1.9) = 1.84.$$

Thus, $x(0.2) \approx 1.62$ and $y(0.2) \approx 1.84$.

Chapter 10

Systems of Nonlinear First-Order Differential Equations

10.1 | Autonomous Systems

1. The corresponding plane autonomous system is

$$x' = y, \quad y' = -9\sin x.$$

If (x, y) is a critical point, $y = 0$ and $-9\sin x = 0$. Therefore $x = \pm n\pi$ and so the critical points are $(\pm n\pi, 0)$ for $n = 0$, 1, 2, ...

3. The corresponding plane autonomous system is

$$x' = y, \quad y' = x^2 - y(1 - x^3).$$

If (x, y) is a critical point, $y = 0$ and so $x^2 - y(1 - x^3) = x^2 = 0$. Therefore $(0, 0)$ is the sole critical point.

5. The corresponding plane autonomous system is

$$x' = y, \quad y' = -x + \epsilon x^3.$$

If (x, y) is a critical point, $y = 0$ and $-x + \epsilon x^3 = 0$. Hence $x(-1 + \epsilon x^2) = 0$ and so $x = 0$, $\sqrt{1/\epsilon}$, $-\sqrt{1/\epsilon}$. The critical points are $(0, 0)$, $(\sqrt{1/\epsilon}, 0)$ and $(-\sqrt{1/\epsilon}, 0)$.

7. From $x + xy = 0$ we have $x(1 + y) = 0$. Therefore $x = 0$ or $y = -1$. If $x = 0$, then, substituting into $-y - xy = 0$, we obtain $y = 0$. Likewise, if $y = -1$, $1 + x = 0$ or $x = -1$. We can conclude that $(0, 0)$ and $(-1, -1)$ are critical points of the system.

9. From $x - y = 0$ we have $y = x$. Substituting into $3x^2 - 4y = 0$ we obtain
   $3x^2 - 4x = x(3x - 4) = 0$. It follows that $(0, 0)$ and $(4/3, 4/3)$ are the critical points of the system.

11. From $x(10 - x - \frac{1}{2}y) = 0$ we obtain $x = 0$ or $x + \frac{1}{2}y = 10$. Likewise $y(16 - y - x) = 0$ implies that $y = 0$ or $x + y = 16$. We therefore have four cases. If $x = 0$, $y = 0$ or $y = 16$. If $x + \frac{1}{2}y = 10$, we can conclude that $y(-\frac{1}{2}y + 6) = 0$ and so $y = 0$, 12. Therefore the critical points of the system are $(0,0)$, $(0,16)$, $(10,0)$, and $(4,12)$.

13. From $x^2 e^y = 0$ we have $x = 0$. Since $e^x - 1 = e^0 - 1 = 0$, the second equation is satisfied for an arbitrary value of y. Therefore any point of the form $(0, y)$ is a critical point.

15. From $x(1 - x^2 - 3y^2) = 0$ we have $x = 0$ or $x^2 + 3y^2 = 1$. If $x = 0$, then substituting into $y(3 - x^2 - 3y^2)$ gives $y(3 - 3y^2) = 0$. Therefore $y = 0$, 1, -1. Likewise $x^2 = 1 - 3y^2$ yields $2y = 0$ so that $y = 0$ and $x^2 = 1 - 3(0)^2 = 1$. The critical points of the system are therefore $(0,0)$, $(0,1)$, $(0,-1)$, $(1,0)$, and $(-1,0)$.

17. (a) From Exercises 8.2, Problem 1, $x = c_1 e^{5t} - c_2 e^{-t}$ and $y = 2c_1 e^{5t} + c_2 e^{-t}$.

(b) From $\mathbf{X}(0) = (-2,2)$ it follows that $c_1 = 0$ and $c_2 = 2$. Therefore $x = -2e^{-t}$ and $y = 2e^{-t}$.

(c)

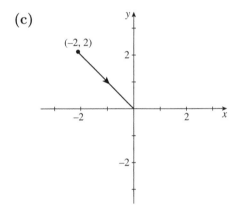

19. (a) From Exercises 8.2, Problem 34, $x = c_1(4\cos 3t - 3\sin 3t) + c_2(4\sin 3t + 3\cos 3t)$ and $y = c_1(5\cos 3t) + c_2(5\sin 3t)$. All solutions are periodic with $p = 2\pi/3$.

(b) From $\mathbf{X}(0) = (4,5)$ it follows that $c_1 = 1$ and $c_2 = 0$. Therefore $x = 4\cos 3t - 3\sin 3t$ and $y = 5\cos 3t$.

(c)

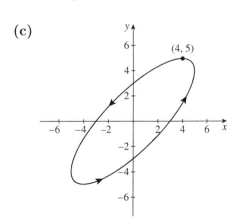

21. (a) From Exercises 8.2, Problem 38, $x = c_1(\sin t - \cos t)e^{4t} + c_2(-\sin t - \cos t)e^{4t}$ and $y = 2c_1(\cos t)e^{4t} + 2c_2(\sin t)e^{4t}$. Because of the presence of e^{4t}, there are no periodic solutions.

(b) From $\mathbf{X}(0) = (-1, 2)$ it follows that $c_1 = 1$ and $c_2 = 0$. Therefore $x = (\sin t - \cos t)e^{4t}$ and $y = 2(\cos t)e^{4t}$.

(c)

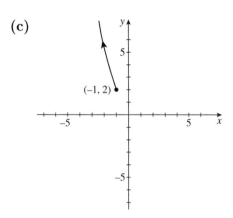

23. Switching to polar coordinates,

$$\frac{dr}{dt} = \frac{1}{r}\left(x\frac{dx}{dt} + y\frac{dy}{dt}\right) = \frac{1}{r}(-xy - x^2r^4 + xy - y^2r^4) = -r^5$$

$$\frac{d\theta}{dt} = \frac{1}{r^2}\left(-y\frac{dx}{dt} + x\frac{dy}{dt}\right) = \frac{1}{r^2}(y^2 + xyr^4 + x^2 - xyr^4) = 1.$$

If we use separation of variables on $\dfrac{dr}{dt} = -r^5$ we obtain

$$r = \left(\frac{1}{4t + c_1}\right)^{1/4} \quad \text{and} \quad \theta = t + c_2.$$

Since $\mathbf{X}(0) = (4, 0)$, $r = 4$ and $\theta = 0$ when $t = 0$. It follows that $c_2 = 0$ and $c_1 = \frac{1}{256}$. The final solution can be written as

$$r = \frac{4}{\sqrt[4]{1024t + 1}}, \qquad \theta = t$$

and so the solution spirals toward the origin as t increases.

25. Switching to polar coordinates,

$$\frac{dr}{dt} = \frac{1}{r}\left(x\frac{dx}{dt} + y\frac{dy}{dt}\right) = \frac{1}{r}[-xy + x^2(1 - r^2) + xy + y^2(1 - r^2)] = r(1 - r^2)$$

$$\frac{d\theta}{dt} = \frac{1}{r^2}\left(-y\frac{dx}{dt} + x\frac{dy}{dt}\right) = \frac{1}{r^2}[y^2 - xy(1 - r^2) + x^2 + xy(1 - r^2)] = 1.$$

Now $dr/dt = r - r^3$ or $(dr/dt) - r = -r^3$ is a Bernoulli differential equation. Following the procedure in Section 2.5 of the text, we let $w = r^{-2}$ so that $w' = -2r^{-3}(dr/dt)$. Therefore

$w' + 2w = 2$, a linear first order differential equation. It follows that $w = 1 + c_1 e^{-2t}$ and so $r^2 = 1/(1 + c_1 e^{-2t})$. The general solution can be written as

$$r = \frac{1}{\sqrt{1 + c_1 e^{-2t}}}, \qquad \theta = t + c_2.$$

If $\mathbf{X}(0) = (1,0)$, $r = 1$ and $\theta = 0$ when $t = 0$. Therefore $c_1 = 0 = c_2$ and so $x = r \cos t = \cos t$ and $y = r \sin t = \sin t$. This solution generates the circle $r = 1$. If $\mathbf{X}(0) = (2,0)$, $r = 2$ and $\theta = 0$ when $t = 0$. Therefore $c_1 = -3/4$, $c_2 = 0$ and so

$$r = \frac{1}{\sqrt{1 - \frac{3}{4} e^{-2t}}}, \qquad \theta = t.$$

This solution spirals toward the circle $r = 1$ as t increases.

27. The system has no critical points, so there are no periodic solutions.

29. The only critical point is $(0,0)$. There appears to be a single periodic solution around $(0,0)$.

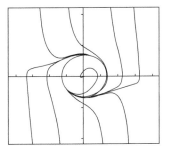

10.2 | Stability of Linear Systems

1. (a) If $\mathbf{X}(0) = \mathbf{X}_0$ lies on the line $y = 2x$, then $\mathbf{X}(t)$ approaches $(0,0)$ along this line. For all other initial conditions, $\mathbf{X}(t)$ approaches $(0,0)$ from the direction determined by the line $y = -x/2$.

(b)

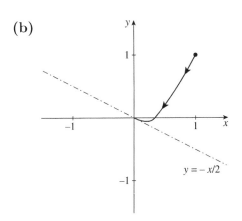

$y = -x/2$

3. **(a)** All solutions are unstable spirals which become unbounded as t increases.

(b)

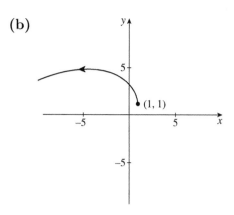

5. **(a)** All solutions approach $(0,0)$ from the direction specified by the line $y = x$.

(b)

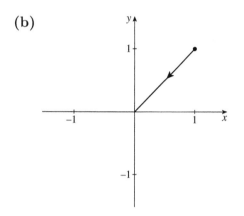

7. **(a)** If $\mathbf{X}(0) = \mathbf{X}_0$ lies on the line $y = 3x$, then $\mathbf{X}(t)$ approaches $(0,0)$ along this line. For all other initial conditions, $\mathbf{X}(t)$ becomes unbounded and $y = x$ serves as the asymptote.

(b)

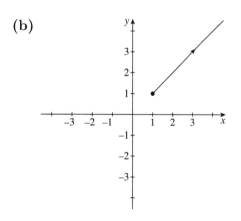

9. Since $\Delta = -41 < 0$, we can conclude from Figure 10.2.12 that $(0,0)$ is a saddle point.

11. Since $\Delta = -19 < 0$, we can conclude from Figure 10.2.12 that $(0,0)$ is a saddle point.

13. Since $\Delta = 1$ and $\tau = -2$, $\tau^2 - 4\Delta = 0$ and so from Figure 10.2.12, $(0,0)$ is a degenerate stable node.

15. Since $\Delta = 0.01$ and $\tau = -0.03$, $\tau^2 - 4\Delta < 0$ and so from Figure 10.2.12, $(0,0)$ is a stable spiral point.

17. $\Delta = 1 - \mu^2$, $\tau = 0$, and so we need $\Delta = 1 - \mu^2 > 0$ for $(0,0)$ to be a center. Therefore $|\mu| < 1$.

19. Note that $\Delta = \mu + 1$ and $\tau = \mu + 1$ and so $\tau^2 - 4\Delta = (\mu+1)^2 - 4(\mu+1) = (\mu+1)(\mu-3)$. It follows that $\tau^2 - 4\Delta < 0$ if and only if $-1 < \mu < 3$. We can conclude that $(0,0)$ will be a saddle point when $\mu < -1$. Likewise $(0,0)$ will be an unstable spiral point when $\tau = \mu+1 > 0$ and $\tau^2 - 4\Delta < 0$. This condition reduces to $-1 < \mu < 3$.

21. $\mathbf{AX}_1 + \mathbf{F} = \mathbf{0}$ implies that $\mathbf{AX}_1 = -\mathbf{F}$ or $\mathbf{X}_1 = -\mathbf{A}^{-1}\mathbf{F}$. Since $\mathbf{X}_p(t) = -\mathbf{A}^{-1}\mathbf{F}$ is a particular solution, it follows from Theorem 10.1.6 that $\mathbf{X}(t) = \mathbf{X}_c(t) + \mathbf{X}_1$ is the general solution to $\mathbf{X}' = \mathbf{AX} + \mathbf{F}$. If $\tau < 0$ and $\Delta > 0$ then $\mathbf{X}_c(t)$ approaches $(0,0)$ by Theorem 10.2.1(a). It follows that $\mathbf{X}(t)$ approaches $\mathbf{X}_1$ as $t \to \infty$.

23. (a) The critical point is $\mathbf{X}_1 = (-3, 4)$.

    (b) From the graph, $\mathbf{X}_1$ appears to be an unstable node or a saddle point.

    (c) Since $\Delta = -1$, $(0,0)$ is a saddle point.

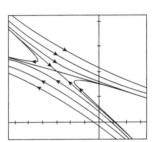

25. (a) The critical point is $\mathbf{X}_1 = (0.5, 2)$.

    (b) From the graph, $\mathbf{X}_1$ appears to be an unstable spiral point.

    (c) Since $\tau = 0.2$, $\Delta = 0.03$, and $\tau 2 - 4\Delta = -0.08$, $(0,0)$ is an unstable spiral point.

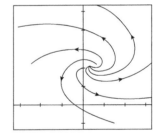

10.3   Linearization and Local Stability

1. Switching to polar coordinates,

$$\frac{dr}{dt} = \frac{1}{r}\left(x\frac{dx}{dt} + y\frac{dy}{dt}\right) = \frac{1}{r}(\alpha x^2 - \beta xy + xy^2 + \beta xy + \alpha y^2 - xy^2) = \frac{1}{r}\alpha r^2 = \alpha r.$$

Therefore $r = ce^{\alpha t}$ and so $r \to 0$ if and only if $\alpha < 0$.

3. The critical points are $x = 0$ and $x = n+1$. Since $g'(x) = k(n+1) - 2kx$, $g'(0) = k(n+1) > 0$ and $g'(n+1) = -k(n+1) < 0$. Therefore $x = 0$ is unstable while $x = n+1$ is asymptotically stable. See Theorem 10.2.

5. The only critical point is $T = T_0$. Since $g'(T) = k$, $g'(T_0) = k > 0$. Therefore $T = T_0$ is unstable by Theorem 10.2.

7. Critical points occur at $x = \alpha, \beta$. Since $g'(x) = k(-\alpha - \beta + 2x)$, $g'(\alpha) = k(\alpha - \beta)$ and $g'(\beta) = k(\beta - \alpha)$. Since $\alpha > \beta$, $g'(\alpha) > 0$ and so $x = \alpha$ is unstable. Likewise $x = \beta$ is asymptotically stable.

9. Critical points occur at $P = a/b$, c but not at $P = 0$. Since $g'(P) = (a - bP) + (P - c)(-b)$,

$$g'(a/b) = (a/b - c)(-b) = -a + bc \quad \text{and} \quad g'(c) = a - bc.$$

Since $a < bc$, $-a + bc > 0$ and $a - bc < 0$. Therefore $P = a/b$ is unstable while $P = c$ is asymptotically stable.

11. The sole critical point is $(1/2, 1)$ and

$$\mathbf{g}'(\mathbf{X}) = \begin{pmatrix} -2y & -2x \\ 2y & 2x - 1 \end{pmatrix}.$$

Computing $\mathbf{g}'((1/2, 1))$ we find that $\tau = -2$ and $\Delta = 2$ so that $\tau^2 - 4\Delta = -4 < 0$. Therefore $(1/2, 1)$ is a stable spiral point.

13. $y' = 2xy - y = y(2x - 1)$. Therefore if (x, y) is a critical point, either $x = 1/2$ or $y = 0$. The case $x = 1/2$ and $y - x^2 + 2 = 0$ implies that $(x, y) = (1/2, -7/4)$. The case $y = 0$ leads to the critical points $(\sqrt{2}, 0)$ and $(-\sqrt{2}, 0)$. We next use the Jacobian matrix

$$\mathbf{g}'(\mathbf{X}) = \begin{pmatrix} -2x & 1 \\ 2y & 2x - 1 \end{pmatrix}$$

to classify these three critical points. For $\mathbf{X} = (\sqrt{2}, 0)$ or $(-\sqrt{2}, 0)$, $\tau = -1$ and $\Delta < 0$. Therefore both critical points are saddle points. For $\mathbf{X} = (1/2, -7/4)$, $\tau = -1$, $\Delta = 7/2$ and so $\tau^2 - 4\Delta = -13 < 0$. Therefore $(1/2, -7/4)$ is a stable spiral point.

15. Since $x^2 - y^2 = 0$, $y^2 = x^2$ and so $x^2 - 3x + 2 = (x - 1)(x - 2) = 0$. It follows that the critical points are $(1, 1)$, $(1, -1)$, $(2, 2)$, and $(2, -2)$. We next use the Jacobian

$$\mathbf{g}'(\mathbf{X}) = \begin{pmatrix} -3 & 2y \\ 2x & -2y \end{pmatrix}$$

to classify these four critical points. For $\mathbf{X} = (1, 1)$, $\tau = -5$, $\Delta = 2$, and so $\tau^2 - 4\Delta = 17 > 0$. Therefore $(1, 1)$ is a stable node. For $\mathbf{X} = (1, -1)$, $\Delta = -2 < 0$ and so $(1, -1)$ is a saddle point. For $\mathbf{X} = (2, 2)$, $\Delta = -4 < 0$ and so we have another saddle point. Finally, if $\mathbf{X} = (2, -2)$, $\tau = 1$, $\Delta = 4$, and so $\tau^2 - 4\Delta = -15 < 0$. Therefore $(2, -2)$ is an unstable spiral point.

17. Since $x' = -2xy = 0$, either $x = 0$ or $y = 0$. If $x = 0$, $y(1 - y^2) = 0$ and so $(0, 0)$, $(0, 1)$, and $(0, -1)$ are critical points. The case $y = 0$ leads to $x = 0$. We next use the Jacobian matrix

$$\mathbf{g}'(\mathbf{X}) = \begin{pmatrix} -2y & -2x \\ -1 + y & 1 + x - 3y^2 \end{pmatrix}$$

to classify these three critical points. For $\mathbf{X} = (0,0)$, $\tau = 1$ and $\Delta = 0$ and so the test is inconclusive. For $\mathbf{X} = (0,1)$, $\tau = -4$, $\Delta = 4$ and so $\tau^2 - 4\Delta = 0$. We can conclude that $(0,1)$ is a stable critical point but we are unable to classify this critical point further in this borderline case. For $\mathbf{X} = (0,-1)$, $\Delta = -4 < 0$ and so $(0,-1)$ is a saddle point.

19. We found the critical points $(0,0)$, $(10,0)$, $(0,16)$ and $(4,12)$ in Problem 11, Section 10.1. Since the Jacobian is

$$\mathbf{g}'(\mathbf{X}) = \begin{pmatrix} 10 - 2x - \frac{1}{2}y & -\frac{1}{2}x \\ -y & 16 - 2y - x \end{pmatrix}$$

we can classify the critical points as follows:

$\mathbf{X}$	τ	Δ	$\tau^2 - 4\Delta$	Conclusion
$(0,0)$	26	160	36	unstable node
$(10,0)$	-4	-60	$-$	saddle point
$(0,16)$	-14	-32	$-$	saddle point
$(4,12)$	-16	24	160	stable node

21. The corresponding plane autonomous system is

$$\theta' = y, \quad y' = (\cos\theta - \frac{1}{2})\sin\theta.$$

Since $|\theta| < \pi$, it follows that critical points are $(0,0)$, $(\pi/3, 0)$ and $(-\pi/3, 0)$. The Jacobian matrix is

$$\mathbf{g}'(\mathbf{X}) = \begin{pmatrix} 0 & 1 \\ \cos 2\theta - \frac{1}{2}\cos\theta & 0 \end{pmatrix}$$

and so at $(0,0)$, $\tau = 0$ and $\Delta = -1/2$. Therefore $(0,0)$ is a saddle point. For $\mathbf{X} = (\pm\pi/3, 0)$, $\tau = 0$ and $\Delta = 3/4$. It is not possible to classify either critical point in this borderline case.

23. The corresponding plane autonomous system is

$$x' = y, \quad y' = x^2 - y(1 - x^3)$$

and the only critical point is $(0,0)$. Since the Jacobian matrix is

$$\mathbf{g}'(\mathbf{X}) = \begin{pmatrix} 0 & 1 \\ 2x + 3x^2y & x^3 - 1 \end{pmatrix},$$

$\tau = -1$ and $\Delta = 0$, and we are unable to classify the critical point in this borderline case.

25. In Problem 5, Section 10.1, we showed that $(0,0)$, $(\sqrt{1/\epsilon}, 0)$ and $(-\sqrt{1/\epsilon}, 0)$ are the critical points. We will use the Jacobian matrix

$$\mathbf{g}'(\mathbf{X}) = \begin{pmatrix} 0 & 1 \\ -1 + 3\epsilon x^2 & 0 \end{pmatrix}$$

to classify these three critical points. For $\mathbf{X} = (0,0)$, $\tau = 0$ and $\Delta = 1$ and we are unable to classify this critical point. For $(\pm\sqrt{1/\epsilon}, 0)$, $\tau = 0$ and $\Delta = -2$ and so both of these critical points are saddle points.

27. The corresponding plane autonomous system is

$$x' = y, \quad y' = -\frac{(\beta + \alpha^2 y^2)x}{1 + \alpha^2 x^2}$$

and the Jacobian matrix is

$$\mathbf{g}'(\mathbf{X}) = \begin{pmatrix} 0 & 1 \\ \dfrac{(\beta + \alpha y^2)(\alpha^2 x^2 - 1)}{(1 + \alpha^2 x^2)^2} & \dfrac{-2\alpha^2 yx}{1 + \alpha^2 x^2} \end{pmatrix}.$$

For $\mathbf{X} = (0,0)$, $\tau = 0$ and $\Delta = \beta$. Since $\beta < 0$, we can conclude that $(0,0)$ is a saddle point.

29. (a) The graphs of $-x + y - x^3 = 0$ and $-x - y + y^2 = 0$ are shown in the figure. The Jacobian matrix is

$$\mathbf{g}'(\mathbf{X}) = \begin{pmatrix} -1 - 3x^2 & 1 \\ -1 & -1 + 2y \end{pmatrix}.$$

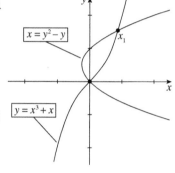

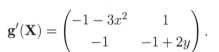

For $\mathbf{X} = (0,0)$, $\tau = -2$, $\Delta = 2$, $\tau^2 - 4\Delta = -4$, and so $(0,0)$ is a stable spiral point.

(b) For $\mathbf{X}_1$, $\Delta = -6.07 < 0$ and so a saddle point occurs at $\mathbf{X}_1$.

31. The differential equation $dy/dx = y'/x' = -2x^3/y$ can be solved by separating variables. It follows that $y^2 + x^4 = c$. If $\mathbf{X}(0) = (x_0, 0)$ where $x_0 > 0$, then $c = x_0^4$ so that $y^2 = x_0^4 - x^4$. Therefore if $-x_0 < x < x_0$, $y^2 > 0$ and so there are two values of y corresponding to each value of x. Therefore the solution $\mathbf{X}(t)$ with $\mathbf{X}(0) = (x_0, 0)$ is periodic and so $(0,0)$ is a center.

33. (a) $x' = 2xy = 0$ implies that either $x = 0$ or $y = 0$. If $x = 0$, then from $1 - x^2 + y^2 = 0$, $y^2 = -1$ and there are no real solutions. If $y = 0$, $1 - x^2 = 0$ and so $(1, 0)$ and $(-1, 0)$ are critical points. The Jacobian matrix is

$$\mathbf{g}'(\mathbf{X}) = \begin{pmatrix} 2y & 2x \\ -2x & 2y \end{pmatrix}$$

and so $\tau = 0$ and $\Delta = 4$ at either $\mathbf{X} = (1, 0)$ or $(-1, 0)$. We obtain no information about these critical points in this borderline case.

(b) The differential equation is

$$\frac{dy}{dx} = \frac{y'}{x'} = \frac{1 - x^2 + y^2}{2xy}$$

or

$$2xy\frac{dy}{dx} = 1 - x^2 + y^2.$$

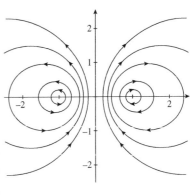

Letting $\mu = y^2/x$, it follows that $d\mu/dx = (1/x^2) - 1$ and so $\mu = -(1/x) - x + 2c$. Therefore $y^2/x = -(1/x) - x + 2c$ which can be put in the form $(x - c)^2 + y^2 = c^2 - 1$. The solution curves are shown and so both $(1, 0)$ and $(-1, 0)$ are centers.

35. The differential equation is $dy/dx = y'/x' = (x^3 - x)/y$ and so $y^2/2 = x^4/4 - x^2/2 + c$ or $y^2 = x^4/2 - x^2 + c_1$. Since $x(0) = 0$ and $y(0) = x'(0) = v_0$, it follows that $c_1 = v_0^2$ and so

$$y^2 = \frac{1}{2}x^4 - x^2 + v_0^2 = \frac{(x^2 - 1)^2 + 2v_0^2 - 1}{2}.$$

The x-intercepts on this graph satisfy

$$x^2 = 1 \pm \sqrt{1 - 2v_0^2}$$

and so we must require that $1 - 2v_0^2 \geq 0$ (or $|v_0| \leq \frac{1}{2}\sqrt{2}$) for real solutions to exist. If $x_0^2 = 1 - \sqrt{1 - 2v_0^2}$ and $-x_0 < x < x_0$, then $(x^2 - 1)^2 + 2v_0^2 - 1 > 0$ and so there are two corresponding values of y. Therefore $\mathbf{X}(t)$ with $\mathbf{X}(0) = (0, v_0)$ is periodic provided that $|v_0| \leq \frac{1}{2}\sqrt{2}$.

37. The corresponding plane autonomous system is

$$x' = y, \quad y' = -\frac{\alpha}{L}x - \frac{\beta}{L}x^3 - \frac{R}{L}y$$

where $x = q$ and $y = q'$. If $\mathbf{X} = (x, y)$ is a critical point, $y = 0$ and $-\alpha x - \beta x^3 = -x(\alpha + \beta x^2) = 0$. If $\beta > 0$, $\alpha + \beta x^2 = 0$ has no real solutions and so $(0, 0)$ is the only critical point. Since

$$\mathbf{g}'(\mathbf{X}) = \begin{pmatrix} 0 & 1 \\ \dfrac{-\alpha - 3\beta x^2}{L} & -\dfrac{R}{L} \end{pmatrix},$$

$\tau = -R/L < 0$ and $\Delta = \alpha/L > 0$. Therefore $(0, 0)$ is a stable critical point. If $\beta < 0$, $(0, 0)$ and $(\pm\hat{x}, 0)$, where $\hat{x}^2 = -\alpha/\beta$ are critical points. At $\mathbf{X}(\pm\hat{x}, 0)$, $\tau = -R/L < 0$ and $\Delta = -2\alpha/L < 0$. Therefore both critical points are saddles.

39. (a) Letting $x = \theta$ and $y = x'$ we obtain the system $x' = y$ and $y' = 1/2 - \sin x$. Since $\sin \pi/6 = \sin 5\pi/6 = 1/2$ we see that $(\pi/6, 0)$ and $(5\pi/6, 0)$ are critical points of the system.

(b) The Jacobian matrix is

$$\mathbf{g}'(\mathbf{X}) = \begin{pmatrix} 0 & 1 \\ -\cos x & 0 \end{pmatrix}$$

and so

$$\mathbf{A}_1 = \mathbf{g}' = ((\pi/6, 0)) = \begin{pmatrix} 0 & 1 \\ -\sqrt{3}/2 & 0 \end{pmatrix} \quad \text{and} \quad \mathbf{A}_2 = \mathbf{g}' = ((5\pi/6, 0)) = \begin{pmatrix} 0 & 1 \\ \sqrt{3}/2 & 0 \end{pmatrix}.$$

Since $\det \mathbf{A}_1 > 0$ and the trace of $\mathbf{A}_1$ is 0, no conclusion can be drawn regarding the critical point $(\pi/6, 0)$. Since $\det \mathbf{A}_2 < 0$, we see that $(5\pi/6, 0)$ is a saddle point.

(c) From the system in part **(A)** we obtain the first-order differential equation

$$\frac{dy}{dx} = \frac{1/2 - \sin x}{y}.$$

Separating variables and integrating we obtain

$$\int y \, dy = \int \left(\frac{1}{2} - \sin x \right) dx$$

and

$$\frac{1}{2}y^2 = \frac{1}{2}x + \cos x + c_1$$

or

$$y^2 = x + 2\cos x + c_2.$$

For x_0 near $\pi/6$, if $\mathbf{X}(0) = (x_0, 0)$ then $c_2 = -x_0 - 2\cos x_0$ and $y^2 = x + 2\cos x - x_0 - 2\cos x_0$. Thus, there are two values of y for each x in a sufficiently small interval around $\pi/6$. Therefore $(\pi/6, 0)$ is a center.

10.4   Autonomous Systems as Mathematical Models

1. We are given that $x(0) = \theta(0) = \pi/3$ and $y(0) = \theta'(0) = w_0$. Since $y^2 = (2g/l)\cos x + c$, $w_0^2 = (2g/l)\cos(\pi/3) + c = g/l + c$ and so $c = w_0^2 - g/l$. Therefore

$$y^2 = \frac{2g}{l}\left(\cos x - \frac{1}{2} + \frac{l}{2g}w_0^2 \right)$$

and the x-intercepts occur where $\cos x = 1/2 - (l/2g)w_0^2$ and so $1/2 - (l/2g)w_0^2$ must be greater than -1 for solutions to exist. This condition is equivalent to $|w_0| < \sqrt{3g/l}$.

3. The corresponding plane autonomous system is

$$x' = y, \quad y' = -g\frac{f'(x)}{1 + [f'(x)]^2} - \frac{\beta}{m}y$$

and

$$\frac{\partial}{\partial x}\left(-g\,\frac{f'(x)}{1+[f'(x)]^2}-\frac{\beta}{m}\,y\right)=-g\,\frac{(1+[f'(x)]^2)f''(x)-f'(x)2f'(x)f''(x)}{(1+[f'(x)]^2)^2}.$$

If $\mathbf{X}_1=(x_1,y_1)$ is a critical point, $y_1=0$ and $f'(x_1)=0$. The Jacobian at this critical point is therefore

$$\mathbf{g}'(\mathbf{X}_1)=\begin{pmatrix} 0 & 1 \\ -gf''(x_1) & -\dfrac{\beta}{m} \end{pmatrix}.$$

5. (a) If $f(x)=x^2/2$, $f'(x)=x$ and so

$$\frac{dy}{dx}=\frac{y'}{x'}=-g\,\frac{x}{1+x^2}\,\frac{1}{y}.$$

We can separate variables to show that $y^2=-g\ln(1+x^2)+c$. But $x(0)=x_0$ and $y(0)=x'(0)=v_0$. Therefore $c=v_0^2+g\ln(1+x_0^2)$ and so

$$y^2=v_0^2-g\ln\left(\frac{1+x^2}{1+x_0^2}\right).$$

Now

$$v_0^2-g\ln\left(\frac{1+x^2}{1+x_0^2}\right)\geq 0 \quad\text{if and only if}\quad x^2\leq e^{v_0^2/g}(1+x_0^2)-1.$$

Therefore, if $|x|\leq [e^{v_0^2/g}(1+x_0^2)-1]^{1/2}$, there are two values of y for a given value of x and so the solution is periodic.

(b) Since $z=x^2/2$, the maximum height occurs at the largest value of x on the cycle. From **(A)**, $x_{\max}=[e^{v_0^2/g}(1+x_0^2)-1]^{1/2}$ and so

$$z_{\max}=\frac{x_{\max}^2}{2}=\frac{1}{2}[e^{v_0^2/g}(1+x_0^2)-1].$$

7. If $x_m<x_1<x_n$, then $F(x_1)>F(x_m)=F(x_n)$. Letting $x=x_1$,

$$G(y)=\frac{c_0}{F(x_1)}=\frac{F(x_m)G(a/b)}{F(x_1)}<G(a/b).$$

Therefore from Property (ii) on page 418 in this section of the text, $G(y)=c_0/F(x_1)$ has two solutions y_1 and y_2 that satisfy $y_1<a/b<y_2$.

9. (a) In the Lotka–Volterra Model the average number of predators is d/c and the average number of prey is a/b. But

$$x'=-ax+bxy-\epsilon_1 x=-(a+\epsilon_1)x+bxy$$
$$y'=-cxy+dy-\epsilon_2 y=-cxy+(d-\epsilon_2)y$$

and so the new critical point in the first quadrant is $(d/c-\epsilon_2/c,\,a/b+\epsilon_1/b)$.

(b) The average number of predators $d/c - \epsilon_2/c$ has decreased while the average number of prey $a/b + \epsilon_1/b$ has increased. The fishery science model is consistent with Volterra's principle.

11. Solving

$$x(20 - 0.4x - 0.3y) = 0$$
$$y(10 - 0.1y - 0.3x) = 0$$

we see that critical points are $(0,0)$, $(0,100)$, $(50,0)$, and $(20,40)$. The Jacobian matrix is

$$\mathbf{g}'(\mathbf{X}) = \begin{pmatrix} 0.08(20 - 0.8x - 0.3y) & -0.024x \\ -0.018y & 0.06(10 - 0.2y - 0.3x) \end{pmatrix}$$

and so

$$\mathbf{A}_1 = \mathbf{g}'((0,0)) = \begin{pmatrix} 1.6 & 0 \\ 0 & 0.6 \end{pmatrix} \qquad \mathbf{A}_2 = \mathbf{g}'((0,100)) = \begin{pmatrix} -0.8 & 0 \\ -1.8 & -0.6 \end{pmatrix}$$

$$\mathbf{A}_3 = \mathbf{g}'((50,0)) = \begin{pmatrix} -1.6 & -1.2 \\ 0 & -0.3 \end{pmatrix} \qquad \mathbf{A}_4 = \mathbf{g}'((20,40)) = \begin{pmatrix} -0.64 & -0.48 \\ -0.72 & -0.24 \end{pmatrix}.$$

Since $\det(\mathbf{A}_1) = \Delta_1 = 0.96 > 0$, $\tau = 2.2 > 0$, and $\tau_1^2 - 4\Delta_1 = 1 > 0$, we see that $(0,0)$ is an unstable node. Since $\det(\mathbf{A}_2) = \Delta_2 = 0.48 > 0$, $\tau = -1.4 < 0$, and $\tau_2^2 - 4\Delta_2 = 0.04 > 0$, we see that $(0,100)$ is a stable node. Since $\det(\mathbf{A}_3) = \Delta_3 = 0.48 > 0$, $\tau = -1.9 < 0$, and $\tau_3^2 - 4\Delta_3 = 1.69 > 0$, we see that $(50,0)$ is a stable node. Since $\det(\mathbf{A}_4) = -0.192 < 0$ we see that $(20,40)$ is a saddle point.

13. For $\mathbf{X} = (K_1, 0)$, $\tau = -r_1 + r_2[1 - (K_1\alpha_{21}/K_2)]$ and $\Delta = -r_1 r_2[1 - (K_1\alpha_{21}/K_2)]$. If we let $c = 1 - K_1\alpha_{21}/K_2$, $\tau^2 - 4\Delta = (cr_2 + r_1)^2 > 0$. Now if $k_1 > K_2/\alpha_{21}$, $c < 0$ and so $\tau < 0$, $\Delta > 0$. Therefore $(K_1, 0)$ is a stable node. If $K_1 < K_2/\alpha_{21}$, $c > 0$ and so $\Delta < 0$. In this case $(K_1, 0)$ is a saddle point.

15. $K_1/\alpha_{12} < K_2 < K_1\alpha_{21}$ and so $\alpha_{12}\alpha_{21} > 1$. Therefore $\Delta = (1 - \alpha_{12}\alpha_{21})\hat{x}\hat{y}\, r_1 r_2/K_1 K_2 < 0$ and so $(\hat{x}, \hat{y})$ is a saddle point.

17. **(a)** The corresponding plane autonomous system is

$$x = y, \quad y' = -\frac{\beta}{m} y|y| - \frac{k}{m} x$$

and so a critical point must satisfy both $y = 0$ and $x = 0$. Therefore $(0,0)$ is the unique critical point.

(b) The Jacobian matrix is

$$\begin{pmatrix} 0 & 1 \\ -\dfrac{k}{m} & -\dfrac{\beta}{m} 2|y| \end{pmatrix}$$

and so $\tau = 0$ and $\Delta = k/m > 0$. Therefore $(0,0)$ is a center, stable spiral point, or an unstable spiral point. Physical considerations suggest that $(0,0)$ must be asymptotically stable and so $(0,0)$ must be a stable spiral point.

19. We have $dy/dx = y'/x' = -f(x)/y$ and so, using separation of variables,

$$\frac{y^2}{2} = -\int_0^x f(\mu)\, d\mu + c \qquad \text{or} \qquad y^2 + 2F(x) = c.$$

We can conclude that for a given value of x there are at most two corresponding values of y. If $(0,0)$ were a stable spiral point there would exist an x with more than two corresponding values of y. Note that the condition $f(0) = 0$ is required for $(0,0)$ to be a critical point of the corresponding plane autonomous system $x' = y$, $y' = -f(x)$.

21. The equation

$$x' = \alpha \frac{y}{1+y} x - x = x\left(\frac{\alpha y}{1+y} - 1\right) = 0$$

implies that $x = 0$ or $y = 1/(\alpha - 1)$. When $\alpha > 0$, $\hat{y} = 1/(\alpha - 1) > 0$. If $x = 0$, then from the differential equation for y', $y = \beta$. On the other hand, if $\hat{y} = 1/(\alpha - 1)$, $\hat{y}/(1 + \hat{y}) = 1/\alpha$ and so $\hat{x}/\alpha - 1/(\alpha - 1) + \beta = 0$. It follows that

$$\hat{x} = \alpha\left(\beta - \frac{1}{\alpha - 1}\right) = \frac{\alpha}{\alpha - 1}[(\alpha - 1)\beta - 1]$$

and if $\beta(\alpha - 1) > 1$, $\hat{x} > 0$. Therefore $(\hat{x}, \hat{y})$ is the unique critical point in the first quadrant. The Jacobian matrix is

$$\mathbf{g}'(\mathbf{X}) = \begin{pmatrix} \alpha\dfrac{y}{y+1} - 1 & \dfrac{\alpha x}{(1+y)^2} \\[2ex] -\dfrac{y}{1+y} & \dfrac{-x}{(1+y)^2} - 1 \end{pmatrix}$$

and for $\mathbf{X} = (\hat{x}, \hat{y})$, the Jacobian can be written in the form

$$\mathbf{g}'((\hat{x}, \hat{y})) = \begin{pmatrix} 0 & \dfrac{(\alpha - 1)^2}{\alpha}\hat{x} \\[2ex] -\dfrac{1}{\alpha} & -\dfrac{(\alpha - 1)^2}{\alpha^2} - 1 \end{pmatrix}.$$

It follows that

$$\tau = -\left[\frac{(\alpha - 1)^2}{\alpha^2}\hat{x} + 1\right] < 0, \quad \Delta = \frac{(\alpha - 1)^2}{\alpha^2}\hat{x}$$

and so $\tau = -(\Delta + 1)$. Therefore $\tau^2 - 4\Delta = (\Delta + 1)^2 - 4\Delta = (\Delta - 1)^2 > 0$. Therefore $(\hat{x}, \hat{y})$ is a stable node.

> ### Chapter 10 in Review

1. True

3. A center or a saddle point

5. False; there are initial conditions for which $\lim\limits_{t\to\infty} \mathbf{X}(t) = (0,0)$.

7. False; this is a borderline case. See Figure 10.3.7 in the text.

9. The system is linear and we identify $\Delta = -\alpha$ and $\tau = \alpha + 1$. Since a critical point will be a center when $\Delta > 0$ and $\tau = 0$ we see that for $\alpha = -1$ critical points will be centered and solutions will be periodic. Not also that when $\alpha = -1$ the system is

$$x' = -x - 2y$$
$$y' = x + y$$

which does have an isolated critical point at $(0,0)$.

11. Switching to polar coordinates,

$$\frac{dr}{dt} = \frac{1}{r}\left(x\frac{dx}{dt} + y\frac{dy}{dt}\right) = \frac{1}{r}(-xy - x^2r^3 + xy - y^2r^3) = -r^4$$

$$\frac{d\theta}{dt} = \frac{1}{r^2}\left(-y\frac{dx}{dt} + x\frac{dy}{dt}\right) = \frac{1}{r^2}(y^2 + xyr^3 + x^2 - xyr^3) = 1.$$

Using separation of variables it follows that $r = \dfrac{1}{\sqrt[3]{3t + c_1}}$ and $\theta = t + c_2$. Since $\mathbf{X}(0) = (1,0)$, $r = 1$ and $\theta = 0$. It follows that $c_1 = 1$, $c_2 = 0$, and so

$$r = \frac{1}{\sqrt[3]{3t + 1}}, \quad \theta = t.$$

As $t \to \infty$, $r \to 0$ and the solution spirals toward the origin.

13. (a) $\tau = 0$, $\Delta = 11 > 0$ and so $(0,0)$ is a center.

(b) $\tau = -2$, $\Delta = 1$, $\tau^2 - 4\Delta = 0$ and so $(0,0)$ is a degenerate stable node.

15. From $x = r\cos\theta$, $y = r\sin\theta$ we have

$$\frac{dx}{dt} = -r\sin\theta\,\frac{d\theta}{dt} + \frac{dr}{dt}\cos\theta$$

$$\frac{dy}{dt} = r\cos\theta\,\frac{d\theta}{dt} + \frac{dr}{dt}\sin\theta.$$

Then $r' = \alpha r$, $\theta' = 1$ gives

$$\frac{dx}{dt} = -r\sin\theta + \alpha r\cos\theta$$

$$\frac{dy}{dt} = r\cos\theta + \alpha r\sin\theta.$$

We see that $r = 0$, which corresponds to $\mathbf{X} = (0,0)$, is a critical point. Solving $r' = \alpha r$ we have $r = c_1 e^{\alpha t}$. Thus, when $\alpha < 0$, $\lim_{t\to\infty} r(t) = 0$ and $(0,0)$ is a stable critical point. When $\alpha = 0$, $r' = 0$ and $r = c_1$. In this case $(0,0)$ is a center, which is stable. Therefore, $(0,0)$ is a stable critical point for the system when $\alpha \leq 0$.

17. Critical points occur at $x = \pm 1$. Since

$$g'(x) = -\frac{1}{2}e^{-x/2}(x^2 - 4x - 1),$$

$g'(1) > 0$ and $g'(-1) < 0$. Therefore $x = 1$ is unstable and $x = -1$ is asymptotically stable.

19. The corresponding plane autonomous system is

$$x' = y, \quad y' = -\frac{\beta}{m}y - \frac{k}{m}(s+x)^3 + g$$

and so the Jacobian is

$$\mathbf{g}'(\mathbf{X}) = \begin{pmatrix} 0 & 1 \\ -\dfrac{3k}{m}(s+x)^2 & -\dfrac{\beta}{m} \end{pmatrix}.$$

For $\mathbf{X} = (0,0)$, $\tau = -\dfrac{\beta}{m} < 0$, $\Delta = \dfrac{3k}{m}s^2 > 0$.

Therefore

$$\tau^2 - 4\Delta = \frac{\beta^2}{m^2} - \frac{12k}{m}s^2 = \frac{1}{m^2}(\beta^2 - 12kms^2).$$

Therefore $(0,0)$ is a stable node if $\beta^2 > 12kms^2$ and a stable spiral point provided $\beta^2 < 12kms^2$, where $ks^3 = mg$.

Chapter 11

Fourier Series

1. $\displaystyle\int_{-2}^{2} xx^2\, dx = \frac{1}{4}x^4\bigg|_{-2}^{2} = 0$

3. $\displaystyle\int_{0}^{2} e^x\left(xe^{-x} - e^{-x}\right) dx = \int_{0}^{2} (x-1)\, dx = \left(\frac{1}{2}x^2 - x\right)\bigg|_{0}^{2} = 0$

5. $\displaystyle\int_{-\pi/2}^{\pi/2} x\cos 2x\, dx = \frac{1}{2}\left(\frac{1}{2}\cos 2x + x\sin 2x\right)\bigg|_{-\pi/2}^{\pi/2} = 0$

7. For $m \neq n$

$$\int_{0}^{\pi/2} \sin(2n+1)x \sin(2m+1)x\, dx = \frac{1}{2}\int_{0}^{\pi/2} \left(\cos 2(n-m)x - \cos 2(n+m+1)x\right)\, dx$$

$$= \frac{1}{4(n-m)}\sin 2(n-m)x\bigg|_{0}^{\pi/2} - \frac{1}{4(n+m+1)}\sin 2(n+m+1)x\bigg|_{0}^{\pi/2} = 0.$$

For $m = n$

$$\int_{0}^{\pi/2} \sin^2(2n+1)x\, dx = \int_{0}^{\pi/2}\left(\frac{1}{2} - \frac{1}{2}\cos 2(2n+1)x\right) dx$$

$$= \frac{1}{2}x\bigg|_{0}^{\pi/2} - \frac{1}{4(2n+1)}\sin 2(2n+1)x\bigg|_{0}^{\pi/2} = \frac{\pi}{4}$$

so that

$$\|\sin(2n+1)x\| = \frac{1}{2}\sqrt{\pi}.$$

9. For $m \neq n$

$$\int_0^\pi \sin nx \sin mx\, dx = \frac{1}{2} \int_0^\pi \left(\cos(n-m)x - \cos(n+m)x \right)\, dx$$

$$= \frac{1}{2(n-m)} \sin(n-m)x \Big|_0^\pi - \frac{1}{2(n+m)} \sin(n+m)x \Big|_0^\pi = 0.$$

For $m = n$

$$\int_0^\pi \sin^2 nx\, dx = \int_0^\pi \left(\frac{1}{2} - \frac{1}{2} \cos 2nx \right) dx = \frac{1}{2} x \Big|_0^\pi - \frac{1}{4n} \sin 2nx \Big|_0^\pi = \frac{\pi}{2}$$

so that

$$\| \sin nx \| = \sqrt{\frac{\pi}{2}}\,.$$

11. For $m \neq n$

$$\int_0^p \cos \frac{n\pi}{p} x \cos \frac{m\pi}{p} x\, dx = \frac{1}{2} \int_0^p \left(\cos \frac{(n-m)\pi}{p} x + \cos \frac{(n+m)\pi}{p} x \right) dx$$

$$= \frac{p}{2(n-m)\pi} \sin \frac{(n-m)\pi}{p} x \Big|_0^p + \frac{p}{2(n+m)\pi} \sin \frac{(n+m)\pi}{p} x \Big|_0^p = 0.$$

For $m = n$

$$\int_0^p \cos^2 \frac{n\pi}{p} x\, dx = \int_0^p \left(\frac{1}{2} + \frac{1}{2} \cos \frac{2n\pi}{p} x \right) dx = \frac{1}{2} x \Big|_0^p + \frac{p}{4n\pi} \sin \frac{2n\pi}{p} x \Big|_0^p = \frac{p}{2}\,.$$

Also

$$\int_0^p 1 \cdot \cos \frac{n\pi}{p} x\, dx = \frac{p}{n\pi} \sin \frac{n\pi}{p} x \Big|_0^p = 0 \quad \text{and} \quad \int_0^p 1^2\, dx = p$$

so that

$$\|1\| = \sqrt{p} \quad \text{and} \quad \left\| \cos \frac{n\pi}{p} x \right\| = \sqrt{\frac{p}{2}}\,.$$

13. Since

$$\int_{-\infty}^\infty e^{-x^2} \cdot 1 \cdot 2x\, dx = 0, \quad \longleftarrow \text{ integrand is an odd function}$$

$$\int_{-\infty}^\infty e^{-x^2} \cdot 1 \cdot (4x^2 - 2)\, dx = 2 \int_0^\infty e^{-x^2} (4x^2 - 2)\, dx \quad \longleftarrow \text{ integrand is an even function}$$

$$= 4 \int_0^\infty x \left(2xe^{-x^2} \right) dx - 4 \int_0^\infty e^{-x^2}\, dx$$

$$= 4 \left(-xe^{-x^2} \Big|_0^\infty + \int_0^\infty e^{-x^2}\, dx \right) - 4 \int_0^\infty e^{-x^2}\, dx$$

$$= -4xe^{-x^2} \Big|_0^\infty + 4 \int_0^\infty e^{-x^2}\, dx - 4 \int_0^\infty e^{-x^2}\, dx = 0$$

and

$$\int_{-\infty}^{\infty} e^{-x^2} \cdot 2x \cdot (4x^2 - 2)\, dx = 0, \quad \longleftarrow \quad \text{integrand is an odd function}$$

the functions are orthogonal.

15. By orthogonality $\int_a^b \phi_0(x)\phi_n(x)dx = 0$ for $n = 1, 2, 3, \ldots$; that is, $\int_a^b \phi_n(x)\,dx = 0$ for $n = 1, 2, 3, \ldots$.

17. Using the fact that ϕ_n and ϕ_m are orthogonal for $n \neq m$ we have

$$\|\phi_m(x) + \phi_n(x)\|^2 = \int_a^b [\phi_m(x) + \phi_n(x)]^2\, dx = \int_a^b \left[\phi_m^2(x) + 2\phi_m(x)\phi_n(x) + \phi_n^2(x)\right]\, dx$$

$$= \int_a^b \phi_m^2(x)\, dx + 2\int_a^b \phi_m(x)\phi_n(x)\, dx + \int_a^b \phi_n^2(x)\, dx$$

$$= \|\phi_m(x)\|^2 + \|\phi_n(x)\|^2.$$

19. The fundamental period is $2\pi/2\pi = 1$.

21. The fundamental period of $\sin x + \sin 2x$ is 2π.

23. The fundamental period of $\sin 3x + \cos 2x$ is 2π since the smallest integer multiples of $2\pi/3$ and $2\pi/2 = \pi$ that are equal are 3 and 2, respectively.

25. (a) For m and n positive integers, we have orthogonality on the interval $[-\pi, \pi]$ because

$$\int_{-\pi}^{\pi} \sin nx \sin mx\, dx = \frac{1}{2}\int_{-\pi}^{\pi} [\cos(m-n)x - \cos(m+n)x]\, dx$$

$$= \frac{1}{2}\left[\frac{\sin(m-n)x}{m-n} - \frac{\sin(m+n)x}{m+n}\right]_{-\pi}^{\pi} = 0, \quad m \neq n$$

(b) The function $f(x) = 1$ is continuous on $[-\pi, \pi]$ but is orthogonal to every function in the orthogonal set:

$$\int_{\pi}^{\pi} 1 \cdot \sin nx\, dx = \frac{\cos \pi - \cos(-\pi)}{n} = 0, \quad n = 1, 2, 3, \ldots$$

Therefore the set $\{\sin nx\}$, $n = 1, 2, 3, \ldots$ is not complete on the interval $[-\pi, \pi]$.

27. First we identify $f_0(x) = 1$, $f_1(x) = x$, $f_2(x) = x^2$, and $f_3(x) = x^3$. Then, we use the formulas from Problem 26. First, we have $\phi_0(x) = f_0(x) = 1$. Then

$$(f_1, \phi_0) = (x, 1) = \int_{-1}^{1} x\, dx = 0 \quad \text{and} \quad (\phi_0, \phi_0) = \int_{-1}^{1} 1\, dx = 2,$$

so

$$\phi_1(x) = f_1(x) - \frac{(f_1, \phi_0)}{(\phi_0, \phi_0)}\phi_0(x) = x - \frac{0}{2}(1) = x.$$

Next

$$(f_2, \phi_0) = (x^2, 1) = \int_{-1}^{1} x^2\, dx = \frac{2}{3}, \quad (f_2, \phi_1) = (x^2, x) = \int_{-1}^{1} x^3\, dx = 0,$$

$$\text{and } (\phi_1, \phi_1) = \int_{-1}^{1} x^2\, dx = \frac{2}{3},$$

so

$$\phi_2(x) = f_2(x) - \frac{(f_2, \phi_0)}{(\phi_0, \phi_0)} \phi_0(x) - \frac{(f_2, \phi_1)}{(\phi_1, \phi_1)} \phi_1(x) = x^2 - \frac{2/3}{2}(1) - \frac{0}{2}(x) = x^2 - \frac{1}{3}.$$

Finally,

$$(f_3, \phi_0) = (x^3, 1) = \int_{-1}^{1} x^3\, dx = 0, \quad (f_3, \phi_1) = (x^3, x) = \int_{-1}^{1} x^4\, dx = \frac{2}{5},$$

and

$$(f_3, \phi_2) = \left(x^3, x^2 - \frac{1}{3}\right) = \int_{-1}^{1}\left(x^5 - \frac{1}{3}x^3\right) dx = 0,$$

so

$$\phi_3(x) = f_3(x) - \frac{(f_3, \phi_0)}{(\phi_0, \phi_0)} \phi_0(x) - \frac{(f_3, \phi_1)}{(\phi_1, \phi_1)} \phi_1(x) - \frac{(f_3, \phi_2)}{(\phi_2, \phi_2)} \phi_2(x)$$

$$= x^3 - 0 - \frac{2/5}{2/3}(x) - 0 = x^3 - \frac{3}{5}x.$$

11.2 | Fourier Series

1. $a_0 = \dfrac{1}{\pi} \displaystyle\int_{-\pi}^{\pi} f(x)\, dx = \dfrac{1}{\pi} \displaystyle\int_{0}^{\pi} 1\, dx = 1$

$a_n = \dfrac{1}{\pi} \displaystyle\int_{-\pi}^{\pi} f(x) \cos \dfrac{n\pi}{pi}x\, dx = \dfrac{1}{\pi} \displaystyle\int_{0}^{\pi} \cos nx\, dx = 0$

$b_n = \dfrac{1}{\pi} \displaystyle\int_{-\pi}^{\pi} f(x) \sin \dfrac{n\pi}{\pi}x\, dx = \dfrac{1}{\pi} \displaystyle\int_{0}^{\pi} \sin nx\, dx = \dfrac{1}{n\pi}(1 - \cos n\pi) = \dfrac{1}{n\pi}[1 - (-1)^n]$

$f(x) = \dfrac{1}{2} + \dfrac{1}{\pi} \displaystyle\sum_{n=1}^{\infty} \dfrac{1 - (-1)^n}{n} \sin nx$

Converges to $\frac{1}{2}$ at $x = 0$.

3. $a_0 = \displaystyle\int_{-1}^{1} f(x)\, dx = \displaystyle\int_{-1}^{0} 1\, dx + \displaystyle\int_{0}^{1} x\, dx = \dfrac{3}{2}$

$a_n = \displaystyle\int_{-1}^{1} f(x) \cos n\pi x\, dx = \displaystyle\int_{-1}^{0} \cos n\pi x\, dx + \displaystyle\int_{0}^{1} x \cos n\pi x\, dx = \dfrac{1}{n^2\pi^2}[(-1)^n - 1]$

$b_n = \displaystyle\int_{-1}^{1} f(x) \sin n\pi x\, dx = \displaystyle\int_{-1}^{0} \sin n\pi x\, dx + \displaystyle\int_{0}^{1} x \sin n\pi x\, dx = -\dfrac{1}{n\pi}$

$$f(x) = \frac{3}{4} + \sum_{n=1}^{\infty} \left[\frac{(-1)^n - 1}{n^2 \pi^2} \cos n\pi x - \frac{1}{n\pi} \sin n\pi x \right]$$

Converges to $\frac{1}{2}$ at $x = 0$.

5. $a_0 = \frac{1}{\pi} \int_{-\pi}^{\pi} f(x)\, dx = \frac{1}{\pi} \int_0^{\pi} x^2\, dx = \frac{1}{3}\pi^2$

$$a_n = \frac{1}{\pi} \int_{-\pi}^{\pi} f(x) \cos nx\, dx = \frac{1}{\pi} \int_0^{\pi} x^2 \cos nx\, dx = \frac{1}{\pi} \left(\frac{x^2}{\pi} \sin nx \Big|_0^{\pi} - \frac{2}{n} \int_0^{\pi} x \sin nx\, dx \right)$$

$$= \frac{2(-1)^n}{n^2}$$

$$b_n = \frac{1}{\pi} \int_0^{\pi} x^2 \sin nx\, dx = \frac{1}{\pi} \left(-\frac{x^2}{n} \cos nx \Big|_0^{\pi} + \frac{2}{n} \int_0^{\pi} x \cos nx\, dx \right)$$

$$= \frac{\pi}{n}(-1)^{n+1} + \frac{2}{n^3 \pi}[(-1)^n - 1]$$

$$f(x) = \frac{\pi^2}{6} + \sum_{n=1}^{\infty} \left[\frac{2(-1)^n}{n^2} \cos nx + \left(\frac{\pi}{n}(-1)^{n+1} + \frac{2[(-1)^n - 1]}{n^3 \pi} \right) \sin nx \right]$$

f is continous on the interval

7. $a_0 = \frac{1}{\pi} \int_{-\pi}^{\pi} f(x)\, dx = \frac{1}{\pi} \int_{-\pi}^{\pi} (x + \pi)\, dx = 2\pi$

$$a_n = \frac{1}{\pi} \int_{-\pi}^{\pi} f(x) \cos nx\, dx = \frac{1}{\pi} \int_{-\pi}^{\pi} (x + \pi) \cos nx\, dx = 0$$

$$b_n = \frac{1}{\pi} \int_{-\pi}^{\pi} f(x) \sin nx\, dx = \frac{2}{n}(-1)^{n+1}$$

$$f(x) = \pi + \sum_{n=1}^{\infty} \frac{2}{n}(-1)^{n+1} \sin nx$$

f is continous on the interval

9. $a_0 = \frac{1}{\pi} \int_{-\pi}^{\pi} f(x)\, dx = \frac{1}{\pi} \int_0^{\pi} \sin x\, dx = \frac{2}{\pi}$

$$a_n = \frac{1}{\pi} \int_{-\pi}^{\pi} f(x) \cos nx\, dx = \frac{1}{\pi} \int_0^{\pi} \sin x \cos nx\, dx = \frac{1}{2\pi} \int_0^{\pi} (\sin(1 + n)x + \sin(1 - n)x)\, dx$$

$$= \frac{1 + (-1)^n}{\pi(1 - n^2)} \quad \text{for } n = 2, 3, 4, \ldots$$

$$a_1 = \frac{1}{2\pi} \int_0^{\pi} \sin 2x\, dx = 0$$

$$b_n = \frac{1}{\pi} \int_{-\pi}^{\pi} f(x) \sin nx\, dx = \frac{1}{\pi} \int_0^{\pi} \sin x \sin nx\, dx$$

$$= \frac{1}{2\pi} \int_0^{\pi} (\cos(1 - n)x - \cos(1 + n)x)\, dx = 0 \quad \text{for } n = 2, 3, 4, \ldots$$

$$b_1 = \frac{1}{2\pi} \int_0^\pi (1 - \cos 2x)\, dx = \frac{1}{2}$$

$$f(x) = \frac{1}{\pi} + \frac{1}{2}\sin x + \sum_{n=2}^\infty \frac{1 + (-1)^n}{\pi(1 - n^2)} \cos nx$$

f is continous on the interval

11. $a_0 = \frac{1}{2} \int_{-2}^2 f(x)\, dx = \frac{1}{2}\left(\int_{-1}^0 -2\, dx + \int_0^1 1\, dx \right) = -\frac{1}{2}$

$a_n = \frac{1}{2} \int_{-2}^2 f(x) \cos \frac{n\pi}{2} x\, dx = \frac{1}{2}\left(\int_{-1}^0 (-2) \cos \frac{n\pi}{2} x\, dx + \int_0^1 \cos \frac{n\pi}{2} x\, dx \right) = -\frac{1}{n\pi} \sin \frac{n\pi}{2}$

$b_n = \frac{1}{2} \int_{-2}^2 f(x) \sin \frac{n\pi}{2} x\, dx = \frac{1}{2}\left(\int_{-1}^0 (-2) \sin \frac{n\pi}{2} x\, dx + \int_0^1 \sin \frac{n\pi}{2} x\, dx \right) = \frac{3}{n\pi}\left(1 - \cos \frac{n\pi}{2}\right)$

$f(x) = -\frac{1}{4} + \sum_{n=1}^\infty \left[-\frac{1}{n\pi} \sin \frac{n\pi}{2} \cos \frac{n\pi}{2} x + \frac{3}{n\pi}\left(1 - \cos \frac{n\pi}{2}\right) \sin \frac{n\pi}{2} x \right]$

Converges to -1 at $x = -1$, $-\frac{1}{2}$ at $x = 0$, and $\frac{1}{2}$ at $x = 1$.

13. $a_0 = \frac{1}{5} \int_{-5}^5 f(x)\, dx = \frac{1}{5}\left(\int_{-5}^0 1\, dx + \int_0^5 (1 + x)\, dx \right) = \frac{9}{2}$

$a_n = \frac{1}{5} \int_{-5}^5 f(x) \cos \frac{n\pi}{5} x\, dx = \frac{1}{5}\left(\int_{-5}^0 \cos \frac{n\pi}{5} x\, dx + \int_0^5 (1 + x) \cos \frac{n\pi}{5} x\, dx \right)$

$= \frac{5}{n^2\pi^2}[(-1)^n - 1]$

$b_n = \frac{1}{5} \int_{-5}^5 f(x) \sin \frac{n\pi}{5} x\, dx = \frac{1}{5}\left(\int_{-5}^0 \sin \frac{n\pi}{5} x\, dx + \int_0^5 (1 + x) \cos \frac{n\pi}{5} x\, dx \right) = \frac{5}{n\pi}(-1)^{n+1}$

$f(x) = \frac{9}{4} + \sum_{n=1}^\infty \left[\frac{5}{n^2\pi^2}[(-1)^n - 1] \cos \frac{n\pi}{5} x + \frac{5}{n\pi}(-1)^{n+1} \sin \frac{n\pi}{5} x \right]$

f is continous on the interval

15. $a_0 = \frac{1}{\pi} \int_{-\pi}^\pi f(x)\, dx = \frac{1}{\pi} \int_{-\pi}^\pi e^x\, dx = \frac{1}{\pi}(e^\pi - e^{-\pi})$

$a_n = \frac{1}{\pi} \int_{-\pi}^\pi f(x) \cos nx\, dx = \frac{(-1)^n(e^\pi - e^{-\pi})}{\pi(1 + n^2)}$

$b_n = \frac{1}{\pi} \int_{-\pi}^\pi f(x) \sin nx\, dx = \frac{1}{\pi} \int_{-\pi}^\pi e^x \sin nx\, dx = \frac{(-1)^n n(e^{-\pi} - e^\pi)}{\pi(1 + n^2)}$

$f(x) = \frac{e^\pi - e^{-\pi}}{2\pi} + \sum_{n=1}^\infty \left[\frac{(-1)^n(e^\pi - e^{-\pi})}{\pi(1 + n^2)} \cos nx + \frac{(-1)^n n(e^{-\pi} - e^\pi)}{\pi(1 + n^2)} \sin nx \right]$

f is continous on the interval

17.

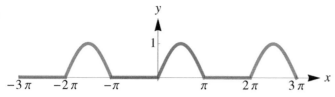

19. The function in Problem 5 is discontinuous at $x = \pi$, so the corresponding Fourier series converges to $\pi^2/2$ at $x = \pi$. That is,

$$\frac{\pi^2}{2} = \frac{\pi^2}{6} + \sum_{n=1}^{\infty} \left[\frac{2(-1)^n}{n^2} \cos n\pi + \left(\frac{\pi}{n}(-1)^{n+1} + \frac{2[(-1)^n - 1]}{n^3\pi} \right) \sin n\pi \right]$$

$$= \frac{\pi^2}{6} + \sum_{n=1}^{\infty} \frac{2(-1)^n}{n^2}(-1)^n = \frac{\pi^2}{6} + \sum_{n=1}^{\infty} \frac{2}{n^2} = \frac{\pi^2}{6} + 2\left(1 + \frac{1}{2^2} + \frac{1}{3^2} + \cdots\right)$$

and

$$\frac{\pi^2}{6} = \frac{1}{2}\left(\frac{\pi^2}{2} - \frac{\pi^2}{6} \right) = 1 + \frac{1}{2^2} + \frac{1}{3^2} + \cdots .$$

At $x = 0$ the series converges to 0 and

$$0 = \frac{\pi^2}{6} + \sum_{n=1}^{\infty} \frac{2(-1)^n}{n^2} = \frac{\pi^2}{6} + 2\left(-1 + \frac{1}{2^2} - \frac{1}{3^2} + \frac{1}{4^2} - \cdots\right)$$

so

$$\frac{\pi^2}{12} = 1 - \frac{1}{2^2} + \frac{1}{3^2} - \frac{1}{4^2} + \cdots .$$

21. The function in Problem 7 is continuous at $x = \pi/2$ so

$$\frac{3\pi}{2} = f\left(\frac{\pi}{2}\right) = \pi + \sum_{n=1}^{\infty} \frac{2}{n}(-1)^{n+1} \sin \frac{n\pi}{2} = \pi + 2\left(1 - \frac{1}{3} + \frac{1}{5} - \frac{1}{7} + \cdots\right)$$

and

$$\frac{\pi}{4} = 1 - \frac{1}{3} + \frac{1}{5} - \frac{1}{7} + \cdots .$$

23. (a) Letting $c_0 = a_0/2$, $c_n = (a_n - ib_n)/2$, and $c_{-n} = (a_n + ib_n)/2$ we have

$$f(x) = \frac{a_0}{2} + \sum_{n=1}^{\infty} \left(a_n \cos \frac{n\pi}{p} x + b_n \sin \frac{n\pi}{p} x \right)$$

$$= c_0 + \sum_{n=1}^{\infty} \left(a_n \frac{e^{in\pi x/p} + e^{-in\pi x/p}}{2} - b_n \frac{ie^{in\pi x/p} - ie^{-in\pi x/p}}{2} \right)$$

$$= c_0 + \sum_{n=1}^{\infty} \left(c_n e^{in\pi x/p} + c_{-n} e^{i(-n)\pi x/p} \right) = \sum_{n=-\infty}^{\infty} c_n e^{in\pi x/p}.$$

(b) From part (a) we have

$$c_n = \frac{1}{2}(a_n - ib_n) = \frac{1}{2p} \int_{-p}^{p} f(x) \left(\cos \frac{n\pi}{p} x - i \sin \frac{n\pi}{p} x \right) dx = \frac{1}{2p} \int_{-p}^{p} f(x) e^{-in\pi x/p} \, dx$$

and

$$c_{-n} = \frac{1}{2}(a_n + ib_n) = \frac{1}{2p}\int_{-p}^{p} f(x)\left(\cos\frac{n\pi}{p}x + i\sin\frac{n\pi}{p}x\right)dx = \frac{1}{2p}\int_{-p}^{p} f(x)e^{in\pi x/p}\,dx$$

for $n = 1, 2, 3, \ldots$. Thus, for $n = \pm 1, \pm 2, \pm 3, \ldots$,

$$c_n = \frac{1}{2p}\int_{-p}^{p} f(x)e^{-in\pi x/p}\,dx.$$

When $n = 0$ the above formula gives

$$c_0 = \frac{1}{2p}\int_{-p}^{p} f(x)\,dx,$$

which is $a_0/2$ where a_0 is (9) in the text. Therefore

$$c_n = \frac{1}{2p}\int_{-p}^{p} f(x)e^{-in\pi x/p}\,dx, \quad n = 0, \pm 1, \pm 2, \ldots.$$

11.3   Fourier Cosine and Sine Series

1. Since $f(-x) = \sin(-3x) = -\sin 3x = -f(x)$, $f(x)$ is an odd function.

3. Since $f(-x) = (-x)^2 - x = x^2 - x$, $f(x)$ is neither even nor odd.

5. Since $f(-x) = e^{|-x|} = e^{|x|} = f(x)$, $f(x)$ is an even function.

7. For $0 < x < 1$, $f(-x) = (-x)^2 = x^2 = -f(x)$, $f(x)$ is an odd function.

9. Since $f(x)$ is not defined for $x < 0$, it is neither even nor odd.

11. Since $f(x)$ is an odd function, we have

$$b_n = \frac{2}{\pi}\int_0^{\pi} 1\cdot\sin nx\,dx = \frac{2}{n\pi}[1 - (-1)^n]$$

Thus

$$f(x) = \sum_{n=1}^{\infty}\frac{2}{n\pi}[1 - (-1)^n]\sin nx.$$

13. Since $f(x)$ is an even function, we expand in a cosine series:

$$a_0 = \frac{2}{\pi}\int_0^{\pi} x\,dx = \pi$$

$$a_n = \frac{2}{\pi}\int_0^{\pi} x\cos nx\,dx = \frac{2}{n^2\pi}[(-1)^n - 1].$$

Thus

$$f(x) = \frac{\pi}{2} + \sum_{n=1}^{\infty}\frac{2}{n^2\pi}[(-1)^n - 1]\cos nx.$$

15. Since $f(x)$ is an even function, we expand in a cosine series:

$$a_0 = 2 \int_0^1 x^2 \, dx = \frac{2}{3}$$

$$a_n = 2 \int_0^1 x^2 \cos n\pi x \, dx = 2 \left(\frac{x^2}{n\pi} \sin n\pi x \Big|_0^1 - \frac{2}{n\pi} \int_0^1 x \sin n\pi x \, dx \right) = \frac{4}{n^2\pi^2}(-1)^n.$$

Thus

$$f(x) = \frac{1}{3} + \sum_{n=1}^{\infty} \frac{4}{n^2\pi^2}(-1)^n \cos n\pi x.$$

17. Since $f(x)$ is an even function, we expand in a cosine series:

$$a_0 = \frac{2}{\pi} \int_0^\pi (\pi^2 - x^2) \, dx = \frac{4}{3}\pi^2$$

$$a_n = \frac{2}{\pi} \int_0^\pi (\pi^2 - x^2) \cos nx \, dx = \frac{2}{\pi} \left(\frac{\pi^2 - x^2}{n} \sin nx \Big|_0^\pi + \frac{2}{n} \int_0^\pi x \sin nx \, dx \right) = \frac{4}{n^2}(-1)^{n+1}.$$

Thus

$$f(x) = \frac{2}{3}\pi^2 + \sum_{n=1}^{\infty} \frac{4}{n^2}(-1)^{n+1} \cos nx \, dx.$$

19. Since $f(x)$ is an odd function, we expand in a sine series:

$$b_n = \frac{2}{\pi} \int_0^\pi (x+1) \sin nx \, dx = \frac{2(\pi+1)}{n\pi}(-1)^{n+1} + \frac{2}{n\pi}.$$

Thus

$$f(x) = \sum_{n=1}^{\infty} \left(\frac{2(\pi+1)}{n\pi}(-1)^{n+1} + \frac{2}{n\pi} \right) \sin nx.$$

21. Since $f(x)$ is an even function, we expand in a cosine series:

$$a_0 = \int_0^1 x \, dx + \int_1^2 1 \, dx = \frac{3}{2}$$

$$a_n = \int_0^1 x \cos \frac{n\pi}{2} x \, dx + \int_1^2 \cos \frac{n\pi}{2} x \, dx = \frac{4}{n^2\pi^2} \left(\cos \frac{n\pi}{2} - 1 \right).$$

Thus

$$f(x) = \frac{3}{4} + \sum_{n=1}^{\infty} \frac{4}{n^2\pi^2} \left(\cos \frac{n\pi}{2} - 1 \right) \cos \frac{n\pi}{2} x.$$

23. Since $f(x)$ is an even function, we expand in a cosine series:

$$a_0 = \frac{2}{\pi} \int_0^\pi \sin x \, dx = \frac{4}{\pi}$$

$$a_n = \frac{2}{\pi} \int_0^\pi \sin x \cos nx \, dx = \frac{1}{\pi} \int_0^\pi \left(\sin(1+n)x + \sin(1-n)x \right) dx$$

$$= \frac{2}{\pi(1-n^2)} \left(1 + (-1)^n \right) \quad \text{for } n = 2, 3, 4, \ldots$$

$$a_1 = \frac{1}{\pi} \int_0^\pi \sin 2x \, dx = 0.$$

Thus

$$f(x) = \frac{2}{\pi} + \sum_{n=2}^\infty \frac{2[1+(-1)^n]}{\pi(1-n^2)} \cos nx.$$

25. $a_0 = 2 \int_0^{1/2} 1 \, dx = 1$

$$a_n = 2 \int_0^{1/2} 1 \cdot \cos n\pi x \, dx = \frac{2}{n\pi} \sin \frac{n\pi}{2}$$

$$b_n = 2 \int_0^{1/2} 1 \cdot \sin n\pi x \, dx = \frac{2}{n\pi} \left(1 - \cos \frac{n\pi}{2} \right)$$

$$f(x) = \frac{1}{2} + \sum_{n=1}^\infty \frac{2}{n\pi} \sin \frac{n\pi}{2} \cos n\pi x$$

$$f(x) = \sum_{n=1}^\infty \frac{2}{n\pi} \left(1 - \cos \frac{n\pi}{2} \right) \sin n\pi x$$

27. $a_0 = \frac{4}{\pi} \int_0^{\pi/2} \cos x \, dx = \frac{4}{\pi}$

$$a_n = \frac{4}{\pi} \int_0^{\pi/2} \cos x \cos 2nx \, dx = \frac{2}{\pi} \int_0^{\pi/2} [\cos(2n+1)x + \cos(2n-1)x] \, dx = \frac{4(-1)^n}{\pi(1-4n^2)}$$

$$b_n = \frac{4}{\pi} \int_0^{\pi/2} \cos x \sin 2nx \, dx = \frac{2}{\pi} \int_0^{\pi/2} [\sin(2n+1)x + \sin(2n-1)x] \, dx = \frac{8n}{\pi(4n^2-1)}$$

$$f(x) = \frac{2}{\pi} + \sum_{n=1}^\infty \frac{4(-1)^n}{\pi(1-4n^2)} \cos 2nx$$

$$f(x) = \sum_{n=1}^\infty \frac{8n}{\pi(4n^2-1)} \sin 2nx$$

29. $a_0 = \dfrac{2}{\pi} \left(\displaystyle\int_0^{\pi/2} x\,dx + \int_{\pi/2}^{\pi} (\pi - x)\,dx \right) = \dfrac{\pi}{2}$

$a_n = \dfrac{2}{\pi} \left(\displaystyle\int_0^{\pi/2} x \cos nx\,dx + \int_{\pi/2}^{\pi} (\pi - x) \cos nx\,dx \right) = \dfrac{2}{n^2\pi} \left(2 \cos \dfrac{n\pi}{2} + (-1)^{n+1} - 1 \right)$

$b_n = \dfrac{2}{\pi} \left(\displaystyle\int_0^{\pi/2} x \sin nx\,dx + \int_{\pi/2}^{\pi} (\pi - x) \sin nx\,dx \right) = \dfrac{4}{n^2\pi} \sin \dfrac{n\pi}{2}$

$f(x) = \dfrac{\pi}{4} + \displaystyle\sum_{n=1}^{\infty} \dfrac{2}{n^2\pi} \left(2 \cos \dfrac{n\pi}{2} + (-1)^{n+1} - 1 \right) \cos nx$

$f(x) = \displaystyle\sum_{n=1}^{\infty} \dfrac{4}{n^2\pi} \sin \dfrac{n\pi}{2} \sin nx$

31. $a_0 = \displaystyle\int_0^1 x\,dx + \int_1^2 1\,dx = \dfrac{3}{2}$

$a_n = \displaystyle\int_0^1 x \cos \dfrac{n\pi}{2} x\,dx = \dfrac{4}{n^2\pi^2} \left(\cos \dfrac{n\pi}{2} - 1 \right)$

$b_n = \displaystyle\int_0^1 x \sin \dfrac{n\pi}{2} x\,dx + \int_1^2 1 \cdot \sin \dfrac{n\pi}{2} x\,dx = \dfrac{4}{n^2\pi^2} \sin \dfrac{n\pi}{2} + \dfrac{2}{n\pi}(-1)^{n+1}$

$f(x) = \dfrac{3}{4} + \displaystyle\sum_{n=1}^{\infty} \dfrac{4}{n^2\pi^2} \left(\cos \dfrac{n\pi}{2} - 1 \right) \cos \dfrac{n\pi}{2} x$

$f(x) = \displaystyle\sum_{n=1}^{\infty} \left(\dfrac{4}{n^2\pi^2} \sin \dfrac{n\pi}{2} + \dfrac{2}{n\pi}(-1)^{n+1} \right) \sin \dfrac{n\pi}{2} x$

33. $a_0 = 2 \displaystyle\int_0^1 (x^2 + x)\,dx = \dfrac{5}{3}$

$a_n = 2 \displaystyle\int_0^1 (x^2 + x) \cos n\pi x\,dx = \left. \dfrac{2(x^2 + x)}{n\pi} \sin n\pi x \right|_0^1 - \dfrac{2}{n\pi} \int_0^1 (2x + 1) \sin n\pi x\,dx$

$\qquad = \dfrac{2}{n^2\pi^2} [3(-1)^n - 1]$

$b_n = 2 \displaystyle\int_0^1 (x^2 + x) \sin n\pi x\,dx = \left. -\dfrac{2(x^2 + x)}{n\pi} \cos n\pi x \right|_0^1 + \dfrac{2}{n\pi} \int_0^1 (2x + 1) \cos n\pi x\,dx$

$\qquad = \dfrac{4}{n\pi}(-1)^{n+1} + \dfrac{4}{n^3\pi^3} [(-1)^n - 1]$

$f(x) = \dfrac{5}{6} + \displaystyle\sum_{n=1}^{\infty} \dfrac{2}{n^2\pi^2} [3(-1)^n - 1] \cos n\pi x$

$f(x) = \displaystyle\sum_{n=1}^{\infty} \left(\dfrac{4}{n\pi}(-1)^{n+1} + \dfrac{4}{n^3\pi^3} [(-1)^n - 1] \right) \sin n\pi x$

35. $a_0 = \dfrac{1}{\pi} \displaystyle\int_0^{2\pi} x^2\, dx = \dfrac{8}{3}\pi^2$

$a_n = \dfrac{1}{\pi} \displaystyle\int_0^{2\pi} x^2 \cos nx\, dx = \dfrac{4}{n^2}$

$b_n = \dfrac{1}{\pi} \displaystyle\int_0^{2\pi} x^2 \sin nx\, dx = -\dfrac{4\pi}{n}$

$f(x) = \dfrac{4}{3}\pi^2 + \displaystyle\sum_{n=1}^{\infty} \left(\dfrac{4}{n^2} \cos nx - \dfrac{4\pi}{n} \sin nx \right)$

37. $a_0 = 2 \displaystyle\int_0^1 (x+1)\, dx = 3$

$a_n = 2 \displaystyle\int_0^1 (x+1) \cos 2n\pi x\, dx = 0$

$b_n = 2 \displaystyle\int_0^1 (x+1) \sin 2n\pi x\, dx = -\dfrac{1}{n\pi}$

$f(x) = \dfrac{3}{2} - \displaystyle\sum_{n=1}^{\infty} \dfrac{1}{n\pi} \sin 2n\pi x$

39. The periodic extensions for the cosine, sine, and Fourier series are shown below:

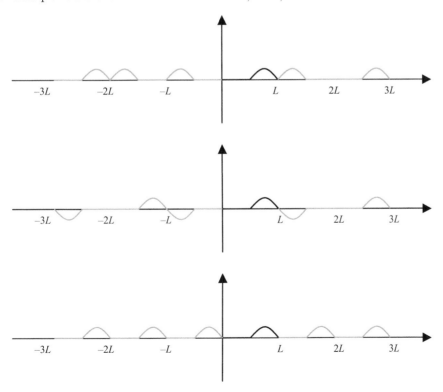

41. The periodic extensions for the cosine, sine, and Fourier series are shown below:

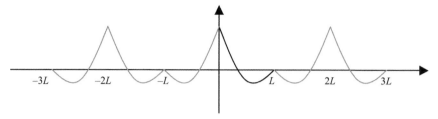

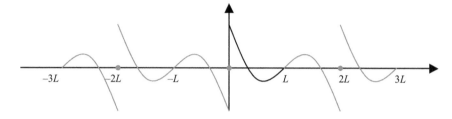

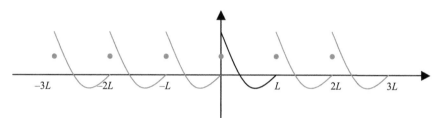

43. We have

$$b_n = \frac{2}{\pi} \int_0^\pi 5 \sin nt \, dt = \frac{10}{n\pi}[1 - (-1)^n]$$

so that

$$f(t) = \sum_{n=1}^\infty \frac{10[1 - (-1)^n]}{n\pi} \sin nt.$$

Substituting the assumption $x_p(t) = \sum_{n=1}^\infty B_n \sin nt$ into the differential equation then gives

$$x_p'' + 10x_p = \sum_{n=1}^\infty B_n(10 - n^2) \sin nt = \sum_{n=1}^\infty \frac{10[1 - (-1)^n]}{n\pi} \sin nt$$

and so $B_n = 10[1 - (-1)^n]/n\pi(10 - n^2)$. Thus

$$x_p(t) = \frac{10}{\pi} \sum_{n=1}^\infty \frac{1 - (-1)^n}{n(10 - n^2)} \sin nt.$$

45. We have

$$a_0 = \frac{2}{\pi} \int_0^\pi \left(2\pi t - t^2\right) dt = \frac{4}{3}\pi^2$$

$$a_n = \frac{2}{\pi} \int_0^\pi \left(2\pi t - t^2\right) \cos nt \, dt = -\frac{4}{n^2}$$

so that

$$f(t) = \frac{2\pi^2}{3} - \sum_{n=1}^\infty \frac{4}{n^2} \cos nt.$$

Substituting the assumption

$$x_p(t) = \frac{A_0}{2} + \sum_{n=1}^{\infty} A_n \cos nt$$

into the differential equation then gives

$$\frac{1}{4} x_p'' + 12 x_p = 6 A_0 + \sum_{n=1}^{\infty} A_n \left(-\frac{1}{4} n^2 + 12 \right) \cos nt = \frac{2\pi^2}{3} - \sum_{n=1}^{\infty} \frac{4}{n^2} \cos nt$$

and $A_0 = \pi^2/9$, $A_n = 16/n^2(n^2 - 48)$. Thus

$$x_p(t) = \frac{\pi^2}{18} + 16 \sum_{n=1}^{\infty} \frac{1}{n^2(n^2 - 48)} \cos nt.$$

47. (a) The general solution is $x(t) = c_1 \cos \sqrt{10}\,t + c_2 \sin \sqrt{10}\,t + x_p(t)$, where

$$x_p(t) = \frac{10}{\pi} \sum_{n=1}^{\infty} \frac{1 - (-1)^n}{n(10 - n^2)} \sin nt.$$

The initial condition $x(0) = 0$ implies $c_1 + x_p(0) = 0$. Since $x_p(0) = 0$, we have $c_1 = 0$ and $x(t) = c_2 \sin \sqrt{10}\,t + x_p(t)$. Then $x'(t) = c_2 \sqrt{10} \cos \sqrt{10}\,t + x_p'(t)$ and $x'(0) = 0$ implies

$$c_2 \sqrt{10} + \frac{10}{\pi} \sum_{n=1}^{\infty} \frac{1 - (-1)^n}{10 - n^2} \cos 0 = 0.$$

Thus

$$c_2 = -\frac{\sqrt{10}}{\pi} \sum_{n=1}^{\infty} \frac{1 - (-1)^n}{10 - n^2}$$

and

$$x(t) = \frac{10}{\pi} \sum_{n=1}^{\infty} \frac{1 - (-1)^n}{10 - n^2} \left[\frac{1}{n} \sin nt - \frac{1}{\sqrt{10}} \sin \sqrt{10}\,t \right].$$

(b) The graph is plotted using eight nonzero terms in the series expansion of $x(t)$.

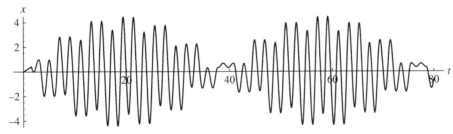

49. (a) We have

$$b_n = \frac{2}{L} \int_0^L \frac{w_0 x}{L} \sin \frac{n\pi}{L} x \, dx = \frac{2 w_0}{n\pi} (-1)^{n+1}$$

so that

$$w(x) = \sum_{n=1}^{\infty} \frac{2 w_0}{n\pi} (-1)^{n+1} \sin \frac{n\pi}{L} x.$$

(b) If we assume $y_p(x) = \sum_{n=1}^{\infty} B_n \sin(n\pi x/L)$ then

$$y_p^{(4)} = \sum_{n=1}^{\infty} \frac{n^4 \pi^4}{L^4} B_n \sin \frac{n\pi}{L} x$$

and so the differential equation $EIy_p^{(4)} = w(x)$ gives

$$B_n = \frac{2w_0(-1)^{n+1}L^4}{EIn^5\pi^5}.$$

Thus

$$y_p(x) = \frac{2w_0 L^4}{EI\pi^5} \sum_{n=1}^{\infty} \frac{(-1)^{n+1}}{n^5} \sin \frac{n\pi}{L} x.$$

51. We note that $w(x)$ is 2π-periodic and even. With $p = \pi$ we find the cosine expansion of

$$f(x) = \begin{cases} w_0, & 0 < x < \pi/2 \\ 0, & \pi/2 < x < \pi \end{cases}$$

We have

$$a_0 = \frac{2}{\pi} \int_0^{\pi} f(x)\,dx = \frac{2}{\pi} \int_0^{\pi/2} w_0\,dx = w_0$$

$$a_n = \frac{2}{\pi} \int_0^{\pi} f(x) \cos nx\,dx = \frac{2}{\pi} \int_0^{\pi/2} w_0 \cos nx\,dx = \frac{2w_0}{n\pi} \sin \frac{n\pi}{2}.$$

Thus,

$$w(x) = \frac{w_0}{2} + \frac{2w_0}{\pi} \sum_{n=1}^{\infty} \frac{1}{n} \sin \frac{n\pi}{2} \cos nx.$$

Now we assume a particular solution of the form $y_p(x) = A_0/2 + \sum_{n=1}^{\infty} A_n \cos nx$. Then $y_p^{(4)}(x) = \sum_{n=1}^{\infty} A_n n^4 \cos nx$ and substituting into the differential equation, we obtain

$$EIy_p^{(4)}(x) + ky_p(x) = \frac{kA_0}{2} + \sum_{n=1}^{\infty} A_n(EIn^4 + k)\cos nx$$

$$= \frac{w_0}{2} + \frac{2w_0}{\pi} \sum_{n=1}^{\infty} \frac{1}{n} \sin \frac{n\pi}{2} \cos nx.$$

Thus

$$A_0 = \frac{w_0}{k} \quad \text{and} \quad A_n = \frac{2w_0}{\pi} \frac{\sin(n\pi/2)}{n(EIn^4 + k)},$$

and

$$y_p(x) = \frac{w_0}{2k} + \frac{2w_0}{\pi} \sum_{n=1}^{\infty} \frac{\sin(n\pi/2)}{n(EIn^4 + k)} \cos nx.$$

53. If $f(x)$ is even then $f(-x) = f(x)$. If $f(x)$ is odd then $f(-x) = -f(x)$. Thus, if $f(x)$ is both even and odd $f(x) = f(-x) = -f(x)$, and $f(x) = 0$.

11.4 Sturm–Liouville Problem

1. For $\lambda \leq 0$ the only solution of the boundary-value problem is $y = 0$. For $\lambda = \alpha^2 > 0$ we have

$$y = c_1 \cos \alpha x + c_2 \sin \alpha x.$$

Now

$$y'(x) = -c_1 \alpha \sin \alpha x + c_2 \alpha \cos \alpha x$$

and $y'(0) = 0$ implies $c_2 = 0$, so

$$y(1) + y'(1) = c_1(\cos \alpha - \alpha \sin \alpha) = 0 \quad \text{or} \quad \cot \alpha = \alpha.$$

The eigenvalues are $\lambda_n = \alpha_n^2$ where α_1, α_2, α_3, ... are the consecutive positive solutions of $\cot \alpha = \alpha$. The corresponding eigenfunctions are $\cos \alpha_n x$ for $n = 1, 2, 3, \ldots$. Using a CAS we find that the first four eigenvalues are approximately 0.7402, 11.7349, 41.4388, and 90.8082 with corresponding approximate eigenfunctions $\cos 0.8603x$, $\cos 3.4256x$, $\cos 6.4373x$, and $\cos 9.5293x$.

3. For $\lambda = 0$ the solution of $y'' = 0$ is $y = c_1 x + c_2$. The condition $y'(0) = 0$ implies $c_1 = 0$, so $\lambda = 0$ is an eigenvalue with corresponding eigenfunction 1.

For $\lambda = -\alpha^2 < 0$ we have $y = c_1 \cosh \alpha x + c_2 \sinh \alpha x$ and $y' = c_1 \alpha \sinh \alpha x + c_2 \alpha \cosh \alpha x$. The condition $y'(0) = 0$ implies $c_2 = 0$ and so $y = c_1 \cosh \alpha x$. Now the condition $y'(L) = 0$ implies $c_1 = 0$. Thus $y = 0$ and there are no negative eigenvalues.

For $\lambda = \alpha^2 > 0$ we have $y = c_1 \cos \alpha x + c_2 \sin \alpha x$ and $y' = -c_1 \alpha \sin \alpha x + c_2 \alpha \cos \alpha x$. The condition $y'(0) = 0$ implies $c_2 = 0$ and so $y = c_1 \cos \alpha x$. Now the condition $y'(L) = 0$ implies $-c_1 \alpha \sin \alpha L = 0$. For $c_1 \neq 0$ this condition will hold when $\alpha L = n\pi$ or $\lambda = \alpha^2 = n^2 \pi^2 / L^2$, where $n = 1, 2, 3, \ldots$. These are the positive eigenvalues with corresponding eigenfunctions $\cos(n\pi x/L)$, $n = 1, 2, 3, \ldots$.

5. The eigenfunctions are $\cos \alpha_n x$ where $\cot \alpha_n = \alpha_n$. Thus

$$\| \cos \alpha_n x \|^2 = \int_0^1 \cos^2 \alpha_n x \, dx = \frac{1}{2} \int_0^1 (1 + \cos 2\alpha_n x) \, dx$$

$$= \frac{1}{2} \left(x + \frac{1}{2\alpha_n} \sin 2\alpha_n x \right) \Bigg|_0^1 = \frac{1}{2} \left(1 + \frac{1}{2\alpha_n} \sin 2\alpha_n \right)$$

$$= \frac{1}{2} \left[1 + \frac{1}{2\alpha_n} (2 \sin \alpha_n \cos \alpha_n) \right] = \frac{1}{2} \left[1 + \frac{1}{\alpha_n} \sin \alpha_n \cot \alpha_n \sin \alpha_n \right]$$

$$= \frac{1}{2} \left[1 + \frac{1}{\alpha_n} (\sin \alpha_n) \alpha_n (\sin \alpha_n) \right] = \frac{1}{2} \left(1 + \sin^2 \alpha_n \right).$$

7. (a) If $\lambda \le 0$ the initial conditions imply $y = 0$. For $\lambda = \alpha^2 > 0$ the general solution of the Cauchy-Euler differential equation is $y = c_1 \cos(\alpha \ln x) + c_2 \sin(\alpha \ln x)$. The condition $y(1) = 0$ implies $c_1 = 0$, so that $y = c_2 \sin(\alpha \ln x)$. The condition $y(5) = 0$ implies $\alpha \ln 5 = n\pi$, $n = 1, 2, 3, \ldots$. Thus, the eigenvalues are $n^2 \pi^2 / (\ln 5)^2$ for $n = 1, 2, 3, \ldots$, with corresponding eigenfunctions $\sin\left[(n\pi / \ln 5) \ln x \right]$.

(b) The self-adjoint form is

$$\frac{d}{dx}[xy'] + \frac{\lambda}{x} y = 0.$$

(c) An orthogonality relation is

$$\int_1^5 \frac{1}{x} \sin\left(\frac{m\pi}{\ln 5} \ln x \right) \sin\left(\frac{n\pi}{\ln 5} \ln x \right) dx = 0, \quad m \ne n.$$

9. We divide by lead coefficient of the differential equation to obtain the form $y'' + \left(-1 + \dfrac{1}{x} \right) y' + \dfrac{n}{x} y = 0$. The integrating factor is then

$$e^{\int (-1 + 1/x)\, dx} = e^{-x + \ln x} = e^{-x} e^{\ln x} = x e^{-x}.$$

Thus, the differential equation is

$$x e^{-x} y'' + (1 - x) e^{-x} y' + n e^{-x} y = 0$$

and the self-adjoint form is

$$\frac{d}{dx}\left[x e^{-x} y' \right] + n e^{-x} y = 0.$$

Identifying the weight function $p(x) = e^{-x}$ and noting that since $r(x) = x e^{-x}$, $r(0) = 0$ and $\lim_{x \to \infty} r(x) = 0$, we have the orthogonality relation

$$\int_0^\infty e^{-x} L_m(x) L_n(x)\, dx = 0, \quad m \ne n.$$

11. (a) The differential equation is

$$(1 + x^2) y'' + 2xy' + \frac{\lambda}{1 + x^2} y = 0.$$

Letting $x = \tan \theta$ we have $\theta = \tan^{-1} x$ and

$$\frac{dy}{dx} = \frac{dy}{d\theta} \frac{d\theta}{dx} = \frac{1}{1 + x^2} \frac{dy}{d\theta}$$

$$\frac{d^2 y}{dx^2} = \frac{d}{dx}\left[\frac{1}{1 + x^2} \frac{dy}{d\theta} \right] = \frac{1}{1 + x^2}\left(\frac{d^2 y}{d\theta^2} \frac{d\theta}{dx} \right) - \frac{2x}{(1 + x^2)^2} \frac{dy}{d\theta}$$

$$= \frac{1}{(1 + x^2)^2} \frac{d^2 y}{d\theta^2} - \frac{2x}{(1 + x^2)^2} \frac{dy}{d\theta}.$$

The differential equation can then be written in terms of $y(\theta)$ as

$$(1+x^2)\left[\frac{1}{(1+x^2)^2}\frac{d^2y}{d\theta^2}-\frac{2x}{(1+x^2)^2}\frac{dy}{d\theta}\right]+2x\left[\frac{1}{1+x^2}\frac{dy}{d\theta}\right]+\frac{\lambda}{1+x^2}y$$

$$=\frac{1}{1+x^2}\frac{d^2y}{d\theta^2}+\frac{\lambda}{1+x^2}y=0$$

or

$$\frac{d^2y}{d\theta^2}+\lambda y=0.$$

The boundary conditions become $y(0)=y(\pi/4)=0$. For $\lambda\le 0$ the only solution of the boundary-value problem is $y=0$. For $\lambda=\alpha^2>0$ the general solution of the differential equation is $y=c_1\cos\alpha\theta+c_2\sin\alpha\theta$. The condition $y(0)=0$ implies $c_1=0$ so $y=c_2\sin\alpha\theta$. Now the condition $y(\pi/4)=0$ implies $c_2\sin\alpha\pi/4=0$. For $c_2\ne 0$ this condition will hold when $\alpha\pi/4=n\pi$ or $\lambda=\alpha^2=16n^2$, where $n=1,2,3,\dots$. These are the eigenvalues with corresponding eigenfunctions $\sin 4n\theta=\sin(4n\tan^{-1}x)$, for $n=1,$ 2, 3, … .

(b) An orthogonality relation is

$$\int_0^1\frac{1}{x^2+1}\sin\left(4m\tan^{-1}x\right)\sin\left(4n\tan^{-1}x\right)dx=0,\quad m\ne n.$$

13. When $\lambda=0$ the differential equation is $r(x)y''+r'(x)y'=0$. By inspection we see that $y=1$ is a solution of the boundary-value problem. Thus, $\lambda=0$ is an eigenvalue.

15. (a) An orthogonality relation is

$$\int_0^1(x_m\cos x_mx-\sin x_mx)(x_n\cos x_nx-\sin x_nx)\,dx=0$$

where $x_m\ne x_n$ are positive solutions of $\tan x=x$.

(b) Referring to Problem 2 we use a CAS to compute

$$\int_0^1(4.4934\cos 4.4934x-\sin 4.4934x)(7.7253\cos 7.7253x-\sin 7.7253x)\,dx=-2.5650\times 10^{-4}$$

$$\approx 0.$$

11.5 Bessel and Legendre Series

1. Identifying $b=3$, we have $\alpha_1=1.2772$, $\alpha_2=2.3385$, $\alpha_3=3.3912$, and $\alpha_4=4.4412$.

3. The boundary condition indicates that we use (15) and (16) in the text. With $b = 2$ we obtain

$$c_i = \frac{2}{4J_1^2(2\alpha_i)} \int_0^2 xJ_0(\alpha_i x)\, dx \qquad \boxed{t = \alpha_i x, \quad dt = \alpha_i\, dx}$$

$$= \frac{1}{2J_1^2(2\alpha_i)} \cdot \frac{1}{\alpha_i^2} \int_0^{2\alpha_i} tJ_0(t)\, dt$$

$$= \frac{1}{2\alpha_i^2 J_1^2(2\alpha_i)} \int_0^{2\alpha_i} \frac{d}{dt}[tJ_1(t)]\, dt \qquad \text{[From (5) in the text]}$$

$$= \frac{1}{2\alpha_i^2 J_1^2(2\alpha_i)} tJ_1(t)\Big|_0^{2\alpha_i} = \frac{1}{\alpha_i J_1(2\alpha_i)}\,.$$

Thus

$$f(x) = \sum_{i=1}^{\infty} \frac{1}{\alpha_i J_1(2\alpha_i)} J_0(\alpha_i x).$$

5. The boundary condition indicates that we use (17) and (18) in the text. With $b = 2$ and $h = 1$ we obtain

$$c_i = \frac{2\alpha_i^2}{(4\alpha_i^2 + 1)J_0^2(2\alpha_i)} \int_0^2 xJ_0(\alpha_i x)\, dx \qquad \boxed{t = \alpha_i x, \quad dt = \alpha_i\, dx}$$

$$= \frac{2\alpha_i^2}{(4\alpha_i^2 + 1)J_0^2(2\alpha_i)} \cdot \frac{1}{\alpha_i^2} \int_0^{2\alpha_i} tJ_0(t)\, dt$$

$$= \frac{2}{(4\alpha_i^2 + 1)J_0^2(2\alpha_i)} \int_0^{2\alpha_i} \frac{d}{dt}[tJ_1(t)]\, dt \qquad \text{[From (5) in the text]}$$

$$= \frac{2}{(4\alpha_i^2 + 1)J_0^2(2\alpha_i)} tJ_1(t)\Big|_0^{2\alpha_i} = \frac{4\alpha_i J_1(2\alpha_i)}{(4\alpha_i^2 + 1)J_0^2(2\alpha_i)}\,.$$

Thus

$$f(x) = 4\sum_{i=1}^{\infty} \frac{\alpha_i J_1(2\alpha_i)}{(4\alpha_i^2 + 1)J_0^2(2\alpha_i)} J_0(\alpha_i x).$$

7. The boundary condition indicates that we use (17) and (18) in the text. With $n = 1$, $b = 4$, and $h = 3$ we obtain

$$c_i = \frac{2\alpha_i^2}{(16\alpha_i^2 - 1 + 9)J_1^2(4\alpha_i)} \int_0^4 xJ_1(\alpha_i x)5x\, dx \qquad \boxed{t = \alpha_i x, \quad dt = \alpha_i\, dx}$$

$$= \frac{5\alpha_i^2}{4(2\alpha_i^2 + 1)J_1^2(4\alpha_i)} \cdot \frac{1}{\alpha_i^3} \int_0^{4\alpha_i} t^2 J_1(t)\, dt$$

$$= \frac{5}{4\alpha_i(2\alpha_i^2 + 1)J_1^2(4\alpha_i)} \int_0^{4\alpha_i} \frac{d}{dt}[t^2 J_2(t)]\, dt \qquad \text{[From (5) in the text]}$$

$$= \frac{5}{4\alpha_i(2\alpha_i^2 + 1)J_1^2(4\alpha_i)} t^2 J_2(t)\Big|_0^{4\alpha_i} = \frac{20\alpha_i J_2(4\alpha_i)}{(2\alpha_i^2 + 1)J_1^2(4\alpha_i)}\,.$$

Thus

$$f(x) = 20 \sum_{i=1}^{\infty} \frac{\alpha_i J_2(4\alpha_i)}{(2\alpha_i^2 + 1)J_1^2(4\alpha_i)} J_1(\alpha_i x).$$

9. The boundary condition indicates that we use (19) and (20) in the text. With $b = 3$ we obtain

$$c_1 = \frac{2}{9} \int_0^3 x x^2 \, dx = \frac{2}{9} \left.\frac{x^4}{4}\right|_0^3 = \frac{9}{2},$$

$$c_i = \frac{2}{9J_0^2(3\alpha_i)} \int_0^3 x J_0(\alpha_i x) x^2 \, dx \qquad \boxed{t = \alpha_i x, \quad dt = \alpha_i \, dx}$$

$$= \frac{2}{9J_0^2(3\alpha_i)} \cdot \frac{1}{\alpha_i^4} \int_0^{3\alpha_i} t^3 J_0(t) \, dt$$

$$= \frac{2}{9\alpha_i^4 J_0^2(3\alpha_i)} \int_0^{3\alpha_i} t^2 \frac{d}{dt}\left[t J_1(t)\right] \, dt \qquad \boxed{\begin{array}{ll} u = t^2 & dv = \frac{d}{dt}\left[t J_1(t)\right]\, dt \\ du = 2t\, dt & v = t J_1(t) \end{array}}$$

$$= \frac{2}{9\alpha_i^4 J_0^2(3\alpha_i)} \left(\left.t^3 J_1(t)\right|_0^{3\alpha_i} - 2 \int_0^{3\alpha_i} t^2 J_1(t) \, dt \right).$$

With $n = 0$ in equation (6) in the text we have $J_0'(x) = -J_1(x)$, so the boundary condition $J_0'(3\alpha_i) = 0$ implies $J_1(3\alpha_i) = 0$. Then

$$c_i = \frac{2}{9\alpha_i^4 J_0^2(3\alpha_i)} \left(-2 \int_0^{3\alpha_i} \frac{d}{dt}\left[t^2 J_2(t)\right] dt \right) = \frac{2}{9\alpha_i^4 J_0^2(3\alpha_i)} \left(\left.-2t^2 J_2(t)\right|_0^{3\alpha_i} \right)$$

$$= \frac{2}{9\alpha_i^4 J_0^2(3\alpha_i)} \left[-18\alpha_i^2 J_2(3\alpha_i) \right] = \frac{-4 J_2(3\alpha_i)}{\alpha_i^2 J_0^2(3\alpha_i)}.$$

Thus

$$f(x) = \frac{9}{2} - 4 \sum_{i=1}^{\infty} \frac{J_2(3\alpha_i)}{\alpha_i^2 J_0^2(3\alpha_i)} J_0(\alpha_i x).$$

11. (a)

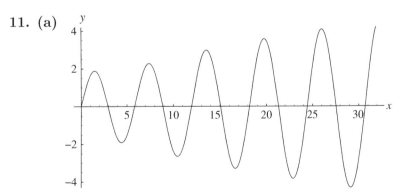

(b) Using **FindRoot** in *Mathematica* we find the roots $x_1 = 2.9496$, $x_2 = 5.8411$, $x_3 = 8.8727$, $x_4 = 11.9561$, and $x_5 = 15.0624$.

(c) Dividing the roots in part **(b)** by 4 we find the eigenvalues $\alpha_1 = 0.7374$, $\alpha_2 = 1.4603$, $\alpha_3 = 2.2182$, $\alpha_4 = 2.9890$, and $\alpha_5 = 3.7656$.

(d) The next five eigenvalues are $\alpha_6 = 4.5451$, $\alpha_7 = 5.3263$, $\alpha_8 = 6.1085$, $\alpha_9 = 6.8915$, and $\alpha_{10} = 7.6749$.

13. Since f is expanded as a series of Bessel functions, $J_1(\alpha_i x)$ and J_1 is an odd function, the series should represent an odd function.

15. We compute

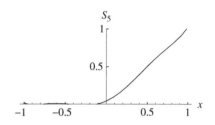

$$c_0 = \frac{1}{2} \int_0^1 x P_0(x)\,dx = \frac{1}{2} \int_0^1 x\,dx = \frac{1}{4}$$

$$c_1 = \frac{3}{2} \int_0^1 x P_1(x)\,dx = \frac{3}{2} \int_0^1 x^2\,dx = \frac{1}{2}$$

$$c_2 = \frac{5}{2} \int_0^1 x P_2(x)\,dx = \frac{5}{2} \int_0^1 \frac{1}{2}(3x^3 - x)\,dx = \frac{5}{16}$$

$$c_3 = \frac{7}{2} \int_0^1 x P_3(x)\,dx = \frac{7}{2} \int_0^1 \frac{1}{2}(5x^4 - 3x^2)\,dx = 0$$

$$c_4 = \frac{9}{2} \int_0^1 x P_4(x)\,dx = \frac{9}{2} \int_0^1 \frac{1}{8}(35x^5 - 30x^3 + 3x)\,dx = -\frac{3}{32}$$

$$c_5 = \frac{11}{2} \int_0^1 x P_5(x)\,dx = \frac{11}{2} \int_0^1 \frac{1}{8}(63x^6 - 70x^4 + 15x^2)\,dx = 0$$

$$c_6 = \frac{13}{2} \int_0^1 x P_6(x)\,dx = \frac{13}{2} \int_0^1 \frac{1}{16}(231x^7 - 315x^5 + 105x^3 - 5x)\,dx = \frac{13}{256}.$$

Thus

$$f(x) = \frac{1}{4} P_0(x) + \frac{1}{2} P_1(x) + \frac{5}{16} P_2(x) - \frac{3}{32} P_4(x) + \frac{13}{256} P_6(x) + \cdots.$$

The figure above is the graph of $S_5(x) = \frac{1}{4}P_0(x) + \frac{1}{2}P_1(x) + \frac{5}{16}P_2(x) - \frac{3}{32}P_4(x) + \frac{13}{256}P_6(x)$.

17. Using $\cos^2 \theta = \frac{1}{2}(\cos 2\theta + 1)$ we have

$$P_2(\cos\theta) = \frac{1}{2}(3\cos^2\theta - 1) = \frac{3}{2}\cos^2\theta - \frac{1}{2} = \frac{3}{4}(\cos 2\theta + 1) - \frac{1}{2} = \frac{3}{4}\cos 2\theta + \frac{1}{4} = \frac{1}{4}(3\cos 2\theta + 1).$$

19. If f is an even function on $(-1, 1)$ then

$$\int_{-1}^1 f(x) P_{2n}(x)\,dx = 2 \int_0^1 f(x) P_{2n}(x)\,dx$$

and

$$\int_{-1}^1 f(x) P_{2n+1}(x)\,dx = 0.$$

Thus

$$c_{2n} = \frac{2(2n)+1}{2} \int_{-1}^{1} f(x)P_{2n}(x)\,dx = \frac{4n+1}{2}\left(2\int_{0}^{1} f(x)P_{2n}(x)\,dx\right)$$

$$= (4n+1)\int_{0}^{1} f(x)P_{2n}(x)\,dx,$$

$c_{2n+1} = 0$, and

$$f(x) = \sum_{n=0}^{\infty} c_{2n}P_{2n}(x).$$

21. From (26) in Problem 19 in the text we find

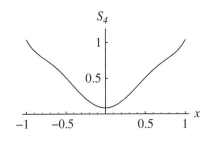

S_4

$$c_0 = \int_{0}^{1} xP_0(x)\,dx = \int_{0}^{1} x\,dx = \frac{1}{2}$$

$$c_2 = 5\int_{0}^{1} xP_2(x)\,dx = 5\int_{0}^{1} \frac{1}{2}(3x^3 - x)\,dx = \frac{5}{8}$$

$$c_4 = 9\int_{0}^{1} xP_4(x)\,dx = 9\int_{0}^{1} \frac{1}{8}(35x^5 - 30x^3 + 3x)\,dx = -\frac{3}{16}$$

and

$$c_6 = 13\int_{0}^{1} xP_6(x)\,dx = 13\int_{0}^{1} \frac{1}{16}(231x^7 - 315x^5 + 105x^3 - 5x)\,dx = \frac{13}{128}.$$

Hence, from (25) in the text,

$$f(x) = \frac{1}{2}P_0(x) + \frac{5}{8}P_2(x) - \frac{3}{16}P_4(x) + \frac{13}{128}P_6 + \cdots.$$

On the interval $-1 < x < 1$ this series represents the function $f(x) = |x|$.

23. Since there is a Legendre polynomial of any specified degree, every polynomial can be represented as a finite linear combination of Legendre polynomials.

Chapter 11 in Review

1. True, since $\int_{-\pi}^{\pi} (x^2 - 1)x^5 \, dx = 0$

3. cosine, since f is even

5. False; the Sturm-Liouville problem,

$$\frac{d}{dx}[r(x)y'] + \lambda p(x)y = 0, \qquad y'(a) = 0, \quad y'(b) = 0,$$

on the interval $[a, b]$, has eigenvalue $\lambda = 0$.

7. The Fourier series will converge to 1, the cosine series to 1, and the sine series to 0 at $x = 0$. Respectively, this is because the rule $(x^2 + 1)$ defining $f(x)$ determines a continuous function on $(-3, 3)$, the even extension of f to $(-3, 0)$ is continuous at 0, and the odd extension of f to $(-3, 0)$ approaches -1 as x approaches 0 from the left.

9. Since the coefficient of y in the differential equation is n^2, the weight function is the integrating factor

$$\frac{1}{a(x)}e^{\int (b/a)\,dx} = \frac{1}{1-x^2}e^{\int -\frac{x}{1-x^2}\,dx} = \frac{1}{1-x^2}e^{\frac{1}{2}\ln(1-x^2)} = \frac{\sqrt{1-x^2}}{1-x^2} = \frac{1}{\sqrt{1-x^2}}$$

on the interval $[-1, 1]$. The orthogonality relation is

$$\int_{-1}^{1} \frac{1}{\sqrt{1-x^2}} T_m(x)T_n(x) \, dx = 0, \quad m \neq n.$$

11. We know from a half-angle formula in trigonometry that $\cos^2 x = \frac{1}{2} + \frac{1}{2}\cos 2x$, which is a cosine series.

13. Since

$$a_0 = \int_{-1}^{0} (-2x) \, dx = 1,$$

$$a_n = \int_{-1}^{0} (-2x)\cos n\pi x \, dx = \frac{2}{n^2\pi^2}[(-1)^n - 1],$$

and

$$b_n = \int_{-1}^{0} (-2x)\sin n\pi x \, dx = \frac{2}{n\pi}(-1)^n$$

for $n = 1, 2, 3, \ldots$ we have

$$f(x) = \frac{1}{2} + \sum_{n=1}^{\infty}\left(\frac{2}{n^2\pi^2}[(-1)^n - 1]\cos n\pi x + \frac{2}{n\pi}(-1)^n \sin n\pi x\right).$$

15. (a) Since

$$a_0 = \frac{2}{1}\int_0^1 e^{-x}\,dx = 2(1 - e^{-1})$$

and

$$a_n = \frac{2}{1}\int_{-1}^1 e^{-x}\cos n\pi x\,dx = \frac{2}{1 + n^2\pi^2}[1 - (-1)^n e^{-1}],$$

for $n = 1, 2, 3, \ldots$, we have the cosine series

$$f(x) = 1 - e^{-1} + 2\sum_{n=1}^{\infty} \frac{1 - (-1)^n e^{-1}}{1 + n^2\pi^2}\cos n\pi x.$$

(b) Since

$$b_n = \frac{2}{1}\int_0^1 e^{-x}\sin n\pi x\,dx = \frac{2n\pi}{1 + n^2\pi^2}[1 - (-1)^n e^{-1}],$$

for $n = 1, 2, 3, \ldots$, we have the sine series

$$f(x) = 2\pi\sum_{n=1}^{\infty} \frac{n[1 - (-1)^n e^{-1}]}{1 + n^2\pi^2}\sin n\pi x.$$

17. The cosine series of f in Problem 15 converges to $F(x)$ on the interval $-1 < x < 1$ since F is the even extension of f to the interval.

19. For $\lambda = \alpha^2 > 0$ a general solution of the given differential equation is

$$y = c_1\cos(3\alpha\ln x) + c_2\sin(3\alpha\ln x)$$

and

$$y' = -\frac{3c_1\alpha}{x}\sin(3\alpha\ln x) + \frac{3c_2\alpha}{x}\cos(3\alpha\ln x).$$

Since $\ln 1 = 0$, the boundary condition $y'(1) = 0$ implies $c_2 = 0$. Therefore

$$y = c_1\cos(3\alpha\ln x).$$

Using $\ln e = 1$ we find that $y(e) = 0$ implies $c_1\cos 3\alpha = 0$ or $3\alpha = (2n-1)\pi/2$, for $n = 1$, 2, 3, ... The eigenvalues are $\lambda = \alpha^2 = (2n-1)^2\pi^2/36$ with corresponding eigenfunctions $\cos[(2n-1)\pi(\ln x)/2]$ for $n = 1, 2, 3, \ldots$.

21. The boundary condition indicates that we use (15) and (16) of Section 11.5 in the text. With $b = 4$ we obtain

$$
c_i = \frac{2}{16 J_1^2(4\alpha_i)} \int_0^4 x J_0(\alpha_i x) f(x)\, dx
$$

$$
= \frac{1}{8 J_1^2(4\alpha_i)} \int_0^2 x J_0(\alpha_i x)\, dx \qquad \boxed{t = \alpha_i x, \quad dt = \alpha_i\, dx}
$$

$$
= \frac{1}{8 J_1^2(4\alpha_i)} \cdot \frac{1}{\alpha_i^2} \int_0^{2\alpha_i} t J_0(t)\, dt
$$

$$
= \frac{1}{8 J_1^2(4\alpha_i)} \int_0^{2\alpha_i} \frac{d}{dt}[t J_1(t)]\, dt \qquad \text{[From (5) in 11.5 in the text]}
$$

$$
= \frac{1}{8 J_1^2(4\alpha_i)} t J_1(t) \Big|_0^{2\alpha_i} = \frac{J_1(2\alpha_i)}{4\alpha_i J_1^2(4\alpha_i)}.
$$

Thus

$$
f(x) = \frac{1}{4} \sum_{i=1}^{\infty} \frac{J_1(2\alpha_i)}{\alpha_i J_1^2(4\alpha_i)} J_0(\alpha_i x.
$$

23. (a)

$$
f_e(x) + f_o(x) = \frac{f(x) + f(-x) + f(x) - f(-x)}{2} = \frac{2 f(x)}{2} = f(x)
$$

(b)

$$
f_e(-x) = \frac{f(-x) + f(-(-x))}{2} = \frac{f(-x) + f(x)}{2} = f_e(x)
$$

$$
f_o(-x) = \frac{f(-x) - f(-(-x))}{2} = \frac{f(-x) - f(x)}{2} = -f_o(x)
$$

25. If we let $u = x + 2p$ then $dx = du$ and we get

$$
\int_0^a f(x)\, dx = \int_{2p}^{a+2p} f(u - 2p)\, du \quad \longleftarrow \quad \text{since } f \text{ is } 2p - \text{periodic, } f(u - 2p) = f(u)
$$

$$
= \int_{2p}^{a+2p} f(u)\, du
$$

$$
= \int_{2p}^{a+2p} f(x)\, dx
$$

Therefore we get

$$\int_a^{a+2p} f(x)\, dx = \int_a^{2p} f(x)\, dx + \int_{2p}^{a+2p} f(x)\, dx$$

$$= \int_a^{2p} f(x)\, dx + \left(\int_0^a f(x)\, dx \right)$$

$$= \int_0^a f(x)\, dx + \int_a^{2p} f(x)\, dx$$

$$= \int_0^{2p} f(x)\, dx$$

Chapter 12

Boundary-Value Problems in Rectangular Coordinates

12.1 | Separable Partial Differential Equations

1. Substituting $u(x, y) = X(x)Y(y)$ into the partial differential equation yields $X'Y = XY'$. Separating variables and using the separation constant $-\lambda$, where $\lambda \neq 0$, we obtain

$$\frac{X'}{X} = \frac{Y'}{Y} = -\lambda.$$

When $\lambda \neq 0$

$$X' + \lambda X = 0 \quad \text{and} \quad Y' + \lambda Y = 0$$

so that

$$X = c_1 e^{-\lambda x} \quad \text{and} \quad Y = c_2 e^{-\lambda y}.$$

A particular product solution of the partial differential equation is

$$u = XY = c_3 e^{-\lambda(x+y)}, \quad \lambda \neq 0.$$

When $\lambda = 0$ the differential equations become $X' = 0$ and $Y' = 0$, so in this case $X = c_4$, $Y = c_5$, and $u = XY = c_6$.

3. Substituting $u(x, y) = X(x)Y(y)$ into the partial differential equation yields $X'Y + XY' = XY$. Separating variables and using the separation constant $-\lambda$ we obtain

$$\frac{X'}{X} = \frac{Y - Y'}{Y} = -\lambda.$$

Then

$$X' + \lambda X = 0 \quad \text{and} \quad Y' - (1 + \lambda)Y = 0$$

so that

$$X = c_1 e^{-\lambda x} \quad \text{and} \quad Y = c_2 e^{(1+\lambda)y}.$$

A particular product solution of the partial differential equation is

$$u = XY = c_3 e^{y + \lambda(y-x)}.$$

5. Substituting $u(x, y) = X(x)Y(y)$ into the partial differential equation yields $xX'Y = yXY'$. Separating variables and using the separation constant $-\lambda$ we obtain

$$\frac{xX'}{X} = \frac{yY'}{Y} = -\lambda.$$

When $\lambda \neq 0$

$$xX' + \lambda X = 0 \qquad \text{and} \qquad yY' + \lambda Y = 0$$

so that

$$X = c_1 x^{-\lambda} \qquad \text{and} \qquad Y = c_2 y^{-\lambda}.$$

A particular product solution of the partial differential equation is

$$u = XY = c_3 (xy)^{-\lambda}.$$

When $\lambda = 0$ the differential equations become $X' = 0$ and $Y' = 0$, so in this case $X = c_4$, $Y = c_5$, and $u = XY = c_6$.

7. Substituting $u(x, y) = X(x)Y(y)$ into the partial differential equation yields $X''Y + X'Y' + XY'' = 0'$, which is not separable.

9. Substituting $u(x, t) = X(x)T(t)$ into the partial differential equation yields $kX''T - XT = XT'$. Separating variables and using the separation constant $-\lambda$ we obtain

$$\frac{kX'' - X}{X} = \frac{T'}{T} = -\lambda.$$

Then

$$X'' + \frac{\lambda - 1}{k}X = 0 \qquad \text{and} \qquad T' + \lambda T = 0.$$

The second differential equation implies $T(t) = c_1 e^{-\lambda t}$. For the first differential equation we consider three cases:

I. If $(\lambda - 1)/k = 0$ then $\lambda = 1$, $X'' = 0$, and $X(x) = c_2 x + c_3$, so

$$u = XT = e^{-t}(A_1 x + A_2).$$

II. If $(\lambda - 1)/k = -\alpha^2 < 0$, then $\lambda = 1 - k\alpha^2$, $X'' - \alpha^2 X = 0$, and $X(x) = c_4 \cosh \alpha x + c_5 \sinh \alpha x$, so

$$u = XT = (A_3 \cosh \alpha x + A_4 \sinh \alpha x)e^{-(1-k\alpha^2)t}.$$

III. If $(\lambda - 1)/k = \alpha^2 > 0$, then $\lambda = 1 + k\alpha^2$, $X'' + \alpha^2 X = 0$, and $X(x) = c_6 \cos \alpha x + c_7 \sin \alpha x$, so

$$u = XT = (A_5 \cos \alpha x + A_6 \sin \alpha x)e^{-(1+\lambda\alpha^2)t}.$$

11. Substituting $u(x,t) = X(x)T(t)$ into the partial differential equation yields $a^2 X''T = XT''$. Separating variables and using the separation constant $-\lambda$ we obtain

$$\frac{X''}{X} = \frac{T''}{a^2 T} = -\lambda.$$

Then

$$X'' + \lambda X = 0 \quad \text{and} \quad T'' + a^2 \lambda T = 0.$$

We consider three cases:

I. If $\lambda = 0$ then $X'' = 0$ and $X(x) = c_1 x + c_2$. Also, $T'' = 0$ and $T(t) = c_3 t + c_4$, so

$$u = XT = (c_1 x + c_2)(c_3 t + c_4).$$

II. If $\lambda = -\alpha^2 < 0$, then $X'' - \alpha^2 X = 0$, and $X(x) = c_5 \cosh \alpha x + c_6 \sinh \alpha x$. Also, $T'' - \alpha^2 a^2 T = 0$ and $T(t) = c_7 \cosh \alpha a t + c_8 \sinh \alpha a t$, so

$$u = XT = (c_5 \cosh \alpha x + c_6 \sinh \alpha x)(c_7 \cosh \alpha a t + c_8 \sinh \alpha a t).$$

III. If $\lambda = \alpha^2 > 0$, then $X'' + \alpha^2 X = 0$, and $X(x) = c_9 \cos \alpha x + c_{10} \sin \alpha x$. Also, $T'' + \alpha^2 a^2 T = 0$ and $T(t) = c_{11} \cos \alpha a t + c_{12} \sin \alpha a t$, so

$$u = XT = (c_9 \cos \alpha x + c_{10} \sin \alpha x)(c_{11} \cos \alpha a t + c_{12} \sin \alpha a t).$$

13. Substituting $u(x,y) = X(x)Y(y)$ into the partial differential equation yields $X''Y + XY'' = 0$. Separating variables and using the separation constant $-\lambda$ we obtain

$$-\frac{X''}{X} = \frac{Y''}{Y} = -\lambda.$$

Then

$$X'' - \lambda X = 0 \quad \text{and} \quad Y'' + \lambda Y = 0.$$

We consider three cases:

I. If $\lambda = 0$ then $X'' = 0$ and $X(x) = c_1 x + c_2$. Also, $Y'' = 0$ and $Y(y) = c_3 y + c_4$ so

$$u = XY = (c_1 x + c_2)(c_3 x + c_4).$$

II. If $\lambda = -\alpha^2 < 0$ then $X'' + \alpha^2 X = 0$ and $X(x) = c_5 \cos \alpha x + c_6 \sin \alpha x$. Also, $Y'' - \alpha^2 Y = 0$ and $Y(y) = c_7 \cosh \alpha y + c_8 \sinh \alpha y$ so

$$u = XY = (c_5 \cos \alpha x + c_6 \sin \alpha x)(c_7 \cosh \alpha y + c_8 \sinh \alpha y).$$

III. If $\lambda = \alpha^2 > 0$ then $X'' - \alpha^2 X = 0$ and $X(x) = c_9 \cosh \alpha x + c_{10} \sinh \alpha x$. Also, $Y'' + \alpha^2 Y = 0$ and $Y(y) = c_{11} \cos \alpha y + c_{12} \sin \alpha y$ so

$$u = XY = (c_9 \cosh \alpha x + c_{10} \sinh \alpha x)(c_{11} \cos \alpha y + c_{12} \sin \alpha y).$$

15. Substituting $u(x, y) = X(x)Y(y)$ into the partial differential equation yields $X''Y + XY'' = XY$. Separating variables and using the separation constant $-\lambda$ we obtain

$$\frac{X''}{X} = \frac{Y - Y''}{Y} = -\lambda.$$

Then

$$X'' + \lambda X = 0 \quad \text{and} \quad Y'' - (1 + \lambda)Y = 0.$$

We consider three cases:

I. If $\lambda = 0$ then $X'' = 0$ and $X(x) = c_1 x + c_2$. Also $Y'' - Y = 0$ and $Y(y) = c_3 \cosh y + c_4 \sinh y$ so

$$u = XY = (c_1 x + c_2)(c_3 \cosh y + c_4 \sinh y).$$

II. If $\lambda = -\alpha^2 < 0$ then $X'' - \alpha^2 X = 0$ and $Y'' + (\alpha^2 - 1)Y = 0$. The solution of the first differential equation is $X(x) = c_5 \cosh \alpha x + c_6 \sinh \alpha x$. The solution of the second differential equation depends on the nature of $\alpha^2 - 1$. We consider three cases:

    (i) If $\alpha^2 - 1 = 0$, or $\alpha^2 = 1$, then $Y(y) = c_7 y + c_8$ and

$$u = XY = (c_5 \cosh \alpha x + c_6 \sinh \alpha x)(c_7 y + c_8).$$

    (ii) If $\alpha^2 - 1 < 0$, or $0 < \alpha^2 < 1$, then $Y(y) = c_9 \cosh \sqrt{1 - \alpha^2}\, y + c_{10} \sinh \sqrt{1 - \alpha^2}\, y$ and

$$u = XY = (c_5 \cosh \alpha x + c_6 \sinh \alpha x)\left(c_9 \cosh \sqrt{1 - \alpha^2}\, y + c_{10} \sinh \sqrt{1 - \alpha^2}\, y\right).$$

    (iii) If $\alpha^2 - 1 > 0$, or $\alpha^2 > 1$, then $Y(y) = c_{11} \cos \sqrt{\alpha^2 - 1}\, y + c_{12} \sin \sqrt{\alpha^2 - 1}\, y$ and

$$u = XY = (c_5 \cosh \alpha x + c_6 \sinh \alpha x)\left(c_{11} \cos \sqrt{\alpha^2 - 1}\, y + c_{12} \sin \sqrt{\alpha^2 - 1}\, y\right).$$

III. If $\lambda = \alpha^2 > 0$, then $X'' + \alpha^2 X = 0$ and $X(x) = c_{13} \cos \alpha x + c_{14} \sin \alpha x$. Also,
$Y'' - (1 + \alpha^2)Y = 0$ and $Y(y) = c_{15} \cosh \sqrt{1 + \alpha^2}\, y + c_{16} \sinh \sqrt{1 + \alpha^2}\, y$ so

$$u = XY = (c_{13} \cos \alpha x + c_{14} \sin \alpha x)\left(c_{15} \cosh \sqrt{1 + \alpha^2}\, y + c_{16} \sinh \sqrt{1 + \alpha^2}\, y\right).$$

17. Identifying $A = B = C = 1$, we compute $B^2 - 4AC = -3 < 0$. The equation is elliptic.

19. Identifying $A = 1$, $B = 6$, and $C = 9$, we compute $B^2 - 4AC = 0$. The equation is parabolic.

21. Identifying $A = 1$, $B = -9$, and $C = 0$, we compute $B^2 - 4AC = 81 > 0$. The equation is hyperbolic.

23. Identifying $A = 1$, $B = 2$, and $C = 1$, we compute $B^2 - 4AC = 0$. The equation is parabolic.

25. Identifying $A = a^2$, $B = 0$, and $C = -1$, we compute $B^2 - 4AC = 4a^2 > 0$. The equation is hyperbolic.

27. Substituting $u(r, t) = R(r)T(t)$ into the partial differential equation yields

$$k\left(R''T + \frac{1}{r}R'T\right) = RT'.$$

Separating variables and using the separation constant $-\lambda$ we obtain

$$\frac{rR'' + R'}{rR} = \frac{T'}{kT} = -\lambda.$$

Then

$$rR'' + R' + \lambda rR = 0 \qquad \text{and} \qquad T' + \lambda kT = 0.$$

Letting $\lambda = \alpha^2$ and writing the first equation as $r^2 R'' + rR' = \alpha^2 r^2 R = 0$ we see that it is a parametric Bessel equation of order 0. As discussed in Chapter 5 of the text, it has solution $R(r) = c_1 J_0(\alpha r) + c_2 Y_0(\alpha r)$. Since a solution of $T' + \alpha^2 kT$ is $T(t) = e^{-k\alpha^2 t}$, we see that a solution of the partial differential equation is

$$u = RT = e^{-k\alpha^2 t}[c_1 J_0(\alpha r) + c_2 Y_0(\alpha r)].$$

29. For $u = A_1 + B_1 x$ we compute $\partial^2 u/\partial x^2 = 0 = \partial u/\partial y$. Then $\partial^2 u/\partial x^2 = 4\,\partial u/\partial y$.

For $u = A_2 e^{\alpha^2 y} \cos 2\alpha x + B_2 e^{\alpha^2 y} \sin 2\alpha x$ we compute

$$\frac{\partial u}{\partial x} = 2\alpha A_2 e^{\alpha^2 y} \sinh 2\alpha x + 2\alpha B_2 e^{\alpha^2 y} \cosh 2\alpha x$$

$$\frac{\partial^2 u}{\partial x^2} = 4\alpha^2 A_2 e^{\alpha^2 y} \cosh 2\alpha x + 4\alpha^2 B_2 e^{\alpha^2 y} \sinh 2\alpha x$$

and

$$\frac{\partial u}{\partial y} = \alpha^2 A_2 e^{\alpha^2 y} \cosh 2\alpha x + \alpha^2 B_2 e^{\alpha^2 y} \sinh 2\alpha x.$$

Then $\partial^2 u/\partial x^2 = 4\,\partial u/\partial y$.

For $u = A_3 e^{-\alpha^2 y} \cosh 2\alpha x + B_3 e^{-\alpha^2 y} \sinh 2\alpha x$ we compute

$$\frac{\partial u}{\partial x} = -2\alpha A_3 e^{-\alpha^2 y} \sin 2\alpha x + 2\alpha B_3 e^{-\alpha^2 y} \cos 2\alpha x$$

$$\frac{\partial^2 u}{\partial x^2} = -4\alpha^2 A_3 e^{-\alpha^2 y} \cos 2\alpha x - 4\alpha^2 B_3 e^{-\alpha^2 y} \sin 2\alpha x$$

and

$$\frac{\partial u}{\partial y} = -\alpha^2 A_3 e^{-\alpha^2 y} \cos 2\alpha x - \alpha^2 B_3 e^{-\alpha^2 y} \sin 2\alpha x.$$

Then $\partial^2 u/\partial x^2 = 4\,\partial u/\partial y$.

31. Assuming $u(x,y) = X(x)Y(y)$ and substituting into $\partial^2 u/\partial x^2 - u = 0$ we get $X''Y - XY = 0$ or $Y(X'' - X) = 0$. This implies $X(x) = c_1 e^x$ or $X(x) = c_2 e^{-x}$. For these choices of X, Y can be any function of y. Two solutions of the partial differential equation are then

$$u_1(x,y) = A(y)e^x \quad \text{and} \quad u_2(x,y) = B(y)e^{-x}.$$

Since the partial differential equation is linear and homogeneous the superposition principle indicates that another solution is

$$u(x,y) = u_1(x,y) + u_2(x,y) = A(y)e^x + B(y)e^{-x}.$$

12.2   Classical PDEs and Boundary-Value Problems

1. $k\dfrac{\partial^2 u}{\partial x^2} = \dfrac{\partial u}{\partial t}, \quad 0 < x < L, \ t > 0$

$u(0,t) = 0, \quad \left.\dfrac{\partial u}{\partial x}\right|_{x=L} = 0, \quad t > 0$

$u(x,0) = f(x), \quad 0 < x < L$

3. $k\dfrac{\partial^2 u}{\partial x^2} = \dfrac{\partial u}{\partial t}, \quad 0 < x < L, \ t > 0$

$u(0,t) = 100, \quad \left.\dfrac{\partial u}{\partial x}\right|_{x=L} = -hu(L,t), \quad t > 0$

$u(x,0) = f(x), \quad 0 < x < L$

5. $k\dfrac{\partial^2 u}{\partial x^2} - hu = \dfrac{\partial u}{\partial t}, \quad 0 < x < L, \ t > 0, \ h$ a constant

$u(0,t) = \sin\dfrac{\pi t}{L}, \quad u(L,t) = 0, \quad t > 0$

$u(x,0) = f(x), \quad 0 < x < L$

7. $a^2\dfrac{\partial^2 u}{\partial x^2} = \dfrac{\partial^2 u}{\partial t^2}, \quad 0 < x < L, \ t > 0$

$u(0,t) = 0, \quad u(L,t) = 0, \quad t > 0$

$u(x,0) = x(L - x), \quad \left.\dfrac{\partial u}{\partial t}\right|_{t=0} = 0, \quad 0 < x < L$

9. $a^2\dfrac{\partial^2 u}{\partial x^2} - 2\beta\dfrac{\partial u}{\partial t} = \dfrac{\partial^2 u}{\partial t^2}, \quad 0 < x < L, \ t > 0$

$u(0,t) = 0, \quad u(L,t) = \sin \pi t, \quad t > 0$

$u(x,0) = f(x), \quad \left.\dfrac{\partial u}{\partial t}\right|_{t=0} = 0, \quad 0 < x < L$

11. $\dfrac{\partial^2 u}{\partial x^2} + \dfrac{\partial^2 u}{\partial y^2} = 0,\quad 0 < x < 4,\ 0 < y < 2$

$\left.\dfrac{\partial u}{\partial x}\right|_{x=0} = 0,\quad u(4, y) = f(y),\quad 0 < y < 2$

$\left.\dfrac{\partial u}{\partial y}\right|_{y=0} = 0,\quad u(x, 2) = 0,\quad 0 < x < 4$

12.3 | Heat Equation

1. Using $u = XT$ and $-\lambda$ as a separation constant we obtain

$$X'' + \lambda X = 0,$$
$$X(0) = 0,$$
$$X(L) = 0,$$

and

$$T' + k\lambda T = 0.$$

This leads to

$$X = c_1 \sin \frac{n\pi}{L} x \qquad \text{and} \qquad T = c_2 e^{-kn^2\pi^2 t/L^2}$$

for $n = 1, 2, 3, \ldots$ so that

$$u = \sum_{n=1}^{\infty} A_n e^{-kn^2\pi^2 t/L^2} \sin \frac{n\pi}{L} x.$$

Imposing

$$u(x, 0) = \sum_{n=1}^{\infty} A_n \sin \frac{n\pi}{L} x$$

gives

$$A_n = \frac{2}{L} \int_0^{L/2} \sin \frac{n\pi}{L} x\, dx = \frac{2}{n\pi}\left(1 - \cos \frac{n\pi}{2}\right)$$

for $n = 1, 2, 3, \ldots$ so that

$$u(x, t) = \frac{2}{\pi} \sum_{n=1}^{\infty} \frac{1 - \cos \frac{n\pi}{2}\, n}{} e^{-kn^2\pi^2 t/L^2} \sin \frac{n\pi}{L} x.$$

3. Using $u = XT$ and $-\lambda$ as a separation constant we obtain

$$X'' + \lambda X = 0,$$
$$X'(0) = 0,$$
$$X'(L) = 0,$$

and

$$T' + k\lambda T = 0.$$

This leads to

$$X = c_1 \cos \frac{n\pi}{L} x \qquad \text{and} \qquad T = c_2 e^{-kn^2\pi^2 t/L^2}$$

for $n = 0, 1, 2, \ldots$ ($\lambda = 0$ is an eigenvalue in this case) so that

$$u = \sum_{n=0}^{\infty} A_n e^{-kn^2\pi^2 t/L^2} \cos \frac{n\pi}{L} x.$$

Imposing

$$u(x, 0) = f(x) = A_0 + \sum_{n=1}^{\infty} A_n \cos \frac{n\pi}{L} x$$

gives

$$u(x, t) = \frac{1}{L} \int_0^L f(x)\, dx + \frac{2}{L} \sum_{n=1}^{\infty} \left(\int_0^L f(x) \cos \frac{n\pi}{L} x\, dx \right) e^{-kn^2\pi^2 t/L^2} \cos \frac{n\pi}{L} x.$$

5. Using $u = XT$ and $-\lambda$ as a separation constant leads to

$$X'' + \lambda X = 0,$$
$$X'(0) = 0,$$
$$X'(L) = 0,$$

and

$$T' + (h + k\lambda)T = 0.$$

Then

$$X = c_1 \cos \frac{n\pi}{L} x \qquad \text{and} \qquad T = c_2 e^{-ht - kn^2\pi^2 t/L^2}$$

for $n = 0, 1, 2, \ldots$ ($\lambda = 0$ is an eigenvalue in this case) so that

$$u = A_0 e^{-ht} + e^{-ht} \sum_{n=1}^{\infty} A_n e^{-kn^2\pi^2 t/L^2} \cos \frac{n\pi}{L} x.$$

Imposing

$$u(x, 0) = f(x) = \sum_{n=0}^{\infty} A_n \cos \frac{n\pi}{L} x$$

gives

$$u(x, t) = \frac{e^{-ht}}{L} \int_0^L f(x)\, dx + \frac{2e^{-ht}}{L} \sum_{n=1}^{\infty} \left(\int_0^L f(x) \cos \frac{n\pi}{L} x\, dx \right) e^{-kn^2\pi^2 t/L^2} \cos \frac{n\pi}{L} x.$$

7. Using $-\lambda$ as the separation constant implies $X'' + \lambda X = 0$ and $T' + k\lambda T = 0$. The boundary conditions are then $X(-L) = X(L)$ and $X'(-L) = X'(L)$.

For $\lambda = 0$, $X(x) = c_1 + c_2 x$. The condition $X(-L) = X(L)$ implies $c_2 = 0$. Therefore an eigenfunction is $X(x) = c_1 \neq 0$. The boundary condition $X'(-L) = X'(L)$ is automatically satisfied.

For $\lambda = -\alpha^2 < 0$, $X(x) = c_3 \cosh \alpha x + c_4 \sinh \alpha x$. From the condition $X(-L) = X(L)$ we obtain

$$c_3 \cosh \alpha L - c_4 \sinh \alpha L = c_3 \cosh \alpha L + c_4 \sinh \alpha L \quad \text{so} \quad 2c_4 \sinh \alpha L = 0.$$

This implies $c_4 = 0$ and so $X(x) = c_3 \cosh \alpha x$. The boundary condition $X'(-L) = X'(L)$ implies

$$-c_3 \alpha \sinh \alpha L = c_3 \sinh \alpha L \quad \text{so} \quad 2\alpha c_3 \sinh \alpha L = 0.$$

Therefore $c_3 = 0$ and $X(x) = 0$.

For $\lambda = \alpha^2 > 0$, $X(x) = c_5 \cos \alpha x + c_6 \sin \alpha x$. The boundary condition $X(-L) = X(L)$ implies

$$c_5 \cos \alpha L - c_6 \sin \alpha L = c_5 \cos \alpha L + c_6 \sin \alpha L \quad \text{so} \quad 2c_6 \sin \alpha L = 0.$$

If $c_6 \neq 0$, then $\alpha = n\pi/L$ for $n = 1, 2, \ldots$. The boundary condition $X'(-L) = X'(L)$ implies

$$-c_5 \alpha \sin \alpha L + c_6 \alpha \cos \alpha L = \alpha c_5 \sin \alpha L + c_6 \alpha \cos \alpha L \quad \text{so} \quad 2\alpha c_5 \sin \alpha L = 0.$$

Then, for $c_5 \neq 0$,

$$\sin \alpha L = 0 \quad \text{and} \quad \alpha = n\pi/L, \; n = 1, 2, \ldots .$$

Thus the coefficients c_5 and c_6 are arbitrary but nonzero. Therefore the eigenvalues are $\lambda_n = (n\pi/L)^2$, $n = 0, 1, 2, \ldots$, and the corresponding eigenfunctions are

$$1, \; \cos \frac{n\pi}{L} x, \; \sin \frac{n\pi}{L} x \quad \text{for} \quad n = 0, 1, 2, \ldots .$$

Forming product solutions with

$$T(t) = \begin{cases} c_7, & \lambda = 0 \\ c_7 e^{-k(n\pi/L)^2 t}, & \lambda > 0, \end{cases}$$

relabeling constants, and summing gives

$$u(x, t) = A_0 + \sum_{k=1}^{\infty} e^{-k(n\pi/L)^2 t} \left(A_n \cos \frac{n\pi}{L} x + B_n \sin \frac{n\pi}{L} x \right).$$

When $t = 0$ we get the full Fourier series of f on $(-L, L)$,

$$f(x) = A_0 + \sum_{k=1}^{\infty} \left(A_n \cos \frac{n\pi}{L} x + B_n \sin \frac{n\pi}{L} x \right).$$

The coefficients are then $A_0 = \frac{1}{2}a_0$, $A_n = a_n$, $B_n = b_n$, or

$$A_0 = \frac{1}{2L} \int_{-L}^{L} f(x)\,dx, \quad A_n = \frac{1}{L} \int_{-L}^{L} f(x) \cos \frac{n\pi}{L}\,dx, \quad B_n = \frac{1}{L} \int_{-L}^{L} f(x) \sin \frac{n\pi}{L}\,dx.$$

9.

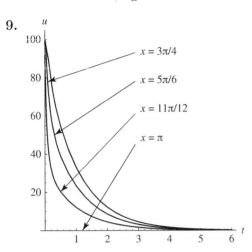

$x = 3\pi/4$

$x = 5\pi/6$

$x = 11\pi/12$

$x = \pi$

12.4 Wave Equation

For Problems 1–10, recall that the solution to the wave equation is given by

$$u(x,t) = \sum_{n=1}^{\infty} \left(A_n \cos \frac{n\pi a}{L} t + B_n \sin \frac{n\pi a}{L} t \right) \sin \frac{n\pi}{L} x$$

where $A_n = \dfrac{2}{L} \displaystyle\int_0^L f(x) \sin \dfrac{n\pi}{L} x\,dx$ *and* $B_n = \dfrac{2}{n\pi a} \displaystyle\int_0^L g(x) \sin \dfrac{n\pi}{L} x\,dx.$

Here, $f(x) = u(x,0)$ *and* $g(x) = \left. \dfrac{\partial u}{\partial t} \right|_{t=0}.$

1. By the discussion in Section 12.4, pages 474 and 475,

$$u(x,t) = \sum_{n=1}^{\infty} \left(A_n \cos \frac{n\pi a}{L} t + B_n \sin \frac{n\pi a}{L} t \right) \sin \frac{n\pi}{L} x$$

The coefficients are given by

$$A_n = \frac{2}{L} \int_0^L f(x) \sin \frac{n\pi}{L} x\,dx = \frac{2}{L} \int_0^L \left(\frac{1}{4} x(L-x) \right) \sin \frac{n\pi}{L} x\,dx = \frac{L^2 \left[1 - (-1)^n \right]}{n^3 \pi^3}$$

$$B_n = \frac{2}{n\pi a} \int_0^L g(x) \sin \frac{n\pi}{L} x\,dx = 0$$

Therefore the solution to the problem is

$$u(x,t) = \sum_{n=1}^{\infty} \left(\frac{L^2 \left[1 - (-1)^n \right]}{n^3 \pi^3} \cdot \cos \frac{n\pi a}{L} t \right) \sin \frac{n\pi}{L} x$$

3. By the discussion in Section 12.4, pages 474 and 475,

$$u(x,t) = \sum_{n=1}^{\infty} (A_n \cos nat + B_n \sin nat) \sin nx$$

The coefficients are given by

$$A_n = \frac{2}{\pi} \int_0^{\pi} f(x) \sin nx \, dx = 0$$

$$B_n = \frac{2}{n\pi a} \int_0^{\pi} g(x) \sin nx \, dx = \frac{1}{n\pi a} \int_0^{\pi} \sin x \sin nx \, dx = 0 \quad \text{for } n = 2, 3, 4, \ldots$$

$$B_1 = \frac{2}{\pi a} \int_0^{\pi} \sin x \sin x \, dx = \frac{1}{a}$$

Therefore the solution to the problem is

$$u(x,t) = B_1 \sin at \sin x = \frac{1}{a} \sin at \sin x$$

5. By the discussion in Section 12.4, pages 474 and 475,

$$u(x,t) = \sum_{n=1}^{\infty} (A_n \cos n\pi at + B_n \sin n\pi at) \sin n\pi x$$

The coefficients are given by

$$A_n = \frac{2}{1} \int_0^1 f(x) \sin n\pi x \, dx = \frac{2}{1} \int_0^1 x(1-x) \sin n\pi x \, dx = \frac{4\left[1-(-1)^n\right]}{n^3 \pi^3}$$

$$B_n = \frac{2}{n\pi a} \int_0^1 g(x) \sin n\pi x \, dx = \frac{2}{n\pi a} \int_0^1 x(1-x) \sin n\pi x \, dx = \frac{4\left[1-(-1)^n\right]}{an^4 \pi^4}$$

Therefore the solution to the problem is

$$u(x,t) = \sum_{n=1}^{\infty} \left(\frac{4\left[1-(-1)^n\right]}{n^3 \pi^3} \cos n\pi at + \frac{4\left[1-(-1)^n\right]}{an^4 \pi^4} \sin n\pi at \right) \sin n\pi x$$

For Problems 7–9, we have $g(x) = 0$ because the string is released from the rest, so $B_n = 0$. So, our general solution is of the form

$$u(x,t) = \sum_{n=1}^{\infty} A_n \cos \frac{n\pi at}{L} \sin \frac{n\pi x}{L}$$

where $A_n = \frac{2}{L} \int_0^L f(x) \sin \frac{n\pi}{L} x \, dx$ as before.

7. By the discussion in Section 12.4, pages 474 and 475,

$$u(x,t) = \sum_{n=1}^{\infty} \left(A_n \cos \frac{n\pi a}{L} t + B_n \sin \frac{n\pi a}{L} t \right) \sin \frac{n\pi}{L} x$$

The coefficients are

$$A_n = \frac{2}{L} \int_0^L f(x) \sin \frac{n\pi}{L} x \, dx = \frac{2}{L} \left[\int_0^{L/2} \frac{2hx}{L} \sin \frac{n\pi}{L} x \, dx + \int_{L/2}^L 2h \left(1 - \frac{x}{L} \right) \sin \frac{n\pi}{L} x \, dx \right]$$

$$= \frac{8h \sin \left(\frac{n\pi}{2} \right)}{n^2 \pi^2}$$

$$B_n = \frac{2}{n\pi a} \int_0^L g(x) \sin \frac{n\pi}{L} x \, dx = 0$$

Therefore the solution to the problem is

$$u(x,t) = \sum_{n=1}^{\infty} \left(\frac{8h \sin \left(\frac{n\pi}{2} \right)}{n^2 \pi^2} \cos \frac{n\pi a}{L} t \right) \sin \frac{n\pi}{L} x$$

9. By the discussion in Section 12.4, pages 474 and 475,

$$u(x,t) = \sum_{n=1}^{\infty} \left(A_n \cos \frac{n\pi a}{L} t + B_n \sin \frac{n\pi a}{L} t \right) \sin \frac{n\pi}{L} x$$

The coefficients are

$$A_n = \frac{2}{L} \int_0^L f(x) \sin \frac{n\pi}{L} x \, dx$$

$$= \frac{2}{L} \left[\int_0^{L/3} \frac{3hx}{L} \sin \frac{n\pi}{L} x \, dx + \int_{L/3}^{2L/3} h \sin \frac{n\pi}{L} x \, dx + \int_{2L/3}^L 3h \left(1 - \frac{x}{L} \right) \sin \frac{n\pi}{L} x \, dx \right]$$

$$= \frac{6h \left[\sin \left(\frac{2n\pi}{3} \right) + \sin \left(\frac{n\pi}{3} \right) \right]}{n^2 \pi^2}$$

$$B_n = \frac{2}{n\pi a} \int_0^L g(x) \sin \frac{n\pi}{L} x \, dx = 0$$

Therefore the solution to the problem is

$$u(x,t) = \sum_{n=1}^{\infty} \left(\frac{6h \left[\sin \left(\frac{2n\pi}{3} \right) + \sin \left(\frac{n\pi}{3} \right) \right]}{n^2 \pi^2} \cos \frac{n\pi a}{L} t \right) \sin \frac{n\pi}{L} x$$

Using the trigonometric identity

$$\sin \frac{2n\pi}{3} = \sin \left(n\pi - \frac{n\pi}{3} \right) = \sin n\pi \cos \frac{n\pi}{3} - \cos n\pi \sin \frac{n\pi}{3} = -(-1)^n \sin \frac{n\pi}{3}$$

we have

$$u(x,t) = \frac{6h}{\pi^2} \sum_{n=1}^{\infty} \frac{1 - (-1)^n}{n^2} \sin \frac{n\pi}{3} \cos \frac{n\pi a}{L} t \sin \frac{n\pi}{L} x.$$

11. By the discussion in Section 12.4, pages 474 and 475, we get

$$\begin{cases} X'' + \lambda X = 0, \\ X'(0) = 0, \\ X'(L) = 0, \end{cases}$$

For $\lambda = 0$,

$$\begin{cases} X'' = 0, \\ X'(0) = 0, \\ X'(L) = 0, \end{cases}$$

Thus $X = c_1 + c_2 x$ and therefore $X = c_1$. Moreover $\{T'' = 0$ implies $T = c_3 + c_4 t$.

Therefore we have $u_0(x, t) = c_1(c_3 + c_4 t) = A_0 + B_0 t$. Using the given conditions and the results from Problem 3 in Section 12.5, the rest of the eigenvalues are $\lambda_n = n^2 \pi^2 / L^2$ with corresponding eigenfunctions $X_n = \cos \frac{n\pi}{L} x$, $n = 1, 2, 3, \ldots$ therefore $T = c_3 \cos \frac{n\pi a}{L} t + c_4 \sin \frac{n\pi a}{L} t$ and we now have

$$u(x, t) = (A_0 + B_0 t) + \sum_{n=1}^{\infty} \left(A_n \cos \frac{n\pi a}{L} t + B_n \sin \frac{n\pi a}{L} t \right) \cos \frac{n\pi}{L} x$$

To complete the problem we need only to find the coefficients. At $t = 0$ we have

$$u(x, 0) = x = A_0 + \sum_{n=1}^{\infty} A_n \cos \frac{n\pi}{L} x \qquad \longleftarrow \text{ A Fourier cosine series for } x$$

where

$$A_0 = \frac{1}{L} \int_0^L x \, dx = \frac{L}{2} \quad \text{and} \quad A_n = \frac{2}{L} \int_0^L x \cos \frac{n\pi}{L} x \, dx = \frac{2L \left[(-1)^n - 1 \right]}{n^2 \pi^2}$$

Similarly at $t = 0$,

$$u_t(x, 0) = 0 = B_0 + \sum_{n=1}^{\infty} \left(B_n \cdot \frac{n\pi a}{L} \right) \cos \frac{n\pi}{L} x$$

and so we get $B_n = 0$ for $n = 0, 1, 2, 3, \ldots$ therefore the solution to the problem is

$$u(x, t) = \frac{L}{2} + \sum_{n=1}^{\infty} \left(\frac{2L \left[(-1)^n - 1 \right]}{n^2 \pi^2} \cos \frac{n\pi a}{L} t \right) \cos \frac{n\pi}{L} x$$

13. From Figure 12.4.6 we expect $u(L/2, t) = 0$ for $t \geq 0$. To prove this we have from Problem 10,

$$u(L/2, t) = \frac{18h}{\pi^2} \sum_{n=1}^{\infty} \frac{1 + (-1)^n}{n^2} \sin \frac{n\pi}{3} \cos \frac{n\pi a}{L} t \sin \frac{n\pi}{2}$$

$$= \begin{cases} 0, & n = 1, 3, 5, \ldots, 2m + 1, \ldots, \text{ we have } 1 + (-1)^{2m+1} = 1 - 1 = 0 \\ \\ 0, & n = 2, 4, 6, \ldots, 2m, \ldots, \text{ we have } \sin \frac{n\pi}{2} = \sin m\pi = 0. \end{cases}$$

15. Using $u = XT$ and $-\lambda$ as a separation constant we obtain

$$X'' + \lambda X = 0,$$
$$X(0) = 0,$$
$$X(\pi) = 0,$$

and

$$T'' + 2\beta T' + \lambda T = 0,$$
$$T'(0) = 0.$$

Solving the differential equations we get

$$X = c_1 \sin nx + c_2 \cos nx \qquad \text{and} \qquad T = e^{-\beta t}\left(c_3 \cos \sqrt{n^2 - \beta^2}\, t + c_4 \sin \sqrt{n^2 - \beta^2}\, t \right)$$

The boundary conditions on X imply $c_2 = 0$ so

$$X = c_1 \sin nx \qquad \text{and} \qquad T = e^{-\beta t}\left(c_3 \cos \sqrt{n^2 - \beta^2}\, t + c_4 \sin \sqrt{n^2 - \beta^2}\, t \right)$$

and

$$u = \sum_{n=1}^{\infty} e^{-\beta t}\left(A_n \cos \sqrt{n^2 - \beta^2}\, t + B_n \sin \sqrt{n^2 - \beta^2}\, t \right) \sin nx.$$

Imposing

$$u(x,0) = f(x) = \sum_{n=1}^{\infty} A_n \sin nx$$

and

$$u_t(x,0) = 0 = \sum_{n=1}^{\infty} \left(B_n \sqrt{n^2 - \beta^2} - \beta A_n \right) \sin nx$$

gives

$$u(x,t) = e^{-\beta t} \sum_{n=1}^{\infty} A_n \left(\cos \sqrt{n^2 - \beta^2}\, t + \frac{\beta}{\sqrt{n^2 - \beta^2}} \sin \sqrt{n^2 - \beta^2}\, t \right) \sin nx,$$

where

$$A_n = \frac{2}{\pi} \int_0^{\pi} f(x) \sin nx \, dx.$$

17. Separating variables in the partial differential equation and using the separation constant $-\lambda = \alpha^4$ gives

$$\frac{X^{(4)}}{X} = -\frac{T''}{a^2 T} = \alpha^4$$

so that

$$X^{(4)} - \alpha^4 X = 0$$
$$T'' + a^2 \alpha^4 T = 0$$

and

$$X = c_1 \cosh \alpha x + c_2 \sinh \alpha x + c_3 \cos \alpha x + c_4 \sin \alpha x$$
$$T = c_5 \cos a\alpha^2 t + c_6 \sin a\alpha^2 t.$$

The boundary conditions translate into $X(0) = X(L) = 0$ and $X''(0) = X''(L) = 0$. From $X(0) = X''(0) = 0$ we find $c_1 = c_3 = 0$. From

$$X(L) = c_2 \sinh \alpha L + c_4 \sin \alpha L = 0$$
$$X''(L) = \alpha^2 c_2 \sinh \alpha L - \alpha^2 c_4 \sin \alpha L = 0$$

we see by subtraction that $c_2 = 0$ and $c_4 \sin \alpha L = 0$. This equation yields the eigenvalues $\alpha = n\pi L$ for $n = 1, 2, 3, \ldots$. The corresponding eigenfunctions are

$$X = c_4 \sin \frac{n\pi}{L} x.$$

Thus

$$u(x,t) = \sum_{n=1}^{\infty} \left(A_n \cos \frac{n^2\pi^2}{L^2} at + B_n \sin \frac{n^2\pi^2}{L^2} at \right) \sin \frac{n\pi}{L} x.$$

From

$$u(x,0) = f(x) = \sum_{n=1}^{\infty} A_n \sin \frac{n\pi}{L} x$$

we obtain

$$A_n = \frac{2}{L} \int_0^L f(x) \sin \frac{n\pi}{L} x \, dx.$$

From

$$\frac{\partial u}{\partial t} = \sum_{n=1}^{\infty} \left(-A_n \frac{n^2\pi^2 a}{L^2} \sin \frac{n^2\pi^2}{L^2} at + B_n \frac{n^2\pi^2 a}{L^2} \cos \frac{n^2\pi^2}{L^2} at \right) \sin \frac{n\pi}{L} x$$

and

$$\frac{\partial u}{\partial t}\bigg|_{t=0} = g(x) = \sum_{n=1}^{\infty} B_n \frac{n^2\pi^2 a}{L^2} \sin \frac{n\pi}{L} x$$

we obtain

$$B_n \frac{n^2\pi^2 a}{L^2} = \frac{2}{L} \int_0^L g(x) \sin \frac{n\pi}{L} x \, dx$$

and

$$B_n = \frac{2L}{n^2\pi^2 a} \int_0^L g(x) \sin \frac{n\pi}{L} x \, dx.$$

19. From (8) in the text we have

$$u(x,t) = \sum_{n=1}^{\infty} \left(A_n \cos \frac{n\pi a}{L} t + B_n \sin \frac{n\pi a}{L} t \right) \sin \frac{n\pi}{L} x.$$

Since $u_t(x,0) = g(x) = 0$ we have $B_n = 0$ and

$$u(x,t) = \sum_{n=1}^{\infty} A_n \cos \frac{n\pi a}{L} t \sin \frac{n\pi}{L} x = \sum_{n=1}^{\infty} A_n \frac{1}{2} \left[\sin \left(\frac{n\pi}{L} x + \frac{n\pi a}{L} t \right) + \sin \left(\frac{n\pi}{L} x - \frac{n\pi a}{L} t \right) \right]$$

$$= \frac{1}{2} \sum_{n=1}^{\infty} A_n \left[\sin \frac{n\pi}{L}(x+at) + \sin \frac{n\pi}{L}(x-at) \right].$$

From

$$u(x,0) = f(x) = \sum_{n=1}^{\infty} A_n \sin \frac{n\pi}{L} x$$

we identify

$$f(x+at) = \sum_{n=1}^{\infty} A_n \sin \frac{n\pi}{L}(x+at)$$

and

$$f(x-at) = \sum_{n=1}^{\infty} A_n \sin \frac{n\pi}{L}(x-at),$$

so that

$$u(x,t) = \frac{1}{2}[f(x+at) + f(x-at)].$$

21. $u(x,t) = \dfrac{1}{2}\left[\sin(x+at) + \sin(x-at) \right] + \dfrac{1}{2a} \displaystyle\int_{x-at}^{x+at} ds$

$$= \frac{1}{2}\left[\sin x \cos at + \cos x \sin at + \sin x \cos at - \cos x \sin at \right] + \frac{1}{2a} s \Big|_{x-at}^{x+at} = \sin x \cos at + t$$

23. $u(x,t) = 0 + \dfrac{1}{2a} \displaystyle\int_{x-at}^{x+at} \sin 2s\, ds = \dfrac{1}{2a}\left[\dfrac{-\cos(2x+2at) + \cos(2x-2at)}{2} \right]$

$$= \frac{1}{4a}\left[-\cos 2x \cos 2at + \sin 2x \sin 2at + \cos 2x \cos 2at + \sin 2x \sin 2at \right]$$

$$= \frac{1}{2a} \sin 2x \sin 2at$$

25. (a)

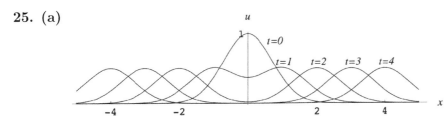

(b)

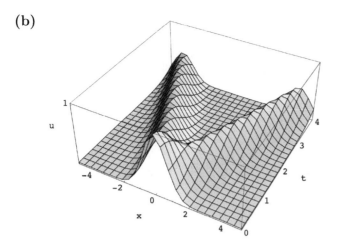

27. (a) With $a = 1$, d'Alembert's solution is

$$u(x,t) = \frac{1}{2} \int_{x-t}^{x+t} g(s)\, ds \qquad \text{where} \qquad g(s) = \begin{cases} 1, & |s| \le 0.1 \\ 0, & |s| > 0.1. \end{cases}$$

Sample plots are shown below.

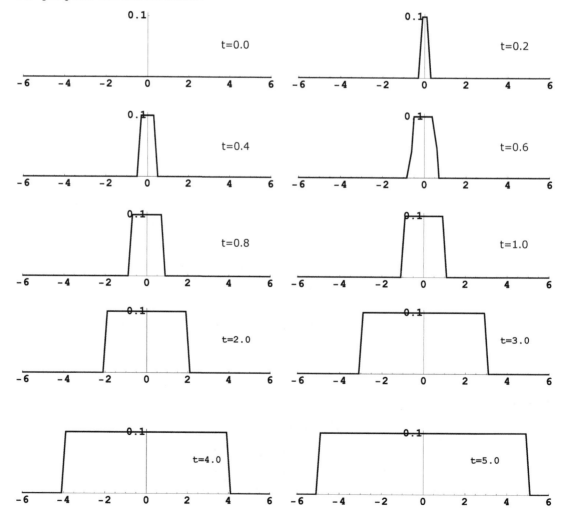

(b) Some frames of the movie are shown in part **(a)**, The string has a roughly rectangular shape with the base on the x-axis increasing in length.

12.5 | Laplace's Equation

1. Using $u = XY$ and $-\lambda$ as a separation constant we obtain

$$X'' + \lambda X = 0,$$

$$X(0) = 0,$$

$$X(a) = 0,$$

and

$$Y'' - \lambda Y = 0,$$

$$Y(0) = 0.$$

With $\lambda = \alpha^2 > 0$ the solutions of the differential equations are

$$X = c_1 \cos \alpha x + c_2 \sin \alpha x \quad \text{and} \quad Y = c_3 \cosh \alpha y + c_4 \sinh \alpha y$$

The boundary and initial conditions imply

$$X = c_2 \sin \frac{n\pi}{a} x \quad \text{and} \quad Y = c_4 \sinh \frac{n\pi}{a} y$$

for $n = 1, 2, 3, \ldots$ so that

$$u = \sum_{n=1}^{\infty} A_n \sin \frac{n\pi}{a} x \sinh \frac{n\pi}{a} y.$$

Imposing

$$u(x, b) = f(x) = \sum_{n=1}^{\infty} A_n \sinh \frac{n\pi b}{a} \sin \frac{n\pi}{a} x$$

gives

$$A_n \sinh \frac{n\pi b}{a} = \frac{2}{a} \int_0^a f(x) \sin \frac{n\pi}{a} x \, dx$$

so that

$$u(x, y) = \sum_{n=1}^{\infty} A_n \sin \frac{n\pi}{a} x \sinh \frac{n\pi}{a} y$$

where

$$A_n = \frac{2}{a} \operatorname{csch} \frac{n\pi b}{a} \int_0^a f(x) \sin \frac{n\pi}{a} x \, dx.$$

3. Using $u = XY$ and $-\lambda$ as a separation constant we obtain

$$X'' + \lambda X = 0,$$

$$X(0) = 0,$$

$$X(a) = 0,$$

and

$$Y'' - \lambda Y = 0,$$

$$Y(b) = 0.$$

With $\lambda = \alpha^2 > 0$ the solutions of the differential equations are

$$X = c_1 \cos \alpha x + c_2 \sin \alpha x \quad \text{and} \quad Y = c_3 \cosh \alpha y + c_4 \sinh \alpha y$$

The boundary and initial conditions imply

$$X = c_2 \sin \frac{n\pi}{a} x \quad \text{and} \quad Y = c_2 \cosh \frac{n\pi}{a} y - c_2 \frac{\cosh \frac{n\pi b}{a}}{\sinh \frac{n\pi b}{a}} \sinh \frac{n\pi}{a} y$$

for $n = 1, 2, 3, \ldots$ so that

$$u = \sum_{n=1}^{\infty} A_n \left(\cosh \frac{n\pi}{a} y - \frac{\cosh \frac{n\pi b}{a}}{\sinh \frac{n\pi b}{a}} \sinh \frac{n\pi}{a} y \right) \sin \frac{n\pi}{a} x.$$

Imposing

$$u(x, 0) = f(x) = \sum_{n=1}^{\infty} A_n \sin \frac{n\pi}{a} x$$

gives

$$A_n = \frac{2}{a} \int_0^a f(x) \sin \frac{n\pi}{a} x \, dx$$

so that

$$u(x, y) = \frac{2}{a} \sum_{n=1}^{\infty} \left(\int_0^a f(x) \sin \frac{n\pi}{a} x \, dx \right) \left(\cosh \frac{n\pi}{a} y - \frac{\cosh \frac{n\pi b}{a}}{\sinh \frac{n\pi b}{a}} \sinh \frac{n\pi}{a} y \right) \sin \frac{n\pi}{a} x.$$

5. Using $u = XY$ and $-\lambda$ as a separation constant we obtain

$$X'' + \lambda X = 0,$$

$$X'(0) = 0,$$

$$X'(a) = 0,$$

and

$$Y'' - \lambda Y = 0,$$

$$Y(b) = 0.$$

With $\lambda = -\alpha^2 < 0$ the solutions of the differential equations are

$$X = c_1 \cosh \alpha x + c_2 \sinh \alpha x \qquad \text{and} \qquad Y = c_3 \cos \alpha y + c_4 \sin \alpha y$$

for $n = 1, 2, 3 \dots$. The boundary and initial conditions imply

$$X = c_2 \sinh n\pi x \qquad \text{and} \qquad Y = c_3 \cos n\pi y$$

for $n = 1, 2, 3, \dots$. Since $\lambda = 0$ is an eigenvalue for the differential equation in X with corresponding eigenfunction x we have

$$u = A_0 x + \sum_{n=1}^{\infty} A_n \sinh n\pi x \cos n\pi y.$$

Imposing

$$u(1, y) = 1 - y = A_0 + \sum_{n=1}^{\infty} A_n \sinh n\pi \cos n\pi y$$

gives

$$A_0 = \int_0^1 (1 - y) \, dy$$

and

$$A_n \sinh n\pi = 2 \int_0^1 (1 - y) \cos n\pi y \, dy = \frac{2[1 - (-1)^n]}{2\pi^2}$$

for $n = 1, 2, 3, \dots$ so that

$$u(x, y) = \frac{1}{2}x + \frac{2}{\pi^2} \sum_{n=1}^{\infty} \frac{1 - (-1)^n}{n^2 \sinh n\pi} \sinh n\pi x \cos n\pi y.$$

7. Using $u = XY$ and $-\lambda$ as a separation constant we obtain

$$X'' + \lambda X = 0,$$

$$X'(0) = X(0)$$

and

$$Y'' - \lambda Y = 0,$$

$$Y(0) = 0,$$

$$Y(\pi) = 0.$$

With $\lambda = \alpha^2 < 0$ the solutions of the differential equations are

$$X = c_1 \cosh \alpha x + c_2 \sinh \alpha x \qquad \text{and} \qquad Y = c_3 \cos \alpha y + c_4 \sin \alpha y$$

The boundary and initial conditions imply

$$Y = c_4 \sin ny \qquad \text{and} \qquad X = c_2(n \cosh nx + \sinh nx)$$

for $n = 1, 2, 3, \ldots$ so that

$$u = \sum_{n=1}^{\infty} A_n(n \cosh nx + \sinh nx) \sin ny.$$

Imposing

$$u(\pi, y) = 1 = \sum_{n=1}^{\infty} A_n(n \cosh n\pi + \sinh n\pi) \sin ny$$

gives

$$A_n(n \cosh n\pi + \sinh n\pi) = \frac{2}{\pi} \int_0^\pi \sin ny \, dy = \frac{2[1 - (-1)^n]}{n\pi}$$

for $n = 1, 2, 3, \ldots$ so that

$$u(x, y) = \frac{2}{\pi} \sum_{n=1}^{\infty} \frac{1 - (-1)^n}{n} \frac{n \cosh nx + \sinh nx}{n \cosh n\pi + \sinh n\pi} \sin ny.$$

9. This boundary-value problem has the form of Problem 1 from the text of this section, with $a = b = 1$, $f(x) = 100$, and $g(x) = 200$. The solution, then, is

$$u(x, y) = \sum_{n=1}^{\infty} (A_n \cosh n\pi y + B_n \sinh n\pi y) \sin n\pi x,$$

where

$$A_n = 2 \int_0^1 100 \sin n\pi x \, dx = 200 \left(\frac{1 - (-1)^n}{n\pi} \right)$$

and

$$B_n = \frac{1}{\sinh n\pi} \left[2 \int_0^1 200 \sin n\pi x \, dx - A_n \cosh n\pi \right]$$

$$= \frac{1}{\sinh n\pi} \left[400 \left(\frac{1 - (-1)^n}{n\pi} \right) - 200 \left(\frac{1 - (-1)^n}{n\pi} \right) \cosh n\pi \right]$$

$$= 200 \left[\frac{1 - (-1)^n}{n\pi} \right] [2 \operatorname{csch} n\pi - \coth n\pi].$$

11. Using $u = XY$ and $-\lambda$ as a separation constant we obtain

$$X'' + \lambda X = 0,$$

$$X(0) = 0,$$

$$X(\pi) = 0,$$

and
$$Y'' - \lambda Y = 0.$$

With $\lambda = \alpha^2 > 0$ the solutions of the differential equations are

$$X = c_1 \cos \alpha x + c_2 \sin \alpha x \qquad \text{and} \qquad Y = c_3 e^{\alpha y} + c_4 e^{-\alpha y}$$

Then the boundedness of u as $y \to \infty$ implies $c_3 = 0$, so $Y = c_4 e^{-ny}$. The boundary conditions at $x = 0$ and $x = \pi$ imply $c_1 = 0$ so $X = c_2 \sin nx$ for $n = 1, 2, 3, \ldots$ and

$$u = \sum_{n=1}^{\infty} A_n e^{-ny} \sin nx.$$

Imposing

$$u(x, 0) = f(x) = \sum_{n=1}^{\infty} A_n \sin nx$$

gives

$$A_n = \frac{2}{\pi} \int_0^{\pi} f(x) \sin nx \, dx$$

so that

$$u(x, y) = \sum_{n=1}^{\infty} \left(\frac{2}{\pi} \int_0^{\pi} f(x) \sin nx \, dx \right) e^{-ny} \sin nx.$$

13. Since the boundary conditions at $y = 0$ and $y = b$ are functions of x we choose to separate Laplace's equation as

$$\frac{X''}{X} = -\frac{Y''}{Y} = -\lambda$$

so that

$$X'' + \lambda X = 0$$

$$Y'' - \lambda Y = 0.$$

Then with $\lambda = \alpha^2$ we have

$$X(x) = c_1 \cos \alpha x + c_2 \sin \alpha x$$

$$Y(y) = c_3 \cosh \alpha y + c_4 \sinh \alpha y.$$

Now $X(0) = 0$ gives $c_1 = 0$ and $X(a) = 0$ implies $\sin \alpha a = 0$ or $\alpha = n\pi/a$ for $n = 1, 2, 3, \ldots$. Thus

$$u_n(x, y) = XY = \left(A_n \cosh \frac{n\pi}{a} y + B_n \sinh \frac{n\pi}{a} y \right) \sin \frac{n\pi}{a} x$$

and

$$u(x, y) = \sum_{n=1}^{\infty} \left(A_n \cosh \frac{n\pi}{a} y + B_n \sinh \frac{n\pi}{a} y \right) \sin \frac{n\pi}{a} x. \tag{1}$$

At $y = 0$ we then have

$$f(x) = \sum_{n=1}^{\infty} A_n \sin \frac{n\pi}{a} x$$

and consequently

$$A_n = \frac{2}{a} \int_0^a f(x) \sin \frac{n\pi}{a} x \, dx. \tag{2}$$

At $y = b$,

$$g(y) = \sum_{n=1}^{\infty} \left(A_n \cosh \frac{n\pi}{a} b + B_n \sinh \frac{n\pi}{b} a \right) \sin \frac{n\pi}{a} x$$

indicates that the entire expression in the parentheses is given by

$$A_n \cosh \frac{n\pi}{a} b + B_n \sinh \frac{n\pi}{a} b = \frac{2}{a} \int_0^a g(x) \sin \frac{n\pi}{a} x \, dx.$$

We can now solve for B_n:

$$B_n \sinh \frac{n\pi}{a} b = \frac{2}{a} \int_0^a g(x) \sin \frac{n\pi}{a} x \, dx - A_n \cosh \frac{n\pi}{a} b$$

$$B_n = \frac{1}{\sinh \frac{n\pi}{a} b} \left(\frac{2}{a} \int_0^a g(x) \sin \frac{n\pi}{a} x \, dx - A_n \cosh \frac{n\pi}{a} b \right). \tag{3}$$

A solution to the given boundary-value problem consists of the series (**1**) with coefficients A_n and B_n given in (**2**) and (**3**), respectively.

15. Referring to the discussion in this section of the text we identify $a = b = \pi$, $f(x) = 0$, $g(x) = 1$, $F(y) = 1$, and $G(y) = 1$. Then $A_n = 0$ and

$$u_1(x, y) = \sum_{n=1}^{\infty} B_n \sinh ny \sin nx$$

where

$$B_n = \frac{2}{\pi \sinh n\pi} \int_0^\pi \sin nx \, dx = \frac{2[1 - (-1)^n]}{n\pi \sinh n\pi}.$$

Next

$$u_2(x, y) = \sum_{n=1}^{\infty} (A_n \cosh nx + B_n \sinh nx) \sin ny$$

where

$$A_n = \frac{2}{\pi} \int_0^\pi \sin ny \, dy = \frac{2[1 - (-1)^n]}{n\pi}$$

and

$$B_n = \frac{1}{\sinh n\pi} \left(\frac{2}{\pi} \int_0^\pi \sin ny \, dy - A_n \cosh n\pi \right)$$

$$= \frac{1}{\sinh n\pi} \left(\frac{2[1 - (-1)^n]}{n\pi} - \frac{2[1 - (-1)^n]}{n\pi} \cosh n\pi \right)$$

$$= \frac{2[1 - (-1)^n]}{n\pi \sinh n\pi} (1 - \cosh n\pi).$$

Now

$$A_n \cosh nx + B_n \sinh nx = \frac{2[1 - (-1)^n]}{n\pi} \left[\cosh nx + \frac{\sinh nx}{\sinh n\pi}(1 - \cosh n\pi) \right]$$

$$= \frac{2[1 - (-1)^n]}{n\pi \sinh n\pi} [\cosh nx \sinh n\pi + \sinh nx - \sinh nx \cosh n\pi]$$

$$= \frac{2[1 - (-1)^n]}{n\pi \sinh n\pi} [\sinh nx + \sinh n(\pi - x)]$$

and

$$u(x, y) = u_1 + u_2 = \frac{2}{\pi} \sum_{n=1}^{\infty} \frac{1 - (-1)^n}{n \sinh n\pi} \sinh ny \sin nx$$

$$+ \frac{2}{\pi} \sum_{n=1}^{\infty} \frac{[1 - (-1)^n][\sinh nx + \sinh n(\pi - x)]}{n \sinh n\pi} \sin ny.$$

17. **(a)**

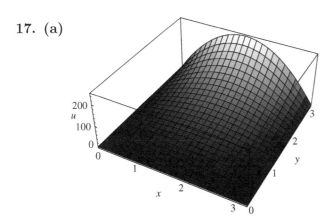

(b) The maximum value occurs at $(\pi/2, \pi)$ and is $f(\pi/2) = 25\pi^2$.

(c) The coefficients are

$$A_n = \frac{2}{\pi} \operatorname{csch} n\pi \int_0^\pi 100x(\pi - x) \sin nx \, dx$$

$$= \frac{200 \operatorname{csch} n\pi}{\pi} \left[\frac{200}{n^3}(1 - (-1)^n) \right] = \frac{400}{n^3 \pi}[1 - (-1)^n] \operatorname{csch} n\pi.$$

See part **(a)** for the graph.

19. Assuming $u(x, y) = X(x)Y(y)$ and substituting into the partial differential equationwe get $X''Y + XY'' = 0$. Separating variables and using $\lambda = \alpha^2$ we get

$$X'' - \alpha^2 X = 0, \quad X'(0) = 0,$$

which implies $X(x) = c_3 \cosh \alpha x$. From

$$Y'' + \alpha^2 Y = 0, \quad Y'(0) = 0, \quad Y'(b) = 0$$

we get $Y(y) = c_1 \cos \alpha y + c_2 \sin \alpha y$ and eigenvalues $\lambda_n = n^2\pi^2/b^2$, $n = 1, 2, 3, \ldots$. The corresponding eigenfunctions are $Y(y) = c_1 \cos(n\pi y/b)$. For $\lambda = 0$ the boundary conditions applied to $X(x) = c_3 + c_4 x$ and $Y(y) = c_1 + c_2 y$ imply $X = c_3$ and $Y = c_1$. Forming products and using the superposition principle then gives

$$u(x, y) = A_0 + \sum_{n=1}^{\infty} A_n \cosh \frac{n\pi}{b} x \cos \frac{n\pi}{b} y.$$

The remaining boundary condition, $u_x(a) = g(y)$ implies

$$g(y) = \frac{\partial u}{\partial x}\bigg|_{x=a} = \sum_{n=1}^{\infty} A_n \frac{n\pi}{b} \sinh \frac{n\pi}{b} x \cos \frac{n\pi}{b} y,$$

and so A_0 remains arbitrary. In order that the series expression for $g(y)$ be a cosine series, the constant term in the series, $a_0/2$, must be 0. Thus, from Section 11.3 in the text,

$$a_0 = \frac{2}{b} \int_0^b g(y)\, dy = 0 \quad \text{so} \quad \int_0^b g(y)\, dy = 0.$$

Also,

$$A_n \frac{n\pi}{b} \sinh \frac{n\pi}{b} a = \frac{2}{b} \int_0^b g(y) \cos \frac{n\pi}{b} y \, dy$$

and

$$A_n = \frac{2}{n\pi \sinh n\pi a/b} \int_0^b g(y) \cos \frac{n\pi}{b} y \, dy.$$

The solution is then

$$u(x, y) = A_0 + \sum_{n=1}^{\infty} A_n \cosh \frac{n\pi}{b} x \cos \frac{n\pi}{b} y,$$

where the A_n are defined above and A_0 is arbitrary. In general, Neumann problems do not have unique solutions.

For a physical interpretation of the compatibility condition $\int_0^b g(y)dy = 0$ see the texts *Elementary Partial Differential Equations* by Paul Berg and James McGregor (Holden-Day) and *Partial Differential Equations of Mathematical Physics* by Tyn Myint-U (North Holland).

21. (a)

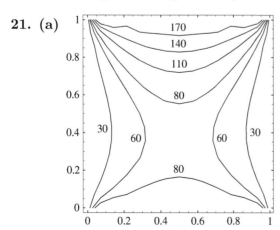

(b)

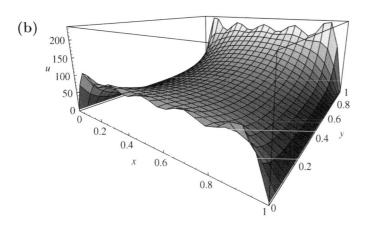

12.6 | Nonhomogeneous Boundary-Value Problems

1. Using $v(x,t) = u(x,t) - 100$ we wish to solve $kv_{xx} = v_t$ subject to $v(0,t) = 0$, $v(1,t) = 0$, and $v(x,0) = -100$. Let $v = XT$ and use $-\lambda$ as a separation constant so that

$$X'' + \lambda X = 0,$$

$$X(0) = 0,$$

$$X(1) = 0,$$

and

$$T' + \lambda kT = 0.$$

This leads to

$$X = c_2 \sin(n\pi x) \qquad \text{and} \qquad T = c_3 e^{-kn^2\pi^2 t}$$

for $n = 1, 2, 3, \ldots$ so that

$$v(x,t) = \sum_{n=1}^{\infty} A_n e^{-kn^2\pi^2 t} \sin n\pi x.$$

Imposing

$$v(x,0) = -100 = \sum_{n=1}^{\infty} A_n \sin n\pi x$$

gives

$$A_n = 2 \int_0^1 (-100) \sin n\pi x \, dx = \frac{-200}{n\pi}[1 - (-1)^n]$$

so that

$$u(x,t) = v(x,t) + 100 = 100 + \frac{200}{\pi} \sum_{n=1}^{\infty} \frac{(-1)^n - 1}{n} e^{-kn^2\pi^2 t} \sin n\pi x.$$

3. If we let $u(x, t) = v(x, t) + \psi(x)$, then we obtain as in Example 1 in the text

$$k\psi'' + r = 0$$

or

$$\psi(x) = -\frac{r}{2k}x^2 + c_1 x + c_2.$$

The boundary conditions become

$$u(0, t) = v(0, t) + \psi(0) = u_0$$

$$u(1, t) = v(1, t) + \psi(1) = u_0.$$

Letting $\psi(0) = \psi(1) = u_0$ we obtain homogeneous boundary conditions in v:

$$v(0, t) = 0 \qquad \text{and} \qquad v(1, t) = 0.$$

Now $\psi(0) = \psi(1) = u_0$ implies $c_2 = u_0$ and $c_1 = r/2k$. Thus

$$\psi(x) = -\frac{r}{2k}x^2 + \frac{r}{2k}x + u_0 = u_0 - \frac{r}{2k}x(x - 1).$$

To determine $v(x, t)$ we solve

$$k\frac{\partial^2 v}{\partial x^2} = \frac{\partial v}{dt}, \quad 0 < x < 1, \ t > 0$$

$$v(0, t) = 0, \quad v(1, t) = 0,$$

$$v(x, 0) = \frac{r}{2k}x(x - 1) - u_0.$$

Separating variables, we find

$$v(x, t) = \sum_{n=1}^{\infty} A_n e^{-kn^2\pi^2 t} \sin n\pi x,$$

where

$$A_n = 2\int_0^1 \left[\frac{r}{2k}x(x - 1) - u_0\right] \sin n\pi x \, dx = 2\left[\frac{u_0}{n\pi} + \frac{r}{kn^3\pi^3}\right][(-1)^n - 1]. \tag{1}$$

Hence, a solution of the original problem is

$$u(x, t) = \psi(x) + v(x, t) = u_0 - \frac{r}{2k}x(x - 1) + \sum_{n=1}^{\infty} A_n e^{-kn^2\pi^2 t} \sin n\pi x,$$

where A_n is defined in (**1**).

5. Substituting $u(x,t) = v(x,t) + \psi(x)$ into the partial differential equation gives

$$k\frac{\partial^2 v}{\partial x^2} + k\psi'' + Ae^{-\beta x} = \frac{\partial v}{\partial t}.$$

This equation will be homogeneous provided ψ satisfies

$$k\psi'' + Ae^{-\beta x} = 0.$$

The solution of this differential equation is obtained by successive integrations:

$$\psi(x) = -\frac{A}{\beta^2 k}e^{-\beta x} + c_1 x + c_2.$$

From $\psi(0) = 0$ and $\psi(1) = 0$ we find

$$c_1 = \frac{A}{\beta^2 k}(e^{-\beta} - 1) \qquad \text{and} \qquad c_2 = \frac{A}{\beta^2 k}.$$

Hence

$$\psi(x) = -\frac{A}{\beta^2 k}e^{-\beta x} + \frac{A}{\beta^2 k}(e^{-\beta} - 1)x + \frac{A}{\beta^2 k} = \frac{A}{\beta^2 k}\left[1 - e^{-\beta x} + (e^{-\beta} - 1)x\right].$$

Now the new problem is

$$k\frac{\partial^2 v}{\partial x^2} = \frac{\partial v}{\partial t}, \quad 0 < x < 1, \quad t > 0,$$

$$v(0,t) = 0, \quad v(1,t) = 0, \quad t > 0,$$

$$v(x,0) = f(x) - \psi(x), \quad 0 < x < 1.$$

Identifying this as the heat equation solved in Section 12.3 in the text with $L = 1$ we obtain

$$v(x,t) = \sum_{n=1}^{\infty} A_n e^{-kn^2\pi^2 t}\sin n\pi x$$

where

$$A_n = 2\int_0^1 [f(x) - \psi(x)]\sin n\pi x\, dx.$$

Thus

$$u(x,t) = \frac{A}{\beta^2 k}\left[1 - e^{-\beta x} + (e^{-\beta} - 1)x\right] + \sum_{n=1}^{\infty} A_n e^{-kn^2\pi^2 t}\sin n\pi x.$$

7. Substituting $u(x,t) = v(x,t) + \psi(x)$ into the partial differential equation gives

$$k\frac{\partial^2 v}{\partial x^2} + k\psi'' - hv - h\psi + hu_0 = \frac{\partial v}{\partial t}.$$

This equation will be homogeneous provided ψ satisfies

$$k\psi'' - h\psi + hu_0 = 0 \qquad \text{or} \qquad k\psi'' - h\psi = -hu_0.$$

This non-homogeneous, linear, second-order, differential equation has solution

$$\psi(x) = c_1 \cosh\sqrt{\frac{h}{k}}\,x + c_2 \sinh\sqrt{\frac{h}{k}}\,x + u_0,$$

where we assume $h > 0$ and $k > 0$. From $\psi(0) = u_0$ and $\psi(1) = 0$ we find $c_1 = 0$ and $c_2 = -u_0/\sinh\sqrt{h/k}$. Thus, the steady-state solution is

$$\psi(x) = -\frac{u_0}{\sinh\sqrt{\frac{h}{k}}}\sinh\sqrt{\frac{h}{k}}\,x + u_0 = u_0\left(1 - \frac{\sinh\sqrt{\frac{h}{k}}\,x}{\sinh\sqrt{\frac{h}{k}}}\right).$$

9. Substituting $u(x, t) = v(x, t) + \psi(x)$ into the partial differential equation gives

$$a^2 \frac{\partial^2 v}{\partial x^2} + a^2\psi'' + Ax = \frac{\partial^2 v}{\partial t^2}.$$

This equation will be homogeneous provided ψ satisfies

$$a^2\psi'' + Ax = 0.$$

The solution of this differential equation is

$$\psi(x) = -\frac{A}{6a^2}x^3 + c_1 x + c_2.$$

From $\psi(0) = 0$ we obtain $c_2 = 0$, and from $\psi(1) = 0$ we obtain $c_1 = A/6a^2$. Hence

$$\psi(x) = \frac{A}{6a^2}(x - x^3).$$

Now the new problem is

$$a^2 \frac{\partial^2 v}{\partial x^2} = \frac{\partial^2 v}{\partial t^2}$$

$$v(0, t) = 0, \quad v(1, t) = 0, \quad t > 0,$$

$$v(x, 0) = -\psi(x), \quad v_t(x, 0) = 0, \quad 0 < x < 1.$$

Identifying this as the wave equation solved in Section 12.4 in the text with $L = 1$, $f(x) = -\psi(x)$, and $g(x) = 0$ we obtain

$$v(x, t) = \sum_{n=1}^{\infty} A_n \cos n\pi at \sin n\pi x$$

where

$$A_n = 2 \int_0^1 [-\psi(x)] \sin n\pi x \, dx = \frac{A}{3a^2} \int_0^1 (x^3 - x) \sin n\pi x \, dx = \frac{2A(-1)^n}{a^2 \pi^3 n^3}.$$

Thus

$$u(x,t) = \frac{A}{6a^2}(x - x^3) + \frac{2A}{a^2 \pi^3} \sum_{n=1}^{\infty} \frac{(-1)^n}{n^3} \cos n\pi a t \sin n\pi x.$$

11. Substituting $u(x,y) = v(x,y) + \psi(y)$ into Laplace's equation we obtain

$$\frac{\partial^2 v}{\partial x^2} + \frac{\partial^2 v}{\partial y^2} + \psi''(y) = 0.$$

This equation will be homogeneous provided ψ satisfies $\psi(y) = c_1 y + c_2$. Considering

$$u(x,0) = v(x,0) + \psi(0) = u_1$$

$$u(x,1) = v(x,1) + \psi(1) = u_0$$

$$u(0,y) = v(0,y) + \psi(y) = 0$$

we require that $\psi(0) = u_1$, $\psi_1 = u_0$ and $v(0,y) = -\psi(y)$. Then $c_1 = u_0 - u_1$ and $c_2 = u_1$. The new boundary-value problem is

$$\frac{\partial^2 v}{\partial x^2} + \frac{\partial^2 v}{\partial y^2} = 0$$

$$v(x,0) = 0, \quad v(x,1) = 0,$$

$$v(0,y) = -\psi(y), \quad 0 < y < 1,$$

where $v(x,y)$ is bounded at $x \to \infty$. This problem is similar to Problem 11 in Section 12.5. The solution is

$$v(x,y) = \sum_{n=1}^{\infty} \left(2 \int_0^1 [-\psi(y) \sin n\pi y] \, dy \right) e^{-n\pi x} \sin n\pi y$$

$$= 2 \sum_{n=1}^{\infty} \left[(u_1 - u_0) \int_0^1 y \sin n\pi y \, dy - u_1 \int_0^1 \sin n\pi y \, dy \right] e^{-n\pi x} \sin n\pi y$$

$$= \frac{2}{\pi} \sum_{n=1}^{\infty} \frac{u_0(-1)^n - u_1}{n} e^{-n\pi x} \sin n\pi y.$$

Thus

$$u(x,y) = v(x,y) + \psi(y) = (u_0 - u_1)y + u_1 + \frac{2}{\pi} \sum_{n=1}^{\infty} \frac{u_0(-1)^n - u_1}{n} e^{-n\pi x} \sin n\pi y.$$

13. From (13) and (14) we have

$$\psi(x,t) = \sin t + x[0 - \sin t] = (1 - x)\sin t,$$

Then the substitution

$$u(x,t) = v(x,t) + \psi(x,t) = v(x,t) + (1 - x)\sin t$$

leads to the boundary-value problem

$$\frac{\partial^2 v}{\partial x^2} + (x - 1)\cos t = \frac{\partial v}{\partial t}, \quad 0 < x < 1, \quad t > 0$$

$$v(0,t) = 0, \quad v(1,t) = 0, \quad t > 0$$

$$v(x,0) = 0, \quad 0 < x < 1.$$

The eigenvalues and eigenfunctions of the Sturm-Liouville problem

$$X'' + \lambda X = 0, \quad X(0) = 0, \quad X(1) = 0$$

are $\lambda_n = \alpha_n^2 = n^2\pi^2$ and $\sin n\pi x$, $n = 1, 2, 3, \ldots$. With $G(x,t) = (x-1)\cos t$ we assume for fixed t that v and G can be written as Fourier sine series:

$$v(x,t) = \sum_{n=1}^{\infty} v_n(t)\sin n\pi x \quad \text{and} \quad G(x,t) = \sum_{n=1}^{\infty} G_n(t)\sin n\pi x.$$

By treating t as a parameter, the coefficients G_n can be computed:

$$G_n(t) = \frac{2}{1}\int_0^1 (x-1)\cos t \sin n\pi x\, dx = 2\cos t \int_0^1 (x-1)\sin n\pi x\, dx = -\frac{2}{n\pi}\cos t.$$

Hence

$$(x-1)\cos t = \sum_{n=1}^{\infty} \frac{-2\cos t}{n\pi}\sin n\pi x.$$

Now, using the series representation for $v(x,t)$, we have

$$\frac{\partial^2 v}{\partial x^2} = \sum_{n=1}^{\infty} v_n(t)(-n^2\pi^2)\sin n\pi x \quad \text{and} \quad \frac{\partial v}{\partial t} = \sum_{n=1}^{\infty} v_n'(t)\sin n\pi x.$$

Writing the partial differential equation as $v_t - v_{xx} = (x-1)\cos t$ and using the above results we have

$$\sum_{n=1}^{\infty} [v_n'(t) + n^2\pi^2 v_n(t)]\sin n\pi x = \sum_{n=1}^{\infty} \frac{-2\cos t}{n\pi}\sin n\pi x.$$

Equating coefficients we get

$$v_n'(t) + n^2\pi^2 v_n(t) = -\frac{2\cos t}{n\pi}.$$

For each n this is a linear first-order differential equation whose general solution is

$$v_n(t) = -\frac{2}{n\pi}\left[\frac{n^2\pi^2\cos t + \sin t}{n^4\pi^4 + 1}\right] + C_n e^{-n^2\pi^2 t}.$$

Thus

$$v(x,t) = \sum_{n=1}^{\infty}\left[-\frac{2n^2\pi^2\cos t + 2\sin t}{n\pi(n^4\pi^4 + 1)} + C_n e^{-n^2\pi^2 t}\right]\sin n\pi x.$$

The initial condition $v(x,0) = 0$ implies

$$\sum_{n=1}^{\infty}\left[-\frac{2n\pi}{n^4\pi^4 + 1} + C_n\right]\sin n\pi x = 0$$

so that $C_n = 2n\pi/(n^4\pi^4 + 1)$. Therefore

$$v(x,t) = \sum_{n=1}^{\infty}\left[-\frac{2n^2\pi^2\cos t + 2\sin t}{n\pi(n^4\pi^4 + 1)} + \frac{2n\pi}{n^4\pi^4 + 1}e^{-n^2\pi^2 t}\right]\sin n\pi x$$

$$= \frac{2}{\pi}\sum_{n=1}^{\infty}\left[\frac{n^2\pi^2 e^{-n^2\pi^2 t} - n^2\pi^2\cos t - \sin t}{n(n^4\pi^4 + 1)}\right]\sin n\pi x$$

and

$$u(x,t) = v(x,t) + \psi(x,t) = (1-x)\sin t + \frac{2}{\pi}\sum_{n=1}^{\infty}\left[\frac{n^2\pi^2 e^{-n^2\pi^2 t} - n^2\pi^2\cos t - \sin t}{n(n^4\pi^4 + 1)}\right]\sin n\pi x.$$

15. From (13) and (14) we have

$$\psi(x,t) = u_0(t) + \frac{X}{L}[u_1(t) - u_0(t)] = x\sin t,$$

Then the substitution

$$u(x,t) = v(x,t) + \psi(x,t) = x\sin t$$

leads to the boundary-value problem

$$\frac{\partial^2 v}{\partial x^2} + x\sin t = \frac{\partial^2 v}{\partial t^2}, \quad 0 < x < 1, \quad t > 0$$

$$v(0,t) = 0, \quad v(1,t) = 0, \quad t > 0$$

$$v(x,0) = 0, \quad \left.\frac{\partial v}{\partial t}\right|_{t=0} = -x, \quad 0 < x < 1.$$

The eigenvalues and eigenfunctions of the Sturm-Liouville problem

$$X'' + \lambda X = 0, \quad X(0) = 0, \quad X(1) = 0$$

are $\lambda_n = \alpha_n^2 = n^2\pi^2$ and $\sin n\pi x$, $n = 1, 2, 3, \ldots$. With $G(x,t) = x\sin t$ we assume for fixed t that v and G can be written as Fourier sine series:

$$v(x,t) = \sum_{n=1}^{\infty} v_n(t) \sin n\pi x \quad \text{and} \quad G(x,t) = \sum_{n=1}^{\infty} G_n(t) \sin n\pi x.$$

By treating t as a parameter, the coefficients G_n can be computed:

$$G_n(t) = \frac{2}{1}\int_0^1 x\sin t \sin n\pi x\, dx = \frac{2}{1}\sin t \int_0^1 x\sin n\pi x\, dx = -2\frac{(-1)^n \sin t}{n\pi}.$$

Hence

$$x\sin t = \sum_{n=1}^{\infty} -2\frac{(-1)^n \sin t}{n\pi}\sin n\pi x.$$

Now, using the series representation for $v(x,t)$, we have

$$\frac{\partial^2 v}{\partial x^2} = \sum_{n=1}^{\infty} v_n(t)(-n^2\pi^2)\sin n\pi x \quad \text{and} \quad \frac{\partial v}{\partial t} = \sum_{n=1}^{\infty} v_n'(t)\sin n\pi x.$$

Writing the partial differential equation as $v_{tt} - v_{xx} = x\sin t$ and using the above results we have

$$\sum_{n=1}^{\infty} [v_n''(t) + n^2\pi^2 v_n(t)]\sin n\pi x = \sum_{n=1}^{\infty} -2\frac{(-1)^n \sin t}{n\pi}\sin n\pi x.$$

Equating coefficients we get

$$v_n''(t) + n^2\pi^2 v_n(t) = -2\frac{(-1)^n \sin t}{n\pi}.$$

For each n the general solution is

$$v_n(t) = A_n \cos n\pi t + B_n \sin n\pi t - 2\frac{(-1)^n}{n\pi\,(n^2\pi^2 - 1)}\sin t$$

Thus

$$v(x,t) = \sum_{n=1}^{\infty} \left[A_n \cos n\pi t + B_n \sin n\pi t - 2\frac{(-1)^n}{n\pi\,(n^2\pi^2 - 1)}\sin t\right]\sin n\pi x.$$

The initial condition $v(x,0) = 0$ implies

$$\sum_{n=1}^{\infty} A_n \sin n\pi x = 0.$$

or $A_n = 0$ for $n = 1, 2, 3, \ldots$. So

$$v(x,t) = \sum_{n=1}^{\infty} \left[B_n \sin n\pi t - 2\frac{(-1)^n}{n\pi\,(n^2\pi^2 - 1)}\sin t\right]\sin n\pi x.$$

and

$$\left.\frac{\partial v}{\partial t}\right|_{t=0} = -x = \sum_{n=1}^{\infty}\left[n\pi B_n - 2\frac{(-1)^n}{n\pi\,(n^2\pi^2-1)}\right]\sin n\pi x.$$

Thinking of $-x$ as a Fourier sine series with coefficients

$$-2\int_0^1 x\sin n\pi x\,dx = 2\frac{(-1)^n}{n\pi}$$

we equate coefficients to obtain

$$n\pi B_n - 2\frac{(-1)^n}{n\pi\,(n^2\pi^2-1)} = 2\frac{(-1)^n}{n\pi}$$

so

$$B_n = 2\frac{(-1)^n}{n\pi} + 2\frac{(-1)^n}{n\pi\,(n^2\pi^2-1)}.$$

Therefore

$$v(x,t) = \sum_{n=1}^{\infty}\left[\left(2\frac{(-1)^n}{n\pi} + 2\frac{(-1)^n}{n\pi\,(n^2\pi^2-1)}\right)\sin n\pi t - 2\frac{(-1)^n}{n\pi\,(n^2\pi^2-1)}\sin t\right]\sin n\pi x$$

and

$$u(x,t) = v(x,t) + \psi(x,t)$$

$$= x\sin t + 2\sum_{n=1}^{\infty}\left[\left(\frac{(-1)^n}{n\pi} + \frac{(-1)^n}{n\pi\,(n^2\pi^2-1)}\right)\sin n\pi t - \frac{(-1)^n}{n\pi\,(n^2\pi^2-1)}\sin t\right]\sin n\pi x.$$

17. Identifying $k = 1$ and $L = \pi$ we see that the eigenfunctions of $X'' + \lambda X = 0$, $X(0) = 0$, $X(\pi) = 0$ are $\sin nx$, $n = 1, 2, 3, \ldots$. Assuming that $u(x,t) = \sum_{n=1}^{\infty}u_n(t)\sin nx$, the formal partial derivatives of u are

$$\frac{\partial^2 u}{\partial x^2} = \sum_{n=1}^{\infty}u_n(t)(-n^2)\sin nx \qquad\text{and}\qquad \frac{\partial u}{\partial t} = \sum_{n=1}^{\infty}u_n'(t)\sin nx.$$

Assuming that $xe^{-3t} = \sum_{n=1}^{\infty}F_n(t)\sin nx$ we have

$$F_n(t) = \frac{2}{\pi}\int_0^\pi xe^{-3t}\sin nx\,dx = \frac{2e^{-3t}}{\pi}\int_0^\pi x\sin nx\,dx = \frac{2e^{-3t}(-1)^{n+1}}{n}.$$

Then

$$xe^{-3t} = \sum_{n=1}^{\infty}\frac{2e^{-3t}(-1)^{n+1}}{n}\sin nx$$

and

$$u_t - u_{xx} = \sum_{n=1}^{\infty}[u_n'(t) + n^2 u_n(t)]\sin nx = xe^{-3t} = \sum_{n=1}^{\infty}\frac{2e^{-3t}(-1)^{n+1}}{n}\sin nx.$$

Equating coefficients we obtain

$$u_n'(t) + n^2 u_n(t) = \frac{2e^{-3t}(-1)^{n+1}}{n}.$$

This is a linear first-order differential equation whose solution is

$$u_n(t) = \frac{2(-1)^{n+1}}{n(n^2-3)}e^{-3t} + C_n e^{-n^2 t}.$$

Thus

$$u(x,t) = \sum_{n=1}^{\infty} \frac{2(-1)^{n+1}}{n(n^2-3)}e^{-3t}\sin nx + \sum_{n=1}^{\infty} C_n e^{-n^2 t}\sin nx$$

and $u(x,0) = 0$ implies

$$\sum_{n=1}^{\infty} \frac{2(-1)^{n+1}}{n(n^2-3)}\sin nx + \sum_{n=1}^{\infty} C_n \sin nx = 0$$

so that $C_n = 2(-1)^n/n(n^2-3)$. Therefore

$$u(x,t) = 2\sum_{n=1}^{\infty} \frac{(-1)^{n+1}}{n(n^2-3)}e^{-3t}\sin nx + 2\sum_{n=1}^{\infty} \frac{(-1)^n}{n(n^2-3)}e^{-n^2 t}\sin nx.$$

19. Identifying $k=1$ and $L=1$ we see that the eigenfunctions of $X'' + \lambda X = 0$, $X(0) = 0$, $X(1) = 0$ are $\sin n\pi x$, $n = 1, 2, 3, \ldots$. Assuming that $u(x,t) = \sum_{n=1}^{\infty} u_n(t)\sin n\pi x$, the formal partial derivatives of u are

$$\frac{\partial^2 u}{\partial x^2} = \sum_{n=1}^{\infty} u_n(t)(-n^2\pi^2)\sin n\pi x \qquad \text{and} \qquad \frac{\partial u}{\partial t} = \sum_{n=1}^{\infty} u_n'(t)\sin n\pi x.$$

Assuming that $-1 + x - x\cos t = \sum_{n=1}^{\infty} F_n(t)\sin n\pi x$ we have

$$F_n(t) = \frac{2}{1}\int_0^1 (-1 + x - x\cos t)\sin n\pi x\, dx = \frac{2[-1 + (-1)^n\cos t]}{n\pi}.$$

Then

$$-1 + x - x\cos t = \frac{2}{\pi}\sum_{n=1}^{\infty} \frac{-1 + (-1)^n\cos t}{n}\sin n\pi x$$

and

$$u_t - u_{xx} = \sum_{n=1}^{\infty} [u_n'(t) + n^2\pi^2 u_n(t)]\sin n\pi x$$

$$= -1 + x - x\cos t = \frac{2}{\pi}\sum_{n=1}^{\infty} \frac{-1 + (-1)^n\cos t}{n}\sin n\pi x.$$

Equating coefficients we obtain

$$u_n'(t) + n^2\pi^2 u_n(t) = \frac{2[-1 + (-1)^n \cos t]}{n\pi}.$$

This is a linear first-order differential equation whose solution is

$$u_n(t) = \frac{2}{n\pi}\left[-\frac{1}{n^2\pi^2} + (-1)^n \frac{n^2\pi^2 \cos t + \sin t}{n^4\pi^4 + 1}\right] + C_n e^{-n^2\pi^2 t}.$$

Thus

$$u(x,t) = \sum_{n=1}^{\infty} \frac{2}{n\pi}\left[-\frac{1}{n^2\pi^2} + (-1)^n \frac{n^2\pi^2 \cos t + \sin t}{n^4\pi^4 + 1}\right]\sin n\pi x + \sum_{n=1}^{\infty} C_n e^{-n^2\pi^2 t} \sin n\pi x$$

and $u(x,0) = x(1-x)$ implies

$$\sum_{n=1}^{\infty} \frac{2}{n\pi}\left[-\frac{1}{n^2\pi^2} + (-1)^n \frac{n^2\pi^2}{n^4\pi^4 + 1} + C_n\right]\sin n\pi x = x(1-x).$$

Hence

$$\frac{2}{n\pi}\left[-\frac{1}{n^2\pi^2} + (-1)^n \frac{n^2\pi^2}{n^4\pi^4 + 1} + C_n\right] = \frac{2}{1}\int_0^1 x(1-x)\sin n\pi x\, dx = 2\left[\frac{1 - (-1)^n}{n^3\pi^3}\right]$$

and

$$C_n = \frac{4 - 2(-1)^n}{n^3\pi^3} - (-1)^n \frac{2n\pi}{n^4\pi^4 + 1}.$$

Therefore

$$u(x,t) = \sum_{n=1}^{\infty} \frac{2}{n\pi}\left[-\frac{1}{n^2\pi^2} + (-1)^n \frac{n^2\pi^2 \cos t + \sin t}{n^4\pi^4 + 1}\right]\sin n\pi x$$

$$+ \sum_{n=1}^{\infty}\left[\frac{4 - 2(-1)^n}{n^3\pi^3} - (-1)^n \frac{2n\pi}{n^4\pi^4 + 1}\right]e^{-n^2\pi^2 t}\sin n\pi x.$$

12.7 Orthogonal Series Expansions

1. Referring to Example 1 in the text we have

$$X(x) = c_1 \cos \alpha x + c_2 \sin \alpha x \qquad \text{and} \qquad T(t) = c_3 e^{-k\alpha^2 t}.$$

From $X'(0) = 0$ (since the left end of the rod is insulated), we find $c_2 = 0$. Then $X(x) = c_1 \cos \alpha x$ and the other boundary condition $X'(1) = -hX(1)$ implies

$$-\alpha \sin \alpha + h \cos \alpha = 0 \qquad \text{or} \qquad \cot \alpha = \frac{\alpha}{h}.$$

Denoting the consecutive positive roots of this latter equation by α_n for $n = 1, 2, 3, \ldots$, we have

$$u(x,t) = \sum_{n=1}^{\infty} A_n e^{-k\alpha_n^2 t} \cos \alpha_n x.$$

From the initial condition $u(x,0) = 1$ we obtain

$$1 = \sum_{n=1}^{\infty} A_n \cos \alpha_n x$$

and

$$A_n = \frac{\displaystyle\int_0^1 \cos \alpha_n x \, dx}{\displaystyle\int_0^1 \cos^2 \alpha_n x \, dx} = \frac{\sin \alpha_n / \alpha_n}{\frac{1}{2}\left[1 + \frac{1}{2\alpha_n} \sin 2\alpha_n\right]} = \frac{2 \sin \alpha_n}{\alpha_n \left[1 + \frac{1}{\alpha_n} \sin \alpha_n \cos \alpha_n\right]}$$

$$= \frac{2 \sin \alpha_n}{\alpha_n \left[1 + \frac{1}{h\alpha_n} \sin \alpha_n (\alpha_n \sin \alpha_n)\right]} = \frac{2h \sin \alpha_n}{\alpha_n [h + \sin^2 \alpha_n]}.$$

The solution is

$$u(x,t) = 2h \sum_{n=1}^{\infty} \frac{\sin \alpha_n}{\alpha_n (h + \sin^2 \alpha_n)} e^{-k\alpha_n^2 t} \cos \alpha_n x.$$

3. Separating variables in Laplace's equation gives

$$X'' + \alpha^2 X = 0$$

$$Y'' - \alpha^2 Y = 0$$

and

$$X(x) = c_1 \cos \alpha x + c_2 \sin \alpha x$$

$$Y(y) = c_3 \cosh \alpha y + c_4 \sinh \alpha y.$$

From $u(0,y) = 0$ we obtain $X(0) = 0$ and $c_1 = 0$. From $u_x(a,y) = -hu(a,y)$ we obtain $X'(a) = -hX(a)$ and

$$\alpha \cos \alpha a = -h \sin \alpha a \qquad \text{or} \qquad \tan \alpha a = -\frac{\alpha}{h}.$$

Let α_n, where $n = 1, 2, 3, \ldots$, be the consecutive positive roots of this equation. From $u(x,0) = 0$ we obtain $Y(0) = 0$ and $c_3 = 0$. Thus

$$u(x,y) = \sum_{n=1}^{\infty} A_n \sinh \alpha_n y \sin \alpha_n x.$$

Now

$$f(x) = \sum_{n=1}^{\infty} A_n \sinh \alpha_n b \sin \alpha_n x$$

and

$$A_n \sinh \alpha_n b = \frac{\int_0^a f(x) \sin \alpha_n x \, dx}{\int_0^a \sin^2 \alpha_n x \, dx}.$$

Since

$$\int_0^a \sin^2 \alpha_n x \, dx = \frac{1}{2} \left[a - \frac{1}{2\alpha_n} \sin 2\alpha_n a \right] = \frac{1}{2} \left[a - \frac{1}{\alpha_n} \sin \alpha_n a \cos \alpha_n a \right]$$

$$= \frac{1}{2} \left[a - \frac{1}{h\alpha_n} (h \sin \alpha_n a) \cos \alpha_n a \right]$$

$$= \frac{1}{2} \left[a - \frac{1}{h\alpha_n} (-\alpha_n \cos \alpha_n a) \cos \alpha_n a \right] = \frac{1}{2h} \left[ah + \cos^2 \alpha_n a \right],$$

we have

$$A_n = \frac{2h}{\sinh \alpha_n b [ah + \cos^2 \alpha_n a]} \int_0^a f(x) \sin \alpha_n x \, dx.$$

5. The boundary-value problem is

$$k \frac{\partial^2 u}{\partial x^2} = \frac{\partial u}{\partial t}, \quad 0 < x < L, \quad t > 0,$$

$$u(0, t) = 0, \quad \left. \frac{\partial u}{\partial x} \right|_{x=L} = 0, \quad t > 0,$$

$$u(x, 0) = f(x), \quad 0 < x < L.$$

Separation of variables leads to

$$X'' + \alpha^2 X = 0$$

$$T' + k\alpha^2 T = 0$$

and

$$X(x) = c_1 \cos \alpha x + c_2 \sin \alpha x$$

$$T(t) = c_3 e^{-k\alpha^2 t}.$$

From $X(0) = 0$ we find $c_1 = 0$. From $X'(L) = 0$ we obtain $\cos \alpha L = 0$ and

$$\alpha = \frac{\pi(2n - 1)}{2L}, \quad n = 1, 2, 3, \dots .$$

Thus

$$u(x, t) = \sum_{n=1}^{\infty} A_n e^{-k(2n-1)^2 \pi^2 t / 4L^2} \sin \left(\frac{2n - 1}{2L} \right) \pi x$$

where

$$A_n = \frac{\int_0^L f(x) \sin\left(\frac{2n-1}{2L}\right) \pi x \, dx}{\int_0^L \sin^2\left(\frac{2n-1}{2L}\right) \pi x \, dx} = \frac{2}{L} \int_0^L f(x) \sin\left(\frac{2n-1}{2L}\right) \pi x \, dx.$$

7. Separation of variables leads to

$$Y'' + \alpha^2 Y = 0$$

$$X'' - \alpha^2 X = 0$$

and

$$Y(y) = c_1 \cos \alpha y + c_2 \sin \alpha y$$

$$X(x) = c_3 \cosh \alpha x + c_4 \sinh \alpha x.$$

From $Y(0) = 0$ we find $c_1 = 0$. From $Y'(1) = 0$ we obtain $\cos \alpha = 0$ and

$$\alpha = \frac{\pi(2n-1)}{2}, \quad n = 1, 2, 3, \dots .$$

Thus

$$Y(y) = c_2 \sin\left(\frac{2n-1}{2}\right) \pi y.$$

From $X'(0) = 0$ we find $c_4 = 0$. Then

$$u(x, y) = \sum_{n=1}^{\infty} A_n \cosh\left(\frac{2n-1}{2}\right) \pi x \sin\left(\frac{2n-1}{2}\right) \pi y$$

where

$$u_0 = u(1, y) = \sum_{n=1}^{\infty} A_n \cosh\left(\frac{2n-1}{2}\right) \pi \sin\left(\frac{2n-1}{2}\right) \pi y$$

and

$$A_n \cosh\left(\frac{2n-1}{2}\right) \pi = \frac{\int_0^1 u_0 \sin\left(\frac{2n-1}{2}\right) \pi y \, dy}{\int_0^1 \sin^2\left(\frac{2n-1}{2}\right) \pi y \, dy} = \frac{4u_0}{(2n-1)\pi}.$$

Thus

$$u(x, y) = \frac{4u_0}{\pi} \sum_{n=1}^{\infty} \frac{1}{(2n-1) \cosh\left(\frac{2n-1}{2}\right) \pi} \cosh\left(\frac{2n-1}{2}\right) \pi x \sin\left(\frac{2n-1}{2}\right) \pi y.$$

9. The boundary-value problem is

$$k\frac{\partial^2 u}{\partial x^2} = \frac{\partial u}{\partial t}, \quad 0 < x < 1, \quad t > 0$$

$$\left.\frac{\partial u}{\partial x}\right|_{x=0} = hu(0,t), \quad \left.\frac{\partial u}{\partial x}\right|_{x=1} = -hu(1,t), \quad h > 0, \quad t > 0,$$

$$u(x,0) = f(x), \quad 0 < x < 1.$$

Referring to Example 1 in the text we have

$$X(x) = c_1 \cos \alpha x + c_2 \sin \alpha x \quad \text{and} \quad T(t) = c_3 e^{-k\alpha^2 t}.$$

Applying the boundary conditions, we obtain

$$X'(0) = hX(0) \quad \text{and} \quad X'(1) = -hX(1)$$

or

$$\alpha c_2 = hc_1$$

$$-\alpha c_1 \sin \alpha + \alpha c_2 \cos \alpha = -hc_1 \cos \alpha - hc_2 \sin \alpha.$$

Choosing $c_1 = \alpha$ and $c_2 = h$ (to satisfy the first equation above) we obtain

$$-\alpha^2 \sin \alpha + h\alpha \cos \alpha = -h\alpha \cos \alpha - h^2 \sin \alpha$$

$$2h\alpha \cos \alpha = (\alpha^2 - h^2) \sin \alpha.$$

The eigenvalues α_n are the consecutive positive roots of

$$\tan \alpha = \frac{2h\alpha}{\alpha^2 - h^2}.$$

Then

$$u(x,t) = \sum_{n=1}^{\infty} A_n e^{-k\alpha_n^2 t} (\alpha_n \cos \alpha_n x + h \sin \alpha_n x)$$

where

$$f(x) = u(x,0) = \sum_{n=1}^{\infty} A_n (\alpha_n \cos \alpha_n x + h \sin \alpha_n x)$$

and

$$A_n = \frac{\displaystyle\int_0^1 f(x)(\alpha_n \cos \alpha_n x + h \sin \alpha_n x)\,dx}{\displaystyle\int_0^1 (\alpha_n \cos \alpha_n x + h \sin \alpha_n x)^2\,dx}$$

$$= \frac{2}{\alpha_n^2 + 2h + h^2} \int_0^1 f(x)(\alpha_n \cos \alpha_n x + h \sin \alpha_n x)\,dx.$$

[Note: the evaluation and simplification of the integral in the denominator requires the use of the relationship $(\alpha^2 - h^2) \sin \alpha = 2h\alpha \cos \alpha$.]

11. Using $u = XT$ and separation constant $-\lambda = \alpha^4$ we find

$$X^{(4)} - \alpha^4 X = 0 \quad \text{and} \quad X(x) = c_1 \cos \alpha x + c_2 \sin \alpha x + c_3 \cosh \alpha x + c_4 \sinh \alpha x.$$

Since $u = XT$ the boundary conditions become

$$X(0) = 0, \quad X'(0) = 0, \quad X''(1) = 0, \quad X'''(1) = 0.$$

Now $X(0) = 0$ implies $c_1 + c_3 = 0$, while $X'(0) = 0$ implies $c_2 + c_4 = 0$. Thus

$$X(x) = c_1 \cos \alpha x + c_2 \sin \alpha x - c_1 \cosh \alpha x - c_2 \sinh \alpha x.$$

The boundary condition $X''(1) = 0$ implies

$$-c_1 \cos \alpha - c_2 \sin \alpha - c_1 \cosh \alpha - c_2 \sinh \alpha = 0$$

while the boundary condition $X'''(1) = 0$ implies

$$c_1 \sin \alpha - c_2 \cos \alpha - c_1 \sinh \alpha - c_2 \cosh \alpha = 0.$$

We then have the system of two equations in two unknowns

$$(\cos \alpha + \cosh \alpha)c_1 + (\sin \alpha + \sinh \alpha)c_2 = 0$$

$$(\sin \alpha - \sinh \alpha)c_1 - (\cos \alpha + \cosh \alpha)c_2 = 0.$$

This homogeneous system will have nontrivial solutions for c_1 and c_2 provided

$$\begin{vmatrix} \cos \alpha + \cosh \alpha & \sin \alpha + \sinh \alpha \\ \sin \alpha - \sinh \alpha & -\cos \alpha - \cosh \alpha \end{vmatrix} = 0$$

or

$$-2 - 2 \cos \alpha \cosh \alpha = 0.$$

Thus, the eigenvalues are determined by the equation $\cos \alpha \cosh \alpha = -1$.

Using a computer to graph $\cosh \alpha$ and $-1/\cos \alpha = -\sec \alpha$ we see that the first two positive eigenvalues occur near 1.9 and 4.7. Applying Newton's method with these initial values we find that the eigenvalues are $\alpha_1 = 1.8751$ and $\alpha_2 = 4.6941$.

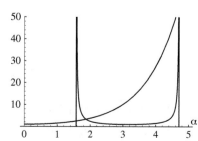

12.8 | Higher Dimensional Problems

1. This boundary-value problem was solved in Example 1 in the text. Identifying $b = c = \pi$ and $f(x, y) = u_0$ we have

$$u(x, y, t) = \sum_{m=1}^{\infty} \sum_{n=1}^{\infty} A_{mn} e^{-k(m^2 + n^2)t} \sin mx \sin ny$$

where

$$A_{mn} = \frac{4}{\pi^2} \int_0^{\pi} \int_0^{\pi} u_0 \sin mx \sin ny \, dx \, dy = \frac{4u_0}{\pi^2} \int_0^{\pi} \sin mx \, dx \int_0^{\pi} \sin ny \, dy$$

$$= \frac{4u_0}{mn\pi^2} [1 - (-1)^m][1 - (-1)^n].$$

3. The conditions $X(0) = 0$ and $Y(0) = 0$ give $c_1 = 0$ and $c_3 = 0$. The conditions $X(\pi) = 0$ and $Y(\pi) = 0$ yield two sets of eigenvalues:

$$\alpha = m, \; m = 1, 2, 3, \ldots \quad \text{and} \quad \beta = n, \; n = 1, 2, 3, \ldots .$$

A product solution of the partial differential equation that satisfies the boundary conditions is

$$u_{mn}(x, y, t) = \left(A_{mn} \cos a\sqrt{m^2 + n^2}\, t + B_{mn} \sin a\sqrt{m^2 + n^2}\, t \right) \sin mx \sin ny.$$

To satisfy the initial conditions we use the superposition principle:

$$u(x, y, t) = \sum_{m=1}^{\infty} \sum_{n=1}^{\infty} \left(A_{mn} \cos a\sqrt{m^2 + n^2}\, t + B_{mn} \sin a\sqrt{m^2 + n^2}\, t \right) \sin mx \sin ny.$$

The initial condition $u_t(x, y, 0) = 0$ implies $B_{mn} = 0$ and

$$u(x, y, t) = \sum_{m=1}^{\infty} \sum_{n=1}^{\infty} A_{mn} \cos a\sqrt{m^2 + n^2}\, t \sin mx \sin ny.$$

At $t = 0$ we have

$$xy(x - \pi)(y - \pi) = \sum_{m=1}^{\infty} \sum_{n=1}^{\infty} A_{mn} \sin mx \sin ny.$$

Using (12) and (13) in the text, it follows that

$$A_{mn} = \frac{4}{\pi^2} \int_0^{\pi} \int_0^{\pi} xy(x - \pi)(y - \pi) \sin mx \sin ny \, dx \, dy$$

$$= \frac{4}{\pi^2} \int_0^{\pi} x(x - \pi) \sin mx \, dx \int_0^{\pi} y(y - \pi) \sin ny \, dy$$

$$= \frac{16}{m^3 n^3 \pi^2} [(-1)^m - 1][(-1)^n - 1].$$

5. The boundary and initial conditions are

$$u(0, y, z) = 0, \quad u(a, y, z) = 0$$
$$u(x, 0, z) = 0, \quad u(x, b, z) = 0$$
$$u(x, y, 0) = 0, \quad u(x, y, c) = f(x, y).$$

The conditions $X(0) = Y(0) = Z(0) = 0$ give $c_1 = c_3 = c_5 = 0$. The conditions $X(a) = 0$ and $Y(b) = 0$ yield two sets of eigenvalues:

$$\alpha = \frac{m\pi}{a}, \ m = 1, 2, 3, \ldots \quad \text{and} \quad \beta = \frac{n\pi}{b}, \ n = 1, 2, 3, \ldots.$$

By the superposition principle

$$u(x, y, t) = \sum_{m=1}^{\infty} \sum_{n=1}^{\infty} A_{mn} \sinh \omega_{mn} z \sin \frac{m\pi}{a} x \sin \frac{n\pi}{b} y$$

where

$$\omega_{mn}^2 = \frac{m^2\pi^2}{a^2} + \frac{n^2\pi^2}{b^2}$$

and

$$A_{mn} = \frac{4}{ab \sinh \omega_{mn} c} \int_0^b \int_0^a f(x, y) \sin \frac{m\pi}{a} x \sin \frac{n\pi}{b} y \, dx \, dy.$$

Chapter 12 in Review

1. Letting $u(x, y) = X(x) + Y(y)$ we have $X'Y' = XY$ and

$$\frac{X'}{X} = YY' = -\lambda.$$

If $\lambda = 0$ then $X' = 0$ and $X(x) = c_1$. Also $Y(y) = 0$ so $u = 0$.

If $\lambda \neq 0$ then $X' + \lambda X = 0$ and $Y + (1/\lambda)Y = 0$. Thus $X(x) = c_1 e^{-\lambda x}$ and $Y(y) = c_2 e^{-y/\lambda}$ so

$$u(x, y) = A e^{(-\lambda x - y/\lambda)}.$$

3. Substituting $u(x, t) = v(x, t) + \psi(x)$ into the partial differential equation we obtain

$$k \frac{\partial^2 v}{\partial x^2} + k\psi''(x) = \frac{\partial v}{\partial t}.$$

This equation will be homogeneous provided ψ satisfies

$$k\psi'' = 0 \quad \text{or} \quad \psi = c_1 x + c_2.$$

Considering

$$u(0, t) = v(0, t) + \psi(0) = u_0$$

we set $\psi(0) = u_0$ so that $\psi(x) = c_1 x + u_0$. Now

$$-\frac{\partial u}{\partial x}\bigg|_{x=\pi} = -\frac{\partial v}{\partial x}\bigg|_{x=\pi} - \psi'(x) = v(\pi, t) + \psi(\pi) - u_1$$

is equivalent to

$$\frac{\partial v}{\partial x}\bigg|_{x=\pi} + v(\pi, t) = u_1 - \psi'(x) - \psi(\pi) = u_1 - c_1 - (c_1\pi + u_0),$$

which will be homogeneous when

$$u_1 - c_1 - c_1\pi - u_0 = 0 \quad \text{or} \quad c_1 = \frac{u_1 - u_0}{1 + \pi}.$$

The steady-state solution is

$$\psi(x) = \left(\frac{u_1 - u_0}{1 + \pi}\right) x + u_0.$$

5. The boundary-value problem is

$$a^2 \frac{\partial^2 u}{\partial x^2} = \frac{\partial^2 u}{\partial t^2}, \quad 0 < x < 1, \quad t > 0,$$

$$u(0, t) = 0, \quad u = (1, t) = 0, \quad t > 0,$$

$$u(x, 0) = 0, \quad \frac{\partial u}{\partial t}\bigg|_{t=0} = g(x), \quad 0 < x < 1.$$

From Section 12.4 in the text we see that $A_n = 0$,

$$B_n = \frac{2}{n\pi a} \int_0^1 g(x) \sin n\pi \, dx = \frac{2}{n\pi a} \int_{1/4}^{3/4} h \sin n\pi x \, dx$$

$$= \frac{2h}{n\pi a} \left(-\frac{1}{n\pi} \cos n\pi x\right)\bigg|_{1/4}^{3/4} = \frac{2h}{n^2\pi^2 a}\left(\cos\frac{n\pi}{4} - \cos\frac{3n\pi}{4}\right)$$

and

$$u(x, t) = \sum_{n=1}^\infty B_n \sin n\pi at \sin n\pi x.$$

7. Using $u = XY$ and $-\lambda$ as a separation constant leads to

$$X'' - \lambda X = 0,$$

$$X(0) = 0,$$

and

$$Y'' + \lambda Y = 0,$$

$$Y(0) = 0,$$

$$Y(\pi) = 0.$$

This leads to

$$Y = c_4 \sin ny \qquad \text{and} \qquad X = c_2 \sinh nx$$

for $n = 1, 2, 3, \ldots$ so that

$$u = \sum_{n=1}^{\infty} A_n \sinh nx \sin ny.$$

Imposing

$$u(\pi, y) = 50 = \sum_{n=1}^{\infty} A_n \sinh n\pi \sin ny$$

gives

$$A_n = \frac{100}{n\pi} \frac{1 - (-1)^n}{\sinh n\pi}$$

so that

$$u(x, y) = \frac{100}{\pi} \sum_{n=1}^{\infty} \frac{1 - (-1)^n}{n \sinh n\pi} \sinh nx \sin ny.$$

9. Using $u = XY$ and $-\lambda$ as a separation constant leads to

$$X'' - \lambda X = 0,$$

and

$$Y'' + \lambda Y = 0,$$
$$Y(0) = 0,$$
$$Y(\pi) = 0.$$

Then

$$X = c_1 e^{nx} + c_2 e^{-nx} \qquad \text{and} \qquad Y = c_3 \cos ny + c_4 \sin ny$$

for $n = 1, 2, 3, \ldots$. Since u must be bounded as $x \to \infty$, we define $c_1 = 0$. Also $Y(0) = 0$ implies $c_3 = 0$ so

$$u = \sum_{n=1}^{\infty} A_n e^{-nx} \sin ny.$$

Imposing

$$u(0, y) = 50 = \sum_{n=1}^{\infty} A_n \sin ny$$

gives

$$A_n = \frac{2}{\pi} \int_0^{\pi} 50 \sin ny \, dy = \frac{100}{n\pi} [1 - (-1)^n]$$

so that

$$u(x, y) = \sum_{n=1}^{\infty} \frac{100}{n\pi} [1 - (-1)^n] e^{-nx} \sin ny.$$

11. (a) The coefficients of the series

$$u(x, 0) = \sum_{n=1}^{\infty} B_n \sin nx$$

are

$$B_n = frac2\pi \int_0^\pi \sin x \sin nx \, dx = \frac{2}{\pi} \int_0^\pi \frac{1}{2} \left[\cos(1-n)x - \cos(1+n)x\right] dx$$

$$= \frac{1}{\pi}\left[\frac{\sin(1-n)x}{1-n}\Bigg|_0^\pi - \frac{\sin(1+n)x}{1+n}\Bigg|_0^\pi\right] = 0 \text{ for } n \neq 1.$$

For $n = 1$,

$$B_1 = \frac{2}{\pi}\int_0^\pi \sin^2 x \, dx = \frac{1}{\pi}\int_0^\pi (1 - \cos 2x)\, dx = 1.$$

Thus

$$u(x, t) = \sum_{n=1}^{\infty} B_n e^{-n^2 t}\sin nx$$

reduces to $u(x, t) = e^{-t}\sin x$ for $n = 1$.

(b) This is like part (a), but in this case, for $n \neq 3$ and $n \neq 5$,

$$B_n = \frac{2}{\pi}\int_0^\pi (100\sin 3x - 30\sin 5x)\sin nx \, dx = 0;$$

while $B_3 = 100$ and $B_5 = -30$. Therefore

$$u(x, t) = 100e^{-9t}\sin 3x - 30e^{-25t}\sin 5x.$$

13. Using $u = XT$ and $-\lambda$, where $\lambda = \alpha^2$, as a separation constant we find

$$X'' + 2X' + \alpha^2 X = 0 \quad \text{and} \quad T'' + 2T' + (1 + \alpha^2)T = 0.$$

Thus for $\alpha > 1$

$$X = c_1 e^{-x}\cos\sqrt{\alpha^2 - 1}\,x + c_2 e^{-x}\sin\sqrt{\alpha^2 - 1}\,x$$
$$T = c_3 e^{-t}\cos\alpha t + c_4 e^{-t}\sin\alpha t.$$

For $0 \leq \alpha \leq 1$ we only obtain $X = 0$. Now the boundary conditions $X(0) = 0$ and $X(\pi) = 0$ give, in turn, $c_1 = 0$ and $\sqrt{\alpha^2 - 1}\pi = n\pi$ or $\alpha^2 = n^2 = n^2 + 1$, $n = 1, 2, 3, \ldots$. The corresponding solutions are $X = c_2 e^{-x}\sin nx$. The initial condition $T'(0) = 0$ implies $c_3 = \alpha c_4$ and so

$$T = c_4 e^{-t}\left[\sqrt{n^2 + 1}\cos\sqrt{n^2 + 1}\,t + \sin\sqrt{n^2 + 1}\,t\right].$$

Using $u = XT$ and the superposition principle, a formal series solution is

$$u(x, t) = e^{-(x+t)}\sum_{n=1}^{\infty} A_n\left[\sqrt{n^2 + 1}\cos\sqrt{n^2 + 1}\,t + \sin\sqrt{n^2 + 1}\,t\right]\sin nx.$$

15. By the discussion in Section 12.6, we assume a solution in the form $u(x,t) = v(x,t) + \psi(x)$. Force this into the differential equation to get $\psi''(x) = 0$ together with the conditions $\psi(0) = 0$ and $\psi'(1) + \psi(1) = u_1$.

Integrating twice and imposing the two conditions leads to $\psi(x) = \frac{1}{2}(u_1 - u_0)x + u_0$. Now

$$u(x,t) = v(x,t) + \psi(x)$$
$$u(0,t) = v(0,t) + \psi(0) = u_0$$
$$v(0,t) + \left[\frac{1}{2}(u_1 - u_0)(0) + u_0\right] = u_0$$
$$v(0,t) = 0$$

Similarly we find that

$$u_x(x,t) = v_x(x,t) + \psi'(x)$$
$$u_x(1,t) = v_x(1,t) + \psi'(1) = u_1 - u(1,t)$$
$$v_x(1,t) + \left[\frac{1}{2}(u_1 - u_0)\right] = u_1 - [v(1,t) + \psi(1)]$$
$$v_x(1,t) = u_1 - [v(1,t) + \psi(1)] - \frac{1}{2}(u_1 - u_0)$$
$$v_x(1,t) = -v(1,t)$$

So now we solve

$$\frac{\partial^2 v}{\partial x^2} = \frac{\partial v}{\partial t}, \quad 0 < x < 1, \ t > 0$$

$$v(0,t) = 0, \quad \left.\frac{\partial v}{\partial x}\right|_{x=1} = -v(1,t), \ t > 0$$

$$v(x,0) = \frac{1}{2}(u_0 - u_1)x, \ 0 < x < 1$$

Using the results of Example 1 from Section 12.7, $v(x,t) = \sum_{n=1}^{\infty} A_n e^{-\alpha_n^2 t} \sin \alpha_n x$ and the coefficients are given by

$$A_n = \frac{\frac{1}{2}(u_0 - u_1) \int_0^1 x \sin \alpha_n x \, dx}{\int_0^1 \sin^2 \alpha_n x \, dx} = \frac{1}{2}(u_0 - u_1)\frac{\frac{\sin \alpha_n - \alpha_n \cos \alpha_n}{\alpha_n^2}}{\frac{1}{2}(1 + \cos^2 \alpha_n)}$$

$$= (u_0 - u_1)\frac{\sin \alpha_n - \alpha_n \cos \alpha_n}{\alpha_n^2 (1 + \cos^2 \alpha_n)}.$$

Since $\tan \alpha_n = -\alpha_n$ we have $\sin \alpha_n = -\alpha_n \cos \alpha_n$ and so

$$A_n = 2(u_1 - u_0)\sum_{n=1}^{\infty} \frac{\cos \alpha_n}{\alpha_n (1 + \cos^2 \alpha_n)} e^{-\alpha_n^2 t} \sin(\alpha_n x)$$

and

$$u(x, t) = u_0 + \frac{1}{2}(u_1 - u_0)x + 2(u_1 - u_0) \sum_{n=1}^{\infty} \frac{\cos \alpha_n}{\alpha_n \left(1 + \cos^2 \alpha_n\right)} e^{-\alpha_n^2 t} \sin\left(\alpha_n x\right)$$

17. Substituting $u(x, y) = v(x, y) + \psi(x)$ into the boundary-value problem yields

$$\frac{\partial^2 v}{\partial x^2} + \frac{\partial^2 v}{\partial y^2} = 0.$$

The condition $\psi''(x) = -2$ has the general solution $\psi(x) = -x^2 + c_1 x + c_2$. The boundary conditions $x = 0$ and $x = \pi$ will be homogeneous if we choose $\psi(0) = 0$. The condition $\psi(\pi) = 0$ gives $c_2 = 0$ and $c_1 = \pi$. Thus

$$\psi(\pi) = -x^2 + \pi x.$$

The boundary-value problem in v is then

$$\frac{\partial^2 v}{\partial x^2} + \frac{\partial^2 v}{\partial y^2} = 0, \quad 0 < x < \pi, \quad 0 < y < \pi$$

$$v(0, y) = 0, \quad v(\pi, y) = 0, \quad 0 < y < \pi$$

$$v(x, 0) = -\psi(x), \quad v(x, \pi) = -\psi(x), \quad 0 < y < \pi.$$

From Problem 16,

$$v(x, y) = \sum_{n=1}^{\infty} \left(A_n \cosh ny + B_n \sinh ny\right) \sin nx$$

$$= \frac{4}{\pi} \sum_{n=1}^{\infty} \frac{[(-1)^n - 1]}{n^3} \left(\frac{\sinh n(\pi - y) + \sinh ny}{\sinh n\pi}\right) \sin nx.$$

Therefore

$$u(x, y) = \psi(x) + v(x, y) = -x^2 + \pi x + \frac{4}{\pi} \sum_{n=1}^{\infty} \frac{[(-1)^n - 1]}{n^3} \left(\frac{\sinh n(\pi - y) + \sinh ny}{\sinh n\pi}\right) \sin nx.$$

19. For $w(x, y) = C \sin \frac{\pi x}{a} \sin \frac{\pi y}{b}$,

$$\frac{\partial^4 w}{\partial x^4} = C \left(\frac{\pi}{a}\right)^4 \sin \frac{\pi x}{a} \sin \frac{\pi y}{b}$$

$$\frac{\partial^4 w}{\partial x^2 \partial y^2} = C \left(\frac{\pi}{a}\right)^2 \left(\frac{\pi}{b}\right)^2 \sin \frac{\pi x}{a} \sin \frac{\pi y}{b}$$

$$\frac{\partial^4 w}{\partial y^4} = C \left(\frac{\pi}{b}\right)^4 \sin \frac{\pi x}{a} \sin \frac{\pi y}{b}$$

Thus

$$\frac{\partial^4 w}{\partial x^4} + 2\frac{\partial^4 w}{\partial x^2 \partial y^2} + \frac{\partial^4 w}{\partial y^4} = C\left[\left(\frac{\pi}{a}\right)^4 + 2\left(\frac{\pi}{a}\right)^2\left(\frac{\pi}{b}\right)^2 + \left(\frac{\pi}{b}\right)^4\right]\sin\frac{\pi x}{a}\sin\frac{\pi y}{b}$$

$$= \frac{q_0}{D}\sin\frac{\pi x}{a}\sin\frac{\pi y}{b}$$

Hence

$$C = \frac{q_0/D}{\pi^4\left(\dfrac{1}{a^2} + \dfrac{1}{b^2}\right)^2} = \frac{q_0 a^4 b^4}{\pi^4 D\left(a^2 + b^2\right)^2}$$

Therefore

$$w(x, y) = \frac{q_0 a^4 b^4}{\pi^4 D\left(a^2 + b^2\right)^2}\sin\frac{\pi x}{a}\sin\frac{\pi y}{b}.$$

Chapter 13

Boundary-Value Problems in Other Coordinate Systems

13.1 | Polar Coordinates

1. We have

$$A_0 = \frac{1}{2\pi} \int_0^\pi u_0 \, d\theta = \frac{u_0}{2}$$

$$A_n = \frac{1}{\pi} \int_0^\pi u_0 \cos n\theta \, d\theta = 0$$

$$B_n = \frac{1}{\pi} \int_0^\pi u_0 \sin n\theta \, d\theta = \frac{u_0}{n\pi}[1 - (-1)^n]$$

and so

$$u(r, \theta) = \frac{u_0}{2} + \frac{u_0}{\pi} \sum_{n=1}^\infty \frac{1 - (-1)^n}{n} r^n \sin n\theta.$$

3. We have

$$A_0 = \frac{1}{2\pi} \int_0^{2\pi} (2\pi\theta - \theta^2) \, d\theta = \frac{2\pi^2}{3}$$

$$A_n = \frac{1}{\pi} \int_0^{2\pi} (2\pi\theta - \theta^2) \cos n\theta \, d\theta = -\frac{4}{n^2}$$

$$B_n = \frac{1}{\pi} \int_0^{2\pi} (2\pi\theta - \theta^2) \sin n\theta \, d\theta = 0$$

and so

$$u(r, \theta) = \frac{2\pi^2}{3} - 4 \sum_{n=1}^\infty \frac{r^n}{n^2} \cos n\theta.$$

5. Proceeding in the usual way by letting $u(r, \theta) = R(r)\Theta(\theta)$ and substituting it into the PDE we get

$$r^2 R'' + rR' - \lambda R = 0 \quad \text{and} \quad \Theta'' + \lambda\Theta = 0$$

As in Example 1 we have $R(r) = c_3 r^n + c_4 r^{-n}$ so that we must take $c_3 = 0$ in order for the solution to remain bounded as $r \to \infty$. We therefore have

$$u(r, \theta) = A_0 + \sum_{n=1}^{\infty} r^{-n} \left(A_n \cos n\theta + B_n \sin n\theta \right)$$

The condition at $r = c$ leads to the coefficients given by

$$A_0 = \frac{1}{2\pi} \int_0^{2\pi} f(\theta) \, d\theta$$

$$A_n = \frac{c^n}{\pi} \int_0^{\pi} f(\theta) \cos n\theta \, d\theta$$

$$B_n = \frac{c^n}{\pi} \int_0^{\pi} f(\theta) \sin n\theta \, d\theta$$

7. Proceeding in the usual way by letting $u(r, \theta) = R(r)\Theta(\theta)$ and substituting it into the PDE we get

$$r^2 R'' + r R' - \lambda R = 0 \quad \text{and} \quad \Theta'' + \lambda \Theta = 0$$

Using the results from Example 1 from Section 11.5, the eigenvalues and eigenfunctions for the boundary-value problem $\Theta'' + \lambda \Theta = 0$, $\Theta'(0) = \Theta'(\pi) = 0$ are $\lambda_n = n^2$ for $n = 0, 1, 2, \ldots$ and $\Theta = c_1 \cos(n\theta)$. Now for $n = 0$ the equation $r^2 R'' + r R' - \lambda R = 0$ has solution $R(r) = c_3 + c_4 \ln r$ which, in order to remain bounded as $r \to 0$, requires that $c_4 = 0$. When $n = 1, 2, 3, \ldots$, the equation $r^2 R'' + r R' - \lambda R = 0$ has solution $R(r) = c_3 r^n + c_4 r^{-n}$ which again requires $c_4 = 0$. Product solutions are therefore

$$u_0(r, \theta) = R_0(r)\Theta_0(\theta) = c_3 c_1 = A_0 \text{ for } n = 0$$

$$u_n(r, \theta) = R_n(r)\Theta_n(\theta) = (c_3 r^n)(c_1 \cos n\theta) = A_n r^n \cos n\theta \text{ for } n = 1, 2, 3, \ldots$$

By the superposition principle

$$u(r, \theta) = A_0 + \sum_{n=1}^{\infty} A_n r^n \cos n\theta$$

From the given condition at $r = 2$, we obtain the coefficients

$$A_0 = \frac{1}{2} \cdot \frac{2}{\pi} \int_0^{\pi/2} u_0 \, d\theta = \frac{u_0}{2}$$

$$A_n = \frac{1}{2^n} \cdot \frac{2}{\pi} \int_0^{\pi/2} u_0 \cos n\theta \, d\theta = \frac{1}{2^n} \cdot \frac{2u_0 \sin\left(\frac{n\pi}{2}\right)}{n\pi}$$

The final solution is therefore

$$u(r, \theta) = \frac{u_0}{2} + \sum_{n=1}^{\infty} \left(\frac{1}{2^n} \cdot \frac{2u_0 \sin\left(\frac{n\pi}{2}\right)}{n\pi} \right) r^n \cos n\theta = \frac{u_0}{2} + \frac{2u_0}{\pi} \sum_{n=1}^{\infty} \left(\frac{\sin\left(\frac{n\pi}{2}\right)}{n} \right) \left(\frac{r}{2}\right)^n \cos n\theta$$

9. Referring to the solution of Problem 6 above we have

$$\Theta(\theta) = c_1 \cos \alpha\theta + c_2 \sin \alpha\theta$$

$$R(r) = c_3 r^\alpha.$$

Applying the boundary conditions $\Theta'(0) = 0$ and $\Theta'(\pi/2) = 0$ we find that $c_2 = 0$ and $\alpha = 2n$ for $n = 0, 1, 2, \ldots$ .Therefore

$$u(r, \theta) = A_0 + \sum_{n=1}^{\infty} A_n r^{2n} \cos 2n\theta.$$

From

$$u(c, \theta) = \begin{cases} 1, & 0 < \theta < \pi/4 \\ 0, & \pi/4 < \theta < \pi/2 \end{cases} = A_0 + \sum_{n=1}^{\infty} A_n c^{2n} \cos 2n\theta$$

we find

$$A_0 = \frac{1}{\pi/2} \int_0^{\pi/4} d\theta = \frac{1}{2}$$

and

$$c^{2n} A_n = \frac{2}{\pi/2} \int_0^{\pi/4} \cos 2n\theta \, d\theta = \frac{2}{n\pi} \sin \frac{n\pi}{2}.$$

Thus

$$u(r, \theta) = \frac{1}{2} + \frac{2}{\pi} \sum_{n=1}^{\infty} \frac{1}{n} \sin \frac{n\pi}{2} \left(\frac{r}{c}\right)^{2n} \cos 2n\theta.$$

11. Proceeding as in Example 1 in the text and again using the periodicity of $u(r, \theta)$, we have

$$\Theta(\theta) = c_1 \cos \alpha\theta + c_2 \sin \alpha\theta$$

where $\alpha = n$ for $n = 0, 1, 2, \ldots$. Then

$$R(r) = c_3 r^n + c_4 r^{-n}.$$

[We do not have $c_4 = 0$ in this case since $0 < a \le r$.] Since $u(b, \theta) = 0$ we have

$$u(r, \theta) = A_0 \ln \frac{r}{b} + \sum_{n=1}^{\infty} \left[\left(\frac{b}{r}\right)^n - \left(\frac{r}{b}\right)^n \right] [A_n \cos n\theta + B_n \sin n\theta].$$

From

$$u(a, \theta) = f(\theta) = A_0 \ln \frac{a}{b} + \sum_{n=1}^{\infty} \left[\left(\frac{b}{a}\right)^n - \left(\frac{a}{b}\right)^n \right] [A_n \cos n\theta + B_n \sin n\theta]$$

we find

$$A_0 \ln \frac{a}{b} = \frac{1}{2\pi} \int_0^{2\pi} f(\theta) \, d\theta,$$

$$\left[\left(\frac{b}{a}\right)^n - \left(\frac{a}{b}\right)^n \right] A_n = \frac{1}{\pi} \int_0^{2\pi} f(\theta) \cos n\theta \, d\theta,$$

and

$$\left[\left(\frac{b}{a}\right)^n - \left(\frac{a}{b}\right)^n\right] B_n = \frac{1}{\pi} \int_0^{2\pi} f(\theta) \sin n\theta \, d\theta.$$

13. Solutions of the separated equations are

$$\Theta(\theta) = c_1, \quad n = 0$$

$$\Theta(\theta) = c_1 \cos n\theta + c_2 \sin n\theta, \quad n = 1, 2, \ldots \quad R(r) \qquad = c)3 + c_2 \ln r, \quad n = 0$$

$$R(r) = c_3 r^n + c_4 r^{-n}, \quad n = 1, 2, \ldots$$

Thus

$$u(r, \theta) = A_0 + B_0 \ln r + \sum_{n=1}^{\infty} \left[(A_n r^n + B_n r^{-n}) \cos n\theta + (C_n r^n + D_n r^{-n}) \sin n\theta\right].$$

When $r = 1$,

$$A_0 + B_0 \ln 1 = \frac{1}{2\pi} \int_0^{2\pi} 75 \sin \theta \, d\theta = 0 \quad \leftarrow \ln 1 = 0$$

$$A_n + B_n = \frac{1}{\pi} \int_0^{2\pi} 75 \sin \theta \cos n\theta \, d\theta = 0, \quad n = 1, 2, \ldots$$

$$C_n + D_n = \frac{1}{\pi} \int_0^{2\pi} 75 \sin \theta \sin n\theta \, d\theta = \begin{cases} 0, & n > 0 \\ 75, & n = 1 \end{cases}, \quad n = 1, 2, \ldots,$$

so

$$A_0 = 0, \quad A_1 + B_1 = 0, \quad C_1 + D_1 = 75,$$

and

$$A_n + B_n = 0, \quad C_n + D_n = 0, \quad \text{for} \quad n > 1.$$

When $r = 2$

$$A_0 + B_0 \ln 2 = \frac{1}{2\pi} \int_0^{2\pi} 60 \cos \theta \, d\theta = 0$$

$$A_n 2^n + B_n 2^{-n} = \frac{1}{\pi} \int_0^{2\pi} 60 \cos \theta \cos n\theta \, d\theta = \begin{cases} 0, & n > 1 \\ 60, & n = 1 \end{cases}$$

$$C_n 2^n + D_n 2^{-n} = \frac{1}{\pi} \int_0^{\infty} 60 \cos \theta \sin n\theta \, d\theta = 0, \quad n = 1, 2, \ldots,$$

so

$$B_0 = 0, \quad 2A_1 + \frac{1}{2} B_1 = 60, \quad 2C_1 + \frac{1}{2} D_1 = 0,$$

and

$$A_n 2^n + B_n 2^{-n} = 0, \quad C_n 2^n + D_n 2^{-n} = 0, \quad \text{for} \quad n > 1.$$

We have $A_0 = 0$ and $B_0 = 0$, and solving the nonhomogeneous systems for $n = 1$,

$$A_1 + B_1 = 0 \qquad\qquad C_1 + D_1 = 75$$

$$2A_1 + \frac{1}{2} B_1 = 60 \qquad\qquad 2C_1 + \frac{1}{2} D_1 = 0$$

yields $A_1 = 40$, $B_1 = -40$, $C_1 = -25$, and $D_1 = 100$. Finally, solving the homogeneous systems

$$A_n + B_n = 0 \qquad\qquad C_n + D_n = 0$$

$$A_n 2^n + B_n 2^{-n} = 0 \qquad\qquad C_n 2^n + D_n 2^{-n} = 0$$

gives $A_n = B_n = C_n = D_n = 0$ for $n > 1$. The solution is then

$$u(r, \theta) = \left(A_1 r + B_1 r^{-1} \right) \cos\theta + \left(C_1 r + D_1 r^{-1} \right) \sin\theta$$

$$= \left(4 - r - 40r^{-1} \right) \cos\theta + \left(-25r + 100r^{-1} \right) \sin\theta$$

$$= 40 \left(r - \frac{1}{r} \right) \cos\theta - 25 \left(r - \frac{4}{r} \right) \sin\theta.$$

15. The homogeneous boundary conditions $\Theta(\theta) = 0$ and $\Theta(\pi) = 0$ imply that $\lambda = 0$ is not an eigenvalue, but, imply for $\Theta(\theta) = c_1 \cos\lambda\theta + c_2 \sin\lambda\theta$, that $c_1 = 0$ and $\lambda_n = n^2$, $n = 1, 2, \ldots$. Then $\Theta(\theta) = c_2 \sin n\theta$, $n = 1, 2, \ldots$. Applying $R(1) = 0$ to $R(r) = c_3 r^n + c_4 r^{-n}$ gives $c_4 = -c_3$ so $R(r) = c_3(r^n - r^{-n})$. Thus $u(r, \theta) = \displaystyle\sum_{n=1}^{\infty} A_n(r^n - r^{-n}) \sin n\theta$ and the boundary condition

$$u(2, \theta) = u_0 = \sum_{n=1}^{\infty} A_n(2^n - 2^{-n}) \sin n\theta$$

implies

$$A_n \left(2^n - 2^{-n} \right) = \frac{2u_0}{\pi} \int_0^\pi \sin n\theta \, d\theta = \frac{2u_0}{\pi} \frac{1 - (-1)^n}{n} \quad \text{or} \quad A_n = \frac{2u_0}{\pi} \frac{1 - (-1)^n}{n\left(2^n - 2^{-n} \right)}.$$

Hence

$$u(r, \theta) = \frac{2u_0}{\pi} \sum_{n=1}^{\infty} \frac{1 - (-1)^n}{n} \frac{r^n - r^{-n}}{2^n - 2^{-n}} \sin n\theta.$$

17. The boundary-value problem is

$$\frac{\partial^2 u}{\partial r^2} + \frac{1}{r} \frac{\partial u}{\partial r} + \frac{1}{r^2} \frac{\partial^2 u}{\partial \theta^2} = 0, \quad 1 < r < 2, \quad 0 < \theta < \pi/2$$

$$\left. \frac{\partial u}{\partial \theta} \right|_{\theta=0} = 0, \quad \left. \frac{\partial u}{\partial \theta} \right|_{\theta=\pi/2} = 0, \quad 1 < r < 2$$

$$u(1, \theta) = 0, \quad u(2, \theta) = f(\theta), \quad 0 < \theta < \pi/2$$

Using $u = R(r)\Theta(\theta)$ and $\Theta'(0) = 0$, $\Theta'(\pi/2) = 0$, $R(1) = 0$ we get $\Theta(\theta) = c_1$ and $R(r) = c_4 \ln r$ for $\lambda = \alpha^2 = 0$ and for $\lambda = -\alpha^2$ we have $\Theta(\theta) = c_1 2n\theta$ and $R(r) = c_3 \left(r^{2n} - r^{-2n}\right)$. Hence

$$u(r, \theta) = A_0 \ln r + \sum_{n=1}^{\infty} A_n \left(r^{2n} - r^{-2n}\right) \cos 2n\theta.$$

For $r = 2$ we have

$$f(\theta) = A_0 \ln 2 + \sum_{n=1}^{\infty} A_n \left(2^{2n} - 2^{-2n}\right) \cos 2n\theta$$

where

$$A_0 \ln 2 = \frac{a_0}{2} \quad \text{gives} \quad A_0 = \frac{2}{\pi \ln 2} \int_0^{\pi/2} f(\theta)\, d\theta$$

and

$$A_n \left(2^{2n} - 2^{-2n}\right) = a_n \quad \text{gives} \quad A_n = \frac{4}{\pi \left(2^{2n} - 2^{-2n}\right)} \int_0^{\pi/2} f(\theta) \cos 2n\theta\, d\theta.$$

19. Let u_1 be the solution of the boundary-value problem

$$\frac{\partial^2}{u_1 \partial r^2} + \frac{1}{r} \frac{\partial u_1}{\partial r} + \frac{1}{r^2} \frac{\partial^2 u_1}{\partial \theta^2} = 0, \quad 0 < \theta < 2\pi, \quad a < r < b$$

$$u_1(a, \theta) = f(\theta), \quad 0 < \theta < 2\pi$$

$$u_1(b, \theta) = 0, \quad 0 < \theta < 2\pi$$

and let u_2 be the solution to the boundary-value problem

$$\frac{\partial^2 u_2}{\partial r^2} + \frac{1}{r} \frac{\partial u_2}{\partial r} + \frac{1}{r^2} \frac{\partial^2 u_2}{\partial \theta^2} = 0, \quad 0 < \theta < 2\pi, \quad a < r < b$$

$$u_2(a, \theta) = 0, \quad 0 < \theta < 2\pi$$

$$u_2(b, \theta) = g(\theta), \quad 0 < \theta < 2\pi$$

Each of these problems can be solved using the methods shown in Problem 9 of this section. Now if $u(r, \theta) = u_1(r, \theta) + u_2(r, \theta)$, then

$$u(a, \theta) = u_1(a, \theta) + u_2(a, \theta) = f(\theta)$$

$$u(b, \theta) = u_1(b, \theta) + u_2(b, \theta) = g(\theta)$$

and $u(r, \theta)$ will be the steady-state temperature of the circular ring with boundary conditions $u(a, \theta) = f(\theta)$ and $u(b, \theta) = g(\theta)$.

21. Using the same reasoning as in Example 1 in the text we obtain

$$u(r, \theta) = A_0 + \sum_{n=1}^{\infty} r^n (A_n \cos n\theta + B_n \sin n\theta).$$

The boundary condition at $r = c$ implies

$$f(\theta) = \sum_{n=1}^{\infty} nc^{n-1}(A_n \cos n\theta + B_n \sin n\theta).$$

Since this condition does not determine A_0, it is an arbitrary constant. However, to be a full Fourier series on $[0, 2\pi]$ we must require that $f(\theta)$ satisfy the condition $A_0 = a_0/2 = 0$ or $\int_0^{2\pi} f(\theta)\, d\theta = 0$. If this integral were not 0, then the series for $f(\theta)$ would contain a nonzero constant, which it obviously does not. With this as a necessary compatibility condition we can then make the identifications

$$nc^{n-1}A_n = a_n \quad \text{and} \quad nc^{n-1}B_n = b_n$$

or

$$A_n = \frac{1}{nc^{n-1}\pi} \int_0^{2\pi} f(\theta) \cos n\theta\, d\theta \quad \text{and} \quad B_n = \frac{1}{nc^{n-1}\pi} \int_0^{2\pi} f(\theta) \sin n\theta\, d\theta.$$

23. (a) From Problem 1 in this section, with $u_0 = 100$,

$$u(r, \theta) = 50 + \frac{100}{\pi} \sum_{n=1}^{\infty} \frac{1 - (-1)^n}{n} r^n \sin n\theta.$$

(b)

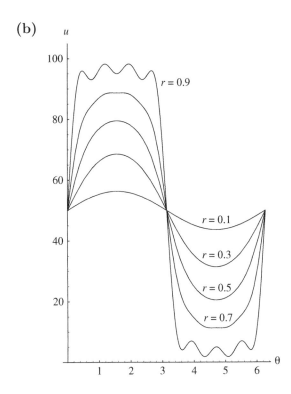

(c) We could use S_5 from part (b) of this problem to compute the approximations, but in a CAS it is just as easy to compute the sum with a much larger number of terms, thereby getting greater accuracy. In this case we use partial sums including the term with r^{99} to find

$$u(0.9, 1.3) \approx 96.5268 \qquad\qquad u(0.9, 2\pi - 1.3) \approx 3.4731$$

$$u(0.7, 2) \approx 87.871 \qquad\qquad u(0.7, 2\pi - 2) \approx 12.129$$

$$u(0.5, 3.5) \approx 36.0744 \qquad\qquad u(0.5, 2\pi - 3.5) \approx 63.9256$$

$$u(0.3, 4) \approx 35.2674 \qquad\qquad u(0.3, 2\pi - 4) \approx 64.7326$$

$$u(0.1, 5.5) \approx 45.4934 \qquad\qquad u(0.1, 2\pi - 5.5) \approx 54.5066$$

(d) At the center of the plate $u(0,0) = 50$. From the graphs in part (b) we observe that the solution curves are symmetric about the point $(\pi, 50)$. In part (c) we observe that the horizontal pairs add up to 100, and hence average 50. This is consistent with the observation about part (b), so it is appropriate to say the average temperature in the plate is $50°$.

13.2 | Polar and Cylindrical Coordinates

1. Referring to the solution of Example 1 in the text we have

$$R(r) = c_1 J_0(\alpha_n r) \qquad \text{and} \qquad T(t) = c_3 \cos a\alpha_n t + c_4 \sin a\alpha_n t$$

where the α_n are the positive roots of $J_0(\alpha c) = 0$. Now, the initial condition $u(r, 0) = R(r)T(0) = 0$ implies $T(0) = 0$ and so $c_3 = 0$. Thus

$$u(r, t) = \sum_{n=1}^{\infty} A_n \sin a\alpha_n t J_0(\alpha_n r) \qquad \text{and} \qquad \frac{\partial u}{\partial t} = \sum_{n=1}^{\infty} a\alpha_n A_n \cos a\alpha_n t J_0(\alpha_n r).$$

From

$$\left. \frac{\partial u}{\partial t} \right|_{t=0} = 1 = \sum_{n=1}^{\infty} a\alpha_n A_n J_0(\alpha_n r)$$

we find

$$a\alpha_n A_n = \frac{2}{c^2 J_1^2(\alpha_n c)} \int_0^c r J_0(\alpha_n r)\, dr \qquad \boxed{x = \alpha_n r, \ dx = \alpha_n\, dr}$$

$$= \frac{2}{c^2 J_1^2(\alpha_n c)} \int_0^{\alpha_n c} \frac{1}{\alpha_n^2} x J_0(x)\, dx$$

$$= \frac{2}{c^2 J_1^2(\alpha_n c)} \int_0^{\alpha_n c} \frac{1}{\alpha_n^2} \frac{d}{dx}[x J_1(x)]\, dx \qquad \boxed{\text{see (4) of Section 11.5 in text}}$$

$$= \frac{2}{c^2 \alpha_n^2 J_1^2(\alpha_n c)} x J_1(x) \Big|_0^{\alpha_n c} = \frac{2}{c\alpha_n J_1(\alpha_n c)}.$$

Then

$$A_n = \frac{2}{ac\alpha_n^2 J_1(\alpha_n c)}$$

and

$$u(r,t) = \frac{2}{ac}\sum_{n=1}^{\infty}\frac{J_0(\alpha_n r)}{\alpha_n^2 J_1(\alpha_n c)}\sin a\alpha_n t.$$

3. Referring to Example 2 in the text we have

$$R(r) = c_1 J_0(\alpha r) + c_2 Y_0(\alpha r)$$
$$Z(z) = c_3 \cosh \alpha z + c_4 \sinh \alpha z$$

where $c_2 = 0$ and $J_0(2\alpha) = 0$ defines the positive eigenvalues $\lambda_n = \alpha_n^2$. From $Z(4) = 0$ we obtain

$$c_3 \cosh 4\alpha_n + c_4 \sinh 4\alpha_n = 0 \qquad \text{or} \qquad c_4 = -c_3 \frac{\cosh 4\alpha_n}{\sinh 4\alpha_n}.$$

Then

$$Z(z) = c_3\left[\cosh \alpha_n z - \frac{\cosh 4\alpha_n}{\sinh 4\alpha_n}\sinh \alpha_n z\right] = c_3 \frac{\sinh 4\alpha_n \cosh \alpha_n z - \cosh 4\alpha_n \sinh \alpha_n z}{\sinh 4\alpha_n}$$

$$= c_3 \frac{\sinh \alpha_n (4-z)}{\sinh 4\alpha_n}$$

and

$$u(r,z) = \sum_{n=1}^{\infty} A_n \frac{\sinh \alpha_n (4-z)}{\sinh 4\alpha_n} J_0(\alpha_n r).$$

From

$$u(r,0) = u_0 = \sum_{n=1}^{\infty} A_n J_0(\alpha_n r)$$

we obtain

$$A_n = \frac{2u_0}{4J_1^2(2\alpha_n)}\int_0^2 r J_0(\alpha_n r)\,dr = \frac{u_0}{\alpha_n J_1(2\alpha_n)}.$$

Thus the temperature in the cylinder is

$$u(r,z) = u_0\sum_{n=1}^{\infty}\frac{\sinh \alpha_n (4-z) J_0(\alpha_n r)}{\alpha_n \sinh(4\alpha_n) J_1(2\alpha_n)}.$$

5. Letting the separation constant be $\lambda = \alpha^2$ and referring to Example 2 in Section 13.2 in the text we have

$$R(r) = c_1 J_0(\alpha r) + c_2 Y_0(\alpha r)$$

$$Z(z) = c_3 \cosh \alpha z + c_4 \sinh \alpha z$$

where $c_2 = 0$ and the positive eigenvalues λ_n are determined by $J_0(2\alpha) = 0$. From $Z'(0) = 0$ we obtain $c_4 = 0$. Then

$$u(r, z) = \sum_{n=1}^{\infty} A_n \cosh \alpha_n z J_0(\alpha_n r).$$

From

$$u(r, 4) = 50 = \sum_{n=1}^{\infty} A_n \cosh 4\alpha_n J_0(\alpha_n r)$$

we obtain (as in Example 1 of Section 13.1)

$$A_n \cosh 4\alpha_n = \frac{2(50)}{4 J_1^2(2\alpha_n)} \int_0^2 r J_0(\alpha_n r)\, dr = \frac{50}{\alpha_n J_1(2\alpha_n)}.$$

Thus the temperature in the cylinder is

$$u(r, z) = 50 \sum_{n=1}^{\infty} \frac{\cosh(\alpha_n z) J_0(\alpha_n r)}{\alpha_n \cosh(4\alpha_n) J_1(2\alpha_n)}.$$

7. Using λ as the separation constant the separated equations are

$$r R'' + R' - r\lambda R = 0 \quad \text{and} \quad Z'' + \lambda Z = 0.$$

The boundary conditions are $Z'(0) = 0$ and $Z'(1) = 0$.

If $\lambda = 0$ the solutions of the ordinary differential equations are

$$R = c_1 + c_2 \ln r \quad \text{and} \quad Z = c_3 + c_4 z.$$

Since $Z'(0) = 0$, $c_4 = 0$. Therefore $Z = c_3$, which satisfies $Z'(1) = 0$. Boundedness at $r = 0$ implies $c_2 = 0$. Therefore $\lambda = 0$ is an eigenvalue with eigenfunction $R = c_1 \neq 0$.

If $\lambda = -\alpha^2 < 0$, the solutions of the ordinary differential equations are

$$R = c_1 J_0(\alpha r) + c_2 Y_0(\alpha r) \quad \text{and} \quad Z = c_3 \cosh \alpha z + c_4 \sinh \alpha z.$$

Since $Z'(0) = 0$, $c_4 = 0$. Therefore $Z = c_3 \cosh \alpha z$. Then $Z'(1) = 0$, so $c_4 \alpha \sinh \alpha = 0$ and $c_4 = 0$. Thus $Z = 0$, and therefore $u = 0$.

If $\lambda = \alpha^2 > 0$, the solutions of the ordinary differential equations are

$$R = c_1 I_0(\alpha r) + c_2 K_0(\alpha r) \quad \text{and} \quad Z = c_3 \cos \alpha z + c_4 \sin \alpha z.$$

Since $Z'(0) = 0$, $c_4 = 0$ and so $Z = c_3 \cos \alpha z$. Now $Z'(1) = 0$ so $c_4 \alpha \sin \alpha = 0$ and $\alpha = n\pi$, $n = 1, 2, 3, \ldots$. The eigenvalues are $\lambda_n = n^2 \pi^2$ and corresponding eigenfunctions are $Z = c_3 \cos n\pi z$. Now, the usual requirement that u be bounded at $r = 0$ implies $c_2 = 0$.

(See Figure 5.3.4 in the text.) Therefore $R = c_1 I_0(\alpha r)$ or $R = c_1 I_0(n\pi r)$. The superposition principle then yields

$$u(r, z) = A_0 + \sum_{n=1}^{\infty} A_n I_0(n\pi r) \cos n\pi z.$$

At $r = 1$

$$u(1, z) = z = A_0 + \sum_{n=1}^{\infty} A_n I_0(n\pi) \cos n\pi z$$

so

$$A_0 = \frac{1}{2} a_0 = \frac{1}{2} \cdot \frac{2}{1} \int_0^1 z \, dz = \frac{1}{2}$$

$$a_n I_0(n\pi) = \frac{2}{1} \int_0^1 z \cos n\pi z \, dz = 2 \frac{(-1)^n - 1}{n^2 \pi^2}$$

and

$$A_n = 2 \frac{(-1)^n - 1}{n^2 \pi^2 I_0(n\pi)}.$$

where we note that $I_0(n\pi)$ has no real zeros. Therefore

$$u(r, z) = \frac{1}{2} + 2 \sum_{n=1}^{\infty} \frac{(-1)^n - 1}{n^2 \pi^2 I_0(n\pi)} I_0(n\pi r) \cos n\pi z.$$

9. Letting $u(r, t) = R(r)T(t)$ and separating variables we obtain

$$\frac{R'' + \frac{1}{r}R'}{R} = \frac{T'}{kT} = -\lambda \qquad \text{and} \qquad R'' + \frac{1}{r}R' + \lambda R = 0, \quad T' + \lambda k T = 0.$$

From the last equation we find $T(t) = e^{-\lambda kt}$. If $\lambda < 0$, $T(t)$ increases without bound as $t \to \infty$. Thus we assume $\lambda = \alpha^2 > 0$. Now

$$R'' + \frac{1}{r}R' + \alpha^2 R = 0$$

is a parametric Bessel equation with solution

$$R(r) = c_1 J_0(\alpha r) + c_2 Y_0(\alpha r).$$

Since Y_0 is unbounded as $r \to 0$ we take $c_2 = 0$. Then $R(r) = c_1 J_0(\alpha r)$ and the boundary condition $u(c, t) = R(c)T(t) = 0$ implies $J_0(\alpha c) = 0$. This latter equation defines the positive eigenvalues $\lambda_n = \alpha_n^2$. Thus

$$u(r, t) = \sum_{n=1}^{\infty} A_n J_0(\alpha_n r) e^{-\alpha_n^2 kt}.$$

From

$$u(r, 0) = f(r) = \sum_{n=1}^{\infty} A_n J_0(\alpha_n r)$$

we find

$$A_n = \frac{2}{c^2 J_1^2(\alpha_n c)} \int_0^c r J_0(\alpha_n r) f(r) \, dr, \quad n = 1, 2, 3, \ldots.$$

11. Referring to Problem 9 we have $T(t) = e^{-\lambda kt}$ and $R(r) = c_1 J_0(\alpha r)$. The boundary condition $hu(1,t) + u_r(1,t) = 0$ implies $hJ_0(\alpha) + \alpha J_0'(\alpha) = 0$ which defines positive eigenvalues $\lambda_n = \alpha_n^2$. Now

$$u(r,t) = \sum_{n=1}^{\infty} A_n J_0(\alpha_n r) e^{-\alpha_n^2 kt}$$

where

$$A_n = \frac{2\alpha_n^2}{(\alpha_n^2 + h^2)J_0^2(\alpha_n)} \int_0^1 r J_0(\alpha_n r) f(r)\, dr.$$

13. Substituting $u(r,t) = v(r,t) + \psi(r)$ into the partial differential equation gives

$$\frac{\partial^2 v}{\partial r^2} + \frac{1}{r}\frac{\partial v}{\partial r} + \psi'' + \frac{1}{r}\psi' = \frac{\partial v}{\partial t}.$$

This equation will be homogeneous provided $\psi'' + \frac{1}{r}\psi' = 0$ or $\psi(r) = c_1 \ln r + c_2$. Since $\ln r$ is unbounded as $r \to 0$ we take $c_1 = 0$. Then $\psi(r) = c_2$ and using $u(2,t) = v(2,t) + \psi(2) = 100$ we set $c_2 = \psi(2) = 100$. Therefore $\psi(r) = 100$. Referring to Problem 5 above, the solution of the boundary-value problem

$$\frac{\partial^2 v}{\partial r^2} + \frac{1}{r}\frac{\partial v}{\partial r} = \frac{\partial v}{\partial t}, \quad 0 < r < 2, \quad t > 0,$$

$$v(2,t) = 0, \quad t > 0,$$

$$v(r,0) = u(r,0) - \psi(r)$$

is

$$v(r,t) = \sum_{n=1}^{\infty} A_n J_0(\alpha_n r) e^{-\alpha_n^2 t}$$

where

$$A_n = \frac{2}{2^2 J_1^2(2\alpha_n)} \int_0^2 r J_0(\alpha_n r)[u(r,0) - \psi(r)]\, dr$$

$$= \frac{1}{2J_1^2(2\alpha_n)}\left[\int_0^1 r J_0(\alpha_n r)[200 - 100]\, dr + \int_1^2 r J_0(\alpha_n r)[100 - 100]\, dr\right]$$

$$= \frac{50}{J_1^2(2\alpha_n)} \int_0^1 r J_0(\alpha_n r)\, dr \qquad \boxed{x = \alpha_n r, \ dx = \alpha_n\, dr}$$

$$= \frac{50}{J_1^2(2\alpha_n)} \int_0^{\alpha_n} \frac{1}{\alpha_n^2} x J_0(x)\, dx$$

$$= \frac{50}{\alpha_n^2 J_1^2(2\alpha_n)} \int_0^{\alpha_n} \frac{d}{dx}[x J_1(x)]\, dx \qquad \boxed{\text{see (5) of Section 11.5 in text}}$$

$$= \frac{50}{\alpha_n^2 J_1^2(2\alpha_n)} (x J_1(x))\Big|_0^{\alpha_n} = \frac{50 J_1(\alpha_n)}{\alpha_n J_1^2(2\alpha_n)}.$$

Thus

$$u(r,t) = v(r,t) + \psi(r) = 100 + 50\sum_{n=1}^{\infty} \frac{J_1(\alpha_n)J_0(\alpha_n r)}{\alpha_n J_1^2(2\alpha_n)} e^{-\alpha_n^2 t}.$$

15. (a) Writing the partial differential equation in the form

$$g\left(x\frac{\partial^2 u}{\partial x^2} + \frac{\partial u}{\partial x}\right) = \frac{\partial^2 u}{\partial t^2}$$

and separating variables we obtain

$$\frac{xX'' + X'}{X} = \frac{T''}{gT} = -\lambda.$$

Letting $\lambda = \alpha^2$ we obtain

$$xX'' + X' + \alpha^2 X = 0 \qquad \text{and} \qquad T'' + g\alpha^2 T = 0.$$

Letting $x = \tau^2/4$ in the first equation we obtain $dx/d\tau = \tau/2$ or $d\tau/dx = 2\tau$. Then

$$\frac{dX}{dx} = \frac{dX}{d\tau}\frac{d\tau}{dx} = \frac{2}{\tau}\frac{dX}{d\tau}$$

and

$$\frac{d^2 X}{dx^2} = \frac{d}{dx}\left(\frac{2}{\tau}\frac{dX}{d\tau}\right) = \frac{2}{\tau}\frac{d}{dx}\left(\frac{dX}{d\tau}\right) + \frac{dX}{d\tau}\frac{d}{dx}\left(\frac{2}{\tau}\right)$$

$$= \frac{2}{\tau}\frac{d}{d\tau}\left(\frac{dX}{d\tau}\right)\frac{d\tau}{dx} + \frac{dX}{d\tau}\frac{d}{d\tau}\left(\frac{2}{\tau}\right)\frac{d\tau}{dx} = \frac{4}{\tau^2}\frac{d^2 X}{d\tau^2} - \frac{4}{\tau^3}\frac{dX}{d\tau}.$$

Thus

$$xX'' + X' + \alpha^2 X = \frac{\tau^2}{4}\left(\frac{4}{\tau^2}\frac{d^2 X}{d\tau^2} - \frac{4}{\tau^3}\frac{dX}{d\tau}\right) + \frac{2}{\tau}\frac{dX}{d\tau} + \alpha^2 X = \frac{d^2 X}{d\tau^2} + \frac{1}{\tau}\frac{dX}{d\tau} + \alpha^2 X = 0.$$

This is a parametric Bessel equation with solution

$$X(\tau) = c_1 J_0(\alpha\tau) + c_2 Y_0(\alpha\tau).$$

(b) To insure a finite solution at $x = 0$ (and thus $\tau = 0$) we set $c_2 = 0$. The condition $u(L, t) = X(L)T(t) = 0$ implies $X\Big|_{x=L} = X\Big|_{\tau=2\sqrt{L}} = c_1 J_0(2\alpha\sqrt{L}) = 0$, which defines positive eigenvalues $\lambda_n = \alpha_n^2$. The solution of $T'' + g\alpha^2 T = 0$ is

$$T(t) = c_3 \cos\left(\alpha_n\sqrt{g}\,t\right) + c_4 \sin\left(\alpha_n\sqrt{g}\,t\right).$$

The boundary condition $u_t(x, 0) = X(x)T'(0) = 0$ implies $c_4 = 0$. Thus

$$u(\tau, t) = \sum_{n=1}^{\infty} A_n \cos\left(\alpha_n\sqrt{g}\,t\right) J_0(\alpha_n\tau).$$

From

$$u(\tau, 0) = f(\tau^2/4) = \sum_{n=1}^{\infty} A_n J_0(\alpha_n\tau)$$

we find

$$A_n = \frac{2}{(2\sqrt{L})^2 J_1^2(2\alpha_n\sqrt{L})} \int_0^{2\sqrt{L}} \tau J_0(\alpha_n\tau) f(\tau^2/4)\, d\tau \qquad \boxed{v = \tau/2, \ dv = d\tau/2}$$

$$= \frac{1}{2L J_1^2(2\alpha_n\sqrt{L})} \int_0^{\sqrt{L}} 2v J_0(2\alpha_n v) f(v^2) 2\, dv$$

$$= \frac{2}{L J_1^2(2\alpha_n\sqrt{L})} \int_0^{\sqrt{L}} v J_0(2\alpha_n v) f(v^2)\, dv.$$

The solution of the boundary-value problem is

$$u(x,t) = \sum_{n=1}^{\infty} A_n \cos\left(\alpha_n\sqrt{g}\,t\right) J_0(2\alpha_n\sqrt{x}\,).$$

17. (a) First we see that

$$\frac{R''\Theta + \dfrac{1}{r} R'\Theta + \dfrac{1}{r^2} R\Theta''}{R\Theta} = \frac{T''}{a^2 T} = -\lambda.$$

This gives $T'' + a^2\lambda T = 0$ and from

$$\frac{R'' + \dfrac{1}{r} R' + \lambda R}{-R/r^2} = \frac{\Theta''}{\Theta} = -\nu$$

we get $\Theta'' + \nu\Theta = 0$ and $r^2 R'' + r R' + (\lambda r^2 - \nu)R = 0$.

(b) With $\lambda = \alpha^2$ and $\nu = \beta^2$ the general solutions of the differential equations in part **(a)** are

$$T = c_1 \cos a\alpha t + c_2 \sin a\alpha t$$
$$\Theta = c_3 \cos \beta\theta + c_4 \sin \beta\theta$$
$$R = c_5 J_\beta(\alpha r) + c_6 Y_\beta(\alpha r).$$

(c) Implicitly we expect $u(r,\theta,t) = u(r,\theta+2\pi,t)$ and so Θ must be 2π-periodic. Therefore $\beta = n$, $n = 0, 1, 2, \ldots$. The corresponding eigenfunctions are 1, $\cos\theta$, $\cos 2\theta$, $\ldots$, $\sin\theta$, $\sin 2\theta$, $\ldots$. Arguing that $u(r,\theta,t)$ is bounded as $r \to 0$ we then define $c_6 = 0$ and so $R = c_3 J_n(\alpha r)$. But $R(c) = 0$ gives $J_n(\alpha c) = 0$; this equation defines the eigenvalues $\lambda_n = \alpha_n^2$. For each n, $\alpha_{ni} = x_{ni}/c$, $i = 1, 2, 3, \ldots$, where x_{ni} are positive roots of $J_n(ac) = 0$. The corresponding eigenfunctions are $J_n(\lambda_{ni} r) = 0$.

(d) $u(r,\theta,t) = \displaystyle\sum_{i=1}^{n} (A_{0i}\cos a\alpha_{0i}t + B_{0i}\sin a\alpha_{0i}t) J_0(\alpha_{0i}r)$

$$+ \sum_{n=1}^{\infty}\sum_{i=1}^{\infty} \Big[(A_{ni}\cos a\alpha_{ni}t + B_{ni}\sin a\alpha_{ni}t)\cos n\theta$$

$$+ (C_{ni}\cos a\alpha_{ni}t + D_{ni}\sin a\alpha_{ni}t)\sin n\theta\Big] J_n(\alpha_{ni}r)$$

19. (a) With $c = 10$ in Example 1 in the text the eigenvalues are $\lambda_n = \alpha_n^2 = x_n^2/100$ where x_n is a positive root of $J_0(x) = 0$. From a CAS we find that $x_1 = 2.4048$, $x_2 = 5.5201$, and $x_3 = 8.6537$, so that the first three eigenvalues are $\lambda_1 = 0.0578$, $\lambda_2 = 0.3047$, and $\lambda_3 = 0.7489$. The corresponding coefficients are

$$A_1 = \frac{2}{100 J_1^2(x_1)} \int_0^{10} r J_0(x_1 r/10)(1 - r/10)\, dr = 0.7845,$$

$$A_2 = \frac{2}{100 J_1^2(x_2)} \int_0^{10} r J_0(x_2 r/10)(1 - r/10)\, dr = 0.0687,$$

and

$$A_3 = \frac{2}{100 J_1^2(x_3)} \int_0^{10} r J_0(x_3 r/10)(1 - r/10)\, dr = 0.0531.$$

Since $g(r) = 0$, $B_n = 0$, $n = 1, 2, 3, \ldots$, and the third partial sum of the series solution is

$$S_3(r, t) = \sum_{n=1}^{\infty} A_n \cos\left(x_n t/10\right) J_0(x_n r/10)$$

$$= 0.7845 \cos\left(0.2405t\right) J_0(0.2405r) + 0.0687 \cos\left(0.5520t\right) J_0(0.5520r)$$

$$+ 0.0531 \cos\left(0.8654t\right) J_0(0.8654r).$$

(b)

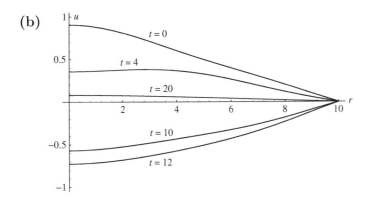

13.3 Spherical Coordinates

1. To compute

$$A_n = \frac{2n + 1}{2c^n} \int_0^{\pi} f(\theta) P_n(\cos\theta) \sin\theta\, d\theta$$

we substitute $x = \cos\theta$ and $dx = -\sin\theta\, d\theta$. Then

$$A_n = \frac{2n + 1}{2c^n} \int_1^{-1} F(x) P_n(x)(-dx) = \frac{2n + 1}{2c^n} \int_{-1}^{1} F(x) P_n(x)\, dx$$

where

$$F(x) = \begin{cases} 0, & -1 < x < 0 \\ 50, & 0 < x < 1 \end{cases}.$$

The coefficients A_n are computed in Example 3 of Section 11.5. Thus

$$u(r, \theta) = \sum_{n=0}^{\infty} A_n r^n P_n(\cos \theta)$$

$$= 50 \left[\frac{1}{2} P_0(\cos \theta) + \frac{3}{4} \left(\frac{r}{c} \right) P_1(\cos \theta) - \frac{7}{16} \left(\frac{r}{c} \right)^3 P_3(\cos \theta) + \frac{11}{32} \left(\frac{r}{c} \right)^5 P_5(\cos \theta) + \cdots \right].$$

3. The coefficients are given by

$$A_n = \frac{2n+1}{2c^n} \int_0^{\pi} \cos \theta \, P_n(\cos \theta) \sin \theta \, d\theta$$

$$= \frac{2n+1}{2c^n} \int_0^{\pi} P_1(\cos \theta) P_n(\cos \theta) \sin \theta \, d\theta \qquad \boxed{x = \cos \theta, \ dx = -\sin \theta \, d\theta}$$

$$= \frac{2n+1}{2c^n} \int_{-1}^{1} P_1(x) P_n(x) \, dx$$

Since $P_n(x)$ and $P_m(x)$ are orthogonal for $m \neq n$, $A_n = 0$ for $n \neq 1$ and

$$A_1 = \frac{2(1)+1}{2c^1} \int_{-1}^{1} P_1(x) P_1(x) \, dx = \frac{3}{2c} \int_{-1}^{1} x^2 \, dx = \frac{1}{c}.$$

Thus

$$u(r, \theta) = \frac{r}{c} P_1(\cos \theta) = \frac{r}{c} \cos \theta.$$

5. Referring to Example 1 in the text we have

$$\Theta = P_n(\cos \theta) \qquad \text{and} \qquad R = c_1 r^n + c_2 r^{-(n+1)}.$$

Since $u(b, \theta) = R(b)\Theta(\theta) = 0$,

$$c_1 b^n + c_2 b^{-(n+1)} = 0 \qquad \text{or} \qquad c_1 = -c_2 b^{-2n-1},$$

and

$$R(r) = -c_2 b^{-2n-1} r^n + c_2 r^{-(n+1)} = c_2 \left(\frac{b^{2n+1} - r^{2n+1}}{b^{2n+1} r^{n+1}} \right).$$

Then

$$u(r, \theta) = \sum_{n=0}^{\infty} A_n \frac{b^{2n+1} - r^{2n+1}}{b^{2n+1} r^{n+1}} P_n(\cos \theta)$$

where

$$\frac{b^{2n+1} - a^{2n+1}}{b^{2n+1} a^{n+1}} A_n = \frac{2n+1}{2} \int_0^{\pi} f(\theta) P_n(\cos \theta) \sin \theta \, d\theta.$$

7. Referring to Example 1 in the text we have

$$r^2 R'' + 2r R' - \lambda R = 0$$

$$\sin \theta \, \Theta'' + \cos \theta \, \Theta' + \lambda \sin \theta \, \Theta = 0.$$

Substituting $x = \cos\theta$, $0 \leq \theta \leq \pi/2$, the latter equation becomes

$$(1 - x^2)\frac{d^2\Theta}{dx^2} - 2x\frac{d\Theta}{dx} + \lambda\Theta = 0, \quad 0 \leq x \leq 1.$$

Taking the solutions of this equation to be the Legendre polynomials $P_n(x)$ corresponding to $\lambda = n(n + 1)$ for $n = 1, 2, 3, \ldots$, we have $\Theta = P_n(\cos\theta)$. Since

$$\left.\frac{\partial u}{\partial\theta}\right|_{\theta=\pi/2} = \Theta'(\pi/2)R(r) = 0$$

we have

$$\Theta'(\pi/2) = -(\sin\pi/2)P_n'(\cos\pi/2) = -P_n'(0) = 0.$$

As noted in the hint, $P_n'(0) = 0$ only if n is even. Thus $\Theta = P_n(\cos\theta)$, $n = 0, 2, 4, \ldots$. As in Example 1, $R(r) = c_1 r^n$. Hence

$$u(r, \theta) = \sum_{n=0}^{\infty} A_{2n} r^{2n} P_{2n}(\cos\theta).$$

At $r = c$,

$$f(\theta) = \sum_{n=0}^{\infty} A_{2n} c^{2n} P_{2n}(\cos\theta).$$

Using Problem 19 in Section 11.5, we obtain

$$c^{2n} A_{2n} = (4n + 1)\int_{\pi/2}^{0} f(\theta)P_{2n}(\cos\theta)(-\sin\theta)\,d\theta$$

and

$$A_{2n} = \frac{4n + 1}{c^{2n}}\int_{0}^{\pi/2} f(\theta)\sin\theta\, P_{2n}(\cos\theta)\,d\theta.$$

9. Checking the hint, we find

$$\frac{1}{r}\frac{\partial^2}{\partial r^2}(ru) = \frac{1}{r}\frac{\partial}{\partial r}\left[r\frac{\partial u}{\partial r} + u\right] = \frac{1}{r}\left[r\frac{\partial^2 u}{\partial r^2} + \frac{\partial u}{\partial r} + \frac{\partial u}{\partial r}\right] = \frac{\partial^2 u}{\partial r^2} + \frac{2}{r}\frac{\partial u}{\partial r}.$$

The partial differential equation then becomes

$$\frac{\partial^2}{\partial r^2}(ru) = r\frac{\partial u}{\partial t}.$$

Now, letting $ru(r, t) = v(r, t) + \psi(r)$, since the boundary condition is nonhomogeneous, we obtain

$$\frac{\partial^2}{\partial r^2}[v(r, t) + \psi(r)] = r\frac{\partial}{\partial t}\left[\frac{1}{r}v(r, t) + \psi(r)\right]$$

or

$$\frac{\partial^2 v}{\partial r^2} + \psi''(r) = \frac{\partial v}{\partial t}.$$

This differential equation will be homogeneous if $\psi''(r) = 0$ or $\psi(r) = c_1 r + c_2$. Now

$$u(r, t) = \frac{1}{r}v(r, t) + \frac{1}{r}\psi(r) \qquad \text{and} \qquad \frac{1}{r}\psi(r) = c_1 + \frac{c_2}{r}.$$

Since we want $u(r, t)$ to be bounded as r approaches 0, we require $c_2 = 0$. Then $\psi(r) = c_1 r$. When $r = 1$

$$u(1, t) = v(1, t) + \psi(1) = v(1, t) + c_1 = 100,$$

and we will have the homogeneous boundary condition $v(1, t) = 0$ when $c_1 = 100$. Consequently, $\psi(r) = 100r$. The initial condition

$$u(r, 0) = \frac{1}{r}v(r, 0) + \frac{1}{r}\psi(r) = \frac{1}{r}v(r, 0) + 100 = 0$$

implies $v(r, 0) = -100r$. We are thus led to solve the new boundary-value problem

$$\frac{\partial^2 v}{\partial r^2} = \frac{\partial v}{\partial t}, \quad 0 < r < 1, \quad t > 0,$$

$$v(1, t) = 0, \quad \lim_{r \to 0} \frac{1}{r}v(r, t) < \infty,$$

$$v(r, 0) = -100r.$$

Letting $v(r, t) = R(r)T(t)$ and using the separation constant $-\lambda$ we obtain

$$R'' + \lambda R = 0 \qquad \text{and} \qquad T' + \lambda T = 0.$$

Using $\lambda = \alpha^2 > 0$ we then have

$$R(r) = c_3 \cos \alpha r + c_4 \sin \alpha r \qquad \text{and} \qquad T(t) = c_5 e^{-\alpha^2 t}.$$

The boundary conditions are equivalent to $R(1) = 0$ and $\lim_{r \to 0} R(r)/r < \infty$. Since

$$\lim_{r \to 0} \frac{\cos \alpha r}{r}$$

does not exist we must have $c_3 = 0$. Then $R(r) = c_4 \sin \alpha r$, and $R(1) = 0$ implies $\alpha = n\pi$ for $n = 1, 2, 3, \ldots$. Thus

$$v_n(r, t) = A_n e^{-n^2 \pi^2 t} \sin n\pi r$$

for $n = 1, 2, 3, \ldots$. Using the condition $\lim_{r \to 0} R(r)/r < \infty$ it is easily shown that there are no eigenvalues for $\lambda = 0$, nor does setting the common constant to $-\lambda = \alpha^2$ when separating variables lead to any solutions. Now, by the superposition principle,

$$v(r, t) = \sum_{n=1}^{\infty} A_n e^{-n^2 \pi^2 t} \sin n\pi r.$$

The initial condition $v(r, 0) = -100r$ implies

$$-100r = \sum_{n=1}^{\infty} A_n \sin n\pi r.$$

This is a Fourier sine series and so

$$A_n = 2 \int_0^1 (-100r \sin n\pi r)\, dr = -200 \left[-\frac{r}{n\pi} \cos n\pi r \Big|_0^1 + \int_0^1 \frac{1}{n\pi} \cos n\pi r\, dr \right]$$

$$= -200 \left[-\frac{\cos n\pi}{n\pi} + \frac{1}{n^2\pi^2} \sin n\pi r \Big|_0^1 \right] = -200 \left[-\frac{(-1)^n}{n\pi} \right] = \frac{(-1)^n 200}{n\pi}.$$

A solution of the problem is thus

$$u(r,t) = \frac{1}{r} v(r,t) + \frac{1}{r} \psi(r) = \frac{1}{r} \sum_{n=1}^\infty (-1)^n \frac{20}{n\pi} e^{-n^2\pi^2 t} \sin n\pi r + \frac{1}{r}(100r)$$

$$= \frac{200}{\pi r} \sum_{n=1}^\infty \frac{(-1)^n}{n} e^{-n^2\pi^2 t} \sin n\pi r + 100.$$

11. We write the differential equation in the form

$$a^2 \frac{1}{r} \frac{\partial^2}{\partial r^2}(ru) = \frac{\partial^2 u}{\partial t^2} \quad \text{or} \quad a^2 \frac{\partial^2}{\partial r^2}(ru) = r \frac{\partial^2 u}{\partial t^2},$$

and then let $v(r,t) = ru(r,t)$. The new boundary-value problem is

$$a^2 \frac{\partial^2 v}{\partial r^2} = \frac{\partial^2 v}{\partial t^2}, \quad 0 < r < c, \quad t > 0$$

$$v(c,t) = 0, \quad t > 0$$

$$v(r,0) = rf(r), \quad \frac{\partial v}{\partial t}\Big|_{t=0} = rg(r).$$

Letting $v(r,t) = R(r)T(t)$ and using the separation constant $-\lambda = -\alpha^2$ we obtain

$$R'' + \alpha^2 R = 0$$
$$T'' + a^2\alpha^2 T = 0$$

and

$$R(r) = c_1 \cos \alpha r + c_2 \sin \alpha r$$
$$T(t) = c_3 \cos a\alpha t + c_4 \sin a\alpha t.$$

Since $u(r,t) = v(r,t)/r$, in order to insure boundedness at $r = 0$ we define $c_1 = 0$. Then $R(r) = c_2 \sin \alpha r$ and the condition $R(c) = 0$ implies $\alpha = n\pi/c$. Thus

$$v(r,t) = \sum_{n=1}^\infty \left(A_n \cos \frac{n\pi a}{c} t + B_n \sin \frac{n\pi a}{c} t \right) \sin \frac{n\pi}{c} r.$$

From

$$v(r,0) = rf(r) = \sum_{n=1}^{\infty} A_n \sin \frac{n\pi}{c} r$$

we see that

$$A_n = \frac{2}{c} \int_0^c rf(r) \sin \frac{n\pi}{c} r\, dr.$$

From

$$\left. \frac{\partial v}{\partial t} \right|_{t=0} = rg(r) = \sum_{n=1}^{\infty} \left(B_n \frac{n\pi a}{c} \right) \sin \frac{n\pi}{c} r$$

we see that

$$B_n = \frac{c}{n\pi a} \cdot \frac{2}{c} \int_0^c rg(r) \sin \frac{n\pi}{c} r\, dr = \frac{2}{n\pi a} \int_0^c rg(r) \sin \frac{n\pi}{c} r\, dr.$$

The solution is

$$u(r,t) = \frac{1}{r} \sum_{n=1}^{\infty} \left(A_n \cos \frac{n\pi a}{c} t + B_n \sin \frac{n\pi a}{c} t \right) \sin \frac{n\pi}{c} r,$$

where A_n and B_n are given above.

13. From (2) in the text $\nabla^2 u + k^2 u = 0$ becomes

$$\frac{\partial^2 u}{\partial r^2} + \frac{2}{r} \frac{\partial u}{\partial r} + \frac{1}{r^2 \sin^2 \theta} \frac{\partial^2 u}{\partial \theta^2} + \frac{1}{r^2} \frac{\partial^2 u}{\partial \theta^2} + \frac{\cot \theta}{r^2} \frac{\partial u}{\partial \theta} = 0$$

or

$$r^2 \frac{\partial^2 u}{\partial r^2} + 2r \frac{\partial u}{\partial r} + k^2 r^2 u = -\frac{1}{\sin^2 \theta} \frac{\partial^2 u}{\partial \phi^2} + \frac{\partial^2 u}{\partial \theta^2} + \cot \theta \frac{\partial u}{\partial \theta}.$$

Since the left hand side of the above equation is strictly a function of r and the right hand side is strictly a function of ϕ and θ, we have

$$r^2 \frac{\partial^2 u}{\partial r^2} + 2r \frac{\partial u}{\partial r} + k^2 r^2 u = 0 \quad \text{and} \quad -\frac{1}{\sin^2 \theta} \frac{\partial^2 u}{\partial \phi^2} + \frac{\partial^2 u}{\partial \theta^2} + \cot \theta \frac{\partial u}{\partial \theta} = 0.$$

Now, assuming that $u(r,\theta,\phi) = R(r)\Theta(\theta)\Phi(\phi)$ we have

$$r^2 R''(r)\Theta(\theta)\Phi(\phi) + 2r R'(r)\Theta(\theta)\Phi(\phi) + k^2 r^2 R(r)\Theta(\theta)\Phi(\phi)$$

$$= -\frac{1}{\sin^2 \theta} R(r)\Theta(\theta)\Phi''(\phi) - R(r)\Theta''(\theta)\Phi(\phi) - \cot \theta\, R'(r)\Theta'(\theta)\Phi(\phi).$$

Thus

$$\frac{r^2 R'' + 2r R' + k^2 r^2 R}{R} = -\frac{1}{\sin^2 \theta} \frac{\Phi''}{\Phi} - \frac{\Theta'' + \cot \theta \Theta'}{\Theta} = n(n+1).$$

This implies that

$$r^2 R'' + 2r R' + \left[k^2 r^2 - n(n+1) \right] R = 0$$

To solve this differential equation identify $r = x$ and $k = \alpha$, and then see Problem 54 in Exercises 6.4.

Chapter 13 in Review

1. We have

$$A_0 = \frac{1}{2\pi} \int_0^\pi u_0 \, d\theta + \frac{1}{2\pi} \int_\pi^{2\pi} (-u_0) \, d\theta = 0$$

$$A_n = \frac{1}{c^n \pi} \int_0^\pi u_0 \cos n\theta \, d\theta + \frac{1}{c^n \pi} \int_\pi^{2\pi} (-u_0) \cos n\theta \, d\theta = 0$$

$$B_n = \frac{1}{c^n \pi} \int_0^\pi u_0 \sin n\theta \, d\theta + \frac{1}{c^n \pi} \int_\pi^{2\pi} (-u_0) \sin n\theta \, d\theta = \frac{2u_0}{c^n n\pi}[1 - (-1)^n]$$

and so

$$u(r, \theta) = \frac{2u_0}{\pi} \sum_{n=1}^\infty \frac{1 - (-1)^n}{n} \left(\frac{r}{c}\right)^n \sin n\theta.$$

3. The conditions $\Theta(0) = 0$ and $\Theta(\pi) = 0$ applied to $\Theta = c_1 \cos \alpha\theta + c_2 \sin \alpha\theta$ give $c_1 = 0$ and $\alpha = n$, $n = 1, 2, 3, \ldots$, respectively. Thus we have the Fourier sine-series coefficients

$$A_n = \frac{2}{\pi} \int_0^\pi u_0(\pi\theta - \theta^2) \sin n\theta \, d\theta = \frac{4u_0}{n^3 \pi}[1 - (-1)^n].$$

Thus

$$u(r, \theta) = \frac{4u_0}{\pi} \sum_{n=1}^\infty \frac{1 - (-1)^n}{n^3} r^n \sin n\theta.$$

5. We solve

$$\frac{\partial^2 u}{\partial r^2} + \frac{1}{r}\frac{\partial u}{\partial r} + \frac{1}{r^2}\frac{\partial^2 u}{\partial \theta^2} = 0, \quad 0 < \theta < \frac{\pi}{4}, \quad \frac{1}{2} < r < 1,$$

$$u(r, 0) = 0, \quad u(r, \pi/4) = 0, \quad \frac{1}{2} < r < 1,$$

$$u(1/2, \theta) = u_0, \quad u_r(1, \theta) = 0, \quad 0 < \theta < \frac{\pi}{4}.$$

Proceeding as in Example 1 in Section 13.1 using the separation constant $\lambda = \alpha^2$ we obtain

$$r^2 R'' + r R' - \lambda R = 0$$
$$\Theta'' + \lambda\Theta = 0$$

with solutions

$$\Theta(\theta) = c_1 \cos \alpha\theta + c_2 \sin \alpha\theta$$
$$R(r) = c_3 r^\alpha + c_4 r^{-\alpha}.$$

Applying the boundary conditions $\Theta(0) = 0$ and $\Theta(\pi/4) = 0$ gives $c_1 = 0$ and $\alpha = 4n$ for $n = 1, 2, 3, \ldots$. From $R_r(1) = 0$ we obtain $c_3 = c_4$.

Therefore

$$u(r, \theta) = \sum_{n=1}^{\infty} A_n \left(r^{4n} + r^{-4n} \right) \sin 4n\theta.$$

From

$$u(1/2, \theta) = u_0 = \sum_{n=1}^{\infty} A_n \left(\frac{1}{2^{4n}} + \frac{1}{2^{-4n}} \right) \sin 4n\theta$$

we find

$$A_n \left(\frac{1}{2^{4n}} + \frac{1}{2^{-4n}} \right) = \frac{2}{\pi/4} \int_0^{\pi/4} u_0 \sin 4n\theta \, d\theta = \frac{2u_0}{n\pi}[1 - (-1)^n]$$

or

$$A_n = \frac{2u_0}{n\pi(2^{4n} + 2^{-4n})}[1 - (-1)^n].$$

Thus the steady-state temperature in the plate is

$$u(r, \theta) = \frac{2u_0}{\pi} \sum_{n=1}^{\infty} \frac{[r^{4n} + r^{-4n}][1 - (-1)^n]}{n[2^{4n} + 2^{-4n}]} \sin 4n\theta.$$

7. Letting $u(r, t) = R(r)T(t)$ and separating variables we obtain

$$\frac{R'' + \frac{1}{r}R' - hR}{R} = \frac{T'}{T} = \lambda$$

so

$$R'' + \frac{1}{r}R' - (\lambda + h)R = 0 \quad \text{and} \quad T' - \lambda T = 0.$$

From the second equation we find $T(t) = c_1 e^{\lambda t}$. If $\lambda > 0$, $T(t)$ increases without bound as $t \to \infty$. Thus we assume $\lambda = -\alpha^2 < 0$. Since $h > 0$ we can take $\mu = -\alpha^2 - h$. Then

$$R'' + \frac{1}{r}R' + \alpha^2 R = 0$$

is a parametric Bessel equation with solution

$$R(r) = c_1 J_0(\alpha r) + c_2 Y_0(\alpha r).$$

Since Y_0 is unbounded as $r \to 0$ we take $c_2 = 0$. Then $R(r) = c_1 J_0(\alpha r)$ and the boundary condition $u(1, t) = R(1)T(t) = 0$ implies $J_0(\alpha) = 0$. This latter equation defines the positive eigenvalues λ_n. Thus

$$u(r, t) = \sum_{n=1}^{\infty} A_n J_0(\alpha_n r) e^{(-\alpha_n^2 - h)t}.$$

From

$$u(r, 0) = 1 = \sum_{n=1}^{\infty} A_n J_0(\alpha_n r)$$

we find

$$A_n = \frac{2}{J_1^2(\alpha_n)} \int_0^1 r J_0(\alpha_n r)\,dr \qquad \boxed{x = \alpha_n r,\ dx = \alpha_n\,dr}$$

$$= \frac{2}{J_1^2(\alpha_n)} \int_0^{\alpha_n} \frac{1}{\alpha_n^2} x J_0(x)\,dx.$$

From recurrence relation (5) in Section 11.5 of the text we have

$$x J_0(x) = \frac{d}{dx}[x J_1(x)].$$

Then

$$A_n = \frac{2}{\alpha_n^2 J_1^2(\alpha_n)} \int_0^{\alpha_n} \frac{d}{dx}[x J_1(x)]\,dx = \frac{2}{\alpha_n^2 J_1^2(\alpha_n)} \left. (x J_1(x)) \right|_0^{\alpha_n} = \frac{2\alpha_1 J_1(\alpha_n)}{\alpha_n^2 J_1^2(\alpha_n)} = \frac{2}{\alpha_n J_1(\alpha_n)}$$

and

$$u(r,t) = 2e^{-ht} \sum_{n=1}^{\infty} \frac{J_0(\alpha_n r)}{\alpha_n J_1(\alpha_n)} e^{-\alpha_n^2 t}$$

9. The boundary-value problem is

$$\frac{\partial^2 u}{\partial r^2} + \frac{1}{r}\frac{\partial u}{\partial r} + \frac{\partial^2 u}{\partial z^2} = 0, \quad 0 < r < 2, \quad 0 < z < 4$$

$$u(2, z) = 50, \quad 0 < z < 4$$

$$\left. \frac{\partial u}{\partial z} \right|_{z=0} = 0, \quad u(r, 4) = 0, \quad 0 < r < 2.$$

We have $u(r, z) = v(r, z) + \psi(r)$ so $\psi'' + (1/r)\psi' = 0$ and $\psi = c_1 \ln r + c_2$. Now, boundedness at $r = 0$ implies $c_2 = 0$. Also, $\psi = c_2$ and $u(r, z) = v(r, z) + \psi(r)$, which implies

$$u(2, z) = v(2, z) + \psi(2) = 50 = c_2,$$

and $\psi(r) = 50$, all of which implies

$$\frac{\partial^2 v}{\partial r^2} + \frac{1}{r}\frac{\partial v}{\partial r} + \frac{\partial^2 v}{\partial z^2} = 0, \quad 0 < r < 2, \quad 0 < z < 4$$

$$v(2, z) = 0, \quad 0 < z < 4$$

$$\left. \frac{\partial v}{\partial z} \right|_{z=0} = 0, \quad v(r, 4) = -50, \quad 0 < r < 2.$$

(See the solution of Problem 5 in Exercises 13.2.) From

$$v(r, z) = -50 \sum_{n=1}^{\infty} \frac{\cosh(\alpha_n z)}{\alpha_n \cosh(4\alpha_n) J_1(2\alpha_n)} J_0(\alpha_n r)$$

we get

$$u(r, z) = 50 - 50 \sum_{n=1}^{\infty} \frac{\cosh(\alpha_n z)}{\alpha_n \cosh(4\alpha_n) J_1(2\alpha_n)} J_0(\alpha_n r).$$

11. Referring to Example 1 in Section 13.3 of the text we have

$$u(r, \theta) = \sum_{n=0}^{\infty} A_n r^n P_n(\cos \theta).$$

For $x = \cos \theta$

$$u(1, \theta) = \begin{cases} 100, & 0 < \theta < \pi/2 \\ -100, & \pi/2 < \theta < \pi \end{cases} = g(x).$$

From Problem 22 in Exercise 11.5 we have

$$u(r, \theta) = 100 \left[\frac{3}{2} r P_1(\cos \theta) - \frac{7}{8} r^3 P_3(\cos \theta) + \frac{11}{16} r^5 P_5(\cos \theta) + \cdots \right].$$

13. We note that the differential equation can be expressed in the form

$$\frac{d}{dx}[xu'] = -\alpha^2 xu.$$

Thus

$$u_n \frac{d}{dx}[xu'_m] = -\alpha_m^2 xu_m u_n$$

and

$$u_m \frac{d}{dx}[xu'_n] = -\alpha_n^2 xu_n u_m.$$

Subtracting we obtain

$$u_n \frac{d}{dx}[xu'_m] - u_m \frac{d}{dx}[xu'_n] = (\alpha_n^2 - \alpha_m^2) xu_m u_n$$

and

$$\int_a^b u_n \frac{d}{dx}[xu'_m]\, dx - \int_a^b u_m \frac{d}{dx}[xu'_n] = (\alpha_n^2 - \alpha_m^2) \int_a^b xu_m u_n\, dx.$$

Using integration by parts this becomes

$$u_n xu'_m \Big|_a^b - \int_a^b xu'_m u'_n\, dx - u_m xu'_n \Big|_a^b + \int_a^b xu'_n u'_m\, dx$$

$$= b[u_n(b)u'_m(b) - u_m(b)u'_n(b)] - a[u_n(a)u'_m(a) - u_m(a)u'_n(a)]$$

$$= (\alpha_n^2 - \alpha_m^2) \int_a^b xu_m u_n\, dx.$$

Since

$$u(x) = Y_0(\alpha a)J_0(\alpha x) - J_0(\alpha a)Y_0(\alpha x)$$

we have

$$u_n(b) = Y_0(\alpha_n a)J_0(\alpha_n b) - J_0(\alpha_n a)Y_0(\alpha_n b) = 0$$

by the definition of the α_n. Similarly $u_m(b) = 0$. Also

$$u_n(a) = Y_0(\alpha a)J_0(\alpha_n a) - J_0(\alpha_n a)Y_0(\alpha_n a) = 0$$

and $u_m(a) = 0$. Therefore

$$\int_a^b x u_m u_n \, dx = \frac{1}{\alpha_n^2 - \alpha_m^2}\left(b[u_n(b)u_m'(b) - u_m(b)u_n'(b)] - a[u_n(a)u_m'(a) - u_m(a)u_n'(a)]\right) = 0$$

and the $u_n(x)$ are orthogonal with respect to the weight function x.

15. We use the superposition principle for Laplace's equation discussed in Section 12.5 and shown schematically in Figure 12.5.3 in the text. That is,

$$\text{Solution } u = \text{ Solution } u_1 \text{ of Problem 1 } + \text{ Solution } u_2 \text{ of Problem 2,}$$

where in Problem 1 the boundary condition on the top and bottom of the cylinder is $u = 0$, while on the lateral surface $r = c$ it is $u = h(z)$, and in Problem 2 the boundary condition on the top of the cylinder $z = L$ is $u = f(r)$, on the bottom $z = 0$ it is $u = g(r)$, and on the lateral surface $r = c$ it is $u = 0$.

Solution for $u_1(r, z)$

Using λ as a separation constant we have

$$\frac{R'' + \frac{1}{r}R'}{R} = -\frac{Z''}{Z} = \lambda,$$

so

$$rR'' + R' - \lambda rR = 0 \qquad \text{and} \qquad Z'' + \lambda Z = 0.$$

The differential equation in Z, together with the boundary conditions $Z(0) = 0$ and $Z(L) = 0$ is a Sturm-Liouville problem. Letting $\lambda = \alpha^2 > 0$ we note that the above differential equation in R is a modified parametric Bessel equation which is discussed in Section 6.3 in the text. Also, we have $Z(z) = c_1 \cos \alpha z + c_2 \sin \alpha z$. The boundary conditions imply $c_1 = 0$ and $\sin \alpha L = 0$. Thus, $\alpha_n = n\pi/L$, $n = 1, 2, 3, \ldots$, so $\lambda_n = n^2\pi^2/L^2$ and

$$R(r) = c_3 I_0\left(\frac{n\pi}{L}r\right) + c_4 K_0\left(\frac{n\pi}{L}r\right).$$

Now boundedness at $r = 0$ implies $c_4 = 0$, so $R(r) = c_3 I_0(n\pi r/L)$ and

$$u_1(r, z) = \sum_{n=1}^{\infty} A_n I_0\left(\frac{n\pi}{L}r\right) \sin\left(\frac{n\pi}{L}z\right).$$

At $r = c$ for $0 < z < L$ we have

$$h(z) = u_1(c, z) = \sum_{n=1}^{\infty} A_n I_0\left(\frac{n\pi}{L}c\right) \sin\left(\frac{n\pi}{L}z\right)$$

which gives

$$A_n = \frac{2}{LI_0(n\pi c/L)} \int_0^L h(z) \sin\left(\frac{n\pi}{L}z\right) dz.$$

$\boxed{\text{Solution for } u_2(r,z)}$

In this case we use $-\lambda$ as a separation constant which leads to

$$\frac{R'' + \frac{1}{r}R'}{R} = -\frac{Z''}{Z} = -\lambda,$$

so

$$rR'' + R' + \lambda r R = 0 \qquad \text{and} \qquad Z'' - \lambda Z = 0.$$

The differential equation in R is a parametric Bessel equation. Using $\lambda = \alpha^2$ we find $R(r) = c_1 J_0(\alpha r) + c_2 Y_0(\alpha r)$. Boundedness at $r = 0$ implies $c_2 = 0$ so $R(r) = c_3 J_0(\alpha r)$. The boundary condition $R(c) = 0$ then gives the defining equation for the eigenvalues: $J_0(\alpha c) = 0$. Let $\lambda_n = \alpha_n^2$ where $\alpha_n c = x_n$ are the roots. The solution of the differential equation in Z is $Z(z) = c_4 \cosh \alpha_n z + c_5 \sinh \alpha_n z$, so

$$u_2(r,z) = \sum_{n=1}^{\infty} (B_n \cosh \alpha_n z + C_n \sinh \alpha_n z) J_0(\alpha_n r).$$

At $z = 0$, for $0 < r < c$, we have

$$f(r) = u_2(r,0) = \sum_{n=1}^{\infty} B_n J_0(\alpha_n r),$$

so

$$B_n = \frac{2}{c^2 J_1^2(\alpha_n c)} \int_0^c r f(r) J_0(\alpha_n r)\, dr.$$

At $z = L$, for $0 < r < c$, we have

$$g(r) = u_2(r,L) = \sum_{n=1}^{\infty} (B_n \cosh \alpha_n L + C_n \sinh \alpha_n L) J_0(\alpha_n r),$$

so

$$B_n \cosh \alpha_n L + C_n \sinh \alpha_n L = \frac{2}{c^2 J_1^2(\alpha_n c)} \int_0^c r g(r) J_0(\alpha_n r)\, dr$$

and

$$C_n = -B_n \frac{\cosh \alpha_n L}{\sinh \alpha_n L} + \frac{2}{c^2 (\sinh \alpha_n L) J_1^2(\alpha_n c)} \int_0^c r g(r) J_0(\alpha_n r)\, dr.$$

By the superposition principle the solution of the original problem is

$$u(r,z) = u_1(r,z) + u_2(r,z).$$

17. Using λ as the separation constant the separated equations are

$$rR'' + R' - r\lambda R = 0 \qquad \text{and} \qquad Z'' + \lambda Z = 0.$$

The boundary conditions are $Z(0) = 0$ and $Z'(1) = 0$.

If $\lambda = 0$ the solutions of the ordinary differential equations are

$$R = c_1 + c_2 \ln r \qquad \text{and} \qquad Z = c_3 + c_4 z.$$

Since $Z(0) = 0$, $c_3 = 0$. Therefore $Z = c_4 z$ and $Z'(1) = 0$ so $Z(1) = c_4 = 0$. The product solution is $u = R(r)Z(z) = 0$. Thus $\lambda = 0$ is not an eigenvalue.

Of $\lambda = -\alpha^2 < 0$, the solutions of the ordinary differential equations are

$$R = c_1 J_0(\alpha r) + c_2 Y_0(\alpha r) \qquad \text{and} \qquad Z = c_3 \cosh \alpha z + c_4 \sinh \alpha z.$$

Since $Z(0) = 0$, $c_3 = 0$. Therefore $Z = c_4 \sinh \alpha z$ and $Z'(1) = 0$ so $c_4 \alpha \cosh \alpha = 0$, which implies $c_4 = 0$. Thus $Z = 0$ and therefore $u = 0$.

If $\lambda = \alpha^2 > 0$, the solution of the ordinary differential equations are

$$R = c_1 I_0(\alpha r) + c_2 K_0(\alpha r) \qquad \text{and} \qquad Z = c_3 \cos \alpha z + c_4 \sin \alpha z.$$

Since $Z(0) = 0$, $c_3 = 0$ and so $Z = c_4 \sin \alpha z$. Now $Z'(1) = 0$ so $c_4 \alpha \cos \alpha = 0$ and $\alpha = (2n-1)\pi/2$, $n = 1, 2, 3, \ldots$. The eigenvalues are $\lambda_n = (2n-1)^2 \pi^2/4$ and corresponding eigenfunctions are $Z = c_4 \sin (2n-1)\pi z/2$. Now, the usual implicit requirement that u be bounded at $r = 0$ implies $c_2 = 0$. (See Figure 6.4.4 in the text.) Therefore $R = c_1 I_0(\alpha r)$ or $R = c_1 I_0 \left((2n-1)\pi r/2 \right)$. The superposition principle yields

$$u(r, z) = \sum_{n=1}^{\infty} A_n I_0 \left(\frac{2n-1}{2} \pi r \right) \sin \frac{2n-1}{2} \pi z.$$

At $r = 1$

$$u(1, z) = u_0 = \sum_{n=1}^{\infty} A_n I_0 \left(\frac{2n-1}{2} \pi \right) \sin \frac{2n-1}{2} \pi z,$$

which is not a Fourier series. Thus

$$A_n I_0 \left(\frac{2n-1}{2} \pi \right) = \frac{\displaystyle\int_0^1 u_0 \sin \frac{2n-1}{2} \pi z \, dz}{\displaystyle\int_0^1 \sin^2 \frac{2n-1}{2} \pi z \, dz} = \frac{4u_0}{(2n-1)\pi}.$$

Therefore

$$u(r, z) = \frac{4u_0}{\pi} \sum_{n=1}^{\infty} \frac{I_0 \left(\frac{2n-1}{2} \pi r \right)}{(2n-1) I_0 \left(\frac{2n-1}{2} \pi \right)} \sin \frac{2n-1}{2} \pi z.$$

19. Using $-\lambda$ as the separation constant, the only value of the separation constant that leads to nontrivial solutions is $\lambda = \alpha^2$. With that we get

$$rR'' + R' - \alpha^2 r R = 0 \qquad \text{or} \qquad R(r) = c_1 J_0(\alpha r) + c_2 Y_0(\alpha r)$$

and $Z'' - \alpha^2 Z = 0$ yields $Z(z) = c_3 e^{-\alpha z} + c_4 e^{\alpha z}$. Boundedness at $r = 0$ implies $c_2 = 0$ so $R(r) = c_1 J_0(\alpha r)$. Then $u(1, z) = 0$ and hence $J_0(\alpha) = 0$. If α_n denotes the positive roots of the last equation, then $Z(z) = c_3 e^{-\alpha_n z} + c_4 e^{\alpha_n z}$. Again, boundedness as $z \to \infty$ implies $c_4 = 0$. Therefore,

$$u_n(r, z) = A_n e^{-\alpha_n z} J_0(\alpha_n r).$$

and

$$u(r, z) = \sum_{k=1}^{\infty} A_n e^{-\alpha_n z} J_0(\alpha_n r).$$

Then

$$u(r, 0) = 1 - r^2 = \sum_{n=1}^{\infty} A_n J_0(\alpha_n r)$$

. Using (16) of Definition 11.5.1 gives

$$A_n = \frac{2}{J_1^2(\alpha_n)} \int_0^1 r \left(1 - r^2\right) J_0(\alpha_n r)\, dr.$$

From Problem 10 of Exercises 1..5, we find the value of the foregoing integral to be

$$A_n = \frac{4 J_2(\alpha_n)}{\alpha_n^2 J_1^2(\alpha_n)}$$

.

Therefore,

$$u(r, z) = 4 \sum_{n=1}^{\infty} \frac{J_2(\alpha_n)}{\alpha_n^2 J_1^2(\alpha_n)} e^{-\alpha_n z} J_0(\alpha_n r)$$

.

We can simplify the last result using the recurrence relation

$$2\nu J_\nu(x) = x J_{\nu+1}(x) + x J_{\nu-1}(x)$$

.

See Problem 28 in Exercises 6.4. With $\nu = 1$ and $x = \alpha_n$ we have

$$2 J_1(\alpha_n) = \alpha_n J_2(\alpha_n) + \alpha_n \overbrace{J_0(\alpha_n)}^{0}$$

$$J_2(\alpha_n) = \frac{2}{\alpha_n} J_1(\alpha_n)$$

$$A_n = \frac{8 u_0}{\alpha_n^3 J_1(\alpha_n)}$$

An alternative form of the answer is

$$u(r, z) = 8 \sum_{n=1}^{\infty} \frac{e^{-\alpha_n z}}{\alpha_n^3 J_1(\alpha_n r)} J_0(\alpha_n r).$$

21. The boundary-value problem in polar coordinates is

$$\frac{\partial^2 u}{\partial r^2} + \frac{1}{r}\frac{\partial u}{\partial r} + \frac{1}{r^2}\frac{\partial^2 u}{\partial \theta^2} = 0, \quad 0 < \theta < 2\pi, \quad 1 < r < 2,$$

$$u(1, \theta) = \sin^2 \theta, \quad \frac{\partial u}{\partial r}\bigg|_{r=2} = 0, \quad 0 < \theta < 2\pi.$$

Letting $u(r, \theta) = R(r)\Theta(\theta)$ and separating variables we obtain

$$r^2 R'' + rR' - \lambda R = 0 \quad \text{and} \quad \Theta'' + \lambda\Theta = 0.$$

For $\lambda = 0$ we have $\Theta'' = 0$ and $r^2 R'' + rR' = 0$. This gives $\Theta = c_1 + c_2\theta$ and $R = c_3 + c_4 \ln r$. The periodicity assumption (as mentioned in Example 1 of Section 13.1 in the text) implies $c_2 = 0$, while the boundary condition $R'(2) = 0$ implies $c_4 = 0$. Thus, for $\lambda = 0$, $u = c_1 c_3 = A_0$. Now, for $\lambda = \alpha^2$, the differential equations become $\Theta'' + \alpha^2\Theta = 0$ and $r^2 R'' + rR' - \alpha^2 R = 0$. The corresponding solutions are $\Theta = c_5 \cos \alpha\theta + c_6 \sin \alpha\theta$ and $R = c_7 r^{\alpha} + c_8 r^{-\alpha}$. In this case the periodicity assumption implies $\alpha = n$, $n = 1, 2, 3, \ldots$, while the boundary condition $R'(2) = 0$ implies $R = c_7(r^n + 4^n r^{-n})$. The product of the solutions is $u_n = (r^n + 4^n r^{-n})(A_n \cos n\theta + B_n \sin n\theta)$ and the superposition principle implies

$$u(r, \theta) = A_0 + \sum_{n=1}^{\infty} (r^n + 4^n r^{-n})(A_n \cos n\theta + B_n \sin n\theta).$$

Using the boundary condition at $r = 1$ we have

$$\sin^2 \theta = \frac{1}{2}(1 - \cos 2\theta) = A_0 + \sum_{n=1}^{\infty} (1 + 4^n)(A_n \cos n\theta + B_n \sin n\theta).$$

From this we conclude that $B_n = 0$ for all integers n, $A_0 = \frac{1}{2}$, $A_1 = 0$, $A_2 = -\frac{1}{34}$, and $A_m = 0$ for $m = 3, 4, 5, \ldots$. Therefore

$$u(r, \theta) = A_0 + A_2 \cos 2\theta = \frac{1}{2} - \frac{1}{34}(r^2 + 16r^{-2})\cos 2\theta = \frac{1}{2} - \left(\frac{1}{34}r^2 + \frac{8}{17}r^{-2}\right)\cos 2\theta.$$

Chapter 14

Integral Transforms

14.1 | Error Function

1. **(a)** The result follows by letting $\tau = u^2$ or $u = \sqrt{\tau}$ in $\operatorname{erf}(\sqrt{t}) = \dfrac{2}{\sqrt{\pi}} \displaystyle\int_0^{\sqrt{t}} e^{-u^2}\, du$.

(b) Using $\mathscr{L}\{t^{-1/2}\} = \dfrac{\sqrt{\pi}}{s^{1/2}}$ and the first translation theorem, it follows from the convolution theorem that

$$\mathscr{L}\left\{\operatorname{erf}(\sqrt{t})\right\} = \frac{1}{\sqrt{\pi}}\, \mathscr{L}\left\{\int_0^t \frac{e^{-\tau}}{\sqrt{\tau}}\, d\tau\right\} = \frac{1}{\sqrt{\pi}}\, \mathscr{L}\{1\}\mathscr{L}\left\{t^{-1/2}e^{-t}\right\} = \frac{1}{\sqrt{\pi}}\, \frac{1}{s}\, \mathscr{L}\left\{t^{-1/2}\right\}\Bigg|_{s \to s+1}$$

$$= \frac{1}{\sqrt{\pi}}\, \frac{1}{s}\, \frac{\sqrt{\pi}}{\sqrt{s+1}} = \frac{1}{s\sqrt{s+1}}.$$

3. By the first translation theorem,

$$\mathscr{L}\left\{e^t \operatorname{erf}(\sqrt{t})\right\} = \mathscr{L}\left\{\operatorname{erf}(\sqrt{t})\right\}\Bigg|_{s \to s-1} = \frac{1}{s\sqrt{s+1}}\Bigg|_{s \to s-1} = \frac{1}{\sqrt{s}\,(s-1)}.$$

5. From Problem 4 we know that

$$\mathscr{L}\left\{e^t \operatorname{erfc}(\sqrt{t})\right\} = \frac{1}{\sqrt{s}(\sqrt{s}+1)}.$$

From this result together with formula 46 from the table of Laplace transforms, we get

$$\mathscr{L}\left\{\frac{1}{\sqrt{\pi t}} - e^t \operatorname{erfc}(\sqrt{t})\right\} = \mathscr{L}\left\{\frac{1}{\sqrt{\pi t}}\right\} - \mathscr{L}\left\{e^t \operatorname{erfc}(\sqrt{t})\right\} = \frac{1}{\sqrt{s}} - \frac{1}{\sqrt{s}(\sqrt{s}+1)} = \frac{1}{\sqrt{s}+1}.$$

7. From entry 3 in Table 14.1.1 and the first translation theorem we have

$$\mathscr{L}\left\{e^{-Gt/C}\operatorname{erf}\left(\frac{x}{2}\sqrt{\frac{RC}{t}}\right)\right\} = \mathscr{L}\left\{e^{-Gt/C}\left[1 - \operatorname{erfc}\left(\frac{x}{2}\sqrt{\frac{RC}{t}}\right)\right]\right\}$$

$$= \mathscr{L}\left\{e^{-Gt/C}\right\} - \mathscr{L}\left\{e^{-Gt/C}\operatorname{erfc}\left(\frac{x}{2}\sqrt{\frac{RC}{t}}\right)\right\}$$

$$= \frac{1}{s + G/C} - \left.\frac{e^{-x\sqrt{RC}\sqrt{s}}}{s}\right|_{s \to s + G/C}$$

$$= \frac{1}{s + G/C} - \frac{e^{-x\sqrt{RC}\sqrt{s+G/C}}}{s + G/C} = \frac{C}{Cs + G}\left(1 - e^{-x\sqrt{RCs+RG}}\right).$$

9. Taking the Laplace transform of both sides of the equation we obtain

$$\mathscr{L}\{y(t)\} = \mathscr{L}\{1\} - \mathscr{L}\left\{\int_0^t \frac{y(\tau)}{\sqrt{t-\tau}}\,d\tau\right\}$$

$$Y(s) = \frac{1}{s} - Y(s)\frac{\sqrt{\pi}}{\sqrt{s}}$$

$$\frac{\sqrt{s}+\sqrt{\pi}}{\sqrt{s}}Y(s) = \frac{1}{s}$$

$$Y(s) = \frac{1}{\sqrt{s}\left(\sqrt{s}+\sqrt{\pi}\right)}.$$

Thus

$$y(t) = \mathscr{L}^{-1}\left\{\frac{1}{\sqrt{s}\left(\sqrt{s}+\sqrt{\pi}\right)}\right\} = e^{\pi t}\operatorname{erfc}(\sqrt{\pi t}). \qquad \boxed{\text{By entry 5 in Table 14.1.1}}$$

11. $\displaystyle\int_a^b e^{-u^2}\,du = \int_a^0 e^{-u^2}\,du + \int_0^b e^{-u^2}\,du = \int_0^b e^{-u^2}\,du - \int_0^a e^{-u^2}\,du$

$$= \frac{\sqrt{\pi}}{2}\operatorname{erf}(b) - \frac{\sqrt{\pi}}{2}\operatorname{erf}(a) = \frac{\sqrt{\pi}}{2}[\operatorname{erf}(b) - \operatorname{erf}(a)]$$

13. Since $\operatorname{erf}(x)/x$ is the indeterminate form $0/0$ as $x \to 0$, we have by L'Hôpital's rule:

$$\lim_{x\to 0}\frac{\operatorname{erf}(x)}{x} = \lim_{x\to 0}\frac{\dfrac{d}{dx}\operatorname{erf}(x)}{\dfrac{d}{dx}x} = \lim_{x\to 0}\frac{\dfrac{2}{\sqrt{\pi}}\dfrac{d}{dx}\displaystyle\int_0^x e^{-t^2}\,dt}{1} = \lim_{x\to 0}\frac{2}{\sqrt{\pi}}e^{-x^2} = \frac{2}{\sqrt{\pi}}\cdot 1 = \frac{2}{\sqrt{\pi}}.$$

15. $\operatorname{erf}(x) + \operatorname{erfc}(x) = 1$

$$\operatorname{erfc}(x) = 1 - \operatorname{erf}(x)$$

$$\operatorname{erfc}(-x) = 1 - \operatorname{erf}(-x)$$

Since $\operatorname{erf}(x)$ is an odd function $\operatorname{erf}(-x) = -\operatorname{erf}(x)$, so

$$\operatorname{erfc}(-x) = 1 - (-\operatorname{erf}(x)) = 1 + \operatorname{erf}(x).$$

14.2 │ Laplace Transform

1. The boundary-value problem is

$$a^2 \frac{\partial^2 u}{\partial x^2} = \frac{\partial^2 u}{\partial t^2}, \quad 0 < x < L, \quad t > 0,$$

$$u(0,t) = 0, \quad u(L,t) = 0, \quad t > 0,$$

$$u(x,0) = A \sin \frac{\pi}{L} x, \quad \frac{\partial u}{\partial t}\bigg|_{t=0} = 0.$$

Transforming the partial differential equation gives

$$\frac{d^2 U}{dx^2} - \left(\frac{s}{a}\right)^2 U = -\frac{s}{a^2} A \sin \frac{\pi}{L} x.$$

Using undetermined coefficients we obtain

$$U(x,s) = c_1 \cosh \frac{s}{a} x + c_2 \sinh \frac{s}{a} x + \frac{As}{s^2 + a^2 \pi^2 / L^2} \sin \frac{\pi}{L} x.$$

The transformed boundary conditions, $U(0,s) = 0$, $U(L,s) = 0$ give in turn $c_1 = 0$ and $c_2 = 0$. Therefore

$$U(x,s) = \frac{As}{s^2 + a^2 \pi^2 / L^2} \sin \frac{\pi}{L} x$$

and

$$u(x,t) = A\mathscr{L}^{-1} \left\{ \frac{s}{s^2 + a^2 \pi^2 / L^2} \right\} \sin \frac{\pi}{L} x = A \cos \frac{a\pi}{L} t \sin \frac{\pi}{L} x.$$

3. The solution of

$$a^2 \frac{d^2 U}{dx^2} - s^2 U = 0$$

is in this case

$$U(x,s) = c_1 e^{-(x/a)s} + c_2 e^{(x/a)s}.$$

Since $\lim_{x\to\infty} u(x,t) = 0$ we have $\lim_{x\to\infty} U(x,s) = 0$. Thus $c_2 = 0$ and

$$U(x,s) = c_1 e^{-(x/a)s}.$$

If $\mathscr{L}\{u(0,t)\} = \mathscr{L}\{f(t)\} = F(s)$ then $U(0,s) = F(s)$. From this we have $c_1 = F(s)$ and

$$U(x,s) = F(s)e^{-(x/a)s}.$$

Hence, by the second translation theorem,

$$u(x,t) = f\left(t - \frac{x}{a}\right) \mathscr{U}\left(t - \frac{x}{a}\right).$$

5. We use

$$U(x, s) = c_1 e^{-(x/a)s} - \frac{g}{s^3} \,.$$

Now

$$\mathcal{L}\{u(0, t)\} = U(0, s) = \frac{A\omega}{s^2 + \omega^2}$$

and so

$$U(0, s) = c_1 - \frac{g}{s^3} = \frac{A\omega}{s^2 + \omega^2} \qquad \text{or} \qquad c_1 = \frac{g}{s^3} + \frac{A\omega}{s^2 + \omega^2} \,.$$

Therefore

$$U(x, s) = \frac{A\omega}{s^2 + \omega^2} e^{-(x/a)s} + \frac{g}{s^3} e^{-(x/a)s} - \frac{g}{s^3}$$

and

$$u(x, t) = A\mathcal{L}^{-1}\left\{ \frac{\omega e^{-(x/a)s}}{s^2 + \omega^2} \right\} + g\mathcal{L}^{-1}\left\{ \frac{e^{-(x/a)s}}{s^3} \right\} - g\mathcal{L}^{-1}\left\{ \frac{1}{s^3} \right\}$$

$$= A \sin\omega\left(t - \frac{x}{a}\right)\mathcal{U}\left(t - \frac{x}{a}\right) + \frac{1}{2}g\left(t - \frac{x}{a}\right)^2 \mathcal{U}\left(t - \frac{x}{a}\right) - \frac{1}{2}gt^2.$$

7. We use

$$U(x, s) = c_1 \cosh\frac{s}{a}x + c_2 \sinh\frac{s}{a}x.$$

Now $U(0, s) = 0$ implies $c_1 = 0$, so $U(x, s) = c_2 \sinh(s/a)x$. The condition $E\, dU/dx\Big|_{x=L} = F_0$
then yields $c_2 = F_0 a/Es \cosh(s/a)L$ and so

$$U(x, s) = \frac{aF_0}{Es}\frac{\sinh(s/a)x}{\cosh(s/a)L} = \frac{aF_0}{Es}\frac{e^{(s/a)x} - e^{-(s/a)x}}{e^{(s/a)L} + e^{-(s/a)L}} = \frac{aF_0}{Es}\frac{e^{(s/a)(x-L)} - e^{-(s/a)(x+L)}}{1 + e^{-2sL/a}}$$

$$= \frac{aF_0}{E}\left[\frac{e^{-(s/a)(L-x)}}{s} - \frac{e^{-(s/a)(3L-x)}}{s} + \frac{e^{-(s/a)(5L-x)}}{s} - \cdots\right]$$

$$- \frac{aF_0}{E}\left[\frac{e^{-(s/a)(L+x)}}{s} - \frac{e^{-(s/a)(3L+x)}}{s} + \frac{e^{-(s/a)(5L+x)}}{s} - \cdots\right]$$

$$= \frac{aF_0}{E}\sum_{n=0}^{\infty}(-1)^n\left[\frac{e^{-(s/a)(2nL+L-x)}}{s} - \frac{e^{-(s/a)(2nL+L+x)}}{s}\right]$$

and

$$u(x, t) = \frac{aF_0}{E}\sum_{n=0}^{\infty}(-1)^n\left[\mathcal{L}^{-1}\left\{\frac{e^{-(s/a)(2nL+L-x)}}{s}\right\} - \mathcal{L}^{-1}\left\{\frac{e^{-(s/a)(2nL+L+x)}}{s}\right\}\right]$$

$$= \frac{aF_0}{E}\sum_{n=0}^{\infty}(-1)^n\left[\left(t - \frac{2nL+L-x}{a}\right)\mathcal{U}\left(t - \frac{2nL+L-x}{a}\right)\right.$$

$$\left. - \left(t - \frac{2nL+L+x}{a}\right)\mathcal{U}\left(t - \frac{2nL+L+x}{a}\right)\right].$$

9. Transforming the partial differential equation gives

$$\frac{d^2U}{dx^2} - s^2U = -sxe^{-x}.$$

Using undetermined coefficients we obtain

$$U(x,s) = c_1e^{-sx} + c_2e^{sx} - \frac{2s}{(s^2-1)^2}e^{-x} + \frac{s}{s^2-1}xe^{-x}.$$

The transformed boundary conditions $\lim_{x\to\infty} U(x,s) = 0$ and $U(0,s) = 0$ give, in turn, $c_2 = 0$ and $c_1 = 2s/(s^2-1)^2$. Therefore

$$U(x,s) = \frac{2s}{(s^2-1)^2}e^{-sx} - \frac{2s}{(s^2-1)^2}e^{-x} + \frac{s}{s^2-1}xe^{-x}.$$

From entries (13) and (26) in the Table of Laplace transforms we obtain

$$u(x,t) = \mathscr{L}^{-1}\left\{\frac{2s}{(s^2-1)^2}e^{-sx} - \frac{2s}{(s^2-1)^2}e^{-x} + \frac{s}{s^2-1}xe^{-x}\right\}$$

$$= 2(t-x)\sinh(t-x)\mathscr{U}(t-x) - te^{-x}\sinh t + xe^{-x}\cosh t.$$

11. We use

$$U(x,s) = c_1e^{-\sqrt{s}\,x} + c_2e^{\sqrt{s}\,x} + \frac{u_1}{s}.$$

The condition $\lim_{x\to\infty} u(x,t) = u_1$ implies $\lim_{x\to\infty} U(x,s) = u_1/s$, so we define $c_2 = 0$. Then

$$U(x,s) = c_1e^{-\sqrt{s}\,x} + \frac{u_1}{s}.$$

From $U(0,s) = u_0/s$ we obtain $c_1 = (u_0 - u_1)/s$. Thus

$$U(x,s) = (u_0 - u_1)\frac{e^{-\sqrt{s}\,x}}{s} + \frac{u_1}{s}$$

and

$$u(x,t) = (u_0-u_1)\mathscr{L}^{-1}\left\{\frac{e^{-x\sqrt{s}}}{s}\right\} + u_1\mathscr{L}^{-1}\left\{\frac{1}{s}\right\} = (u_0-u_1)\,\text{erfc}\left(\frac{x}{2\sqrt{t}}\right) + u_1.$$

13. We use

$$U(x,s) = c_1e^{-\sqrt{s}\,x} + c_2e^{\sqrt{s}\,x} + \frac{u_0}{s}.$$

The condition $\lim_{x\to\infty} u(x,t) = u_0$ implies $\lim_{x\to\infty} U(x,s) = u_0/s$, so we define $c_2 = 0$. Then

$$U(x,s) = c_1e^{-\sqrt{s}\,x} + \frac{u_0}{s}.$$

The transform of the remaining boundary conditions gives

$$\frac{dU}{dx}\bigg|_{x=0} = U(0,s).$$

This condition yields $c_1 = -u_0/s(\sqrt{s}+1)$. Thus

$$U(x,s) = -u_0 \frac{e^{-\sqrt{s}\,x}}{s(\sqrt{s}+1)} + \frac{u_0}{s}$$

and

$$u(x,t) = -u_0 \mathscr{L}^{-1}\left\{\frac{e^{-x\sqrt{s}}}{s(\sqrt{s}+1)}\right\} + u_0 \mathscr{L}^{-1}\left\{\frac{1}{s}\right\}$$

$$= u_0 e^{x+t} \operatorname{erfc}\left(\sqrt{t}+\frac{x}{2\sqrt{t}}\right) - u_0 \operatorname{erfc}\left(\frac{x}{2\sqrt{t}}\right) + u_0 \qquad \boxed{\text{By entry (6) in Table 14.1.1}}$$

15. We use
$$U(x,s) = c_1 e^{-\sqrt{s}\,x} + c_2 e^{\sqrt{s}\,x}.$$

The condition $\lim_{x\to\infty} u(x,t) = 0$ implies $\lim_{x\to\infty} U(x,s) = 0$, so we define $c_2 = 0$. Hence

$$U(x,s) = c_1 e^{-\sqrt{s}\,x}.$$

The transform of $u(0,t) = f(t)$ is $U(0,s) = F(s)$. Therefore

$$U(x,s) = F(s)e^{-\sqrt{s}\,x}$$

and

$$u(x,t) = \mathscr{L}^{-1}\left\{F(s)e^{-x\sqrt{s}}\right\} = \frac{x}{2\sqrt{\pi}}\int_0^t \frac{f(t-\tau)e^{-x^2/4\tau}}{\tau^{3/2}}\,d\tau.$$

17. Transforming the partial differential equation gives

$$\frac{d^2 U}{dx^2} - sU = -60.$$

Using undetermined coefficients we obtain

$$U(x,s) = c_1 e^{-\sqrt{s}\,x} + c_2 e^{\sqrt{s}\,x} + \frac{60}{s}.$$

The condition $\lim_{x\to\infty} u(x,t) = 60$ implies $\lim_{x\to\infty} U(x,s) = 60/s$, so we define $c_2 = 0$. The transform of the remaining boundary condition gives

$$U(0,s) = \frac{60}{s} + \frac{40}{s}e^{-2s}.$$

This condition yields $c_1 = \frac{40}{s}e^{-2s}$. Thus

$$U(x,s) = \frac{60}{s} + 40e^{-2s}\frac{e^{-\sqrt{s}\,x}}{s}.$$

Using the Table of Laplace transforms and the second translation theorem we obtain

$$u(x,t) = \mathscr{L}^{-1}\left\{\frac{60}{s} + 40e^{-2s}\frac{e^{-\sqrt{s}\,x}}{s}\right\} = 60 + 40\operatorname{erfc}\left(\frac{x}{2\sqrt{t-2}}\right)\mathscr{U}(t-2).$$

19. Transforming the partial differential equation gives

$$\frac{d^2U}{dx^2} - sU = 0$$

and so

$$U(x, s) = c_1 e^{-\sqrt{s}\,x} + c_2 e^{\sqrt{s}\,x}.$$

The condition $\lim\limits_{x \to -\infty} u(x, t) = 0$ implies $\lim\limits_{x \to -\infty} U(x, s) = 0$, so we define $c_1 = 0$. The transform of the remaining boundary condition gives

$$\left. \frac{dU}{dx} \right|_{x=1} = \frac{100}{s} - U(1, s).$$

This condition yields

$$c_2 \sqrt{s}\, e^{\sqrt{s}} = \frac{100}{s} - c_2 e^{\sqrt{s}}$$

from which it follows that

$$c_2 = \frac{100}{s(\sqrt{s} + 1)} e^{-\sqrt{s}}.$$

Thus

$$U(x, s) = 100 \frac{e^{-(1-x)\sqrt{s}}}{s(\sqrt{s} + 1)}.$$

Using the Table of Laplace transforms we obtain

$$u(x, t) = 100 \mathscr{L}^{-1} \left\{ \frac{e^{-(1-x)\sqrt{s}}}{s(\sqrt{s} + 1)} \right\} = 100 \left[-e^{1-x+t} \operatorname{erfc}\left(\sqrt{t} + \frac{1-x}{\sqrt{t}} \right) + \operatorname{erfc}\left(\frac{1-x}{2\sqrt{t}} \right) \right].$$

21. The solution of

$$\frac{d^2U}{dx^2} - sU = -u_0 - u_0 \sin \frac{\pi}{L} x$$

is

$$U(x, s) = c_1 \cosh\left(\sqrt{s}\, x \right) + c_2 \sinh\left(\sqrt{s}\, x \right) + \frac{u_0}{s} + \frac{u_0}{s + \pi^2/L^2} \sin \frac{\pi}{L} x.$$

The transformed boundary conditions $U(0, s) = u_0/s$ and $U(L, s) = u_0/s$ give, in turn, $c_1 = 0$ and $c_2 = 0$. Therefore

$$U(x, s) = \frac{u_0}{s} + \frac{u_0}{s + \pi^2/L^2} \sin \frac{\pi}{L} x$$

and

$$u(x, t) = u_0 \mathscr{L}^{-1} \left\{ \frac{1}{s} \right\} + u_0 \mathscr{L}^{-1} \left\{ \frac{1}{s + \pi^2/L^2} \right\} \sin \frac{\pi}{L} x = u_0 + u_0 e^{-\pi^2 t/L^2} \sin \frac{\pi}{L} x.$$

23. We use

$$U(x,s) = c_1 \cosh\sqrt{\frac{s}{k}}\,x + c_2 \sinh\sqrt{\frac{s}{k}}\,x + \frac{u_0}{s}.$$

The transformed boundary conditions $\left.\dfrac{dU}{dx}\right|_{x=0} = 0$ and $U(1,s) = 0$ give, in turn, $c_2 = 0$ and $c_1 = -u_0/s\cosh\sqrt{s/k}$. Therefore

$$U(x,s) = \frac{u_0}{s} - \frac{u_0 \cosh\sqrt{s/k}\,x}{s\cosh\sqrt{s/k}} = \frac{u_0}{s} - u_0\frac{e^{\sqrt{s/k}\,x} + e^{-\sqrt{s/k}\,x}}{s(e^{\sqrt{s/k}} + e^{-\sqrt{s/k}})}$$

$$= \frac{u_0}{s} - u_0\frac{e^{\sqrt{s/k}\,(x-1)} + e^{-\sqrt{s/k}\,(x+1)}}{s(1 + e^{-2\sqrt{s/k}})}$$

$$= \frac{u_0}{s} - u_0\left[\frac{e^{-\sqrt{s/k}\,(1-x)}}{s} - \frac{e^{-\sqrt{s/k}\,(3-x)}}{s} + \frac{e^{-\sqrt{s/k}\,(5-x)}}{s} - \cdots\right]$$

$$- u_0\left[\frac{e^{-\sqrt{s/k}\,(1+x)}}{s} - \frac{e^{-\sqrt{s/k}\,(3+x)}}{s} + \frac{e^{-\sqrt{s/k}\,(5+x)}}{s} - \cdots\right]$$

$$= \frac{u_0}{s} - u_0\sum_{n=0}^{\infty}(-1)^n\left[\frac{e^{-(2n+1-x)\sqrt{s}/\sqrt{k}}}{s} + \frac{e^{-(2n+1+x)\sqrt{s}/\sqrt{k}}}{s}\right]$$

and

$$u(x,t) = u_0\mathscr{L}^{-1}\left\{\frac{1}{s}\right\} - u_0\sum_{n=0}^{\infty}(-1)^n\left[\mathscr{L}^{-1}\left\{\frac{e^{-(2n+1-x)\sqrt{s}/\sqrt{k}}}{s}\right\} - \mathscr{L}^{-1}\left\{\frac{e^{-(2n+1+x)\sqrt{s}/\sqrt{k}}}{s}\right\}\right]$$

$$= u_0 - u_0\sum_{n=0}^{\infty}(-1)^n\left[\operatorname{erfc}\left(\frac{2n+1-x}{2\sqrt{kt}}\right) - \operatorname{erfc}\left(\frac{2n+1+x}{2\sqrt{kt}}\right)\right].$$

25. We use

$$U(x,s) = c_1 e^{-\sqrt{RCs+RG}\,x} + c_2 e^{\sqrt{RCs+RG}} + \frac{Cu_0}{Cs+G}.$$

The condition $\lim_{x\to\infty}\partial u/\partial x = 0$ implies $\lim_{x\to\infty}dU/dx = 0$, so we define $c_2 = 0$. Applying $U(0,s) = 0$ to

$$U(x,s) = c_1 e^{-\sqrt{RCsRG}\,x} + \frac{Cu_0}{Cs+G}$$

gives $c_1 = -Cu_0/(Cs+G)$. Therefore

$$U(x,s) = -Cu_0\frac{e^{-\sqrt{RCs+RG}\,x}}{Cs+G} + \frac{Cu_0}{Cs+G}$$

and

$$u(x,t) = u_0 \mathcal{L}^{-1}\left\{\frac{1}{s+G/C}\right\} - u_0 \mathcal{L}^{-1}\left\{\frac{e^{-x\sqrt{RC}\sqrt{s+G/C}}}{s+G/C}\right\}$$

$$= u_0 e^{-Gt/C} - u_0 e^{-Gt/C}\,\text{erfc}\left(\frac{x\sqrt{RC}}{2\sqrt{t}}\right)$$

$$= u_0 e^{-Gt/C}\left[1 - \text{erfc}\left(\frac{x}{2}\sqrt{\frac{RC}{t}}\right)\right]$$

$$= u_0 e^{-Gt/C}\,\text{erf}\left(\frac{x}{2}\sqrt{\frac{RC}{t}}\right).$$

27. Using the Laplace transform with respect to t of the partial differential equationgives

$$\frac{d^2 U}{dr^2} + \frac{2}{r}\frac{dU}{dr} = sU - \overbrace{u(r,0)}^{0},$$

where $\mathcal{L}\{u(r,t)\} = U(r,s)$. The ordinary differential equation is equivalent to

$$rU'' + 2U' - srU = 0.$$

If we let $v(r,s) = rU(r,s)$ then differentiation with respect to r gives $v'' = rU'' + 2U'$. The transformed ordinary differential equation becomes

$$v'' - sv = 0$$

$$v = c_1 e^{-\sqrt{s}r} + c_2 e^{\sqrt{s}r}$$

$$U(r,s) = c_1\frac{e^{-\sqrt{s}r}}{r} + c_2\frac{e^{\sqrt{s}r}}{r}.$$

The usual argument about boundedness as $r \to \infty$ implies $c_2 = 0$. Therefore $U(r,s) = c_1 e^{-\sqrt{s}r}/r$. Now the Laplace transform of $u(1,t) = 100$ is

$$U(1,s) = \frac{100}{s} = c_1 e^{-\sqrt{s}}$$

so

$$c_1 = 100\frac{e^{\sqrt{s}}}{s}$$

and

$$U(r,s) = \frac{100}{r}\frac{e^{-\sqrt{s}(r-1)}}{s}$$

Thus

$$u(r,t) = \frac{100}{r}\mathcal{L}^{-1}\left\{\frac{e^{-(r-1)\sqrt{s}}}{s}\right\}.$$

From the third entry in Table 14.1.1 we get

$$u(r,t) = \frac{100}{r}\,\text{erfc}\left(\frac{r-1}{2\sqrt{t}}\right).$$

29. We use

$$U(x,s) = c_1 e^{-\sqrt{s/k}\,x} + c_2 e^{\sqrt{s/k}\,x}.$$

Now $\lim\limits_{x\to\infty} u(x,t) = 0$ implies $\lim\limits_{x\to\infty} U(x,s) = 0$, so we define $c_2 = 0$. Then

$$U(x,s) = c_1 e^{-\sqrt{s/k}\,x}.$$

Finally, from $U(0,s) = u_0/s$ we obtain $c_1 = u_0/s$. Thus

$$U(x,s) = u_0\,\frac{e^{-\sqrt{s/k}\,x}}{s}$$

and

$$u(x,t) = u_0\mathscr{L}^{-1}\left\{\frac{e^{-\sqrt{s/k}\,x}}{s}\right\} = u_0\mathscr{L}^{-1}\left\{\frac{e^{-(x/\sqrt{k})\sqrt{s}}}{s}\right\} = u_0\,\text{erfc}\left(\frac{x}{2\sqrt{kt}}\right).$$

Since $\text{erfc}(0) = 1$,

$$\lim_{t\to\infty} u(x,t) = \lim_{t\to\infty} u_0\,\text{erfc}(x/2\sqrt{kt}) = u_0.$$

31.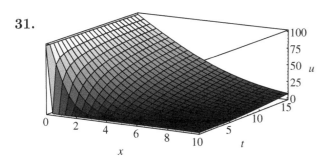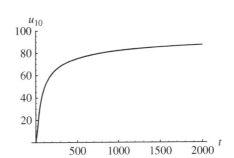

33. **(a)** Letting $C(x,s) = \mathscr{L}\{c(x,t)\}$ we obtain

$$\frac{d^2C}{dx^2} - \frac{s}{k}C = 0 \qquad \text{subject to} \qquad \left.\frac{dC}{dx}\right|_{x=0} = -A.$$

The solution of this initial-value problem is

$$C(x,s) = A\sqrt{k}\,\frac{e^{-(x/\sqrt{k})\sqrt{s}}}{\sqrt{s}},$$

so that

$$c(x,t) = A\sqrt{\frac{k}{\pi t}}\,e^{-x^2/4kt}.$$

(b)

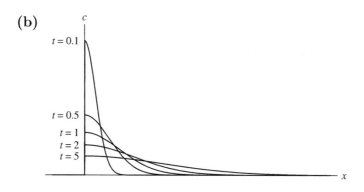

(c) $\displaystyle\int_0^\infty c(x,t)\,dx = Ak\ \mathrm{erf}\left(\frac{x}{2\sqrt{kt}}\right)\Big|_0^\infty = Ak(1-0) = Ak$

14.3   Fourier Integral

1. From formulas (5) and (6) in the text,

$$A(\alpha) = \int_{-1}^0 (-1)\cos \alpha x\,dx + \int_0^1 (2)\cos \alpha x\,dx = -\frac{\sin \alpha}{\alpha} + 2\frac{\sin \alpha}{\alpha} = \frac{\sin \alpha}{\alpha}$$

and

$$B(\alpha) = \int_{-1}^0 (-1)\sin \alpha x\,dx + \int_0^1 (2)\sin \alpha x\,dx = \frac{1-\cos \alpha}{\alpha} - 2\frac{\cos \alpha - 1}{\alpha} = \frac{3(1-\cos \alpha)}{\alpha}.$$

Hence

$$f(x) = \frac{1}{\pi}\int_0^\infty \frac{\sin \alpha \cos \alpha x + 3(1-\cos \alpha)\sin \alpha x}{\alpha}\,d\alpha.$$

3. From formulas (5) and (6) in the text,

$$A(\alpha) = \int_0^3 x\cos \alpha x\,dx = \frac{x\sin \alpha x}{\alpha}\Big|_0^3 - \frac{1}{\alpha}\int_0^3 \sin \alpha x\,dx$$

$$= \frac{3\sin 3\alpha}{\alpha} + \frac{\cos \alpha x}{\alpha^2}\Big|_0^3 = \frac{3\alpha \sin 3\alpha + \cos 3\alpha - 1}{\alpha^2}$$

and

$$B(\alpha) = \int_0^3 x\sin \alpha x\,dx = -\frac{x\cos \alpha x}{\alpha}\Big|_0^3 + \frac{1}{\alpha}\int_0^3 \cos \alpha x\,dx$$

$$= -\frac{3\cos 3\alpha}{\alpha} + \frac{\sin \alpha x}{\alpha^2}\Big|_0^3 = \frac{\sin 3\alpha - 3\alpha \cos 3\alpha}{\alpha^2}.$$

Hence

$$f(x) = \frac{1}{\pi} \int_0^\infty \frac{(3\alpha \sin 3\alpha + \cos 3\alpha - 1)\cos \alpha x + (\sin 3\alpha - 3\alpha \cos 3\alpha)\sin \alpha x}{\alpha^2} \, d\alpha$$

$$= \frac{1}{\pi} \int_0^\infty \frac{3\alpha(\sin 3\alpha \cos \alpha x - \cos 3\alpha \sin \alpha x) + \cos 3\alpha \cos \alpha x + \sin 3\alpha \sin \alpha x - \cos \alpha x}{\alpha^2} \, d\alpha$$

$$= \frac{1}{\pi} \int_0^\infty \frac{3\alpha \sin \alpha(3 - x) + \cos \alpha(3 - x) - \cos \alpha x}{\alpha^2} \, d\alpha.$$

5. From formula (5) in the text,

$$A(\alpha) = \int_0^\infty e^{-x} \cos \alpha x \, dx.$$

Recall $\mathscr{L}\{\cos kt\} = s/(s^2 + k^2)$. If we set $s = 1$ and $k = \alpha$ we obtain

$$A(\alpha) = \frac{1}{1 + \alpha^2}.$$

Now

$$B(\alpha) = \int_0^\infty e^{-x} \sin \alpha x \, dx.$$

Recall $\mathscr{L}\{\sin kt\} = k/(s^2 + k^2)$. If we set $s = 1$ and $k = \alpha$ we obtain

$$B(\alpha) = \frac{\alpha}{1 + \alpha^2}.$$

Hence

$$f(x) = \frac{1}{\pi} \int_0^\infty \frac{\cos \alpha x + \alpha \sin \alpha x}{1 + \alpha^2} \, d\alpha.$$

7. The function is odd. Thus from formula (11) in the text

$$B(\alpha) = 5 \int_0^1 \sin \alpha x \, dx = \frac{5(1 - \cos \alpha)}{\alpha}.$$

Hence from formula (10) in the text,

$$f(x) = \frac{10}{\pi} \int_0^\infty \frac{(1 - \cos \alpha)\sin \alpha x}{\alpha} \, d\alpha.$$

9. The function is even. Thus from formula (9) in the text

$$A(\alpha) = \int_0^\pi x \cos \alpha x \, dx = \frac{x \sin \alpha x}{\alpha}\Big|_0^\pi - \frac{1}{\alpha} \int_0^\pi \sin \alpha x \, dx$$

$$= \frac{\pi \alpha \sin \pi \alpha}{\alpha} + \frac{1}{\alpha^2} \cos \alpha x \Big|_0^\pi = \frac{\pi \alpha \sin \pi \alpha + \cos \pi \alpha - 1}{\alpha^2}.$$

Hence from formula (8) in the text

$$f(x) = \frac{2}{\pi} \int_0^\infty \frac{(\pi \alpha \sin \pi \alpha + \cos \pi \alpha - 1)\cos \alpha x}{\alpha^2} \, d\alpha.$$

11. The function is odd. Thus from formula (11) in the text

$$B(\alpha) = \int_0^\infty (e^{-x} \sin x) \sin \alpha x \, dx$$

$$= \frac{1}{2} \int_0^\infty e^{-x} [\cos(1-\alpha)x - \cos(1+\alpha)x] \, dx$$

$$= \frac{1}{2} \int_0^\infty e^{-x} \cos(1-\alpha)x \, dx - \frac{1}{2} \int_0^\infty e^{-x} \cos(1+\alpha)x, dx.$$

Now recall

$$\mathscr{L}\{\cos kt\} = \int_0^\infty e^{-st} \cos kt \, dt = s/(s^2 + k^2).$$

If we set $s = 1$, and in turn, $k = 1 - \alpha$ and then $k = 1 + \alpha$, we obtain

$$B(\alpha) = \frac{1}{2} \frac{1}{1+(1-\alpha)^2} - \frac{1}{2} \frac{1}{1+(1+\alpha)^2} = \frac{1}{2} \frac{(1+\alpha)^2 - (1-\alpha)^2}{[1+(1-\alpha)^2][1+(1+\alpha)^2]}.$$

Simplifying the last expression gives

$$B(\alpha) = \frac{2\alpha}{4 + \alpha^4}.$$

Hence from formula (10) in the text

$$f(x) = \frac{4}{\pi} \int_0^\infty \frac{\alpha \sin \alpha x}{4 + \alpha^4} \, d\alpha.$$

13. For the cosine integral,

$$A(\alpha) = \int_0^\infty e^{-kx} \cos \alpha x \, dx = \frac{k}{k^2 + \alpha^2}.$$

Hence

$$f(x) = \frac{2}{\pi} \int_0^\infty \frac{k \cos \alpha x}{k^2 + \alpha^2} \, d\alpha = \frac{2k}{\pi} \int_0^\infty \frac{\cos \alpha x}{k^2 + \alpha^2} \, d\alpha.$$

For the sine integral,

$$B(\alpha) = \int_0^\infty e^{-kx} \sin \alpha x \, dx = \frac{\alpha}{k^2 + \alpha^2}.$$

Hence

$$f(x) = \frac{2}{\pi} \int_0^\infty \frac{\alpha \sin \alpha x}{k^2 + \alpha^2} \, d\alpha.$$

15. For the cosine integral,

$$A(\alpha) = \int_0^\infty x e^{-2x} \cos \alpha x \, dx.$$

But we know

$$\mathscr{L}\{t \cos kt\} = -\frac{d}{ds} \frac{s}{(s^2 + k^2)} = \frac{s^2 - k^2}{(s^2 + k^2)^2}.$$

If we set $s = 2$ and $k = \alpha$ we obtain

$$A(\alpha) = \frac{4 - \alpha^2}{(4 + \alpha^2)^2}.$$

Hence

$$f(x) = \frac{2}{\pi} \int_0^\infty \frac{(4 - \alpha^2) \cos \alpha x}{(4 + \alpha^2)^2} \, d\alpha.$$

For the sine integral,

$$B(\alpha) = \int_0^\infty x e^{-2x} \sin \alpha x \, dx.$$

From Problem 12, we know

$$\mathscr{L}\{t \sin kt\} = \frac{2ks}{(s^2 + k^2)^2}.$$

If we set $s = 2$ and $k = \alpha$ we obtain

$$B(\alpha) = \frac{4\alpha}{(4 + \alpha^2)^2}.$$

Hence

$$f(x) = \frac{8}{\pi} \int_0^\infty \frac{\alpha \sin \alpha x}{(4 + \alpha^2)^2} \, d\alpha.$$

17. By formula (8) in the text

$$f(x) = \frac{2}{\pi} \int_0^\infty e^{-\alpha} \cos \alpha x \, d\alpha = \frac{2}{\pi} \frac{1}{1 + x^2}, \quad x > 0.$$

19. (a) From formula (7) in the text with $x = 2$, we have

$$\frac{1}{2} = \frac{2}{\pi} \int_0^\infty \frac{\sin \alpha \cos \alpha}{\alpha} \, d\alpha = \frac{1}{\pi} \int_0^\infty \frac{\sin 2\alpha}{\alpha} \, d\alpha.$$

If we let $\alpha = x$ we obtain

$$\int_0^\infty \frac{\sin 2x}{x} \, dx = \frac{\pi}{2}.$$

(b) If we now let $2x = kt$ where $k > 0$, then $dx = (k/2) \, dt$ and the integral in part **(a)** becomes

$$\int_0^\infty \frac{\sin kt}{kt/2} (k/2) \, dt = \int_0^\infty \frac{\sin kt}{t} \, dt = \frac{\pi}{2}.$$

21. (a) From the identity

$$\sin A \cos B = \frac{1}{2}[\sin (A + B) + \sin (A - B)]$$

we have

$$\sin \alpha \cos \alpha x = \frac{1}{2}[\sin (\alpha + \alpha x) + \sin (\alpha - \alpha x)]$$

$$= \frac{1}{2}[\sin \alpha(1 + x) + \sin \alpha(1 - x)]$$

$$= \frac{1}{2}[\sin \alpha(x + 1) - \sin \alpha(x - 1)].$$

Then

$$\frac{2}{\pi} \int_0^\infty \frac{\sin \alpha \cos \alpha x}{\alpha} \, d\alpha = \frac{1}{\pi} \int_0^\infty \frac{\sin \alpha(x+1) - \sin \alpha(x-1)}{\alpha} \, d\alpha.$$

(b) Noting that

$$F_b = \frac{1}{\pi} \int_0^b \frac{\sin \alpha(x+1) - \sin \alpha(x-1)}{\alpha} \, d\alpha$$

$$= \frac{1}{\pi} \left[\int_0^b \frac{\sin \alpha(x+1)}{\alpha} \, d\alpha - \int_0^b \frac{\sin \alpha(x-1)}{\alpha} \, d\alpha \right]$$

and letting $t = \alpha(x+1)$ so that $dt = (x+1)\,d\alpha$ in the first integral and $t = \alpha(x-1)$ so that $dt = (x-1)\,d\alpha$ in the second integral we have

$$F_b = \frac{1}{\pi} \left[\int_0^{b(x+1)} \frac{\sin t}{t} \, dt - \int_0^{b(x-1)} \frac{\sin t}{t} \, dt \right].$$

Since $\mathrm{Si}\,(x) = \int_0^x [(\sin t)/t]\,dt$, this becomes

$$F_b = \frac{1}{\pi}[\mathrm{Si}\,(b(x+1)) - \mathrm{Si}\,(b(x-1))].$$

(c) In *Mathematica* we define

$$\mathbf{f[b\_] := (1/Pi)(SinIntegral[b(x+1)] - SinIntegral[b(x-1)]}$$

Graphs of $F_b(x)$ for $b = 4,\ 6,\ 15,$ and 75 are shown below.

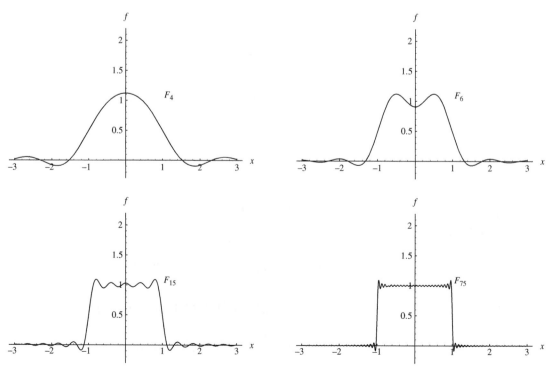

14.4 | Fourier Transforms

For the boundary-value problems in this section it is sometimes useful to note that the identities

$$e^{i\alpha} = \cos\alpha + i\sin\alpha \qquad and \qquad e^{-i\alpha} = \cos\alpha - i\sin\alpha$$

imply

$$e^{i\alpha} + e^{-i\alpha} = 2\cos\alpha \qquad and \qquad e^{i\alpha} - e^{-i\alpha} = 2i\sin\alpha.$$

1. Using the Fourier transform, the partial differential equation becomes

$$\frac{dU}{dt} + k\alpha^2 U = 0 \qquad \text{and so} \qquad U(\alpha, t) = ce^{-k\alpha^2 t}.$$

Now

$$\mathscr{F}\{u(x,0)\} = U(\alpha, 0) = \mathscr{F}\left\{e^{-|x|}\right\}.$$

We have

$$\mathscr{F}\left\{e^{-|x|}\right\} = \int_{-\infty}^{\infty} e^{-|x|}e^{i\alpha x}\,dx = \int_{-\infty}^{\infty} e^{-|x|}(\cos\alpha x + i\sin\alpha x)\,dx = \int_{-\infty}^{\infty} e^{-|x|}\cos\alpha x\,dx.$$

The integral

$$\int_{-\infty}^{\infty} e^{-|x|}\sin\alpha x\,dx = 0$$

since the integrand is an odd function of x. Continuing we obtain

$$\mathscr{F}\left\{e^{-|x|}\right\} = 2\int_{0}^{\infty} e^{-x}\cos\alpha x\,dx = \frac{2}{1+\alpha^2}.$$

But $U(\alpha, 0) = c = 2/(1+\alpha^2)$ gives

$$U(\alpha, t) = \frac{2e^{-k\alpha^2 t}}{1+\alpha^2}$$

and so

$$u(x,t) = \frac{2}{2\pi}\int_{-\infty}^{\infty} \frac{e^{-k\alpha^2 t}e^{-i\alpha x}}{1+\alpha^2}\,d\alpha = \frac{1}{\pi}\int_{-\infty}^{\infty} \frac{e^{-k\alpha^2 t}}{1+\alpha^2}(\cos\alpha x - i\sin\alpha x)\,d\alpha$$

$$= \frac{1}{\pi}\int_{-\infty}^{\infty} \frac{e^{-k\alpha^2 t}\cos\alpha x}{1+\alpha^2}\,d\alpha = \frac{2}{\pi}\int_{0}^{\infty} \frac{e^{-k\alpha^2 t}\cos\alpha x}{1+\alpha^2}\,d\alpha.$$

3. Using the Fourier sine transform, the partial differential equation becomes

$$\frac{dU}{dt} + k\alpha^2 U = k\alpha u_0.$$

The general solution of this linear equation is

$$U(\alpha, t) = ce^{-k\alpha^2 t} + \frac{u_0}{\alpha}.$$

But $U(\alpha, 0) = 0$ implies $c = -u_0/\alpha$ and so

$$U(\alpha, t) = u_0 \frac{1 - e^{-k\alpha^2 t}}{\alpha}$$

and

$$u(x, t) = \frac{2u_0}{\pi} \int_0^\infty \frac{1 - e^{-k\alpha^2 t}}{\alpha} \sin \alpha x \, d\alpha.$$

5. Using the Fourier sine transform we find

$$U(\alpha, t) = ce^{-k\alpha^2 t}.$$

Now

$$\mathscr{F}_S\{u(x, 0)\} = U(\alpha, 0) = \int_0^1 \sin \alpha x \, dx = \frac{1 - \cos \alpha}{\alpha}.$$

From this we find $c = (1 - \cos \alpha)/\alpha$ and so

$$U(\alpha, t) = \frac{1 - \cos \alpha}{\alpha} e^{-k\alpha^2 t}$$

and

$$u(x, t) = \frac{2}{\pi} \int_0^\infty \frac{1 - \cos \alpha}{\alpha} e^{-k\alpha^2 t} \sin \alpha x \, d\alpha.$$

7. Using the Fourier cosine transform we find

$$U(\alpha, t) = ce^{-k\alpha^2 t}.$$

Now

$$\mathscr{F}_C\{u(x, 0)\} = \int_0^1 \cos \alpha x \, dx = \frac{\sin \alpha}{\alpha} = U(\alpha, 0).$$

From this we obtain $c = (\sin \alpha)/\alpha$ and so

$$U(\alpha, t) = \frac{\sin \alpha}{\alpha} e^{-k\alpha^2 t}$$

and

$$u(x, t) = \frac{2}{\pi} \int_0^\infty \frac{\sin \alpha}{\alpha} e^{-k\alpha^2 t} \cos \alpha x \, d\alpha.$$

9. (a) Using the Fourier transform we obtain

$$U(\alpha, t) = c_1 \cos \alpha a t + c_2 \sin \alpha a t.$$

If we write

$$\mathscr{F}\{u(x, 0)\} = \mathscr{F}\{f(x)\} = F(\alpha)$$

and

$$\mathscr{F}\{u_t(x, 0)\} = \mathscr{F}\{g(x)\} = G(\alpha)$$

we first obtain $c_1 = F(\alpha)$ from $U(\alpha, 0) = F(\alpha)$ and then $c_2 = G(\alpha)/\alpha a$ from

$$\left. dU/dt \right|_{t=0} = G(\alpha). \text{ Thus}$$

$$U(\alpha, t) = F(\alpha) \cos \alpha a t + \frac{G(\alpha)}{\alpha a} \sin \alpha a t$$

and

$$u(x, t) = \frac{1}{2\pi} \int_{-\infty}^{\infty} \left(F(\alpha) \cos \alpha a t + \frac{G(\alpha)}{\alpha a} \sin \alpha a t \right) e^{-i\alpha x} \, d\alpha.$$

(b) If $g(x) = 0$ then $c_2 = 0$ and

$$u(x, t) = \frac{1}{2\pi} \int_{-\infty}^{\infty} F(\alpha) \cos \alpha a t e^{-i\alpha x} \, d\alpha$$

$$= \frac{1}{2\pi} \int_{-\infty}^{\infty} F(\alpha) \left(\frac{e^{\alpha a t i} + e^{-\alpha a t i}}{2} \right) e^{-i\alpha x} \, d\alpha$$

$$= \frac{1}{2} \left[\frac{1}{2\pi} \int_{-\infty}^{\infty} F(\alpha) e^{-i(x-at)\alpha} \, d\alpha + \frac{1}{2\pi} \int_{-\infty}^{\infty} F(\alpha) e^{-i(x+at)\alpha} \, d\alpha \right]$$

$$= \frac{1}{2} \left[f(x - at) + f(x + at) \right].$$

11. Using the Fourier cosine transform we obtain

$$U(x, \alpha) = c_1 \cosh \alpha x + c_2 \sinh \alpha x.$$

Now the Fourier cosine transforms of $u(0, y) = e^{-y}$ and $u(\pi, y) = 0$ are, respectively, $U(0, \alpha) = 1/(1 + \alpha^2)$ and $U(\pi, \alpha) = 0$. The first of these conditions gives $c_1 = 1/(1 + \alpha^2)$. The second condition gives

$$c_2 = -\frac{\cosh \alpha \pi}{(1 + \alpha^2) \sinh \alpha \pi}.$$

Hence

$$U(x, \alpha) = \frac{\cosh \alpha x}{1 + \alpha^2} - \frac{\cosh \alpha \pi \sinh \alpha x}{(1 + \alpha^2) \sinh \alpha \pi} = \frac{\sinh \alpha \pi \cosh \alpha x - \cosh \alpha \pi \sinh \alpha x}{(1 + \alpha^2) \sinh \alpha \pi}$$

$$= \frac{\sinh \alpha (\pi - x)}{(1 + \alpha^2) \sinh \alpha \pi}$$

and

$$u(x, y) = \frac{2}{\pi} \int_0^{\infty} \frac{\sinh \alpha (\pi - x)}{(1 + \alpha^2) \sinh \alpha \pi} \cos \alpha y \, d\alpha.$$

13. Using the Fourier cosine transform with respect to x gives

$$U(\alpha, y) = c_1 e^{-\alpha y} + c_2 e^{\alpha y}.$$

Since we expect $u(x, y)$ to be bounded as $y \to \infty$ we define $c_2 = 0$. Thus

$$U(\alpha, y) = c_1 e^{-\alpha y}.$$

Now

$$\mathscr{F}_C\{u(x,0)\} = \int_0^1 50 \cos \alpha x \, dx = 50 \frac{\sin \alpha}{\alpha}$$

and so

$$U(\alpha, y) = 50 \frac{\sin \alpha}{\alpha} e^{-\alpha y}$$

and

$$u(x,y) = \frac{100}{\pi} \int_0^\infty \frac{\sin \alpha}{\alpha} e^{-\alpha y} \cos \alpha x \, d\alpha.$$

15. We use the Fourier sine transform with respect to x to obtain

$$U(\alpha, y) = c_1 \cosh \alpha y + c_2 \sinh \alpha y.$$

The transforms of $u(x,0) = f(x)$ and $u(x,2) = 0$ give, in turn, $U(\alpha, 0) = F(\alpha)$ and $U(\alpha, 2) = 0$. The first condition gives $c_1 = F(\alpha)$ and the second condition then yields

$$c_2 = -\frac{F(\alpha) \cosh 2\alpha}{\sinh 2\alpha}.$$

Hence

$$U(\alpha, y) = F(\alpha) \cosh \alpha y - \frac{F(\alpha) \cosh 2\alpha \sinh \alpha y}{\sinh 2\alpha}$$

$$= F(\alpha) \frac{\sinh 2\alpha \cosh \alpha y - \cosh 2\alpha \sinh \alpha y}{\sinh 2\alpha}$$

$$= F(\alpha) \frac{\sinh \alpha(2 - y)}{\sinh 2\alpha}$$

and

$$u(x,y) = \frac{2}{\pi} \int_0^\infty F(\alpha) \frac{\sinh \alpha(2 - y)}{\sinh 2\alpha} \sin \alpha x \, d\alpha.$$

17. We solve two boundary-value problems:

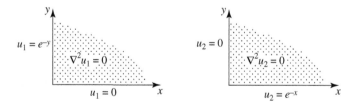

Using the Fourier sine transform with respect to y gives

$$u_1(x,y) = \frac{2}{\pi} \int_0^\infty \frac{\alpha e^{-\alpha x}}{1 + \alpha^2} \sin \alpha y \, d\alpha.$$

The Fourier sine transform with respect to x yields the solution to the second problem:

$$u_2(x,y) = \frac{2}{\pi} \int_0^\infty \frac{\alpha e^{-\alpha y}}{1 + \alpha^2} \sin \alpha x \, d\alpha.$$

We define the solution of the original problem to be

$$u(x,y) = u_1(x,y) + u_2(x,y) = \frac{2}{\pi} \int_0^\infty \frac{\alpha}{1 + \alpha^2} \left[e^{-\alpha x} \sin \alpha y + e^{-\alpha y} \sin \alpha x \right] d\alpha.$$

19. Using the Fourier transform, the partial differential equation equation becomes

$$\frac{dU}{dt} + k\alpha^2 U = 0 \qquad \text{and so} \qquad U(\alpha, t) = ce^{-k\alpha^2 t}.$$

Now

$$\mathscr{F}\{u(x,0)\} = U(\alpha, 0) = \sqrt{\pi}\, e^{-\alpha^2/4}$$

by the given result. This gives $c = \sqrt{\pi}\, e^{-\alpha^2/4}$ and so

$$U(\alpha, t) = \sqrt{\pi}\, e^{-\left(\frac{1}{4} + kt\right)\alpha^2}.$$

Using the given Fourier transform again we obtain

$$u(x,t) = \sqrt{\pi}\, \mathscr{F}^{-1}\{e^{-(1+4kt)\alpha^2/4}\} = \frac{1}{\sqrt{1+4kt}}\, e^{-x^2/(1+4kt)}.$$

21. Using the Fourier transform with respect to x gives

$$U(\alpha, y) = c_1 \cosh \alpha y + c_2 \sinh \alpha y.$$

The transform of the boundary condition $\left.\partial u/\partial y\right|_{y=0} = 0$ is $\left.dU/dy\right|_{y=0} = 0$. This condition gives $c_2 = 0$. Hence

$$U(\alpha, y) = c_1 \cosh \alpha y.$$

Now by the given information the transform of the boundary condition $u(x,1) = e^{-x^2}$ is $U(\alpha, 1) = \sqrt{\pi}\, e^{-\alpha^2/4}$. This condition then gives $c_1 = \sqrt{\pi}\, e^{-\alpha^2/4} \cosh \alpha$. Therefore

$$U(\alpha, y) = \sqrt{\pi}\, \frac{e^{-\alpha^2/4} \cosh \alpha y}{\cosh \alpha}$$

and

$$U(x,y) = \frac{1}{2\sqrt{\pi}} \int_{-\infty}^{\infty} \frac{e^{-\alpha^2/4} \cosh \alpha y}{\cosh \alpha} e^{-i\alpha x}\, d\alpha = \frac{1}{2\sqrt{\pi}} \int_{-\infty}^{\infty} \frac{e^{-\alpha^2/4} \cosh \alpha y}{\cosh \alpha} \cos \alpha x\, d\alpha$$

$$= \frac{1}{\sqrt{\pi}} \int_0^{\infty} \frac{e^{-\alpha^2/4} \cosh \alpha y}{\cosh \alpha} \cos \alpha x\, d\alpha.$$

23. Using the definition of f and the solution in Problem 20 we obtain

$$u(x,t) = \frac{u_0}{2\sqrt{k\pi t}} \int_{-1}^{1} e^{-(x-\tau)^2/4kt}\, d\tau.$$

If $v = (x-\tau)/2\sqrt{kt}$, then $d\tau = -2\sqrt{kt}\, du$ and the integral becomes

$$v(x,t) = \frac{u_0}{\sqrt{\pi}} \int_{(x-1)/2\sqrt{kt}}^{(x+1)/2\sqrt{kt}} e^{-v^2}\, dv.$$

Using the result in Problem 11 of Exercises 14.1 in the text, we have

$$u(x,t) = \frac{u_0}{2}\left[\operatorname{erf}\left(\frac{x+1}{2\sqrt{kt}}\right) - \operatorname{erf}\left(\frac{x-1}{2\sqrt{kt}}\right)\right].$$

25. We use the Fourier cosine transform in the variable z. The transform of the partial differential equation is

$$\frac{d^2U}{dr^2} + \frac{1}{r}\frac{dU}{dr} - \alpha^2 U = 0$$

which implies that

$$U(r, \alpha) = c_1 I_0(\alpha r) + c_2 K(\alpha r)$$

Assuming boundedness at $r = 0$ forces $c_2 = 0$, so $U(r, \alpha) = c_1 I_0(\alpha r)$. The transform of the remaining boundary condition is

$$U(1, \alpha) = \int_0^1 \cos(\alpha z)\, dz = \frac{\sin \alpha}{\alpha} = c_1 I_0(\alpha)$$

which gives $c_1 = \dfrac{\sin \alpha}{\alpha I_0(\alpha)}$ and $U(r, \alpha) = \dfrac{\sin \alpha}{\alpha I_0(\alpha r)} I_0(\alpha r)$. Thus

$$u(r, z) = \frac{2}{\pi}\int_0^\infty \frac{I_0(\alpha r)}{\alpha I_0(\alpha)} \sin \alpha \cos \alpha z \, dz.$$

27. (a) If

$$\int_0^\infty f(x)\cos \alpha x \, dx = F(\alpha) \quad \text{and} \quad F(\alpha) = \begin{cases} 1 - \alpha, & 0 \le \alpha \le 1 \\ 0, & \alpha > 1, \end{cases}$$

then by (10) of this section

$$f(x) = \frac{2}{\pi}\int_0^\infty F(\alpha)\cos \alpha x \, d\alpha$$

$$= \frac{2}{\pi}\int_0^1 F(\alpha)\cos \alpha x \, d\alpha + \frac{2}{\pi}\int_1^\infty F(\alpha)\cos \alpha x \, d\alpha$$

$$= \frac{2}{\pi}\int_0^1 (1 - \alpha)\cos \alpha x \, d\alpha + \frac{2}{\pi}\int_1^\infty 0 \cdot \cos \alpha x \, d\alpha.$$

Integration by parts then gives

$$f(x) = \frac{2(1 - \cos x)}{\pi x^2}.$$

(b) Therefore we know

$$\int_0^\infty \frac{2(1 - \cos x)}{\pi x^2}\cos \alpha x \, dx = \begin{cases} 1 - \alpha, & 0 \le \alpha \le 1 \\ 0, & \alpha > 1. \end{cases}$$

For $\alpha = 0$,

$$\frac{2}{\pi}\int_0^\infty \frac{1 - \cos x}{x^2}\, dx = 1 \quad \text{or} \quad \int_0^\infty \frac{1 - \cos x}{x^2}\, dx = \frac{\pi}{2}.$$

From the identity

$$\sin^2 \frac{x}{2} = \frac{1}{2}(1 - \cos x)$$

we have

$$1 - \cos x = 2 \sin^2 \frac{x}{2}$$

or

$$2 \int_0^\infty \frac{\sin^2 \frac{x}{2}}{x^2} \, dx = \frac{\pi}{2}$$

Now, if $t = x/2$ we have $x = 2t$ and $dx = 2 \, dt$. This implies that the foregoing integral is

$$\int_0^\infty \frac{\sin^2 t}{t^2} \, dt = \frac{\pi}{2} \qquad \text{or} \qquad \int_0^\infty \frac{\sin^2 x}{x^2} \, dx = \frac{\pi}{2}.$$

Chapter 14 in Review

1. The partial differential equation and the boundary conditions indicate that the Fourier cosine transform is appropriate for the problem. We find in this case

$$u(x, y) = \frac{2}{\pi} \int_0^\infty \frac{\sinh \alpha y}{\alpha(1 + \alpha^2) \cosh \alpha \pi} \cos \alpha x \, d\alpha.$$

3. The Laplace transform gives

$$U(x, s) = c_1 e^{-\sqrt{s+h}\, x} + c_2 e^{\sqrt{s+h}\, x} + \frac{u_0}{s + h}.$$

The condition $\lim_{x \to \infty} \partial u / \partial x = 0$ implies $\lim_{x \to \infty} dU/dx = 0$ and so we define $c_2 = 0$. Thus

$$U(x, s) = c_1 e^{-\sqrt{s+h}\, x} + \frac{u_0}{s + h}.$$

The condition $U(0, s) = 0$ then gives $c_1 = -u_0/(s + h)$ and so

$$U(x, s) = \frac{u_0}{s + h} - u_0 \frac{e^{-\sqrt{s+h}\, x}}{s + h}.$$

With the help of the first translation theorem we then obtain

$$u(x, t) = u_0 \mathscr{L}^{-1} \left\{ \frac{1}{s + h} \right\} - u_0 \mathscr{L}^{-1} \left\{ \frac{e^{-\sqrt{s+h}\, x}}{s + h} \right\} = u_0 e^{-ht} - u_0 e^{-ht} \operatorname{erfc}\left(\frac{x}{2\sqrt{t}} \right)$$

$$= u_0 e^{-ht} \left[1 - \operatorname{erfc}\left(\frac{x}{2\sqrt{t}} \right) \right] = u_0 e^{-ht} \operatorname{erf}\left(\frac{x}{2\sqrt{t}} \right).$$

5. The Laplace transform gives

$$U(x, s) = c_1 e^{-\sqrt{s}\, x} + c_2 e^{\sqrt{s}\, x}.$$

The condition $\lim_{x \to \infty} u(x, t) = 0$ implies $\lim_{x \to \infty} U(x, s) = 0$ and so we define $c_2 = 0$. Thus

$$U(x, s) = c_1 e^{-\sqrt{s}\, x}.$$

The transform of the remaining boundary condition is $U(0, s) = 1/s^2$. This gives $c_1 = 1/s^2$. Hence

$$U(x, s) = \frac{e^{-\sqrt{s}\,x}}{s^2} \qquad \text{and} \qquad u(x, t) = \mathscr{L}^{-1}\left\{\frac{1}{s}\frac{e^{-\sqrt{s}\,x}}{s}\right\}.$$

Using

$$\mathscr{L}^{-1}\left\{\frac{1}{s}\right\} = 1 \qquad \text{and} \qquad \mathscr{L}^{-1}\left\{\frac{e^{-\sqrt{s}\,x}}{s}\right\} = \text{erfc}\left(\frac{x}{2\sqrt{t}}\right),$$

it follows from the convolution theorem that

$$u(x, t) = \int_0^t \text{erfc}\left(\frac{x}{2\sqrt{\tau}}\right) d\tau.$$

7. The Fourier transform gives the solution

$$u(x, t) = \frac{u_0}{2\pi}\int_{-\infty}^{\infty}\left(\frac{e^{i\alpha\pi} - 1}{i\alpha}\right)e^{-i\alpha x}e^{-k\alpha^2 t}\,d\alpha$$

$$= \frac{u_0}{2\pi}\int_{-\infty}^{\infty}\frac{e^{i\alpha(\pi - x)} - e^{-i\alpha x}}{i\alpha}e^{-k\alpha^2 t}\,d\alpha$$

$$= \frac{u_0}{2\pi}\int_{-\infty}^{\infty}\frac{\cos\alpha(\pi - x) + i\sin\alpha(\pi - x) - \cos\alpha x + i\sin\alpha x}{i\alpha}e^{-k\alpha^2 t}\,d\alpha.$$

Since the imaginary part of the integrand of the last integral is an odd function of α, we obtain

$$u(x, t) = \frac{u_0}{2\pi}\int_{-\infty}^{\infty}\frac{\sin\alpha(\pi - x) + \sin\alpha x}{\alpha}e^{-k\alpha^2 t}\,d\alpha.$$

9. We solve the two problems

$$\frac{\partial^2 u_1}{\partial x^2} + \frac{\partial^2 u_1}{\partial y^2} = 0, \quad x > 0, \quad y > 0,$$

$$u_1(0, y) = 0, \quad y > 0,$$

$$u_1(x, 0) = \begin{cases} 100, & 0 < x < 1 \\ 0, & x > 1 \end{cases}$$

and

$$\frac{\partial^2 u_2}{\partial x^2} + \frac{\partial^2 u_2}{\partial y^2} = 0, \quad x > 0, \quad y > 0,$$

$$u_2(0, y) = \begin{cases} 50, & 0 < y < 1 \\ 0, & y > 1 \end{cases}$$

$$u_2(x, 0) = 0.$$

Using the Fourier sine transform with respect to x we find

$$u_1(x, y) = \frac{200}{\pi} \int_0^\infty \left(\frac{1 - \cos \alpha}{\alpha} \right) e^{-\alpha y} \sin \alpha x \, d\alpha.$$

Using the Fourier sine transform with respect to y we find

$$u_2(x, y) = \frac{100}{\pi} \int_0^\infty \left(\frac{1 - \cos \alpha}{\alpha} \right) e^{-\alpha x} \sin \alpha y \, d\alpha.$$

The solution of the problem is then

$$u(x, y) = u_1(x, y) + u_2(x, y).$$

11. The Fourier sine transform with respect to x and undetermined coefficients give

$$U(\alpha, y) = c_1 \cosh \alpha y + c_2 \sinh \alpha y + \frac{A}{\alpha}.$$

The transforms of the boundary conditions are

$$\frac{dU}{dy}\bigg|_{y=0} = 0 \qquad \text{and} \qquad \frac{dU}{dy}\bigg|_{y=\pi} = \frac{B\alpha}{1 + \alpha^2}.$$

The first of these conditions gives $c_2 = 0$ and so

$$U(\alpha, y) = c_1 \cosh \alpha y + \frac{A}{\alpha}.$$

The second transformed boundary condition yields $c_1 = B/(1 + \alpha^2) \sinh \alpha \pi$. Therefore

$$U(\alpha, y) = \frac{B \cosh \alpha y}{(1 + \alpha^2) \sinh \alpha \pi} + \frac{A}{\alpha}$$

and

$$u(x, y) = \frac{2}{\pi} \int_0^\infty \left(\frac{B \cosh \alpha y}{(1 + \alpha^2) \sinh \alpha \pi} + \frac{A}{\alpha} \right) \sin \alpha x \, d\alpha.$$

13. Using the Fourier transform gives

$$U(\alpha, t) = c_1 e^{-k\alpha^2 t}.$$

Now

$$u(\alpha, 0) = \int_0^\infty e^{-x} e^{i\alpha x} \, dx = \frac{e^{(i\alpha - 1)x}}{i\alpha - 1}\bigg|_0^\infty = 0 - \frac{1}{i\alpha - 1} = \frac{1}{1 - i\alpha} = c_1$$

so

$$U(\alpha, t) = \frac{1 + i\alpha}{1 + \alpha^2} e^{-k\alpha^2 t}$$

and

$$u(x, t) = \frac{1}{2\pi} \int_{-\infty}^\infty \frac{1 + i\alpha}{1 + \alpha^2} e^{-k\alpha^2 t} e^{-i\alpha x} \, d\alpha.$$

Since
$$\frac{1+i\alpha}{1+\alpha^2}(\cos\alpha x - i\sin\alpha x) = \frac{\cos\alpha x + \alpha\sin\alpha x}{1+\alpha^2} + \frac{i(\alpha\cos\alpha x - \sin\alpha x)}{1+\alpha^2}$$

and the integral of the product of the second term with $e^{-k\alpha^2 t}$ is 0 (it is an odd function), we have
$$u(x,t) = \frac{1}{2\pi}\int_{-\infty}^{\infty}\frac{\cos\alpha x + \alpha\sin\alpha x}{1+\alpha^2}e^{-k\alpha^2 t}\,d\alpha.$$

15. The Fourier cosine transform of the partial differential equation gives
$$\frac{dU}{dt} + k\alpha^2 U = 0 \qquad \text{so} \qquad U(\alpha,t) = c_1 e^{-k\alpha^2 t}.$$

The transform of the initial condition gives
$$U(\alpha,0) = \frac{\alpha}{\alpha^2+1} = c_1$$
$$U(\alpha,t) = \frac{\alpha}{\alpha^2+1}e^{-k\alpha^2 t}$$

and so
$$u(x,t) = \frac{2}{\pi}\int_0^{\infty}\frac{\alpha e^{-k\alpha^2 t}}{\alpha^2+1}\cos\alpha x\,d\alpha.$$

17. The transform of the partial differential equation with respect to t is
$$\frac{d^2 U}{dx^2} - sU = 0$$
$$U(x,s) = c_1 e^{-\sqrt{s}\,x} + c_2 e^{\sqrt{s}\,x}$$

The transform of the boundary condition $\lim_{x\to\infty} u(x,t) = 0$ is $\lim_{x\to\infty} U(,x,s) = 0$. Hence $c_2 = 0$ and
$$U(x,s) = c_1 e^{-\sqrt{s}\,x}$$

By writing the boundary condition $x = 0$ as
$$u(0,t) = u_0 - u_0\mathscr{U}(t-1)$$

its transform is
$$U(0,s) = \frac{u_0}{s} - \frac{u_0}{s}e^{-s}$$
$$c_1 = \frac{u_0}{s} - \frac{u_0}{s}e^{-s}$$
$$U(x,s) = u_0\frac{e^{-\sqrt{s}\,x}}{s} - u_0\frac{e^{-\sqrt{s}\,x}}{s}e^{-s}$$
$$u(x,t) = u_0\mathscr{L}^{-1}\left\{\frac{e^{-\sqrt{s}\,x}}{s}\right\} - u_0\mathscr{L}^{-1}\left\{\frac{e^{-\sqrt{s}\,x}}{s}e^{-s}\right\}$$

by entry 3 of Table 14.1.1 and the inverse form of the second translation theorem that:

$$u(x,t) = u_0 \operatorname{erfc}\left(\frac{x}{2\sqrt{t}}\right) - u_0 \operatorname{erfc}\left(\frac{x}{2\sqrt{t-1}}\right)\mathscr{U}(t-1)$$

or

$$u(x,t) = \begin{cases} u_0 \operatorname{erfc}\left(\dfrac{x}{2\sqrt{t}}\right), & 0 < t < 1 \\[2ex] u_0 \operatorname{erfc}\left(\dfrac{x}{2\sqrt{t}}\right) - u_0 \operatorname{erfc}\left(\dfrac{x}{2\sqrt{t-1}}\right), & t > 1. \end{cases}$$

19. Using the boundary condition $u(0,y) = 0$ the Fourier sine transform of

$$\frac{\partial^2 u}{\partial x^2} + \frac{\partial^2 u}{\partial y^2} = hu, \quad x < 0, \quad 0 < y < \pi$$

with respect to x is

$$-\alpha^2 U + \frac{d^2 U}{dy^2} = hU$$

$$\frac{d^2 U}{dy^2} - \left(\alpha^2 + h\right) U = 0$$

Therefore

$$U(\alpha, y) = c_1 \cosh\sqrt{\alpha^2 + h}\, y + c_2 \sinh\sqrt{\alpha^2 + h}\, y$$

The condition $U(\alpha, 0) = 0$ gives $c_1 = 0$, so $U(\alpha, y) = c_2 \sinh\sqrt{\alpha^2 + h}\, y$. The condition $U(\alpha, \pi) = F(\alpha)$ where

$$F(\alpha) = \int_0^\infty f(x) \sin \alpha x\, dx$$

gives

$$U(\alpha, \pi) = c_2 \sinh\sqrt{\alpha^2 + h}\, \pi = F(\alpha) \quad \text{or} \quad c_2 = \frac{F(\alpha)}{\sinh\sqrt{\alpha^2 + h}\, \pi}$$

Thus

$$U(\alpha, y) = \frac{F(\alpha) \sinh\sqrt{\alpha^2 + h}\, y}{\sinh\sqrt{\alpha^2 + h}\, \pi}$$

and

$$u(x,y) = \frac{2}{\pi} \int_0^\infty \frac{F(\alpha) \sinh\sqrt{\alpha^2 + h}\, y}{\sinh\sqrt{\alpha^2 + h}\, \pi} \sin \alpha x\, d\alpha.$$

21. [This is Problem 16 in Exercises 14.2 with $f(t) = -100$.] The Laplace transform with respect to t of the partial differential equation gives

$$\frac{d^2 U}{dx^2} - sU = 0 \quad \text{so} \quad U(x,s) = c_1 e^{-\sqrt{s}\,x} + c_2 e^{\sqrt{s}\,x}.$$

The boundary condition $\lim_{x\to\infty} u(x,t) = 0$ implies $\lim_{x\to\infty} U(x,s) = 0$, so we take $c_2 = 0$. Then

$$U(x,s) = c_1 e^{-\sqrt{s}\,x}.$$

The transform of the boundary condition at $x = 0$ implies

$$\left.\frac{dU}{dx}\right|_{x=0} = -\frac{100}{s}.$$

Also,

$$U'(0, s) = -\sqrt{s}\, c_1 = -\frac{100}{s}$$

so

$$c_1 = \frac{100}{s\sqrt{s}}$$

and

$$U(x, s) = \frac{100}{s\sqrt{s}} e^{-\sqrt{s}\,x}.$$

From entry 4 of Table 14.1.1 with the identification $a = x$ we get

$$u(x, t) = 200\sqrt{\frac{t}{\pi}}\, e^{-x^2/4t} - 100x\, \text{erfc}\left(\frac{x}{2\sqrt{t}}\right).$$

Had we used the convolution theorem in the inverse of

$$U(x, \alpha) = \frac{100}{s} \cdot \frac{1}{\sqrt{s}} e^{-\sqrt{s}\,x}$$

then from the inverse form of the convolution theorem, (4) of Section 4.4,

$$u(x, t) = \frac{100}{\sqrt{\pi}} \int_0^t \frac{e^{-x^2/4(t-\tau)}}{\sqrt{t-\tau}}\, d\tau.$$

Chapter 15

Numerical Solutions of Partial Differential Equations

15.1 | Laplace's Equation

1. The figure shows the values of $u(x, y)$ along the boundary. We need to determine u_{11} and u_{21}. The system is

$$u_{21} + 2 + 0 + 0 - 4u_{11} = 0 \qquad \text{or} \qquad -4u_{11} + u_{21} = -2$$
$$1 + 2 + u_{11} + 0 - 4u_{21} = 0 \qquad \qquad u_{11} - 4u_{21} = -3.$$

Solving we obtain $u_{11} = 11/15$ and $u_{21} = 14/15$.

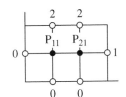

3. The figure shows the values of $u(x, y)$ along the boundary. We need to determine u_{11}, u_{21}, u_{12}, and u_{22}. By symmetry $u_{11} = u_{21}$ and $u_{12} = u_{22}$. The system is

$$u_{21} + u_{12} + 0 + 0 - 4u_{11} = 0$$
$$0 + u_{22} + u_{11} + 0 - 4u_{21} = 0 \qquad \text{or} \qquad 3u_{11} + u_{12} = 0$$
$$u_{22} + \sqrt{3}/2 + 0 + u_{11} - 4u_{12} = 0 \qquad \qquad u_{11} - 3u_{12} = -\frac{\sqrt{3}}{2}.$$
$$0 + \sqrt{3}/2 + u_{12} + u_{21} - 4u_{22} = 0$$

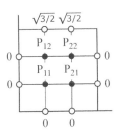

Solving we obtain $u_{11} = u_{21} = \sqrt{3}/16$ and $u_{12} = u_{22} = 3\sqrt{3}/16$.

5. The figure shows the values of $u(x, y)$ along the boundary. For Gauss-Seidel the coefficients of the unknowns u_{11}, u_{21}, u_{31}, u_{12}, u_{22}, u_{32}, u_{13}, u_{23}, u_{33} are shown in the matrix

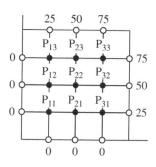

$$\begin{bmatrix} 0 & .25 & 0 & .25 & 0 & 0 & 0 & 0 & 0 \\ .25 & 0 & .25 & 0 & .25 & 0 & 0 & 0 & 0 \\ 0 & .25 & 0 & 0 & 0 & .25 & 0 & 0 & 0 \\ .25 & 0 & 0 & 0 & .25 & 0 & .25 & 0 & 0 \\ 0 & .25 & 0 & .25 & 0 & .25 & 0 & .25 & 0 \\ 0 & 0 & .25 & 0 & .25 & 0 & 0 & 0 & .25 \\ 0 & 0 & 0 & .25 & 0 & 0 & 0 & .25 & 0 \\ 0 & 0 & 0 & 0 & .25 & 0 & .25 & 0 & .25 \\ 0 & 0 & 0 & 0 & 0 & .25 & 0 & .25 & 0 \end{bmatrix}$$

The constant terms in the equations are 0, 0, 6.25, 0, 0, 12.5, 6.25, 12.5, 37.5. We use 25 as the initial guess for each variable. Then $u_{11} = 6.25$, $u_{21} = u_{12} = 12.5$, $u_{31} = u_{13} = 18.75$, $u_{22} = 25$, $u_{32} = u_{23} = 37.5$, and $u_{33} = 56.25$

7. (a) Using the difference approximations for u_{xx} and u_{yy} we obtain

$$u_{xx} + u_{yy} = \frac{1}{h^2}(u_{i+1,j} + u_{i,j+1} + u_{i-1,j} + u_{i,j-1} - 4u_{ij}) = f(x, y)$$

so that

$$u_{i+1,j} + u_{i,j+1} + u_{i-1,j} + u_{i,j-1} - 4u_{ij} = h^2 f(x, y).$$

(b) By symmetry, as shown in the figure, we need only solve for u_1, u_2, u_3, u_4, and u_5. The difference equations are

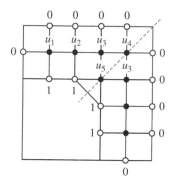

$$u_2 + 0 + 0 + 1 - 4u_1 = \frac{1}{4}(-2)$$

$$u_3 + 0 + u_1 + 1 - 4u_2 = \frac{1}{4}(-2)$$

$$u_4 + 0 + u_2 + u_5 - 4u_3 = \frac{1}{4}(-2)$$

$$0 + 0 + u_3 + u_3 - 4u_4 = \frac{1}{4}(-2)$$

$$u_3 + u_3 + 1 + 1 - 4u_5 = \frac{1}{4}(-2)$$

or

$$u_1 = 0.25u_2 + 0.375$$

$$u_2 = 0.25u_1 + 0.25u_3 + 0.375$$

$$u_3 = 0.25u_2 + 0.25u_4 + 0.25u_5 + 0.125$$

$$u_4 = 0.5u_3 + 0.125$$

$$u_5 = 0.5u_3 + 0.625.$$

Using Gauss-Seidel iteration we find $u_1 = 0.5427$, $u_2 = 0.6707$, $u_3 = 0.6402$, $u_4 = 0.4451$, and $u_5 = 0.9451$.

15.2 | Heat Equation

1. We identify $c = 1$, $a = 2$, $T = 1$, $n = 8$, and $m = 40$. Then $h = 2/8 = 0.25$, $k = 1/40 = 0.025$, and $\lambda = 2/5 = 0.4$.

TIME	X = 0.25	X = 0.50	X = 0.75	X = 1.00	X = 1.25	X = 1.50	X = 1.75
0.000	1.0000	1.0000	1.0000	1.0000	0.0000	0.0000	0.0000
0.025	0.6000	1.0000	1.0000	0.6000	0.4000	0.0000	0.0000
0.050	0.5200	0.8400	0.8400	0.6800	0.3200	0.1600	0.0000
0.075	0.4400	0.7120	0.7760	0.6000	0.4000	0.1600	0.0640
0.100	0.3728	0.6288	0.6800	0.5904	0.3840	0.2176	0.0768
0.125	0.3261	0.5469	0.6237	0.5437	0.4000	0.2278	0.1024
0.150	0.2840	0.4893	0.5610	0.5182	0.3886	0.2465	0.1116
0.175	0.2525	0.4358	0.5152	0.4835	0.3836	0.2494	0.1209
0.200	0.2248	0.3942	0.4708	0.4562	0.3699	0.2517	0.1239
0.225	0.2027	0.3571	0.4343	0.4275	0.3571	0.2479	0.1255
0.250	0.1834	0.3262	0.4007	0.4021	0.3416	0.2426	0.1242
0.275	0.1672	0.2989	0.3715	0.3773	0.3262	0.2348	0.1219
0.300	0.1530	0.2752	0.3448	0.3545	0.3101	0.2262	0.1183
0.325	0.1407	0.2541	0.3209	0.3329	0.2943	0.2166	0.1141
0.350	0.1298	0.2354	0.2990	0.3126	0.2787	0.2067	0.1095
0.375	0.1201	0.2186	0.2790	0.2936	0.2635	0.1966	0.1046
0.400	0.1115	0.2034	0.2607	0.2757	0.2488	0.1865	0.0996
0.425	0.1036	0.1895	0.2438	0.2589	0.2347	0.1766	0.0945
0.450	0.0965	0.1769	0.2281	0.2432	0.2211	0.1670	0.0896
0.475	0.0901	0.1652	0.2136	0.2283	0.2083	0.1577	0.0847
0.500	0.0841	0.1545	0.2002	0.2144	0.1961	0.1487	0.0800
0.525	0.0786	0.1446	0.1876	0.2014	0.1845	0.1402	0.0755
0.550	0.0736	0.1354	0.1759	0.1891	0.1735	0.1320	0.0712
0.575	0.0689	0.1269	0.1650	0.1776	0.1632	0.1243	0.0670
0.600	0.0645	0.1189	0.1548	0.1668	0.1534	0.1169	0.0631
0.625	0.0605	0.1115	0.1452	0.1566	0.1442	0.1100	0.0594
0.650	0.0567	0.1046	0.1363	0.1471	0.1355	0.1034	0.0559
0.675	0.0532	0.0981	0.1279	0.1381	0.1273	0.0972	0.0525
0.700	0.0499	0.0921	0.1201	0.1297	0.1196	0.0914	0.0494
0.725	0.0468	0.0864	0.1127	0.1218	0.1124	0.0859	0.0464
0.750	0.0439	0.0811	0.1058	0.1144	0.1056	0.0807	0.0436
0.775	0.0412	0.0761	0.0994	0.1074	0.0992	0.0758	0.0410
0.800	0.0387	0.0715	0.0933	0.1009	0.0931	0.0712	0.0385
0.825	0.0363	0.0671	0.0876	0.0948	0.0875	0.0669	0.0362
0.850	0.0341	0.0630	0.0823	0.0890	0.0822	0.0628	0.0340
0.875	0.0320	0.0591	0.0772	0.0836	0.0772	0.0590	0.0319
0.900	0.0301	0.0555	0.0725	0.0785	0.0725	0.0554	0.0300
0.925	0.0282	0.0521	0.0681	0.0737	0.0681	0.0521	0.0282
0.950	0.0265	0.0490	0.0640	0.0692	0.0639	0.0489	0.0265
0.975	0.0249	0.0460	0.0601	0.0650	0.0600	0.0459	0.0249
1.000	0.0234	0.0432	0.0564	0.0610	0.0564	0.0431	0.0233

3. We identify $c = 1$, $a = 2$, $T = 1$, $n = 8$, and $m = 40$. Then $h = 2/8 = 0.25$, $k = 1/40 = 0.025$, and $\lambda = 2/5 = 0.4$.

TIME	X = 0.25	X = 0.50	X = 0.75	X = 1.00	X = 1.25	X = 1.50	X = 1.75
0.000	1.0000	1.0000	1.0000	1.0000	0.0000	0.0000	0.0000
0.025	0.7074	0.9520	0.9566	0.7444	0.2545	0.0371	0.0053
0.050	0.5606	0.8499	0.8685	0.6633	0.3303	0.1034	0.0223
0.075	0.4684	0.7473	0.7836	0.6191	0.3614	0.1529	0.0462
0.100	0.4015	0.6577	0.7084	0.5837	0.3753	0.1871	0.0684
0.125	0.3492	0.5821	0.6428	0.5510	0.3797	0.2101	0.0861
0.150	0.3069	0.5187	0.5857	0.5199	0.3778	0.2247	0.0990
0.175	0.2721	0.4652	0.5359	0.4901	0.3716	0.2329	0.1078
0.200	0.2430	0.4198	0.4921	0.4617	0.3622	0.2362	0.1132
0.225	0.2186	0.3809	0.4533	0.4348	0.3507	0.2358	0.1160
0.250	0.1977	0.3473	0.4189	0.4093	0.3378	0.2327	0.1166
0.275	0.1798	0.3181	0.3881	0.3853	0.3240	0.2275	0.1157
0.300	0.1643	0.2924	0.3604	0.3626	0.3097	0.2208	0.1136
0.325	0.1507	0.2697	0.3353	0.3412	0.2953	0.2131	0.1107
0.350	0.1387	0.2495	0.3125	0.3211	0.2808	0.2047	0.1071
0.375	0.1281	0.2313	0.2916	0.3021	0.2666	0.1960	0.1032
0.400	0.1187	0.2150	0.2725	0.2843	0.2528	0.1871	0.0989
0.425	0.1102	0.2002	0.2549	0.2675	0.2393	0.1781	0.0946
0.450	0.1025	0.1867	0.2387	0.2517	0.2263	0.1692	0.0902
0.475	0.0955	0.1743	0.2236	0.2368	0.2139	0.1606	0.0858
0.500	0.0891	0.1630	0.2097	0.2228	0.2020	0.1521	0.0814
0.525	0.0833	0.1525	0.1967	0.2096	0.1906	0.1439	0.0772
0.550	0.0779	0.1429	0.1846	0.1973	0.1798	0.1361	0.0731
0.575	0.0729	0.1339	0.1734	0.1856	0.1696	0.1285	0.0691
0.600	0.0683	0.1256	0.1628	0.1746	0.1598	0.1214	0.0653
0.625	0.0641	0.1179	0.1530	0.1643	0.1506	0.1145	0.0617
0.650	0.0601	0.1106	0.1438	0.1546	0.1419	0.1080	0.0582
0.675	0.0564	0.1039	0.1351	0.1455	0.1336	0.1018	0.0549
0.700	0.0530	0.0976	0.1270	0.1369	0.1259	0.0959	0.0518
0.725	0.0497	0.0917	0.1194	0.1288	0.1185	0.0904	0.0488
0.750	0.0467	0.0862	0.1123	0.1212	0.1116	0.0852	0.0460
0.775	0.0439	0.0810	0.1056	0.1140	0.1050	0.0802	0.0433
0.800	0.0413	0.0762	0.0993	0.1073	0.0989	0.0755	0.0408
0.825	0.0388	0.0716	0.0934	0.1009	0.0931	0.0711	0.0384
0.850	0.0365	0.0674	0.0879	0.0950	0.0876	0.0669	0.0362
0.875	0.0343	0.0633	0.0827	0.0894	0.0824	0.0630	0.0341
0.900	0.0323	0.0596	0.0778	0.0841	0.0776	0.0593	0.0321
0.925	0.0303	0.0560	0.0732	0.0791	0.0730	0.0558	0.0302
0.950	0.0285	0.0527	0.0688	0.0744	0.0687	0.0526	0.0284
0.975	0.0268	0.0496	0.0647	0.0700	0.0647	0.0495	0.0268
1.000	0.0253	0.0466	0.0609	0.0659	0.0608	0.0465	0.0252

(x, y)	exact	approx	abs error
(0.25, 0.1)	0.3794	0.4015	0.0221
(1, 0.5)	0.1854	0.2228	0.0374
(1.5, 0.8)	0.0623	0.0755	0.0132

5. We identify $c = 1$, $a = 2$, $T = 1$, $n = 8$, and $m = 20$. Then $h = 2/8 = 0.25$, $k = 1/20 = 0.05$, and $\lambda = 4/5 = 0.8$.

TIME	X = 0.25	X = 0.50	X = 0.75	X = 1.00	X = 1.25	X = 1.50	X = 1.75
0.00	1.0000	1.0000	1.0000	1.0000	0.0000	0.0000	0.0000
0.05	0.5265	0.8693	0.8852	0.6141	0.3783	0.0884	0.0197
0.10	0.3972	0.6551	0.7043	0.5883	0.3723	0.1955	0.0653
0.15	0.3042	0.5150	0.5844	0.5192	0.3812	0.2261	0.1010
0.20	0.2409	0.4171	0.4901	0.4620	0.3636	0.2385	0.1145
0.25	0.1962	0.3452	0.4174	0.4092	0.3391	0.2343	0.1178
0.30	0.1631	0.2908	0.3592	0.3624	0.3105	0.2220	0.1145
0.35	0.1379	0.2482	0.3115	0.3208	0.2813	0.2056	0.1077
0.40	0.1181	0.2141	0.2718	0.2840	0.2530	0.1876	0.0993
0.45	0.1020	0.1860	0.2381	0.2514	0.2265	0.1696	0.0904
0.50	0.0888	0.1625	0.2092	0.2226	0.2020	0.1523	0.0816
0.55	0.0776	0.1425	0.1842	0.1970	0.1798	0.1361	0.0732
0.60	0.0681	0.1253	0.1625	0.1744	0.1597	0.1214	0.0654
0.65	0.0599	0.1104	0.1435	0.1544	0.1418	0.1079	0.0582
0.70	0.0528	0.0974	0.1268	0.1366	0.1257	0.0959	0.0518
0.75	0.0466	0.0860	0.1121	0.1210	0.1114	0.0851	0.0460
0.80	0.0412	0.0760	0.0991	0.1071	0.0987	0.0754	0.0408
0.85	0.0364	0.0672	0.0877	0.0948	0.0874	0.0668	0.0361
0.90	0.0322	0.0594	0.0776	0.0839	0.0774	0.0592	0.0320
0.95	0.0285	0.0526	0.0687	0.0743	0.0686	0.0524	0.0284
1.00	0.0252	0.0465	0.0608	0.0657	0.0607	0.0464	0.0251

(x, y)	exact	approx	abs error
(0.25, 0.1)	0.3794	0.3972	0.0178
(1, 0.5)	0.1854	0.2226	0.0372
(1.5, 0.8)	0.0623	0.0754	0.0131

7. (a) We identify $c = 15/88 \approx 0.1705$, $a = 20$, $T = 10$, $n = 10$, and $m = 10$. Then $h = 2$, $k = 1$, and $\lambda = 15/352 \approx 0.0426$.

TIME	X = 2.00	X = 4.00	X = 6.00	X = 8.00	X = 10.00	X = 12.00	X = 14.00	X = 16.00	X = 18.00
0.00	30.0000	30.0000	30.0000	30.0000	30.0000	30.0000	30.0000	30.0000	30.0000
1.00	28.7733	29.9749	29.9995	30.0000	30.0000	30.0000	29.9995	29.9749	28.7733
2.00	27.6450	29.9037	29.9970	29.9999	30.0000	29.9999	29.9970	29.9037	27.6450
3.00	26.6051	29.7938	29.9911	29.9997	30.0000	29.9997	29.9911	29.7938	26.6051
4.00	25.6452	29.6517	29.9805	29.9991	29.9999	29.9991	29.9805	29.6517	25.6452
5.00	24.7573	29.4829	29.9643	29.9981	29.9998	29.9981	29.9643	29.4829	24.7573
6.00	23.9347	29.2922	29.9421	29.9963	29.9996	29.9963	29.9421	29.2922	23.9347
7.00	23.1711	29.0836	29.9134	29.9936	29.9992	29.9936	29.9134	29.0836	23.1711
8.00	22.4612	28.8606	29.8782	29.9898	29.9986	29.9898	29.8782	28.8606	22.4612
9.00	27.7999	28.6263	29.8362	29.9848	29.9977	29.9848	29.8362	28.6263	21.7999
10.00	21.1829	28.3831	29.7878	29.9782	29.9964	29.9782	29.7878	28.3831	21.1829

(b) We identify $c = 15/88 \approx 0.1705$, $a = 50$, $T = 10$, $n = 10$, and $m = 10$. Then $h = 5$, $k = 1$, and $\lambda = 3/440 \approx 0.0068$.

TIME	X = 5.00	X = 10.00	X = 15.00	X = 20.00	X = 25.00	X = 30.00	X = 35.00	X = 40.00	X = 45.00
0.00	30.0000	30.0000	30.0000	30.0000	30.0000	30.0000	30.0000	30.0000	30.0000
1.00	29.7968	29.9993	30.0000	30.0000	30.0000	30.0000	30.0000	29.9993	29.7968
2.00	29.5964	29.9973	30.0000	30.0000	30.0000	30.0000	30.0000	29.9973	29.5964
3.00	29.3987	29.9939	30.0000	30.0000	30.0000	30.0000	30.0000	29.9939	29.3987
4.00	29.2036	29.9893	29.9999	30.0000	30.0000	30.0000	29.9999	29.9893	29.2036
5.00	29.0112	29.9834	29.9998	30.0000	30.0000	30.0000	29.9998	29.9834	29.0112
6.00	28.8212	29.9762	29.9997	30.0000	30.0000	30.0000	29.9997	29.9762	28.8213
7.00	28.6339	29.9679	29.9995	30.0000	30.0000	30.0000	29.9995	29.9679	28.6339
8.00	28.4490	29.9585	29.9992	30.0000	30.0000	30.0000	29.9993	29.9585	28.4490
9.00	28.2665	29.9479	29.9989	30.0000	30.0000	30.0000	29.9989	29.9479	28.2665
10.00	28.0864	29.9363	29.9986	30.0000	30.0000	30.0000	29.9986	29.9363	28.0864

(c) We identify $c = 50/27 \approx 1.8519$, $a = 20$, $T = 10$, $n = 10$, and $m = 10$. Then $h = 2$, $k = 1$, and $\lambda = 25/54 \approx 0.4630$.

TIME	X = 2.00	X = 4.00	X = 6.00	X = 8.00	X = 10.00	X = 12.00	X = 14.00	X = 16.00	X = 18.00
0.00	18.0000	32.0000	42.0000	48.0000	50.0000	48.0000	42.0000	32.0000	18.0000
1.00	16.4489	30.1970	40.1561	46.1495	48.1486	46.1495	40.1561	30.1970	16.4489
2.00	15.3312	28.5348	38.3465	44.3067	46.3001	44.3067	38.3465	28.5348	15.3312
3.00	14.4216	27.0416	36.6031	42.4847	44.4619	42.4847	36.6031	27.0416	14.4216
4.00	13.6371	25.6867	34.9416	40.6988	42.6453	40.6988	34.9416	25.6867	13.6371
5.00	12.9378	24.4419	33.3628	38.9611	40.8634	38.9611	33.3628	24.4419	12.9378
6.00	12.3012	23.2863	31.8624	37.2794	39.1273	37.2794	31.8624	23.2863	12.3012
7.00	11.7137	22.2051	30.4350	35.6578	37.4446	35.6578	30.4350	22.2051	11.7137
8.00	11.1659	21.1877	29.0757	34.0984	35.8202	34.0984	29.0757	21.1877	11.1659
9.00	10.6517	20.2261	27.7799	32.6014	34.2567	32.6014	27.7799	20.2261	10.6517
10.00	10.1665	19.3143	26.5439	31.1662	32.7549	31.1662	26.5439	19.3143	10.1665

(d) We identify $c = 260/159 \approx 1.6352$, $a = 100$, $T = 10$, $n = 10$, and $m = 10$. Then $h = 10$, $k = 1$, and $\lambda = 13/795 \approx 00164$.

TIME	X = 10.00	X = 20.00	X = 30.00	X = 40.00	X = 50.00	X = 60.00	X = 70.00	X = 80.00	X = 90.00
0.00	8.0000	16.0000	24.0000	32.0000	40.0000	32.0000	24.0000	16.0000	8.0000
1.00	8.0000	16.0000	24.0000	31.9979	39.7425	31.9979	24.0000	16.0000	8.0000
2.00	8.0000	16.0000	23.9999	31.9918	39.4932	31.9918	23.9999	16.0000	8.0000
3.00	8.0000	16.0000	23.9997	31.9820	39.2517	31.9820	23.9997	16.0000	8.0000
4.00	8.0000	16.0000	23.9993	31.9687	39.0176	31.9687	23.9993	16.0000	8.0000
5.00	8.0000	16.0000	23.9987	31.9520	38.7905	31.9520	23.9987	16.0000	8.0000
6.00	8.0000	15.9999	23.9978	31.9323	38.5701	31.9323	23.9978	15.9999	8.0000
7.00	8.0000	15.9999	23.9966	31.9097	38.3561	31.9097	23.9966	15.9999	8.0000
8.00	8.0000	15.9998	23.9951	31.8844	38.1483	31.8844	23.9951	15.9998	8.0000
9.00	8.0000	15.9997	23.9931	31.8566	37.9463	31.8566	23.9931	15.9997	8.0000
10.00	8.0000	15.9996	23.9908	31.8265	37.7499	31.8265	23.9908	15.9996	8.0000

9. (a) We identify $c = 15/88 \approx 0.1705$, $a = 20$, $T = 10$, $n = 10$, and $m = 10$. Then $h = 2$, $k = 1$, and $\lambda = 15/352 \approx 0.0426$.

TIME	X = 2.00	X = 4.00	X= 6.00	X = 8.00	X = 10.00	X = 12.00	X = 14.00	X = 16.00	X = 18.00
0.00	30.0000	30.0000	30.0000	30.0000	30.0000	30.0000	30.0000	30.0000	30.0000
1.00	28.7733	29.9749	29.9995	30.0000	30.0000	30.0000	29.9998	29.9916	29.5911
2.00	27.6450	29.9037	29.9970	29.9999	30.0000	30.0000	29.9990	29.9679	29.2150
3.00	26.6051	29.7938	29.9911	29.9997	30.0000	29.9999	29.9970	29.9313	28.8684
4.00	25.6452	29.6517	29.9805	29.9991	30.0000	29.9997	29.9935	29.8839	28.5484
5.00	24.7573	29.4829	29.9643	29.9981	29.9999	29.9994	29.9881	29.8276	28.2524
6.00	23.9347	29.2922	29.9421	29.9963	29.9997	29.9988	29.9807	29.7641	27.9782
7.00	23.1711	29.0836	29.9134	29.9936	29.9995	29.9979	29.9711	29.6945	27.7237
8.00	22.4612	28.8606	29.8782	29.9899	29.9991	29.9966	29.9594	29.6202	27.4870
9.00	27.7999	28.6263	29.8362	29.9848	29.9985	29.9949	29.9454	29.5421	27.2666
10.00	21.1829	28.3831	29.7878	29.9783	29.9976	29.9927	29.9293	29.4610	27.0610

(b) We identify $c = 15/88 \approx 0.1705$, $a = 50$, $T = 10$, $n = 10$, and $m = 10$. Then $h = 5$, $k = 1$, and $\lambda = 3/440 \approx 0.0068$.

TIME	X = 5.00	X = 10.00	X= 15.00	X = 20.00	X = 25.00	X = 30.00	X = 35.00	X = 40.00	X = 45.00
0.00	30.0000	30.0000	30.0000	30.0000	30.0000	30.0000	30.0000	30.0000	30.0000
1.00	29.7968	29.9993	30.0000	30.0000	30.0000	30.0000	30.0000	29.9998	29.9323
2.00	29.5964	29.9973	30.0000	30.0000	30.0000	30.0000	30.0000	29.9991	29.8655
3.00	29.3987	29.9939	30.0000	30.0000	30.0000	30.0000	30.0000	29.9980	29.7996
4.00	29.2036	29.9893	29.9999	30.0000	30.0000	30.0000	30.0000	29.9964	29.7345
5.00	29.0112	29.9834	29.9998	30.0000	30.0000	30.0000	29.9999	29.9945	29.6704
6.00	28.8212	29.9762	29.9997	30.0000	30.0000	30.0000	29.9999	29.9921	29.6071
7.00	28.6339	29.9679	29.9995	30.0000	30.0000	30.0000	29.9998	29.9893	29.5446
8.00	28.4490	29.9585	29.9992	30.0000	30.0000	30.0000	29.9997	29.9862	29.4830
9.00	28.2665	29.9479	29.9989	30.0000	30.0000	30.0000	29.9996	29.9827	29.4222
10.00	28.0864	29.9363	29.9986	30.0000	30.0000	30.0000	29.9995	29.9788	29.3621

(c) We identify $c = 50/27 \approx 1.8519$, $a = 20$, $T = 10$, $n = 10$, and $m = 10$. Then $h = 2$, $k = 1$, and $\lambda = 25/54 \approx 0.4630$.

TIME	X = 2.00	X = 4.00	X= 6.00	X = 8.00	X = 10.00	X = 12.00	X = 14.00	X = 16.00	X = 18.00
0.00	18.0000	32.0000	42.0000	48.0000	50.0000	48.0000	42.0000	32.0000	18.0000
1.00	16.4489	30.1970	40.1562	46.1502	48.1531	46.1773	40.3274	31.2520	22.9449
2.00	15.3312	28.5350	38.3477	44.3130	46.3327	44.4671	39.0872	31.5755	24.6930
3.00	14.4219	27.0429	36.6090	42.5113	44.5759	42.9362	38.1976	31.7478	25.4131
4.00	13.6381	25.6913	34.9606	40.7728	42.9127	41.5716	37.4340	31.7086	25.6986
5.00	12.9409	24.4545	33.4091	39.1182	41.3519	40.3240	36.7033	31.5136	25.7663
6.00	12.3088	23.3146	31.9546	37.5566	39.8880	39.1565	35.9745	31.2134	25.7128
7.00	11.7294	22.2589	30.5939	36.0884	38.5109	38.0470	35.2407	30.8434	25.5871
8.00	11.1946	21.2785	29.3217	34.7092	37.2109	36.9834	34.5032	30.4279	25.4167
9.00	10.6987	20.3660	28.1318	33.4130	35.9801	35.9591	33.7660	29.9836	25.2181
10.00	10.2377	19.5150	27.0178	32.1929	34.8117	34.9710	33.0338	29.5224	25.0019

(d) We identify $c = 260/159 \approx 1.6352$, $a = 100$, $T = 10$, $n = 10$, and $m = 10$. Then $h = 10$, $k = 1$, and $\lambda = 13/795 \approx 00164$.

TIME	X = 10.00	X = 20.00	X= 30.00	X = 40.00	X = 50.00	X = 60.00	X = 70.00	X = 80.00	X = 90.00
0.00	8.0000	16.0000	24.0000	32.0000	40.0000	32.0000	24.0000	16.0000	8.0000
1.00	8.0000	16.0000	24.0000	31.9979	39.7425	31.9979	24.0000	16.0026	8.3218
2.00	8.0000	16.0000	23.9999	31.9918	39.4932	31.9918	24.0000	16.0102	8.6333
3.00	8.0000	16.0000	23.9997	31.9820	39.2517	31.9820	24.0001	16.0225	8.9350
4.00	8.0000	16.0000	23.9993	31.9687	39.0176	31.9687	24.0002	16.0392	9.2272
5.00	8.0000	16.0000	23.9987	31.9520	38.7905	31.9521	24.0003	16.0599	9.5103
6.00	8.0000	15.9999	23.9978	31.9323	38.5701	31.9324	24.0005	16.0845	9.7846
7.00	8.0000	15.9999	23.9966	31.9097	38.3561	31.9098	24.0008	16.1126	10.0506
8.00	8.0000	15.9998	23.9951	31.8844	38.1483	31.8846	24.0012	16.1441	10.3084
9.00	8.0000	15.9997	23.9931	31.8566	37.9463	31.8569	24.0017	16.1786	10.5585
10.00	8.0000	15.9996	23.9908	31.8265	37.7499	31.8270	24.0023	16.2160	10.8012

11. (a) The differential equation is $k\dfrac{\partial^2 u}{\partial x^2} = \dfrac{\partial u}{\partial t}$ where $k = K/\gamma\rho$. If we let $u(x, t) = v(x, t) + \psi(x)$, then

$$\frac{\partial^2 u}{\partial x^2} = \frac{\partial^2 v}{\partial x^2} + \psi'' \quad \text{and} \quad \frac{\partial u}{\partial t} = \frac{\partial v}{\partial t}.$$

Substituting into the differential equation gives

$$k\frac{\partial^2 v}{\partial x^2} + k\psi'' = \frac{\partial v}{\partial t}.$$

Requiring $k\psi'' = 0$ we have $\psi(x) = c_1 x + c_2$. The boundary conditions become

$$u(0, t) = v(0, t) + \psi(0) = 20 \quad \text{and} \quad u(20, t) = v(20, t) + \psi(20) = 30.$$

Letting $\psi(0) = 20$ and $\psi(20) = 30$ we obtain the homogeneous boundary conditions in v: $v(0, t) = v(20, t) = 0$. Now $\psi(0) = 20$ and $\psi(20) = 30$ imply that $c_1 = 1/2$ and $c_2 = 20$. The steady-state solution is $\psi(x) = \frac{1}{2}x + 20$.

(b) To use the Crank-Nicholson method we identify $c = 375/212 \approx 1.7689$, $a = 20$, $T = 400$, $n = 5$, and $m = 40$. Then $h = 4$, $k = 10$, and $\lambda = 1875/1696 \approx 1.1055$.

TIME	X = 4.00	X = 8.00	X = 12.00	X = 16.00
0.00	50.0000	50.0000	50.0000	50.0000
10.00	32.7433	44.2679	45.4228	38.2971
20.00	29.9946	36.2354	38.3148	35.8160
30.00	26.9487	32.1409	34.0874	32.9644
40.00	25.2691	29.2562	31.2704	31.2580
50.00	24.1178	27.4348	29.4296	30.1207
60.00	23.3821	26.2339	28.2356	29.3810
70.00	22.8995	25.4560	27.4554	28.8998
80.00	22.5861	24.9481	26.9482	28.5859
90.00	22.3817	24.6176	26.6175	28.3817
100.00	22.2486	24.4022	26.4023	28.2486
110.00	22.1619	24.2620	26.2620	28.1619
120.00	22.1055	24.1707	26.1707	28.1055
130.00	22.0687	24.1112	26.1112	28.0687
140.00	22.0447	24.0725	26.0724	28.0447
150.00	22.0291	24.0472	26.0472	28.0291
160.00	22.0190	24.0307	26.0307	28.0190
170.00	22.0124	24.0200	26.0200	28.0124
180.00	22.0081	24.0130	26.0130	28.0081
190.00	22.0052	24.0085	26.0085	28.0052
200.00	22.0034	24.0055	26.0055	28.0034
210.00	22.0022	24.0036	26.0036	28.0022
220.00	22.0015	24.0023	26.0023	28.0015
230.00	22.0009	24.0015	26.0015	28.0009
240.00	22.0006	24.0010	26.0010	28.0006
250.00	22.0004	24.0007	26.0007	28.0004
260.00	22.0003	24.0004	26.0004	28.0003
270.00	22.0002	24.0003	26.0003	28.0002
280.00	22.0001	24.0002	26.0002	28.0001
290.00	22.0001	24.0001	26.0001	28.0001
300.00	22.0000	24.0001	26.0001	28.0000
310.00	22.0000	24.0001	26.0001	28.0000
320.00	22.0000	24.0000	26.0000	28.0000
330.00	22.0000	24.0000	26.0000	28.0000
340.00	22.0000	24.0000	26.0000	28.0000
350.00	22.0000	24.0000	26.0000	28.0000

We observe that the approximate steady-state temperatures agree exactly with the corresponding values of $\psi(x)$.

15.3 Wave Equation

1. (a) Identifying $h = 1/4$ and $k = 1/10$ we see that $\lambda = 2/5$.

TIME	X = 0.25	X = 0.5	X = 0.75
0.00	0.1875	0.2500	0.1875
0.10	0.1775	0.2400	0.1775
0.20	0.1491	0.2100	0.1491
0.30	0.1066	0.1605	0.1066
0.40	0.0556	0.0938	0.0556
0.50	0.0019	0.0148	0.0019
0.60	−0.0501	−0.0682	−0.0501
0.70	−0.0970	−0.1455	−0.0970
0.80	−0.1361	−0.2072	−0.1361
0.90	−0.1648	−0.2462	−0.1648
1.00	−0.1802	−0.2591	−0.1802

(b) Identifying $h = 2/5$ and $k = 1/10$ we see that $\lambda = 1/4$.

TIME	X = 0.4	X = 0.8	X = 1.2	X = 1.6
0.00	0.0032	0.5273	0.5273	0.0032
0.10	0.0194	0.5109	0.5109	0.0194
0.20	0.0652	0.4638	0.4638	0.0652
0.30	0.1318	0.3918	0.3918	0.1318
0.40	0.2065	0.3035	0.3035	0.2065
0.50	0.2743	0.2092	0.2092	0.2743
0.60	0.3208	0.1190	0.1190	0.3208
0.70	0.3348	0.0413	0.0413	0.3348
0.80	0.3094	−0.0180	−0.0180	0.3094
0.90	0.2443	−0.0568	−0.0568	0.2443
1.00	0.1450	−0.0768	−0.0768	0.1450

(c) Identifying $h = 1/10$ and $k = 1/25$ we see that $\lambda = 2\sqrt{2}/5$.

TIME	X = 0.1	X = 0.2	X = 0.3	X = 0.4	X = 0.5	X = 0.6	X = 0.7	X = 0.8	X = 0.9
0.00	0.0000	0.0000	0.0000	0.0000	0.0000	0.5000	0.5000	0.5000	0.5000
0.04	0.0000	0.0000	0.0000	0.0000	0.0800	0.4200	0.5000	0.5000	0.4200
0.08	0.0000	0.0000	0.0000	0.0256	0.2432	0.2568	0.4744	0.4744	0.2312
0.12	0.0000	0.0000	0.0082	0.1126	0.3411	0.1589	0.3792	0.3710	0.0462
0.16	0.0000	0.0026	0.0472	0.2394	0.3076	0.1898	0.2108	0.1663	−0.0496
0.20	0.0008	0.0187	0.1334	0.3264	0.2146	0.2651	0.0215	−0.0933	−0.0605
0.24	0.0071	0.0657	0.2447	0.3159	0.1735	0.2463	−0.1266	−0.3056	−0.0625
0.28	0.0299	0.1513	0.3215	0.2371	0.2013	0.0849	−0.2127	−0.3829	−0.1223
0.32	0.0819	0.2525	0.3168	0.1737	0.2033	−0.1345	−0.2580	−0.3223	−0.2264
0.36	0.1623	0.3197	0.2458	0.1657	0.0877	−0.2853	−0.2843	−0.2104	−0.2887
0.40	0.2412	0.3129	0.1727	0.1583	−0.1223	−0.3164	−0.2874	−0.1473	−0.2336
0.44	0.2657	0.2383	0.1399	0.0658	−0.3046	−0.2761	−0.2549	−0.1565	−0.0761
0.48	0.1965	0.1410	0.1149	−0.1216	−0.3593	−0.2381	−0.1977	−0.1715	0.0800
0.52	0.0466	0.0531	0.0225	−0.3093	−0.2992	−0.2260	−0.1451	−0.1144	0.1300
0.56	−0.1161	−0.0466	−0.1662	−0.3876	−0.2188	−0.2114	−0.1085	0.0111	0.0602
0.60	−0.2194	−0.2069	−0.3875	−0.3411	−0.1901	−0.1662	−0.0666	0.1140	−0.0446
0.64	−0.2485	−0.4290	−0.5362	−0.2611	−0.2021	−0.0969	0.0012	0.1084	−0.0843
0.68	−0.2559	−0.6276	−0.5625	−0.2503	−0.1993	−0.0298	0.0720	0.0068	−0.0354
0.72	−0.3003	−0.6865	−0.5097	−0.3230	−0.1585	0.0156	0.0893	−0.0874	0.0384
0.76	−0.3722	−0.5652	−0.4538	−0.4029	−0.1147	0.0289	0.0265	−0.0849	0.0596
0.80	−0.3867	−0.3464	−0.4172	−0.4068	−0.1172	−0.0046	−0.0712	−0.0005	0.0155
0.84	−0.2647	−0.1633	−0.3546	−0.3214	−0.1763	−0.0954	−0.1249	0.0665	−0.0386
0.88	−0.0254	−0.0738	−0.2202	−0.2002	−0.2559	−0.2215	−0.1079	0.0385	−0.0468
0.92	0.2064	−0.0157	−0.0325	−0.1032	−0.3067	−0.3223	−0.0804	−0.0636	−0.0127
0.96	0.3012	0.1081	0.1380	−0.0487	−0.2974	−0.3407	−0.1250	−0.1548	0.0092
1.00	0.2378	0.3032	0.2392	−0.0141	−0.2223	−0.2762	−0.2481	−0.1840	−0.0244

3. (a) Identifying $h = 1/5$ and $k = 0.5/10 = 0.05$ we see that $\lambda = 0.25$.

TIME	X = 0.2	X = 0.4	X = 0.6	X = 0.8
0.00	0.5878	0.9511	0.9511	0.5878
0.05	0.5808	0.9397	0.9397	0.5808
0.10	0.5599	0.9059	0.9059	0.5599
0.15	0.5256	0.8505	0.8505	0.5256
0.20	0.4788	0.7748	0.7748	0.4788
0.25	0.4206	0.6806	0.6806	0.4206
0.30	0.3524	0.5701	0.5701	0.3524
0.35	0.2757	0.4460	0.4460	0.2757
0.40	0.1924	0.3113	0.3113	0.1924
0.45	0.1046	0.1692	0.1692	0.1046
0.50	0.0142	0.0230	0.0230	0.0142

(b) Identifying $h = 1/5$ and $k = 0.5/20 = 0.025$ we see that $\lambda = 0.125$.

TIME	X = 0.2	X = 0.4	X = 0.6	X = 0.8
0.00	0.5878	0.9511	0.9511	0.5878
0.03	0.5860	0.9482	0.9482	0.5860
0.05	0.5808	0.9397	0.9397	0.5808
0.08	0.5721	0.9256	0.9256	0.5721
0.10	0.5599	0.9060	0.9060	0.5599
0.13	0.5445	0.8809	0.8809	0.5445
0.15	0.5257	0.8507	0.8507	0.5257
0.18	0.5039	0.8153	0.8153	0.5039
0.20	0.4790	0.7750	0.7750	0.4790
0.23	0.4513	0.7302	0.7302	0.4513
0.25	0.4209	0.6810	0.6810	0.4209
0.28	0.3879	0.6277	0.6277	0.3879
0.30	0.3527	0.5706	0.5706	0.3527
0.33	0.3153	0.5102	0.5102	0.3153
0.35	0.2761	0.4467	0.4467	0.2761
0.38	0.2352	0.3806	0.3806	0.2352
0.40	0.1929	0.3122	0.3122	0.1929
0.43	0.1495	0.2419	0.2419	0.1495
0.45	0.1052	0.1701	0.1701	0.1052
0.48	0.0602	0.0974	0.0974	0.0602
0.50	0.0149	0.0241	0.0241	0.0149

5. We identify $c = 24944.4$, $k = 0.00020045$ seconds $= 0.20045$ milliseconds, and $\lambda = 0.5$. Time in the table is expressed in milliseconds.

TIME	X = 10	X = 20	X = 30	X = 40	X = 50
0.00000	0.1000	0.2000	0.3000	0.2000	0.1000
0.20045	0.1000	0.2000	0.2750	0.2000	0.1000
0.40089	0.1000	0.1938	0.2125	0.1938	0.1000
0.60134	0.0984	0.1688	0.1406	0.1688	0.0984
0.80178	0.0898	0.1191	0.0828	0.1191	0.0898
1.00223	0.0661	0.0531	0.0432	0.0531	0.0661
1.20268	0.0226	−0.0121	0.0085	−0.0121	0.0226
1.40312	−0.0352	−0.0635	−0.0365	−0.0635	−0.0352
1.60357	−0.0913	−0.1011	−0.0950	−0.1011	−0.0913
1.80401	−0.1271	−0.1347	−0.1566	−0.1347	−0.1271
2.00446	−0.1329	−0.1719	−0.2072	−0.1719	−0.1329
2.20491	−0.1153	−0.2081	−0.2402	−0.2081	−0.1153
2.40535	−0.0920	−0.2292	−0.2571	−0.2292	−0.0920
2.60580	−0.0801	−0.2230	−0.2601	−0.2230	−0.0801
2.80624	−0.0838	−0.1903	−0.2445	−0.1903	−0.0838
3.00669	−0.0932	−0.1445	−0.2018	−0.1445	−0.0932
3.20713	−0.0921	−0.1003	−0.1305	−0.1003	−0.0921
3.40758	−0.0701	−0.0615	−0.0440	−0.0615	−0.0701
3.60803	−0.0284	−0.0205	0.0336	−0.0205	−0.0284
3.80847	0.0224	0.0321	0.0842	0.0321	0.0224
4.00892	0.0700	0.0953	0.1087	0.0953	0.0700
4.20936	0.1064	0.1555	0.1265	0.1555	0.1064
4.40981	0.1285	0.1962	0.1588	0.1962	0.1285
4.61026	0.1354	0.2106	0.2098	0.2106	0.1354
4.81070	0.1273	0.2060	0.2612	0.2060	0.1273
5.01115	0.1070	0.1955	0.2851	0.1955	0.1070
5.21159	0.0821	0.1853	0.2641	0.1853	0.0821
5.41204	0.0625	0.1689	0.2038	0.1689	0.0625
5.61249	0.0539	0.1347	0.1260	0.1347	0.0539
5.81293	0.0520	0.0781	0.0526	0.0781	0.0520
6.01338	0.0436	0.0086	−0.0080	0.0086	0.0436
6.21382	0.0156	−0.0564	−0.0604	−0.0564	0.0156
6.41427	−0.0343	−0.1043	−0.1107	−0.1043	−0.0343
6.61472	−0.0931	−0.1364	−0.1578	−0.1364	−0.0931
6.81516	−0.1395	−0.1630	−0.1942	−0.1630	−0.1395
7.01561	−0.1568	−0.1915	−0.2150	−0.1915	−0.1568
7.21605	−0.1436	−0.2173	−0.2240	−0.2173	−0.1436
7.41650	−0.1129	−0.2263	−0.2297	−0.2263	−0.1129
7.61695	−0.0824	−0.2078	−0.2336	−0.2078	−0.0824
7.81739	−0.0625	−0.1644	−0.2247	−0.1644	−0.0625
8.01784	−0.0526	−0.1106	−0.1856	−0.1106	−0.0526
8.21828	−0.0440	−0.0611	−0.1091	−0.0611	−0.0440
8.41873	−0.0287	−0.0192	−0.0085	−0.0192	−0.0287
8.61918	−0.0038	0.0229	0.0867	0.0229	−0.0038
8.81962	0.0287	0.0743	0.1500	0.0743	0.0287
9.02007	0.0654	0.1332	0.1755	0.1332	0.0654
9.22051	0.1027	0.1858	0.1799	0.1858	0.1027
9.42096	0.1352	0.2160	0.1872	0.2160	0.1352
9.62140	0.1540	0.2189	0.2089	0.2189	0.1540
9.82185	0.1506	0.2030	0.2356	0.2030	0.1506
10.02230	0.1226	0.1822	0.2461	0.1822	0.1226

Chapter 15 in Review

1. Using the figure we obtain the system

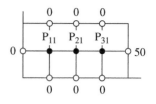

$$u_{21} + 0 + 0 + 0 - 4u_{11} = 0$$

$$u_{31} + 0 + u_{11} + 0 - 4u_{21} = 0$$

$$50 + 0 + u_{21} + 0 - 4u_{31} = 0.$$

By Gauss-Elimination then,

$$
\begin{bmatrix}
-4 & 1 & 0 & | & 0 \\
1 & -4 & 1 & | & 0 \\
0 & 1 & -4 & | & -50
\end{bmatrix}
\xrightarrow[\text{operations}]{\text{row}}
\begin{bmatrix}
1 & -4 & 1 & | & 0 \\
0 & 1 & -4 & | & -50 \\
0 & 0 & 1 & | & 13.3928
\end{bmatrix}.
$$

The solution is $u_{11} = 0.8929$, $u_{21} = 3.5714$, $u_{31} = 13.3928$.

3. (a)

TIME	X = 0.0	X = 0.2	X = 0.4	X = 0.6	X = 0.8	X = 1.0
0.00	0.0000	0.2000	0.4000	0.6000	0.8000	0.0000
0.01	0.0000	0.2000	0.4000	0.6000	0.5500	0.0000
0.02	0.0000	0.2000	0.4000	0.5375	0.4250	0.0000
0.03	0.0000	0.2000	0.3844	0.4750	0.3469	0.0000
0.04	0.0000	0.1961	0.3609	0.4203	0.2922	0.0000
0.05	0.0000	0.1883	0.3346	0.3734	0.2512	0.0000

(b)

TIME	X = 0.0	X = 0.2	X = 0.4	X = 0.6	X = 0.8	X = 1.0
0.00	0.0000	0.2000	0.4000	0.6000	0.8000	1.0000
0.01	0.0000	0.2000	0.4000	0.6000	0.8000	0.0000
0.02	0.0000	0.2000	0.4000	0.6000	0.5500	0.0000
0.03	0.0000	0.2000	0.4000	0.5375	0.4250	0.0000
0.04	0.0000	0.2000	0.3844	0.4750	0.3469	0.0000
0.05	0.0000	0.1961	0.3609	0.4203	0.2922	0.0000

(c) The table in part **(b)** is the same as the table in part **(a)** shifted downward one row.